ELEMENTS OF MODERN ALGEBRA

FIFTH EDITION

ELEMENTS OF MODERN ALGEBRA

Jimmie Gilbert

Linda Gilbert

University of South Carolina Spartanburg

Brooks/Cole
Thomson Learning™

Pacific Grove • Albany • Belmont • Boston • Cincinnati • Johannesburg • London • Madrid
Melbourne • Mexico City • New York • Scottsdale • Singapore • Tokyo • Toronto

Publisher: *Gary Ostedt*
Marketing Team: *Caroline Croley, Debra Johnston*
Editorial Assistant: *Carol Benedict*
Production Editor: *Janet Hill*
Production Service and Composition: *WestWords, Inc.*
Print Buyer: *Vena Dyer*
Cover Design: *Roger Knox*

Cover Photo: *Hideharu Naito/Photonica*
Interior Illustration: *George Nichols, Heather Theurer*
Photo Researcher: *Terry Powell*
Interior Design: *John Edeen*
Cover and Text Printing/Binding: *R. R. Donnelley & Sons, Crawfordsville*

For more information, contact:
BROOKS/COLE
511 Forest Lodge Road
Pacific Grove, CA 93950 USA
www.brookscole.com

Printed in the United States of America

10   9   8   7   6   5   4   3

Library of Congress Cataloging-in-Publication Data
Gilbert, Jimmie, [date]
    Elements of modern algebra / Jimmie Gilbert, Linda Gilbert. — 5th ed.
       p.   cm.
    Includes bibliographical references and index.
    ISBN 0-534-37351-8
    1. Algebra, Abstract. I. Gilbert, Linda. II. Title.
QA162.G527 1999
512'.02—dc21   99-32433

Photo Credits: p 52 The Royal Society; p 108 The Photographer's Window; p 169 CORBIS/Dorling Kindersley Ltd., London; p 152 Smithsonian Institution; p 206 Public Domain Images; p 237 Public Domain Images; p 264 Stock Montage, Inc.; p 289 The Royal Irish Academy; p 334 Smithsonian Institution.

Dedicated to Georgia Phillips
and the memory of Ella Gilbert

PREFACE

As the earlier editions were, this book is intended as a text for an introductory course in algebraic structures (groups, rings, fields, and so forth). Such a course is often used to bridge the gap from manipulative to theoretical mathematics and to help prepare secondary mathematics teachers for their careers. Some flexibility is provided by including more material than would normally be taught in one course, and a dependency diagram of the chapters/sections (Figure P-1) is included at the end of this preface. Several sections, including a new one on cryptography, are marked "optional" and may be skipped by instructors who prefer to spend more time on later topics.

The reviewers for this fifth edition were opposed to major changes in the text. The consensus was that changes should be along the lines of "fine tuning." Consistent with the advice we received, the following list of changes from the fourth edition contains nothing drastic.

▶ An optional new section, Introduction to Cryptography, presents another application of modular arithmetic in Chapter 2.

▶ Feedback from earlier editions has made it clear that a large majority of users favored use of the terms *onto* and *one-to-one* over the corresponding terms *surjective* and *injective*. In this edition, we have overcome inertia and now give preference to *onto* and *one-to-one,* though the other terms are still included and used on rare occasions.

▶ The basic material on mapping composition is now included in Section 1.2, and Section 1.3 has been changed into an optional investigation of the relationships between two mappings and their compositions. Omitting the new Section 1.3 slightly reduces the time required to reach the topic of groups.

▶ In an attempt to improve clarity, minor rewriting has been done in 16 sections of this edition. Responding to a reviewer's suggestion, changes in notation from script letters to italics or bold print have been made in several places. Among these changes, the real numbers and the complex numbers are now denoted by **R** and **C**, respectively.

▶ The section on mathematical induction has been expanded in this edition, with more emphasis on the method of proof by complete induction.

▶ A modest number (122) of new problems are included in this edition.

▶ In response to a suggestion from one of our most loyal adopters, we have included Euclid's theorem on the infinitude of primes as Theorem 2.19.

The following *user-friendly* features are retained from the fourth edition:

▶ **Descriptive labels and titles** are placed on definitions and theorems to indicate their content and relevance.

▶ **Strategy boxes** that give guidance and explanation about techniques of proof are included. This feature forms a component of the bridge that enables students to become more proficient in their proof construction skills.

▶ **Symbolic marginal notes** such as "$(p \land q) \Rightarrow r$" and "$\sim p \Leftarrow (\sim q \land \sim r)$" are used to help students analyze the logic in the proofs of theorems without interrupting the natural flow of the proof.

▶ A **reference system** provides guideposts to continuations and interconnections of exercises throughout the text. As an example, consider Exercise 19 in Section 4.4. The marginal notation "Sec. 3.1, #28 ≫" indicates that this exercise is *connected* to Exercise 28 in the *earlier* Section 3.1. The marginal notation "Sec. 4.5, #7 ≪" indicates that this exercise has a *continuation* in Exercise 7 in the *later* Section 4.5.

▶ An **appendix** on the basics of logic and methods of proof is included.

▶ A **biographical sketch** of a great mathematician whose contributions are relevant to that material concludes each chapter.

▶ A **gradual introduction and development** of concepts is used, proceeding from the simplest structures to the more complex.

▶ An **abundance of examples** that are designed to develop the student's intuition are included.

▶ Enough **exercises** to allow instructors to make different assignments of approximately the same difficulty are included.

▶ **Exercise sets** are designed to develop the student's maturity and ability to construct proofs, and contain many problems that are elementary or of a computational nature.

▶ A **summary of key words and phrases** is included at the end of each chapter.

▶ A **list of special notations** used in the book is on the front endpapers.

▶ **Group tables** for the most common examples are on the back endpapers.

▶ An **updated bibliography** is included.

Groups appear in the text before rings. The standard topics in elementary group theory are included, and the last two sections in Chapter 4 provide an optional sample of more advanced work in finite abelian groups.

Several users of the text have inquired as to what material we teach in our courses. Our basic goal in a single course is to reach the end of Section 5.3 (The Field of Quotients of an Integral Domain), omitting the last two sections of Chapter 4 along the way. We would also omit other optional sections if class meetings are in short supply. The sections on applications naturally lend themselves well to outside student projects involving additional writing and/or library research.

For the most part, the problems in an exercise set are arranged in order of difficulty, with easier problems first, but exceptions to this arrangement occur if it violates logical order. If one problem is needed or useful in another problem, the more basic problem appears first. When teaching from this text, both authors use a ground rule that any previous result, including prior exercises, may be used in constructing a proof. Whether to adopt this ground rule is, of course, completely optional.

Some users have indicated that they omit Chapter 7 (Real and Complex Numbers) because their students are already familiar with it. Others cover Chapter 8 (Polynomials) before Chapter 7. These and other options are diagrammed in Figure P-1 at the end of this preface.

The treatment of the set $\mathbf{Z}_n$ of congruence classes modulo n is a unique feature of this text, in that it threads throughout most of the book. The first contact with $\mathbf{Z}_n$ is early in Chapter 2, where it appears as a set of equivalence classes. Binary operations of addition and multiplication are defined in $\mathbf{Z}_n$ at a later point in that chapter. Both the additive and multiplicative structures are drawn upon for examples in Chapters 3 and 4. The development of $\mathbf{Z}_n$ continues in Chapter 5, where it appears in its familiar context as a ring. This development culminates in Chapter 6 with the final description of $\mathbf{Z}_n$ as a quotient ring of the integers by the principal ideal (n).

A minimal amount of mathematical maturity is assumed in the text; a major goal is to develop mathematical maturity. The material is presented in a theorem-proof format, with definitions and major results easily located with a user-friendly format. The treatment is rigorous and self-contained, in keeping with the objectives of training the student in the techniques of algebra and of providing a bridge to higher-level mathematical courses.

ACKNOWLEDGMENTS

We are grateful to the following people for their reviews and helpful comments with regard to the first four editions:

Lateef A. Adelani, *Harris-Stowe College*

Philip C. Almes, *Wayland Baptist University*

Edwin F. Baumgartner, *Le Moyne College*

Bruce M. Bemis, *Westminster College*

Steve Benson, *St. Olaf College*

Louise M. Berard, *Wilkes College*

Thomas D. Bishop, *Arkansas State University*

James C. Bradford, *Abilene Christian University*

Shirley Branan, *Birmingham Southern College*

Gordon Brown, *University of Colorado, Boulder*

Harmon C. Brown, *Harding University*

Marshall Cates, *California State University, Los Angeles*

Patrick Costello, *Eastern Kentucky University*

Richard Cowan, *Shorter College*

Elwyn H. Davis, *Pittsburg State University*

David J. DeVries, *Georgia College*

Paul J. Fairbanks, *Bridgewater State College*

Howard Frisinger, *Colorado State University*

Nickolas Heerema, *Florida State University*

Edward K. Hinson, *University of New Hampshire*

David L. Johnson, *Lehigh University*

William J. Keane, *Boston College*

Robert E. Kennedy, *Central Missouri State University*

William F. Keigher, *Rutgers University*

Andre E. Kezdy, *University of Louisville*

Stanley M. Lukawecki, *Clemson University*

Joan S. Morrison, *Goucher College*

Carl R. Spitznagel, *John Carroll University*

Ralph C. Steinlage, *University of Dayton*

James J. Tattersall, *Providence College*

Mark L. Teply, *University of Wisconsin-Milwaukee*

Krishnanand Verma, *University of Minnesota, Duluth*

Robert P. Webber, *Longwood College*

Diana Y. Wei, *Norfolk State University*

Carroll G. Wells, *Western Kentucky University*

Burdette C. Wheaton, *Mankato State University*

Henry Wyzinski, *Indiana University Northwest*

We also express our thanks to these people for their valuable suggestions in reviews of the fifth edition:

Joel Brawley, *Clemson University*
John D. Elwin, *San Diego State University*
Robert E. Kennedy, *Central Missouri State University*
Richard J. Painter, *Colorado State University*
James J. Tattersall, *Providence College*

We are indebted to the following people for suggesting improvements and corrections for this and earlier editions:

Brian D. Beasley, *Presbyterian College*
David M. Bloom, *Brooklyn College of the City University of New York*
David J. DeVries, *Georgia College*
J. Taylor Hollist, *State University of New York at Oneonta*

Our sincere appreciation goes to Gary Ostedt and Carol Benedict for the excellent editorial guidance they provided for this revision. Also, special thanks go to Janet Hill and Richard Saunders for an extremely smooth and efficient production.

Jimmie Gilbert
Linda Gilbert

Chapters/Sections Dependency Diagram

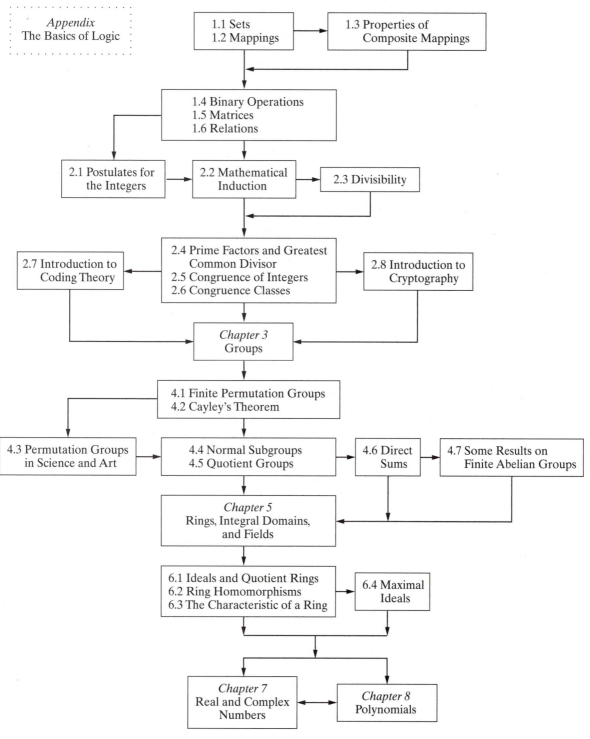

Figure P-1

CONTENTS

FUNDAMENTALS

INTRODUCTION

This chapter presents the fundamental concepts of set, mapping, binary operation, and relation. It also contains a section on matrices, which will serve as a basis for examples and exercises from time to time in the remainder of the text. Much of the material in this chapter may be familiar from earlier courses. If that is the case, appropriate omissions can be made to expedite the study of later topics.

1.1 SETS

Abstract algebra had its beginnings in attempts to solve mathematical problems such as the solution of polynomial equations by radicals and geometric constructions with straightedge and compass. From the solutions of specific problems, general techniques evolved that could be used to solve problems of the same type, and treatments were generalized to deal with whole classes of problems rather than individual ones.

In our study of abstract algebra, we shall make use of our knowledge of the various number systems. At the same time, in many cases we wish to examine how certain properties are consequences of other known properties. This sort of examination deepens our understanding of the system. As we proceed, we shall be careful to distinguish between the properties we have assumed and made available for use, and those that must be deduced from these properties. We must accept without definition some terms that are basic objects in our mathematical systems. Initial assumptions about each system are formulated using these undefined terms.

One such undefined term is **set**. We think of a set as a collection of objects about which it is possible to determine whether or not a particular object is a member of the set. Sets are usually denoted by capital letters and are sometimes described by a list of their elements, as illustrated in the following examples.

Example I   We write

$$A = \{0, 1, 2, 3\}$$

to indicate that the set A contains the elements 0, 1, 2, 3, and no other elements. The notation $\{0, 1, 2, 3\}$ is read as "the set with elements 0, 1, 2, and 3."   ∎

Example 2   The set B, consisting of all the nonnegative integers, is written

$$B = \{0, 1, 2, 3, \dots\}.$$

The three dots $\dots$, called an *ellipsis*, means that the pattern established before the dots continues indefinitely. The notation $\{0, 1, 2, 3, \dots\}$ is read as "the set with elements 0, 1, 2, 3, and so on."   ∎

As in Examples 1 and 2, it is customary to avoid repetition when listing the elements of a set. Another way of describing sets is called *set-builder notation*. Set-builder notation uses braces to enclose a property that is the qualification for membership in the set.

Example 3   The set B in Example 2 can be described using set-builder notation as

$$B = \{x \,|\, x \text{ is a nonnegative integer}\}.$$

The vertical slash is shorthand for "such that," and we read "B is the set of all x such that x is a nonnegative integer."   ∎

There is also a shorthand notation for "is an element of." We write "$x \in A$" to mean "x is an element of the set A." We write "$x \notin A$" to mean "x is not an element of the set A." For the set A in Example 1, we can write

$$2 \in A \quad \text{and} \quad 7 \notin A.$$

Definition 1.1   SUBSET

Let A and B be sets. Then A is called a **subset** of B if and only if every element of A is an element of B. Either of the notations $A \subseteq B$ or $B \supseteq A$ indicates that A is a subset of B.

The notation $A \subseteq B$ is read "A is a subset of B" or "A is contained in B." Also, $B \supseteq A$ is read as "B contains A." The symbol $\in$ is reserved for elements, whereas the symbol $\subseteq$ is reserved for subsets.

Example 4   We write

$$a \in \{a, b, c, d\} \quad \text{or} \quad \{a\} \subseteq \{a, b, c, d\}.$$

However,

$$a \subseteq \{a, b, c, d\} \quad \text{and} \quad \{a\} \in \{a, b, c, d\}$$

are both *incorrect* uses of set notation.   ∎

Definition 1.2   EQUALITY OF SETS

Two sets are **equal** if and only if they contain exactly the same elements.

The sets A and B are equal, and we write $A = B$, if each member of A is also a member of B, and if each member of B is also a member of A. Typically, a proof that two sets are equal is presented in two parts. The first shows that $A \subseteq B$, and the second that $B \subseteq A$. We then conclude that $A = B$. We shall have an example of this type of proof shortly.

Definition 1.3   PROPER SUBSET

If A and B are sets, then A is a **proper subset** of B if and only if $A \subseteq B$ and $A \neq B$.

We sometimes write $A \subset B$ to denote that A is a proper subset of B.

Example 5   The following statements illustrate the notation for proper subsets and equality of sets.

$$\{1, 2, 4\} \subset \{1, 2, 3, 4, 5\} \qquad \{a, c\} = \{c, a\}$$ ∎

There are two basic operations, *union* and *intersection*, that are used to combine sets. These operations are defined as follows.

Definition 1.4   UNION, INTERSECTION

> If A and B are sets, the **union** of A and B is the set $A \cup B$ (read "A union B"), given by
>
> $$A \cup B = \{x \mid x \in A \text{ or } x \in B\}.$$
>
> The **intersection** of A and B is the set $A \cap B$ (read "A intersection B"), given by
>
> $$A \cap B = \{x \mid x \in A \text{ and } x \in B\}.$$

The union of two sets A and B is the set whose elements are either in A or in B or in both A and B. The intersection of sets A and B is the set of those elements common to both A and B.

Example 6   Suppose $A = \{2, 4, 6\}$ and $B = \{4, 5, 6, 7\}$. Then

$$A \cup B = \{2, 4, 5, 6, 7\}$$

and

$$A \cap B = \{4, 6\}. \qquad \blacksquare$$

The operations of union and intersection of two sets have some properties that are analogous to properties of addition and multiplication of numbers.

Example 7   It is easy to see that for any sets A and B, $A \cup B = B \cup A$:

$$\begin{aligned}
A \cup B &= \{x \mid x \in A \text{ or } x \in B\} \\
&= \{x \mid x \in B \text{ or } x \in A\} \\
&= B \cup A.
\end{aligned}$$

Because of the fact that $A \cup B = B \cup A$, we say that the operation union has the **commutative property**. It is just as easy to show that $A \cap B = B \cap A$, and we say also that the operation intersection has the **commutative property**. $\blacksquare$

It is easy to find sets that have no elements at all in common. For example, the sets

$$A = \{1, -1\} \quad \text{and} \quad B = \{0, 2, 3\}$$

have no elements in common. Hence, there are no elements in their intersection, $A \cap B$, and we say that the intersection is *empty*. Thus, it is logical to introduce the *empty set*.

Definition 1.5   EMPTY SET, DISJOINT SETS

> The **empty set** is the set that has no elements, and the empty set is denoted by $\varnothing$ or $\{\ \}$. Two sets A and B are called **disjoint** if and only if $A \cap B = \varnothing$.

The sets $\{1, -1\}$ and $\{0, 2, 3\}$ are disjoint, since

$$\{1, -1\} \cap \{0, 2, 3\} = \varnothing.$$

There is only one empty set $\varnothing$, and $\varnothing$ is a subset of every set. For a set A with n elements (n a nonnegative integer), we can write out all the subsets of A. For example, if

$$A = \{a, b, c\},$$

then the subsets of A are

$$\varnothing, \{a\}, \{b\}, \{c\}, \{a,b\}, \{a,c\}, \{b,c\}, A.$$

Definition 1.6 **POWER SET**

> For any set A, the **power set** of A, denoted by $\mathscr{P}(A)$, is the set of all subsets of A and is written
>
> $$\mathscr{P}(A) = \{X \mid X \subseteq A\}.$$

Example 8 For $A = \{a, b, c\}$, the power set of A is

$$\mathscr{P}(A) = \{\varnothing, \{a\}, \{b\}, \{c\}, \{a,b\}, \{a,c\}, \{b,c\}, A\}. \qquad \blacksquare$$

It is often helpful to draw a picture or diagram of the sets under discussion. When we do this, we assume that all the sets we are dealing with, along with all possible unions and intersections of those sets, are subsets of some **universal set**, denoted by U. In Figure 1-1, we let two overlapping circles represent the two sets A and B. The sets A and B are subsets of the universal set U, represented by the rectangle. Hence, the circles are contained in the rectangle. The intersection of A and B, $A \cap B$, is the crosshatched region where the two circles overlap. This type of pictorial representation is called a **Venn diagram**.

Figure 1-1

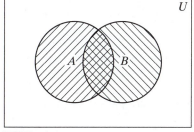

$\boxed{/\!/\!/}$: A

$\boxed{\backslash\!\backslash\!\backslash}$: B

$\boxed{\times\!\times\!\times}$: $A \cap B$

Another special subset is defined next.

Definition 1.7   COMPLEMENT

> For arbitrary subsets A and B of the universal set U, the **complement** of B in A is
>
> $$A - B = \{x \in U \mid x \in A \text{ and } x \notin B\}.$$

The special notation A' is reserved for a particular complement, $U - A$:

$$A' = U - A = \{x \in U \mid x \notin A\}.$$

We read A' as simply "the complement of A" rather than as "the complement of A in U."

Example 9   Let

$$U = \{x \mid x \text{ is an integer}\}$$
$$A = \{x \mid x \text{ is an even integer}\}$$
$$B = \{x \mid x \text{ is a positive integer}\}.$$

Then

$$B - A = \{x \mid x \text{ is a positive odd integer}\}$$
$$= \{1, 3, 5, 7, \ldots\}$$
$$A - B = \{x \mid x \text{ is a nonpositive even integer}\}$$
$$= \{0, -2, -4, -6, \ldots\}$$
$$A' = \{x \mid x \text{ is an odd integer}\}$$
$$= \{\ldots, -3, -1, 1, 3, \ldots\}$$
$$B' = \{x \mid x \text{ is a nonpositive integer}\}$$
$$= \{0, -1, -2, -3, \ldots\}. \qquad \blacksquare$$

Example 10   The overlapping circles representing the sets A and B separate the interior of the rectangle representing U into four regions, labeled 1, 2, 3, and 4, in the Venn diagram in Figure 1-2. Each region represents a particular subset of U.

Figure 1-2

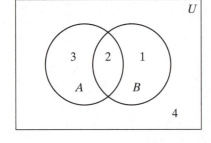

Region 1: $B - A$
Region 2: $A \cap B$
Region 3: $A - B$
Region 4: $(A \cup B)'$ $\blacksquare$

Many of the examples and exercises in this book involve familiar systems of numbers, and we adopt the following standard notations for some of these systems:

$\mathbf{Z}$ denotes the set of all **integers**;
$\mathbf{Z}^+$ denotes the set of all **positive integers**;
$\mathbf{Q}$ denotes the set of all **rational numbers**;

R denotes the set of all **real numbers**;
C denotes the set of all **complex numbers**.

We recall that a **complex number** is defined as a number of the form $a + bi$, where a and b are real numbers and $i = \sqrt{-1}$. Also, a real number x is **rational** if and only if x can be written as a quotient of integers that has a nonzero denominator. That is,

$$\mathbf{Q} = \left\{ \frac{m}{n} \;\middle|\; m \in \mathbf{Z}, n \in \mathbf{Z}, \text{ and } n \neq 0 \right\}.$$

The relationships that the number systems in the preceding paragraph have to each other is indicated by the Venn diagram in Figure 1-3.

Figure 1-3

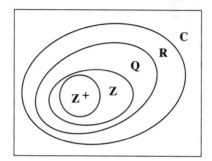

$$\mathbf{Z}^+ \subset \mathbf{Z} \subset \mathbf{Q} \subset \mathbf{R} \subset \mathbf{C}$$

Our work in this book usually assumes a knowledge of the various number systems that would be known from a precalculus or college algebra course. Some exceptions occur when we wish to examine how certain properties are consequences of other properties in a particular system. Exceptions of this kind occur with the integers in Chapter 2 and the complex numbers in Chapter 7, and these exceptions are clearly indicated when they occur.

The operations of union and intersection can be applied repeatedly. For instance, we might form the intersection of A and B, obtaining $A \cap B$, and then form the intersection of this set with a third set C: $(A \cap B) \cap C$.

Example 11   The sets $(A \cap B) \cap C$ and $A \cap (B \cap C)$ are equal, since

$$
\begin{aligned}
(A \cap B) \cap C &= \{x \mid x \in A \text{ and } x \in B\} \cap C \\
&= \{x \mid x \in A \text{ and } x \in B \text{ and } x \in C\} \\
&= A \cap \{x \mid x \in B \text{ and } x \in C\} \\
&= A \cap (B \cap C).
\end{aligned}
$$

In analogy with the associative property

$$x + (y + z) = (x + y) + z$$

for addition of numbers, we say that the operation of intersection is **associative**. When we work with numbers, we drop the parentheses for convenience and write

$$x + y + z = x + (y + z) = (x + y) + z.$$

Similarly, for sets A, B, and C, we write

$$A \cap B \cap C = A \cap (B \cap C) = (A \cap B) \cap C. \qquad \blacksquare$$

Just as simply, we can show (see Exercise 19 in this section) that the union of sets is an associative operation. We write

$$A \cup B \cup C = A \cup (B \cup C) = (A \cup B) \cup C.$$

Example 12   A separation of a nonempty set A into mutually disjoint nonempty subsets is called a **partition** of the set A. If

$$A = \{a, b, c, d, e, f\},$$

then one partition of A is

$$X_1 = \{a, d\}, \qquad X_2 = \{b, c, f\}, \qquad X_3 = \{e\},$$

since

$$A = X_1 \cup X_2 \cup X_3$$

with $X_1 \neq \varnothing$, $X_2 \neq \varnothing$, $X_3 \neq \varnothing$, and

$$X_1 \cap X_2 = \varnothing, \qquad X_1 \cap X_3 = \varnothing, \qquad X_2 \cap X_3 = \varnothing.$$

The concept of a partition is fundamental to many of the topics encountered later in this book. $\qquad \blacksquare$

The operations of intersection, union, and forming complements can be combined in all sorts of ways, and several nice equalities can be obtained that relate some of these results. For example, it can be shown that

$$A \cap (B \cup C) = (A \cap B) \cup (A \cap C)$$

and that

$$A \cup (B \cap C) = (A \cup B) \cap (A \cup C).$$

Because of the resemblance between these equations and the familiar distributive property $x(y + z) = xy + xz$ for numbers, we call these equations **distributive properties**.

We shall prove the first of these distributive laws in the next example and leave the last one as an exercise. To prove the first law, we shall show that $A \cap (B \cup C) \subseteq (A \cap B) \cup (A \cap C)$ and that $(A \cap B) \cup (A \cap C) \subseteq A \cap (B \cup C)$. This illustrates the point made earlier in the discussion of equality of sets, immediately after Definition 1.2.

The symbol $\Rightarrow$ is shorthand for "implies," and $\Leftarrow$ is shorthand for "is implied by." We use them in the next example.

Example 13   To prove

$$A \cap (B \cup C) = (A \cap B) \cup (A \cap C),$$

we first let $x \in A \cap (B \cup C)$. Now

$$
\begin{aligned}
x \in A \cap (B \cup C) &\Rightarrow x \in A \quad \text{and} \quad x \in (B \cup C) \\
&\Rightarrow x \in A, \quad \text{and} \quad x \in B \quad \text{or} \quad x \in C \\
&\Rightarrow x \in A \quad \text{and} \quad x \in B, \quad \text{or} \quad x \in A \quad \text{and} \quad x \in C \\
&\Rightarrow x \in A \cap B, \quad \text{or} \quad x \in A \cap C \\
&\Rightarrow x \in (A \cap B) \cup (A \cap C).
\end{aligned}
$$

Thus, $A \cap (B \cup C) \subseteq (A \cap B) \cup (A \cap C)$.

Conversely, suppose $x \in (A \cap B) \cup (A \cap C)$. Then

$$
\begin{aligned}
x \in (A \cap B) \cup (A \cap C) &\Rightarrow x \in A \cap B, \quad \text{or} \quad x \in A \cap C \\
&\Rightarrow x \in A \quad \text{and} \quad x \in B, \quad \text{or} \quad x \in A \quad \text{and} \quad x \in C \\
&\Rightarrow x \in A, \quad \text{and} \quad x \in B \quad \text{or} \quad x \in C \\
&\Rightarrow x \in A \quad \text{and} \quad x \in (B \cup C) \\
&\Rightarrow x \in A \cap (B \cup C).
\end{aligned}
$$

Therefore, $(A \cap B) \cup (A \cap C) \subseteq A \cap (B \cup C)$, and we have shown that $A \cap (B \cup C) = (A \cap B) \cup (A \cap C)$.

It should be evident that the second part of the proof can be obtained from the first simply by reversing the steps. That is, when each $\Rightarrow$ is replaced by $\Leftarrow$, a valid implication results. In fact, then, we could obtain a proof of both parts by replacing $\Rightarrow$ with $\Leftrightarrow$, where $\Leftrightarrow$ is short for "if and only if." Thus,

$$
\begin{aligned}
x \in A \cap (B \cup C) &\Leftrightarrow x \in A \quad \text{and} \quad x \in (B \cup C) \\
&\Leftrightarrow x \in A, \quad \text{and} \quad x \in B \quad \text{or} \quad x \in C \\
&\Leftrightarrow x \in A \quad \text{and} \quad x \in B, \quad \text{or} \quad x \in A \quad \text{and} \quad x \in C \\
&\Leftrightarrow x \in A \cap B, \quad \text{or} \quad x \in A \cap C \\
&\Leftrightarrow x \in (A \cap B) \cup (A \cap C). \qquad \blacksquare
\end{aligned}
$$

STRATEGY ▶

In proving an equality of sets S and T, we can often use the technique of showing that $S \subseteq T$ and then check to see if the steps are reversible. In many cases, the steps are indeed reversible, and we obtain the other part of the proof easily. However, this method should not obscure the fact that there are still two parts to the argument: $S \subseteq T$ and $T \subseteq S$.

There are some interesting relations between complements and unions or intersections. For example, it is true that

$$
(A \cap B)' = A' \cup B'.
$$

This statement is one of two that are known as **De Morgan's Laws**. De Morgan's other law is the statement that

$$
(A \cup B)' = A' \cap B'.
$$

Stated somewhat loosely in words, the first law says that the complement of an intersection is the union of the individual complements. The second similarly says that the complement of a union is the intersection of the individual complements.

EXERCISES I.I

1. For each set A, describe A by indicating a property that is a qualification for membership in A.
   - **a.** $A = \{0, 2, 4, 6, 8, 10\}$
   - **b.** $A = \{1, -1\}$
   - **c.** $A = \{-1, -2, -3, \ldots\}$
   - **d.** $A = \{1, 4, 9, 16, 25, \ldots\}$

2. Decide whether or not each statement is true for $A = \{2, 7, 11\}$ and $B = \{1, 2, 9, 10, 11\}$.
   - **a.** $2 \subseteq A$
   - **b.** $\{11, 2, 7\} \subseteq A$
   - **c.** $2 = A \cap B$
   - **d.** $\{7, 11\} \in A$
   - **e.** $A \subseteq B$
   - **f.** $\{7, 11, 2\} = A$

3. Decide whether or not each statement is true, where A and B are arbitrary sets.
   - **a.** $B \cup A \subseteq A$
   - **b.** $B \cap A \subseteq A \cup B$
   - **c.** $\varnothing \subseteq A$
   - **d.** $0 \in \varnothing$
   - **e.** $\varnothing \in \{\varnothing\}$
   - **f.** $\varnothing \subseteq \{\varnothing\}$
   - **g.** $\{\varnothing\} \subseteq \varnothing$
   - **h.** $\{\varnothing\} = \varnothing$
   - **i.** $\varnothing \in \varnothing$
   - **j.** $\varnothing \subseteq \varnothing$

4. Decide whether or not each of the following is true for all sets A, B, and C.
   - **a.** $A \cap A' = \varnothing$
   - **b.** $A \cap \varnothing = A \cup \varnothing$
   - **c.** $A \cap (B \cup C) = A \cup (B \cap C)$
   - **d.** $A \cup (B' \cap C') = A \cup (B \cup C)'$
   - **e.** $A \cup (B \cap C) = (A \cup B) \cap C$
   - **f.** $(A \cap B) \cup C = A \cap (B \cup C)$
   - **g.** $A \cup (B \cap C) = (A \cap C) \cup (B \cap C)$
   - **h.** $A \cap (B \cup C) = (A \cup B) \cap (A \cup C)$

5. Evaluate each of the following sets, where

$$U = \{0, 1, 2, 3, \ldots, 10\}$$
$$A = \{0, 1, 2, 3, 4, 5\}$$
$$B = \{0, 2, 4, 6, 8, 10\}$$
$$C = \{2, 3, 5, 7\}.$$

   - **a.** $A \cup B$
   - **b.** $A \cap C$
   - **c.** $A' \cup B$
   - **d.** $A \cap B \cap C$
   - **e.** $A' \cap B \cap C$
   - **f.** $A \cup (B \cap C)$
   - **g.** $A \cap (B \cup C)$
   - **h.** $(A \cup B')'$
   - **i.** $A - B$
   - **j.** $B - A$
   - **k.** $A - (B - C)$
   - **l.** $C - (B - A)$
   - **m.** $(A - B) \cap (C - B)$
   - **n.** $(A - B) \cap (A - C)$

6. Write each of the following as either A, A', U, or $\varnothing$, where A is an arbitrary subset of the universal set U.
   - **a.** $A \cap A$
   - **b.** $A \cup A$
   - **c.** $A \cap A'$
   - **d.** $A \cup A'$
   - **e.** $A \cup \varnothing$
   - **f.** $A \cap \varnothing$
   - **g.** $A \cap U$
   - **h.** $A \cup U$
   - **i.** $U \cup A'$
   - **j.** $A - \varnothing$
   - **k.** $\varnothing'$
   - **l.** U'
   - **m.** $(A')'$
   - **n.** $\varnothing - A$

7. Write out the power set, $\mathcal{P}(A)$, for each set A.
   - **a.** $A = \{a\}$
   - **b.** $A = \{0, 1\}$
   - **c.** $A = \{a, b, c\}$
   - **d.** $A = \{1, 2, 3, 4\}$

Sec. 3.1, #51, 52, 53 ≪

8. Describe two partitions of each of the following sets.
   - **a.** $\{x \mid x \text{ is an integer}\}$
   - **b.** $\{a, b, c, d\}$
   - **c.** $\{1, 5, 9, 11, 15\}$
   - **d.** $\{x \mid x \text{ is a complex number}\}$

9. Write out all the different partitions of the given set A.
   - **a.** $A = \{1, 2, 3\}$
   - **b.** $A = \{1, 2, 3, 4\}$

10. If $n \in \mathbf{Z}^+$ and the set A has n elements, how many elements does the power set $\mathcal{P}(A)$ have?

11. Give an example of sets A and B such that $\mathcal{P}(A \cup B) \neq \mathcal{P}(A) \cup \mathcal{P}(B)$.

12. State the most general conditions on the subsets A and B of U under which the given equality holds.

    a. $A \cap B = A$

    b. $A \cup B' = A$

    c. $A \cup B = A$

    d. $A \cap B' = A$

    e. $A \cap B = U$

    f. $A' \cap B' = \varnothing$

    g. $A \cup \varnothing = U$

    h. $A' \cap U = \varnothing$

13. Let $\mathbf{Z}$ denote the set of all integers, and let

$$A = \{x \mid x = 3p - 2 \text{ for some } p \in \mathbf{Z}\},$$
$$B = \{x \mid x = 3q + 1 \text{ for some } q \in \mathbf{Z}\}.$$

    Prove that $A = B$.

14. Let $\mathbf{Z}$ denote the set of all integers, and let

$$C = \{x \mid x = 3r - 1 \text{ for some } r \in \mathbf{Z}\},$$
$$D = \{x \mid x = 3s + 2 \text{ for some } s \in \mathbf{Z}\}.$$

    Prove that $C = D$.

In Exercises 15–32, prove each statement.

15. $A \cap B \subseteq A \cup B$

16. $(A')' = A$

17. If $A \subseteq B$ and $B \subseteq C$, then $A \subseteq C$.

18. $A \subseteq B$ if and only if $B' \subseteq A'$.

19. $A \cup (B \cup C) = (A \cup B) \cup C$

20. $(A \cup B)' = A' \cap B'$

21. $(A \cap B)' = A' \cup B'$

22. $A \cup (B \cap C) = (A \cup B) \cap (A \cup C)$

23. $A \cap (A' \cup B) = A \cap B$

24. $A \cup (A' \cap B) = A \cup B$

25. $A \cup (A \cap B) = A \cap (A \cup B)$

26. If $A \subseteq B$, then $A \cup C \subseteq B \cup C$.

27. If $A \subseteq B$, then $A \cap C \subseteq B \cap C$.

28. $B - A = B \cap A'$

29. $A \cap (B - A) = \varnothing$

30. $A \cup (B - A) = A \cup B$

31. $(A \cup B) - C = (A - C) \cup (B - C)$

32. $(A - B) \cup (A \cap B) = A$

33. Express $(A \cup B) - (A \cap B)$ in terms of unions and intersections that involve A, A', B, and B'.

34. Let the operation of addition be defined on subsets A and B of U by $A + B = (A \cup B) - (A \cap B)$. Use a Venn diagram with labeled regions to illustrate each of the following statements.

    a. $A + B = (A - B) \cup (B - A)$

    b. $A + (B + C) = (A + B) + C$

    c. $A \cap (B + C) = (A \cap B) + (A \cap C)$.

1.2 MAPPINGS

The concept of a function is fundamental to nearly all areas of mathematics. The term *function* is the one most widely used for the concept that we have in mind, but it has become traditional to use the terms *mapping* and *transformation* in algebra. It is likely that these words are used because they express an intuitive feel for the association between the elements involved. The basic idea is that correspondences of a certain type exist between the elements of two sets. There is to be a rule of association between the elements of a first set and those of a second set. The association is to be such that for each element in the first set, there is one and only one associated element in the second set. This rule of association leads to a natural pairing of the elements that are to correspond, and then to the formal statement in Definition 1.9.

By an **ordered pair** of elements we mean a pairing (a, b), where there is to be a distinction between the pair (a, b) and the pair (b, a), if a and b are different. That is, there is to be a first position and a second position such that $(a, b) = (c, d)$ if and only if both $a = c$ and $b = d$. This ordering is altogether different from listing the elements of a set, for there the order of listing is of no consequence at all. The sets $\{1, 2\}$ and $\{2, 1\}$ have exactly the same elements, and $\{1, 2\} = \{2, 1\}$. When we speak of ordered pairs, however, $(1, 2)$ and $(2, 1)$ are not considered to be equal. With these ideas in mind, we make the following definition.

Definition 1.8   CARTESIAN PRODUCT

> For two nonempty sets A and B, the **Cartesian product** $A \times B$ is the set of all ordered pairs (a, b) of elements $a \in A$ and $b \in B$. That is,
>
> $$A \times B = \{(a, b) \mid a \in A \text{ and } b \in B\}.$$

Example 1   If $A = \{1, 2\}$ and $B = \{3, 4, 5\}$, then

$$A \times B = \{(1, 3), (1, 4), (1, 5), (2, 3), (2, 4), (2, 5)\}.$$

We observe that the order in which the sets appear is important. In this example,

$$B \times A = \{(3, 1), (3, 2), (4, 1), (4, 2), (5, 1), (5, 2)\},$$

so $A \times B$ and $B \times A$ are quite distinct from each other.   ■

We can now make our formal definition of a mapping.

Definition 1.9   MAPPING, IMAGE

> Let A and B be nonempty sets. A subset f of $A \times B$ is a **mapping** from A to B if and only if for each $a \in A$, there is a unique (one and only one) element $b \in B$ such that $(a, b) \in f$. If f is a mapping from A to B and the pair (a, b) belongs to f, we write $b = f(a)$ and call b the **image** of a under f.

Figure 1-4 illustrates the pairing between a and $f(a)$. A mapping f from A to B is the same as a function from A to B, and the image of $a \in A$ under f is the same as the value of the function f at a. Two mappings f from A to B and g from A to B are **equal** if and only if $f(x) = g(x)$ for all $x \in A$.

Figure 1-4

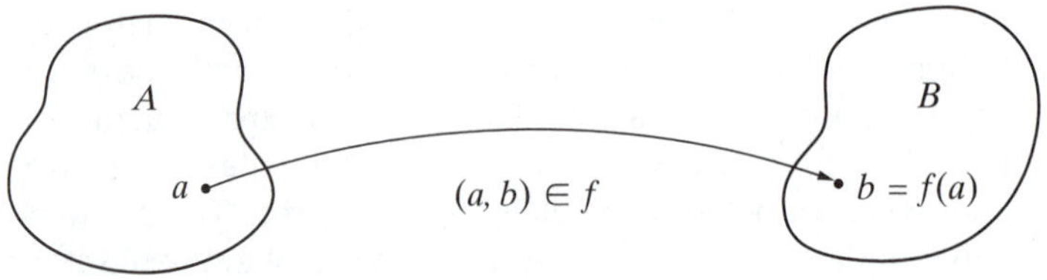

Example 2 Let $A = \{-2, 1, 2\}$, and let $B = \{1, 4, 9\}$. The set f given by

$$f = \{(-2, 4), (1, 1), (2, 4)\}$$

is a mapping from A to B, since for each $a \in A$ there is a unique element $b \in B$ such that $(a, b) \in f$. As is frequently the case, this mapping can be efficiently described by giving the rule for the image under f. In this case, $f(a) = a^2$, $a \in A$. This mapping is pictured in Figure 1-5.

Figure 1-5

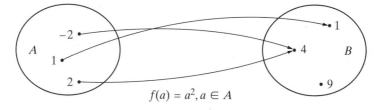

$$f(a) = a^2, a \in A$$

When it is possible to describe a mapping by giving a simple rule for the image of an element, it is certainly desirable to do so. We must keep in mind, however, that the set A, the set B, and the rule must all be known before the mapping is determined. If f is a mapping from A to B, we write $f: A \rightarrow B$ or $A \xrightarrow{f} B$ to indicate this.

Definition 1.10 DOMAIN, CODOMAIN, RANGE

> Let f be a mapping from A to B. The set A is called the **domain** of f, and B is called the **codomain** of f. The **range** of f is the set
>
> $$C = \{y \mid y \in B \text{ and } y = f(x) \text{ for some } x \in A\}.$$
>
> The range of f is denoted by $f(A)$.

Example 3 Let $A = \{-2, 1, 2\}$ and $B = \{1, 4, 9\}$, and let f be the mapping described in the previous example:

$$f = \{(a, b) \mid f(a) = a^2, a \in A\}.$$

The domain of f is A, the codomain of f is B, and the range of f is $\{1, 4\} \subset B$.   ∎

If $f: A \rightarrow B$, the notation used in Definition 1.10 can be extended as follows to arbitrary subsets $S \subseteq A$.

Definition 1.11 IMAGE, INVERSE IMAGE

> If $f: A \rightarrow B$ and $S \subseteq A$, then
>
> $$f(S) = \{y \mid y \in B \text{ and } y = f(x) \text{ for some } x \in S\}.$$
>
> The set $f(S)$ is called the **image** of S under f. For any subset T of B, the **inverse image** of T is denoted by $f^{-1}(T)$ and defined by
>
> $$f^{-1}(T) = \{x \mid x \in A \text{ and } f(x) \in T\}.$$

We note that the image $f(A)$ is the same as the range of f. Also, both notations $f(S)$ and $f^{-1}(T)$ in Definition 1.11 denote *sets*, not values of a mapping. We illustrate these notations in the next example.

Example 4   Let $f: A \to B$ as in Example 3. If $S = \{1, 2\}$, then $f(S) = \{1, 4\}$ as shown in Figure 1-6.

Figure 1-6

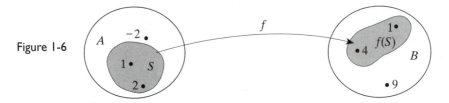

With $T = \{4, 9\}$, $f^{-1}(T)$ is given by $f^{-1}(T) = \{-2, 2\}$ as shown in Figure 1-7.

Figure 1-7

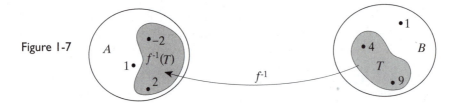

■

Among the various mappings from a nonempty set A to a nonempty set B, there are some that have properties worthy of special designation. We make the following definition.

Definition 1.12   **ONTO, SURJECTIVE**

> Let $f: A \to B$. Then f is called **onto**, or **surjective**, if and only if $B = f(A)$. Alternately, an onto mapping f is called a mapping from A **onto B**.

We begin our discussion of *onto* mappings by first describing what is meant by a mapping that does not satisfy the requirement in Definition 1.12. To show that a given mapping $f: A \to B$ is **not onto**, we need only find a single element b in B for which no $x \in A$ exists such that $f(x) = b$. Such an element b and the sets A, B, and $f(A)$ are diagrammed in Figure 1-8.

Figure 1-8

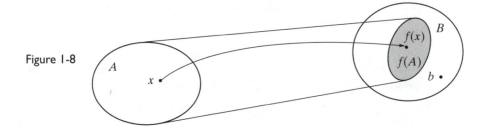

Example 5   Suppose we have $f: A \to B$ where $A = \{-1, 0, 1\}$, $B = \{4, -4\}$, and $f = \{(-1, 4), (0, 4), (1, 4)\}$. The mapping f is not onto, since there is no $a \in A$ such that $f(a) = -4 \in B$. ∎

STRATEGY ▶

According to our definition, a mapping f from A to B is onto if and only if every element of B is the image of at least one element in A. A standard way to demonstrate that $f: A \to B$ is onto is to take an arbitrary element b in B and show (usually by some kind of formula) that there exists an element $a \in A$ such that $b = f(a)$.

Example 6   Let $f: \mathbf{Z} \to \mathbf{Z}$, where $\mathbf{Z}$ is the set of integers. If f is defined by

$$f = \{(a, 2 - a) \mid a \in \mathbf{Z}\},$$

then we write $f(a) = 2 - a, a \in \mathbf{Z}$.

To show that f is onto (surjective), we choose an arbitrary element $b \in \mathbf{Z}$. Then there exists $2 - b \in \mathbf{Z}$ such that

$$(2 - b, b) \in f$$

since $f(2 - b) = 2 - (2 - b) = b$, and hence f is onto. ∎

Definition 1.13   **ONE-TO-ONE, INJECTIVE**

Let $f: A \to B$. Then f is called **one-to-one**, or **injective**, if and only if different elements of A always have different images under f.

Analogous to our treatment of the onto property, we first examine the situation when a mapping fails to have the one-to-one property. To show that f is **not one-to-one**, we need only find two elements $a_1 \in A$ and $a_2 \in A$ such that $a_1 \neq a_2$ and $f(a_1) = f(a_2)$. A pair of elements with this property is shown in Figure 1-9.

Figure 1-9

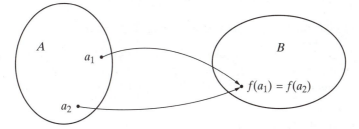

STRATEGY ▶

The preceeding discussion illustrates how *only one exception is needed to show that a given statement is false*. An example that provides such an exception is referred to as a **counterexample**.

Example 7 Suppose we reconsider the mapping $f: A \to B$ from Example 5 where $A = \{-1, 0, 1\}$, $B = \{4, -4\}$, and $f = \{(-1, 4), (0, 4), (1, 4)\}$. We see that f is not one-to-one, since

$$f(-1) = f(0) = 4 \quad \text{but} \quad -1 \neq 0.$$ ∎

A mapping $f: A \to B$ is one-to-one if and only if it has the property that $a_1 \neq a_2$ in A always implies that $f(a_1) \neq f(a_2)$ in B. This is just a precise statement of the fact that different elements always have different images. The trouble with this statement is that it is formulated in terms of unequal quantities, whereas most of the manipulations in mathematics deal with equalities. For this reason, we take the logically equivalent *contrapositive* statement "$f(a_1) = f(a_2)$ always implies $a_1 = a_2$" as our working form of the definition.

STRATEGY ▶

We usually show that f is one-to-one by assuming that $f(a_1) = f(a_2)$ and proving that this implies that $a_1 = a_2$.

This strategy is used to show that the mapping in Example 6 is one-to-one.

Example 8 Suppose $f: \mathbf{Z} \to \mathbf{Z}$ is defined by

$$f = \{(a, 2 - a) | a \in \mathbf{Z}\}.$$

To show that f is one-to-one (injective), we assume that for $a_1 \in \mathbf{Z}$ and $a_2 \in \mathbf{Z}$,

$$f(a_1) = f(a_2).$$

Then we have

$$2 - a_1 = 2 - a_2,$$

and this implies that $a_1 = a_2$. Thus, f is one-to-one. ∎

Definition 1.14 **ONE-TO-ONE CORRESPONDENCE, BIJECTION**

Let $f: A \to B$. The mapping f is called **bijective** if and only if f is both surjective and injective. A bijective mapping from A to B is called a **one-to-one correspondence** from A to B, or a **bijection** from A to B.

Example 9 The mapping $f: \mathbf{Z} \to \mathbf{Z}$ defined in Example 6 by

$$f = \{(a, 2 - a) | a \in \mathbf{Z}\}$$

is both onto and one-to-one. Thus, f is a one-to-one correspondence. ∎

Just after Example 10 in Section 1.1, the symbols $\mathbf{Z}$, $\mathbf{Z}^+$, $\mathbf{Q}$, $\mathbf{R}$, and $\mathbf{C}$ were introduced as standard notations for some of the number systems. Another set of numbers that we use often enough to justify a special notation is the set of all even integers. The set **E** of all even integers includes 0 and all negative even integers, $-2, -4, -6, \ldots$, as well as the positive even integers, $2, 4, 6, \ldots$. Thus,

$$\mathbf{E} = \{\ldots, -6, -4, -2, 0, 2, 4, 6, \ldots\},$$

and we define n to be an **even integer** if and only if $n = 2k$ for some integer k. An integer n is defined to be an **odd integer** if and only if $n = 2k + 1$ for some integer k and the set of all odd integers is the complement of **E** in **Z**:

$$\mathbf{Z} - \mathbf{E} = \{\ldots, -5, -3, -1, 1, 3, 5, \ldots\}.$$

Note that we could also define an odd integer by using the expression $n = 2j - 1$ for some integer j.

The next two examples show that a mapping may be onto but not one-to-one, or it may be one-to-one but not onto.

Example 10   In this example, we encounter a mapping that is onto but not one-to-one. Let $h\colon \mathbf{Z} \to \mathbf{Z}$ be defined by

$$h(x) = \begin{cases} \dfrac{x - 2}{2} & \text{if } x \text{ is even} \\[2ex] \dfrac{x - 3}{2} & \text{if } x \text{ is odd.} \end{cases}$$

To attempt a proof that h is onto, let b be an arbitrary element in **Z** and consider the equation $h(x) = b$. There are two possible values for $h(x)$, depending on whether x is even or odd. Considering both of these values, we have

$$\frac{x - 2}{2} = b \quad \text{for } x \text{ even,} \quad \text{or} \quad \frac{x - 3}{2} = b \quad \text{for } x \text{ odd.}$$

Solving each of these equations separately for x yields

$$x = 2b + 2 \quad \text{for } x \text{ even,} \quad \text{or} \quad x = 2b + 3 \quad \text{for } x \text{ odd.}$$

We note that $2b + 2 = 2(b + 1)$ is an even integer for every choice of b in **Z** and that $2b + 3 = 2(b + 1) + 1$ is an odd integer for every choice of b in **Z**. Thus, there are two values, $2b + 2$ and $2b + 3$, for x in **Z** such that

$$h(2b + 2) = b \quad \text{and} \quad h(2b + 3) = b.$$

This proves that h is onto. Since $2b + 2 \neq 2b + 3$ and $h(2b + 2) = h(2b + 3)$, we have also proved that h is not one-to-one.   ∎

Example 11   Consider now the mapping $f\colon \mathbf{Z} \to \mathbf{Z}$ defined by

$$f(x) = 2x + 1.$$

To attempt a proof that f is onto, consider an arbitrary element b in **Z**. We have

$$\begin{aligned} f(x) = b &\Leftrightarrow 2x + 1 = b \\ &\Leftrightarrow 2x \qquad = b - 1, \end{aligned}$$

and the equation $2x = b - 1$ has a solution x in $\mathbf{Z}$ if and only if $b - 1$ is an even integer—that is, if and only if b is an odd integer. Thus, only odd integers are in the range of f, and therefore f is not onto.

The proof that f is one-to-one is straightforward:

$$f(m) = f(n) \Rightarrow 2m + 1 = 2n + 1$$
$$\Rightarrow \quad 2m = 2n$$
$$\Rightarrow \quad m = n.$$

Thus, f is one-to-one even though it is not onto. ∎

In Section 3.1 and other places in our work, we need to be able to apply two mappings in succession, one after the other. In order for this successive application to be possible, the mappings involved must be compatible, as required in the next definition.

Definition 1.15 **COMPOSITE MAPPING**

Let $g: A \to B$ and $f: B \to C$. The **composite mapping** $f \circ g$ is the mapping from A to C defined by

$$(f \circ g)(x) = f(g(x))$$

for all $x \in A$.

The process of forming the composite mapping is called **composition of mappings** and the result $f \circ g$ is sometimes called the **composition** of g and f. Readers familiar with calculus will recognize this as the setting for the *chain rule* of derivatives.

The composite mapping $f \circ g$ is diagrammed in Figure 1-10. Note that the domain of f must contain the range of g before the composition $f \circ g$ is defined.

Figure 1-10

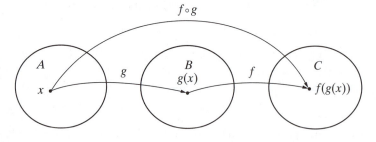

Example 12 Let $\mathbf{Z}$ be the set of integers, A the set of nonnegative integers, and B the set of nonpositive integers. Suppose the mappings g and f are defined as

$$g: \mathbf{Z} \to A, \quad g(x) = x^2$$
$$f: A \to B, \quad f(x) = -x.$$

Then the composition $f \circ g$ is a mapping from $\mathbf{Z}$ to B with

$$(f \circ g)(x) = f(g(x)) = f(x^2) = -x^2.$$

Note that $f \circ g$ is not onto, since $-3 \in B$, but there is no integer x such that

$$(f \circ g)(x) = -x^2 = -3.$$

Also, $f \circ g$ is not one-to-one, since

$$(f \circ g)(-2) = -(-2)^2 = -4 = (f \circ g)(2)$$

and

$$-2 \neq 2. \qquad \blacksquare$$

In connection with the composition of mappings, a word of caution about notation is in order. Some mathematicians use the notation xf to indicate the image of x under f. That is, both notations xf and $f(x)$ represent the value of f at x. When the xf notation is used, mappings are applied from left to right, and the composite mapping $f \circ g$ is defined by the equation $x(f \circ g) = (xf)g$. We consistently use the $f(x)$ notation in this book, but the xf notation is found in some other texts on algebra.

When the composite mapping can be formed, we have an operation defined that is **associative**. If $h\colon A \to B, g\colon B \to C$, and $f\colon C \to D$, then

$$\begin{aligned}
((f \circ g) \circ h)(x) &= (f \circ g)(h(x)) \\
&= f[g(h(x))] \\
&= f((g \circ h)(x)) \\
&= (f \circ (g \circ h))(x)
\end{aligned}$$

for all $x \in A$. Thus, the compositions $(f \circ g) \circ h$ and $f \circ (g \circ h)$ are the same mapping from A to D.

EXERCISES 1.2

1. For the given sets, form the indicated Cartesian product.
   a. $A \times B; A = \{a, b\}, B = \{0, 1\}$
   b. $B \times A; A = \{a, b\}, B = \{0, 1\}$
   c. $A \times B; A = \{2, 4, 6, 8\}, B = \{2\}$
   d. $B \times A; A = \{1, 5, 9\}, B = \{-1, 1\}$
   e. $B \times A; A = B = \{1, 2, 3\}$

2. For each of the following mappings, state the domain, the codomain, and the range, where $f\colon \mathbf{E} \to \mathbf{Z}$.
   a. $f(x) = x/2, x \in \mathbf{E}$ b. $f(x) = x, x \in \mathbf{E}$
   c. $f(x) = |x|, x \in \mathbf{E}$ d. $f(x) = x + 1, x \in \mathbf{E}$

3. For each of the following mappings, write out $f(S)$ and $f^{-1}(T)$ for the given S and T, where $f\colon \mathbf{Z} \to \mathbf{Z}$.
   a. $f(x) = |x|; S = \mathbf{Z} - \mathbf{E}, T = \{1, 3, 4\}$
   b. $f(x) = \begin{cases} x + 1 & \text{if } x \text{ is even} \\ x & \text{if } x \text{ is odd}; \end{cases}$ $S = \{0, 1, 5, 9\}, T = \mathbf{Z} - \mathbf{E}$
   c. $f(x) = x^2; S = \{-2, -1, 0, 1, 2\}, T = \{2, 7, 11\}$

4. For each of the following mappings $f\colon \mathbf{Z} \to \mathbf{Z}$, determine if the mapping is onto and if it is one-to-one. Justify all negative answers.
   a. $f(x) = 2x$ b. $f(x) = 3x$

c. $f(x) = x + 3$

d. $f(x) = x^3$

e. $f(x) = |x|$

f. $f(x) = x - |x|$

g. $f(x) = \begin{cases} x & \text{if } x \text{ is even} \\ 2x - 1 & \text{if } x \text{ is odd} \end{cases}$

h. $f(x) = \begin{cases} x & \text{if } x \text{ is even} \\ x - 1 & \text{if } x \text{ is odd} \end{cases}$

i. $f(x) = \begin{cases} x & \text{if } x \text{ is even} \\ \dfrac{x - 1}{2} & \text{if } x \text{ is odd} \end{cases}$

j. $f(x) = \begin{cases} x - 1 & \text{if } x \text{ is even} \\ 2x & \text{if } x \text{ is odd} \end{cases}$

5. For each of the following mappings $f: \mathbf{R} \to \mathbf{R}$, determine if the mapping is onto and if it is one-to-one. Justify all negative answers. (Compare these results with the corresponding parts of Exercise 4.)

   a. $f(x) = 2x$

   b. $f(x) = 3x$

   c. $f(x) = x + 3$

   d. $f(x) = x^3$

   e. $f(x) = |x|$

   f. $f(x) = x - |x|$

6. For the given subsets A and B of $\mathbf{Z}$, let $f(x) = 2x$ and determine whether $f: A \to B$ is onto and if it is one-to-one. Justify all negative answers.

   a. $A = \mathbf{Z}, B = \mathbf{E}$

   b. $A = \mathbf{E}, B = \mathbf{E}$

7. For the given subsets A and B of $\mathbf{Z}$, let $f(x) = |x|$ and determine whether $f: A \to B$ is onto and if it is one-to-one. Justify all negative answers.

   a. $A = \mathbf{Z}, B = \mathbf{Z}^+ \cup \{0\}$

   b. $A = \mathbf{Z}^+, B = \mathbf{Z}$

   c. $A = \mathbf{Z}^+, B = \mathbf{Z}^+$

8. For each of the following parts, give an example of a mapping from $\mathbf{E}$ to $\mathbf{E}$ that satisfies the given conditions.

   a. one-to-one and onto

   b. one-to-one and not onto

   c. onto and not one-to-one

   d. not one-to-one and not onto

9. For the given $f: \mathbf{Z} \to \mathbf{Z}$, determine whether f is onto or one-to-one and prove that your conclusions are correct.

   a. $f(x) = \begin{cases} 2x + 1 & \text{if } x \text{ is even} \\ \dfrac{x + 1}{2} & \text{if } x \text{ is odd} \end{cases}$

   b. $f(x) = \begin{cases} \dfrac{x}{2} & \text{if } x \text{ is even} \\ \dfrac{x - 3}{2} & \text{if } x \text{ is odd} \end{cases}$

   c. $f(x) = \begin{cases} 3x & \text{if } x \text{ is even} \\ 2x & \text{if } x \text{ is odd} \end{cases}$

   d. $f(x) = \begin{cases} 2x - 1 & \text{if } x \text{ is even} \\ 2x & \text{if } x \text{ is odd} \end{cases}$

10. Let $A = \mathbf{R} - \{0\}$ and $B = \mathbf{R}$. For the given $f: A \to B$, determine if f is onto or one-to-one and prove that your decisions are correct.

    a. $f(x) = \dfrac{x - 1}{x}$

    b. $f(x) = \dfrac{2x - 1}{x}$

    c. $f(x) = \dfrac{x}{x^2 + 1}$

    d. $f(x) = \dfrac{2x - 1}{x^2 + 1}$

11. For the given $f: A \to B$, determine whether f is onto or one-to-one. Prove that your conclusions are correct.

    a. $A = \mathbf{Z} \times \mathbf{Z}, B = \mathbf{Z} \times \mathbf{Z}, f(x, y) = (y, x)$

    b. $A = \mathbf{Z} \times \mathbf{Z}, B = \mathbf{Z}, f(x, y) = x + y$

    c. $A = \mathbf{Z} \times \mathbf{Z}, B = \mathbf{Z}, f(x, y) = x$

    d. $A = \mathbf{Z}, B = \mathbf{Z} \times \mathbf{Z}, f(x) = (x, 1)$

12. **a.** Show that the mapping f given in Example 2 is neither onto nor one-to-one.

    b. For this mapping f, show that if $S = \{1, 2\}$, then $f^{-1}(f(S)) \neq S$.

    c. For this same f and $T = \{4, 9\}$, show that $f(f^{-1}(T)) \neq T$.

13. Let $g: \mathbf{Z} \to \mathbf{Z}$ be given by $g(x) = \begin{cases} x & \text{if } x \text{ is even} \\ \dfrac{x+1}{2} & \text{if } x \text{ is odd.} \end{cases}$

    a. For $S = \{3, 4\}$, find $g(S)$ and $g^{-1}(g(S))$.

    b. For $T = \{5, 6\}$, find $g^{-1}(T)$ and $g(g^{-1}(T))$.

14. Let $f: \mathbf{Z} \to \mathbf{Z}$ be given by $f(x) = \begin{cases} 2x - 1 & \text{if } x \text{ is even} \\ 2x & \text{if } x \text{ is odd.} \end{cases}$

    a. For $S = \{0, 1, 2\}$, find $f(S)$ and $f^{-1}(f(S))$.

    b. For $T = \{-1, 1, 4\}$, find $f^{-1}(T)$ and $f(f^{-1}(T))$.

15. Let $f: \mathbf{Z} \to \mathbf{Z}$ and $g: \mathbf{Z} \to \mathbf{Z}$ be defined as follows. In each case, compute $(f \circ g)(x)$ for arbitrary $x \in \mathbf{Z}$.

    a. $f(x) = 2x$, $g(x) = \begin{cases} x & \text{if } x \text{ is even} \\ 2x - 1 & \text{if } x \text{ is odd} \end{cases}$

    b. $f(x) = 2x, g(x) = x^3$

    c. $f(x) = x + |x|$, $g(x) = \begin{cases} \dfrac{x}{2} & \text{if } x \text{ is even} \\ -x & \text{if } x \text{ is odd} \end{cases}$

    d. $f(x) = \begin{cases} \dfrac{x}{2} & \text{if } x \text{ is even} \\ x + 1 & \text{if } x \text{ is odd} \end{cases}$ $g(x) = \begin{cases} x - 1 & \text{if } x \text{ is even} \\ 2x & \text{if } x \text{ is odd} \end{cases}$

    e. $f(x) = x^2, g(x) = x - |x|$

16. Let f and g be defined in the various parts of Exercise 15. In each part, compute $(g \circ f)(x)$ for arbitrary $x \in \mathbf{Z}$.

In Exercises 17–20, suppose m and n are positive integers, A is a set with exactly m elements, and B is a set with exactly n elements.

17. How many mappings are there from A to B ?

18. If $m = n$, how many one-to-one correspondences are there from A to B ?

19. If $m \le n$, how many one-to-one mappings are there from A to B ?

20. If $m \ge n$, how many onto mappings are there from A to B ?

21. Let a and b be constant integers with $a \ne 0$, and let the mapping $f: \mathbf{Z} \to \mathbf{Z}$ be defined by $f(x) = ax + b$.

    a. Prove that f is one-to-one.

    b. Prove that f is onto if and only if $a = 1$ or $a = -1$.

22. Let $f: A \to B$, where A and B are nonempty.

    a. Prove that $f(S_1 \cup S_2) = f(S_1) \cup f(S_2)$ for all subsets S_1 and S_2 of A.

    b. Prove that $f(S_1 \cap S_2) \subseteq f(S_1) \cap f(S_2)$ for all subsets S_1 and S_2 of A.

    c. Give an example where there are subsets S_1 and S_2 of A such that

$$f(S_1 \cap S_2) \ne f(S_1) \cap f(S_2).$$

    d. Prove that $f(S_1) - f(S_2) \subseteq f(S_1 - S_2)$ for all subsets S_1 and S_2 of A.

    e. Give an example where there are subsets S_1 and S_2 of A such that

$$f(S_1) - f(S_2) \ne f(S_1 - S_2).$$

23. Let $f: A \rightarrow B$, where A and B are nonempty, and let T_1 and T_2 be subsets of B.
    a. Prove that $f^{-1}(T_1 \cup T_2) = f^{-1}(T_1) \cup f^{-1}(T_2)$.
    b. Prove that $f^{-1}(T_1 \cap T_2) = f^{-1}(T_1) \cap f^{-1}(T_2)$.
    c. Prove that $f^{-1}(T_1) - f^{-1}(T_2) = f^{-1}(T_1 - T_2)$.
    d. Prove that if $T_1 \subseteq T_2$, then $f^{-1}(T_1) \subseteq f^{-1}(T_2)$.

24. Let $g: A \rightarrow B$ and $f: B \rightarrow C$. Prove that $(f \circ g)^{-1}(T) = g^{-1}(f^{-1}(T))$ for any subset of T of C.

25. Let $f: A \rightarrow B$, where A and B are nonempty. Prove that f has the property that $f^{-1}(f(S)) = S$ for every subset S of A if and only if f is one-to-one. (Compare with Exercise 12b.)

26. Let $f: A \rightarrow B$, where A and B are nonempty. Prove that f has the property that $f(f^{-1}(T)) = T$ for every subset T of B if and only if f is onto. (Compare with Exercise 12c.)

1.3 PROPERTIES OF COMPOSITE MAPPINGS (OPTIONAL)

In many cases, we will be dealing with mappings of a set into itself; that is, the domain and codomain of the mappings are the same. In these cases, the mappings $f \circ g$ and $g \circ f$ are both defined, and the question arises as to whether $f \circ g$ and $g \circ f$ are equal. That is, is mapping composition commutative when the domain and codomain are equal? The following example shows that the answer is no.

Example I Let $\mathbf{Z}$ be the set of all integers, and let the mappings $f: \mathbf{Z} \rightarrow \mathbf{Z}$ and $g: \mathbf{Z} \rightarrow \mathbf{Z}$ be defined for each $n \in \mathbf{Z}$ by

$$f(n) = 2n$$

$$g(n) = \begin{cases} \dfrac{n}{2} & \text{if } n \text{ is even} \\ 4 & \text{if } n \text{ is odd.} \end{cases}$$

In this case, the composition mappings $f \circ g$ and $g \circ f$ are both defined. We have, on the one hand,

$$\begin{aligned} (g \circ f)(n) &= g(f(n)) \\ &= g(2n) \\ &= n, \end{aligned}$$

so $(g \circ f)(n) = n$ for all $n \in \mathbf{Z}$. On the other hand,

$$(f \circ g)(n) = f(g(n))$$

$$= \begin{cases} f\left(\dfrac{n}{2}\right) = n & \text{if } n \text{ is even} \\ f(4) = 8 & \text{if } n \text{ is odd,} \end{cases}$$

so $f \circ g \neq g \circ f$. Thus, mapping composition is not commutative. ∎

In the next example we use the same functions $f, g, g \circ f$, and $f \circ g$ as in Example 1. For each of them, we determine if the mapping is onto and if it is one-to-one.

Example 2 Let f and g be the same as in Example 1. We see that f is one-to-one since

$$f(m) = f(n) \;\Rightarrow\; 2m = 2n$$
$$\Rightarrow\;\; m = n.$$

To show that f is not onto, consider the equation $f(n) = 1$:

$$f(n) = 1 \;\Rightarrow\; 2n = 1$$
$$\Rightarrow\;\; n = \tfrac{1}{2},$$

and $\tfrac{1}{2}$ is not an element of **Z**. Thus, 1 is not in the range of f.

We see that g is not one-to-one since

$$g(3) = 4 \quad \text{and} \quad g(5) = 4.$$

However, we can show that g is onto. For any $m \in$ **Z**, the integer $2m$ is in **Z** and

$$g(2m) = \frac{2m}{2} \; \text{since } 2m \text{ is even}$$
$$= m.$$

Thus every $m \in$ **Z** is in the range of g, and g is onto.

Using the computed values from Example 1, we have

$$(g \circ f)(n) = n$$

and

$$(f \circ g) = \begin{cases} n & \text{if } n \text{ is even} \\ 8 & \text{if } n \text{ is odd.} \end{cases}$$

The value $(g \circ f)(n) = n$ shows that $g \circ f$ is both onto and one-to-one. Since

$$(f \circ g)(1) = 8 \quad \text{and} \quad (f \circ g)(3) = 8,$$

$f \circ g$ is not one-to-one. Since $(f \circ g)(n)$ is always an even integer, there is no $n \in$ **Z** such that

$$(f \circ g)(n) = 5,$$

and hence $f \circ g$ is not onto.

Summarizing our results, we have that

f is one-to-one and not onto;
g is onto and not one-to-one;
$g \circ f$ is both onto and one-to-one;
$f \circ g$ is neither onto nor one-to-one.   ∎

Considerations such as those in Example 2 raise the question of how the one-to-one and onto properties of the mappings f, g, and $f \circ g$ are related. One general statement concerning these relationships is given in the next theorem, and others can be found in the exercises.

STRATEGY ▶

> To show that $f \circ g$ is onto in the proof of the next theorem, we use the standard procedure described on p. 15: We take an arbitrary $c \in C$ and prove that there exists an $a \in A$ such that $(f \circ g)(a) = c$.

Theorem 1.16 **COMPOSITION OF ONTO MAPPINGS**

Let $g: A \to B$ and $f: B \to C$. If f and g are both onto, then $f \circ g$ is onto.

$(p \wedge q) \Rightarrow r$ **Proof**   Suppose f and g satisfy the stated conditions. The composition $f \circ g$ maps A to C. Suppose $c \in C$. Since f is onto, there exists $b \in B$ such that

$$f(b) = c.$$

Since g is onto, every element in B is an image under g. In particular, for the specific b such that $f(b) = c$, there exists $a \in A$ such that

$$g(a) = b.$$

Hence, for $c \in C$, there exists $a \in A$ such that

$$(f \circ g)(a) = f(g(a)) = f(b) = c,$$

and $f \circ g$ is onto.

The mappings in Example 3 provide a combination of properties that is different from the one in Example 2.

Example 3   Let $f: \mathbf{Z} \to \mathbf{Z}$ and $g: \mathbf{Z} \to \mathbf{Z}$ be defined as follows:

$$f(x) = \begin{cases} x & \text{if } x \text{ is even} \\ \dfrac{x-1}{2} & \text{if } x \text{ is odd,} \end{cases}$$

$$g(x) = 4x.$$

We shall determine which of the mappings, f, g, $f \circ g$, and $g \circ f$ are onto, and also which of these mappings are one-to-one.

For arbitrary $n \in \mathbf{Z}$, $2n + 1$ is odd in $\mathbf{Z}$, and $f(2n + 1) = n$. Thus, f is onto. We have $f(2) = 2$ and also $f(5) = 2$, so f is not one-to-one.

Since $g(x)$ is always a multiple of 4, there is no $x \in \mathbf{Z}$ such that $g(x) = 3$. Hence, g is not onto. However,

$$g(x) = g(z) \Rightarrow 4x = 4z$$
$$\Rightarrow x = z,$$

so g is one-to-one.

Now

$$(f \circ g)(x) = f(g(x)) = f(4x) = 4x.$$

This means that $(f \circ g)(x) = g(x)$ for all $x \in \mathbf{Z}$. Therefore, $f \circ g = g$ is not onto and is one-to-one.

Computing $(g \circ f)(x)$, we obtain

$$(g \circ f)(x) = g(f(x))$$

$$= \begin{cases} g(x) & \text{if } x \text{ is even} \\ g\left(\dfrac{x-1}{2}\right) & \text{if } x \text{ is odd} \end{cases} = \begin{cases} 4x & \text{if } x \text{ is even} \\ 2(x-1) & \text{if } x \text{ is odd.} \end{cases}$$

Since $(g \circ f)(x)$ is never odd, there is no x such that $(g \circ f)(x) = 1$, and $g \circ f$ is not onto. Also, since $(g \circ f)(2) = 8$ and $(g \circ f)(5) = 8$, $g \circ f$ is not one-to-one.

We can summarize our results as follows:

f is onto and not one-to-one;
g is one-to-one and not onto;
$f \circ g$ is one-to-one and not onto;
$g \circ f$ is neither onto nor one-to-one.   ∎

EXERCISES 1.3

1. For each of the following pairs $f: \mathbf{Z} \to \mathbf{Z}$ and $g: \mathbf{Z} \to \mathbf{Z}$, decide if $f \circ g$ is onto or one-to-one and justify all negative answers.

   a. $f(x) = 2x$, $g(x) = \begin{cases} x & \text{if } x \text{ is even} \\ 2x - 1 & \text{if } x \text{ is odd} \end{cases}$

   b. $f(x) = 2x$, $g(x) = x^3$

   c. $f(x) = x + |x|$, $g(x) = \begin{cases} \dfrac{x}{2} & \text{if } x \text{ is even} \\ -x & \text{if } x \text{ is odd} \end{cases}$

   d. $f(x) = \begin{cases} \dfrac{x}{2} & \text{if } x \text{ is even} \\ x + 1 & \text{if } x \text{ is odd} \end{cases}$, $g(x) = \begin{cases} x - 1 & \text{if } x \text{ is even} \\ 2x & \text{if } x \text{ is odd} \end{cases}$

   e. $f(x) = x^2$, $g(x) = x - |x|$

2. For each pair f, g given in Exercise 1, decide if $g \circ f$ is onto or one-to-one, and justify all negative answers.

3. Give an example of mappings f and g such that one of f or g is not onto but $f \circ g$ is onto.

4. Give an example of mappings f and g, different from those in Example 3, such that one of f or g is not one-to-one but $f \circ g$ is one-to-one.

5. **a.** Give an example of mappings f and g, different from those in Example 2, where f is one-to-one, g is onto, and $f \circ g$ is not one-to-one.
   b. Give an example of mappings f and g, different from those in Example 2, where f is one-to-one, g is onto, and $f \circ g$ is not onto.

6. **a.** Give an example of mappings f and g, where f is onto, g is one-to-one, and $f \circ g$ is not one-to-one.
   b. Give an example of mappings f and g, different from those in Example 3, where f is onto, g is one-to-one, and $f \circ g$ is not onto.

7. Suppose $f, g,$ and h are all mappings of a set A into itself.
   a. Prove that if g is onto and $f \circ g = h \circ g,$ then $f = h.$
   b. Prove that if f is one-to-one and $f \circ g = f \circ h,$ then $g = h.$
8. a. Find mappings $f, g,$ and h of a set A into itself such that $f \circ g = h \circ g$ and $f \neq h.$
   b. Find mappings $f, g,$ and h of a set A into itself such that $f \circ g = f \circ h$ and $g \neq h.$
9. Let $g: A \to B$ and $f: B \to C.$ Prove that $f \circ g$ is one-to-one if each of f and g is one-to-one.
10. Let $g: A \to B$ and $f: B \to C.$ Prove that f is onto if $f \circ g$ is onto.
11. Let $g: A \to B$ and $f: B \to C.$ Prove that g is one-to-one if $f \circ g$ is one-to-one.
12. Let $f: A \to B$ and $g: B \to A.$ Prove that f is one-to-one and onto if $f \circ g$ is one-to-one and $g \circ f$ is onto.
13. Let $f: A \to B$ and $g: B \to C.$ For an arbitrary subset S of $C,$ prove that $(f \circ g)^{-1}(S) = g^{-1}(f^{-1}(S)).$

1.4   BINARY OPERATIONS

We are familiar with the operations of addition, subtraction, and multiplication on real numbers. These are examples of *binary operations*. When we speak of a binary operation on a set, we have in mind a process that combines two elements of the set to produce a third element of the set. This third element, the result of the operation on the first two, must be unique. That is, there must be one and only one result from the combination. Also, it must always be possible to combine the two elements, no matter which two are chosen. This discussion is admittedly a bit vague, in that the terms "process" and "combine" are somewhat indefinite. To eliminate this vagueness, we make the following formal definition.

Definition 1.17   BINARY OPERATION

> A **binary operation** on a nonempty set A is a mapping f from $A \times A$ to $A.$

It is conventional in mathematics to assume that when a formal definition is made, it is automatically biconditional. That is, it is understood to be an "if and only if" statement, without this being written out explicitly. In Definition 1.17, for example, it is understood as part of the definition that f is a binary operation on the nonempty set A if and only if f is a mapping from $A \times A$ to $A.$ Throughout the remainder of this book, we will adhere to this convention when we make definitions.

We now have a precise definition of the term *binary operation*, but some of the feel for the concept may have been lost. However, the definition gives us what we want. Suppose f is a mapping from $A \times A$ to $A.$ Then $f(x, y)$ is defined for every ordered pair (x, y) of elements of $A,$ and the image $f(x, y)$ is unique. In other words, any two elements x and y of A can be combined to obtain a unique third element of A by finding the value $f(x, y).$ The result of performing the binary operation on x and y is $f(x, y),$ and the only thing unfamiliar about this is the notation for the result. We are accustomed to indicating results of binary operations by symbols such as $x + y$ and $x - y.$ We can use a similar notation and write $x * y$ in place of $f(x, y).$ Thus, $x * y$ represents the result of an arbitrary binary operation $*$ on $A,$ just as $f(x, y)$ represents the value of an arbitrary mapping from $A \times A$ to $A.$

Example 1 Two examples of binary operations on **Z** are the mappings from **Z** $\times$ **Z** to **Z**, defined as follows:

1. $x * y = x + y - 1$, for $(x, y) \in$ **Z** $\times$ **Z**.
2. $x * y = 1 + xy$, for $(x, y) \in$ **Z** $\times$ **Z**. ∎

Example 2 The operation of forming the intersection $A \cap B$ of subsets A and B of a universal set U is a binary operation on the collection of all subsets of U. This is also true of the operation of forming the union. ∎

Since we are dealing with ordered pairs in connection with a binary operation, the results $x * y$ and $y * x$ may well be different.

Definition 1.18 **COMMUTATIVITY, ASSOCIATIVITY**

> If $*$ is a binary operation on the nonempty set A, then $*$ is called **commutative** if $x * y = y * x$ for all x and y in A. If $x * (y * z) = (x * y) * z$ for all x, y, z in A, we say that the binary operation is **associative**.

Example 3 The usual binary operations of addition and multiplication on the integers are both commutative and associative. However, the binary operation of subtraction on the integers does not have either of these properties. For example, $5 - 7 \neq 7 - 5$, and $9 - (8 - 3) \neq (9 - 8) - 3$. ∎

Suppose we consider the two binary operations given in Example 1.

Example 4 The binary operation $*$ defined on **Z** by

$$x * y = x + y - 1$$

is commutative, since

$$x * y = x + y - 1 = y + x - 1 = y * x.$$

Note that $*$ is also associative, since

$$
\begin{aligned}
x * (y * z) &= x * (y + z - 1) \\
&= x + (y + z - 1) - 1 \\
&= x + y + z - 2
\end{aligned}
$$

and

$$
\begin{aligned}
(x * y) * z &= (x + y - 1) * z \\
&= (x + y - 1) + z - 1 \\
&= x + y + z - 2.
\end{aligned}
$$
∎

Example 5 The binary operation $*$ defined on **Z** by

$$x * y = 1 + xy$$

is commutative, since

$$x * y = 1 + xy = 1 + yx = y * x.$$

To check whether $*$ is associative, we compute

$$x * (y * z) = x * (1 + yz) = 1 + x(1 + yz) = 1 + x + xyz$$

and

$$(x * y) * z = (1 + xy) * z = 1 + (1 + xy)z = 1 + z + xyz.$$

Thus, we can demonstrate that $*$ is not associative by choosing x, y, and z in $\mathbf{Z}$ with $x \neq z$. Using $x = 1, y = 2, z = 3$, we get

$$1 * (2 * 3) = 1 * (1 + 6) = 1 * 7 = 1 + 7 = 8$$

and

$$(1 * 2) * 3 = (1 + 2) * 3 = 3 * 3 = 1 + 9 = 10.$$

Hence, $*$ is not associative on $\mathbf{Z}$.   ∎

The commutative and associative properties are properties of the binary operation itself. In contrast, the property described in the next definition depends on the set under consideration as well as the binary operation.

Definition 1.19   CLOSURE

> Suppose that $*$ is a binary operation on a nonempty set A, and let $B \subseteq A$. If $x * y$ is an element of B for all $x \in B$ and $y \in B$, then B is **closed** with respect to $*$.

In the special case where $B = A$ in Definition 1.19, the property of being closed is automatic, since the result $x * y$ is required to be in A by the definition of a binary operation on A.

Example 6   Consider the binary operation $*$ defined on the set of integers $\mathbf{Z}$ by

$$x * y = |x| + |y|, \quad (x, y) \in \mathbf{Z} \times \mathbf{Z}.$$

The set B of negative integers is not closed with respect to $*$, since $x = -1 \in B$ and $y = -2 \in B$, but

$$x * y = (-1) * (-2) = |-1| + |-2| = 3 \notin B.$$ ∎

Example 7   The definition of an odd integer that was stated in Section 1.2 can be used to prove that the set S of all odd integers is closed under multiplication.

Let x and y be arbitrary odd integers. According to the definition of an odd integer, this means that $x = 2m + 1$ for some integer m and $y = 2n + 1$ for some integer n. Forming the product, we obtain

$$xy = (2m + 1)(2n + 1)$$
$$= 4mn + 2m + 2n + 1$$

$$= 2(2mn + m + n) + 1$$
$$= 2k + 1,$$

where $k = 2mn + m + n \in \mathbf{Z}$, and therefore xy is an odd integer. Hence, the set S of all odd integers is closed with respect to multiplication. ∎

Definition 1.20 **IDENTITY ELEMENT**

> Let $*$ be a binary operation on the nonempty set A. An element e in A is called an **identity element** with respect to the binary operation $*$ if e has the property that
>
> $$e * x = x * e = x$$
>
> for all $x \in A$.

The phase "with respect to the binary operation" is critical in this definition because the particular binary operation being considered is all-important. This is pointed out in the next example.

Example 8   The integer 1 is an identity with respect to the operation of multiplication ($1 \cdot x = x \cdot 1 = x$), but not with respect to the operation of addition ($1 + x \neq x$). ∎

Example 9   The element 1 is the identity element with respect to the binary operation $*$, given by

$$x * y = x + y - 1, \qquad (x, y) \in \mathbf{Z} \times \mathbf{Z},$$

since

$$x * 1 = x + 1 - 1 = x \quad \text{and} \quad 1 * x = 1 + x - 1 = x. \qquad ∎$$

Example 10   There is no identity element with respect to the binary operation $*$ defined by

$$x * y = 1 + xy, \qquad (x, y) \in \mathbf{Z} \times \mathbf{Z},$$

since there is no fixed integer z such that

$$x * z = z * x = 1 + xz = x, \quad \text{for all } x \in \mathbf{Z}. \qquad ∎$$

Whenever a set has an identity element with respect to a binary operation on the set, it is then in order to raise the question of inverses.

Definition 1.21 **RIGHT INVERSE, LEFT INVERSE, INVERSE**

> Suppose that e is an identity element for the binary operation $*$ on the set A, and let $a \in A$. If there exists an element $b \in A$ such that $a * b = e$, then b is called a **right inverse** of a with respect to this operation. Similarly, if $b * a = e$, then b is called a **left inverse** of a. If both of $a * b = e$ and $b * a = e$ hold, then b is called an **inverse of** a and a is called an **invertible** element of A.

Sometimes an inverse is referred to as a *two-sided inverse*, to emphasize that both of the required equations hold.

Example 11   Each element $x \in \mathbf{Z}$ has a two-sided inverse $(-x + 2) \in \mathbf{Z}$ with respect to the binary operation $*$ given by

$$x * y = x + y - 1, \qquad (x, y) \in \mathbf{Z} \times \mathbf{Z},$$

since

$$x * (-x + 2) = (-x + 2) * x = -x + 2 + x - 1 = 1 = e.$$ ∎

Definition 1.22   **PERMUTATION**

> A one-to-one correspondence from a set A to itself is called a **permutation** on A. For any nonempty set A, we adopt the notation $\mathcal{S}(A)$ as standard for the set of all permutations on A. The set of all mappings from A to A will be denoted by $\mathcal{M}(A)$.

From the discussion at the end of Section 1.2, we know that composition of mappings is an associative binary operation on $\mathcal{M}(A)$. The **identity mapping** I_A is defined by

$$I_A(x) = x \quad \text{for all } x \in A.$$

For any f in $\mathcal{M}(A)$,

$$(I_A \circ f)(x) = I_A(f(x)) = f(x)$$

and

$$(f \circ I_A)(x) = f(I_A(x)) = f(x),$$

so $I_A \circ f = f \circ I_A = f$. That is, I_A is an identity element for mapping composition.

Example 12   In Example 1 of Section 1.3, we defined the mappings $f: \mathbf{Z} \to \mathbf{Z}$ and $g: \mathbf{Z} \to \mathbf{Z}$ by

$$f(n) = 2n$$

and

$$g(n) = \begin{cases} \dfrac{n}{2} & \text{if } n \text{ is even} \\ 4 & \text{if } n \text{ is odd.} \end{cases}$$

For these mappings, $(g \circ f)(n) = n$ for all $n \in \mathbf{Z}$, so $g \circ f = I_\mathbf{Z}$ and g is a left inverse for f. Note, however, that

$$(f \circ g)(n) = \begin{cases} n & \text{if } n \text{ is even} \\ 8 & \text{if } n \text{ is odd.} \end{cases}$$

So $f \circ g \neq I_\mathbf{Z}$, and g is not a right inverse for f. ∎

Example 12 furnishes some insight into the next two lemmas.[†]

STRATEGY ▶ Each of these lemmas makes a statement of the form "*p* if and only if *q*." For this kind of statement, there are two things to be proved:

1. ($p \Leftarrow q$) The "if" part, where we assume *q* is true and prove that *p* must then be true, and
2. ($p \Rightarrow q$) The "only if" part, where we assume that *p* is true and prove that *q* must then be true.

Lemma 1.23 LEFT INVERSES AND THE ONE-TO-ONE PROPERTY

Let *A* be a nonempty set, and let $f: A \to A$. Then *f* is one-to-one if and only if *f* has a left inverse.

$p \Leftarrow q$ **Proof** Assume first that *f* has a left inverse *g*, and suppose that $f(a_1) = f(a_2)$. Since $g \circ f = I_A$, we have

$$a_1 = I_A(a_1) = (g \circ f)(a_1) = g(f(a_1)) = g(f(a_2))$$
$$= (g \circ f)(a_2) = I_A(a_2) = a_2.$$

Thus, $f(a_1) = f(a_2)$ implies $a_1 = a_2$, and *f* is one-to-one.

$p \Rightarrow q$ Conversely, now assume that *f* is one-to-one. We shall define a left inverse *g* of *f*. Let a_0 represent an arbitrarily chosen but fixed element in *A*. For each *x* in *A*, $g(x)$ is defined by this rule:

1. If there is an element *y* in *A* such that $f(y) = x$, then $g(x) = y$.
2. If no such element *y* exists in *A*, then $g(x) = a_0$.

When the first part of the rule applies, the element *y* is unique because *f* is one-to-one ($f(y_1) = x = f(y_2) \Rightarrow y_1 = y_2 = g(x)$). Thus, $g(x)$ is unique in this case. When the second part of the rule applies, $g(x) = a_0$ is surely unique, and *g* is a mapping from *A* to *A*. For all *a* in *A*, we have

$$(g \circ f)(a) = g(f(a)) = a$$

because $x = f(a)$ requires $g(x) = a$. Thus, *g* is a left inverse for *f*.

There is a connection between the onto property and right inverses that is similar to the one between the one-to-one property and left inverses. This connection is stated in Lemma 1.24, and its proof involves using the **Axiom of Choice**. In one of its simplest forms, this axiom states that it is possible to make a choice of an element from each of the sets in a nonempty collection of nonempty sets. We assume the Axiom of Choice in this text, and it should be noted that this is an *assumption*.

[†]A **lemma** is a proposition whose main purpose is to help prove another proposition.

Lemma 1.24   **RIGHT INVERSES AND THE ONTO PROPERTY**

Let A be a nonempty set, and $f: A \to A$. Then f is an onto mapping if and only if f has a right inverse.

$p \Leftarrow q$ **Proof**   Assume that f has a right inverse g, and let a_0 by an arbitrarily chosen element of A. Now $g(a_0)$ is an element of A, and

$$f(g(a_0)) = (f \circ g)(a_0)$$
$$= I_A(a_0) \qquad \text{since } g \text{ is a right inverse of } f$$
$$= a_0.$$

Thus, a_0 is the image of $g(a_0)$ under f, and this proves that f is onto if f has a right inverse.

$p \Rightarrow q$ Let us assume now that f is onto, and we shall define a right inverse of f as follows. Let a_0 be an arbitrary element of A. Since f is onto, there exists at least one element x of A such that $f(x) = a_0$. Choose[†] one of these elements, say x_0, and define $g(a_0)$ by

$$g(a_0) = x_0.$$

For each a_0 in A, we have a unique value $g(a_0)$ such that

$$(f \circ g)(a_0) = f(g(a_0))$$
$$= f(x_0)$$
$$= a_0 \qquad \text{by the choice of } x_0.$$

Therefore, $f \circ g = I_A$, and g is a right inverse of f.

Lemmas 1.23 and 1.24 enable us to prove the following important theorem.

Theorem 1.25   **INVERSES AND PERMUTATIONS**

Let $f: A \to A$. Then f is invertible if and only if f is a permutation on A.

$p \Rightarrow q$ **Proof**   If f has an inverse g, then $g \circ f = I_A$ and $f \circ g = I_A$. Note that $g \circ f = I_A$ implies that f is one-to-one by Lemma 1.23, and $f \circ g = I_A$ implies that f is onto by Lemma 1.24. Thus, f is a permutation on A.

$p \Leftarrow q$ Now suppose that f is a permutation on A. Then f has a left inverse g by Lemma 1.23 and a right inverse h by Lemma 1.24. We have $g \circ f = I_A$ and $f \circ h = I_A$, so

$$g = g \circ I_A = g \circ (f \circ h) = (g \circ f) \circ h = I_A \circ h = h.$$

That is, $g = h$, and f has an inverse.

The last theorem shows that the members of the set $\mathcal{S}(A)$ are special in that each of them is invertible.

[†]The Axiom of Choice implies that this is possible.

STRATEGY ▶

Exercise 9 of this section requests a proof that the inverse of an element with respect to an associative binary operation is unique. A standard way to prove the uniqueness of an entity is to assume that two such entities exist and then prove the two to be equal.

The unique inverse of a permutation f is denoted by f^{-1}. It is left as an exercise to prove that f^{-1} is a permutation on A.

There is one other property of the set $S(A)$ that is significant. Whenever f and g are in $S(A)$, then $f \circ g$ is also in $S(A)$. (See Exercise 17 of this section.) Thus, $S(A)$ is *closed* under mapping composition.

Some of the preceding results are illustrated in the following example.

Example 13   From Example 11 of Section 1.2, we know that the mapping $f \colon \mathbf{Z} \to \mathbf{Z}$ defined by

$$f(x) = 2x + 1$$

is one-to-one and not onto. According to Lemmas 1.23 and 1.24, f has a left inverse but fails to have a right inverse. The two-part rule for g in the proof of Lemma 1.23 can be used as a guide in defining a left inverse of the f under consideration here.

The first part of the rule reads as follows: If there is an element y in $\mathbf{Z}$ such that $f(y) = x$, then $g(x) = y$. Since we have $f(x) = 2x + 1$ here, the equation $f(y) = x$ requires that x be odd and that $2y + 1 = x$. Solving this equation for y, we obtain

$$y = \frac{x - 1}{2}.$$

Thus, the equation $g(x) = y$ becomes

$$g(x) = \frac{x - 1}{2} \quad \text{for } x \text{ odd}$$

in this instance.

According to the second part of the rule for g in the proof of Lemma 1.23, we may choose an arbitrary fixed a_0 in $\mathbf{Z}$ and define $g(x) = a_0$ when x is not in the range of f. Choosing $a_0 = 4$ gives us a left inverse g of f defined as follows:

$$g(x) = \begin{cases} \dfrac{x - 1}{2} & \text{if } x \text{ is odd} \\ 4 & \text{if } x \text{ is even.} \end{cases}$$ ■

EXERCISES 1.4

1. Decide if the given set B is closed with respect to the binary operation defined on the set of integers $\mathbf{Z}$. If B is not closed, exhibit elements $x \in B$ and $y \in B$, such that $x * y \notin B$.
   a. $x * y = xy$, $\quad B = \{-1, -2, -3, \dots\}$

b. $x * y = x - y, \quad B = \{1, 2, 3, \ldots\}$

c. $x * y = x^2 + y^2, \quad B = \{1, 2, 3, \ldots\}$

d. $x * y = \operatorname{sgn} x + \operatorname{sgn} y, \quad B = \{-1, 0, 1\}$ where $\operatorname{sgn} x = \begin{cases} 1 & \text{if } x > 0 \\ 0 & \text{if } x = 0 \\ -1 & \text{if } x < 0. \end{cases}$

2. In each part following, a rule is given that determines a binary operation $*$ on the set $\mathbf{Z}$ of all integers. Determine in each case if the operation is commutative or associative and if there is an identity element.

   a. $x * y = x + xy$
   b. $x * y = x$
   c. $x * y = x + 2y$
   d. $x * y = 3(x + y)$
   e. $x * y = x - y$
   f. $x * y = x + xy + y$

3. Let S be a set of four elements given by $S = \{A, B, C, D\}$. In the following table, all of the elements of S are listed in a row at the top and in a column at the left. The result $x * y$ is found in the row that starts with x at the left and in the column that has y at the top. For example, $B * C = B$ and $B * D = A$.

$*$	A	B	C	D
A	B	C	A	B
B	C	D	B	A
C	A	B	C	D
D	A	B	D	D

   a. Is the binary operation $*$ commutative? Why?
   b. Determine whether there is an identity element in S for $*$.
   c. If there is an identity element, which elements have inverses?

Sec. 1.5, #8 ≪

4. Prove that the set of all odd integers is not closed with respect to addition.

5. Use the definition of an even integer that was stated in Section 1.2 to prove that the set $\mathbf{E}$ of all even integers is closed with respect to
   a. addition
   b. multiplication.

6. Assume that $*$ is an associative binary operation on the nonempty set A. Prove that

$$a * [b * (c * d)] = [a * (b * c)] * d$$

   for all $a, b, c,$ and d in A.

7. Assume that $*$ is a binary operation on a nonempty set A, and suppose that $*$ is both commutative and associative. Use the definitions of the commutative and associative properties to show that

$$[(a * b) * c] * d = (d * c) * (a * b)$$

   for all $a, b, c,$ and d in A.

8. Let $*$ be a binary operation on the nonempty set A. Prove that if A contains an identity element with respect to $*$, the identity element is unique. (*Hint:* Assume that both e_1 and e_2 are identity elements for $*$ and then prove that $e_1 = e_2$.)

9. Assume that $*$ is an associative binary operation on A with an identity element. Prove that the inverse of an element is unique when it exists.

10. For each of the following mappings $f: \mathbf{Z} \rightarrow \mathbf{Z}$, exhibit a right inverse of f with respect to mapping composition whenever one exists.

   a. $f(x) = 2x$

   b. $f(x) = 3x$

   c. $f(x) = x + 2$

   d. $f(x) = 1 - x$

   e. $f(x) = x^3$

   f. $f(x) = x^2$

   g. $f(x) = \begin{cases} x & \text{if } x \text{ is even} \\ 2x - 1 & \text{if } x \text{ is odd} \end{cases}$

   h. $f(x) = \begin{cases} x & \text{if } x \text{ is even} \\ x - 1 & \text{if } x \text{ is odd} \end{cases}$

   i. $f(x) = |x|$

   j. $f(x) = x - |x|$

   k. $f(x) = \begin{cases} x & \text{if } x \text{ is even} \\ \dfrac{x - 1}{2} & \text{if } x \text{ is odd} \end{cases}$

   l. $f(x) = \begin{cases} x - 1 & \text{if } x \text{ is even} \\ 2x & \text{if } x \text{ is odd} \end{cases}$

   m. $f(x) = \begin{cases} \dfrac{x}{2} & \text{if } x \text{ is even} \\ x + 2 & \text{if } x \text{ is odd} \end{cases}$

   n. $f(x) = \begin{cases} x + 1 & \text{if } x \text{ is even} \\ \dfrac{x + 1}{2} & \text{if } x \text{ is odd} \end{cases}$

11. For each of the mappings f given in Exercise 10, determine whether f has a left inverse. Exhibit a left inverse whenever one exists.

12. If n is a positive integer and the set A has n elements, how many elements are in the set $S(A)$ of all permutations on A?

13. Let $f: A \rightarrow A$, where A is nonempty. Prove that f has a left inverse if and only if $f^{-1}(f(S)) = S$ for every subset S of A.

14. Let $f: A \rightarrow A$, where A is nonempty. Prove that f has a right inverse if and only if $f(f^{-1}(T)) = T$ for every subset T of A.

15. Prove that if f is a permutation on A, then f^{-1} is a permutation on A.

16. Prove that if f is a permutation on A, then $(f^{-1})^{-1} = f$.

17. Let f and g be permutations on A. Prove that $f \circ g$ is one-to-one and onto.

18. Let f and g be permutations on A. Prove that $(f \circ g)^{-1} = g^{-1} \circ f^{-1}$.

19. Let f and g be mappings from A to A. Prove that if $f \circ g$ is invertible, then f is onto and g is one-to-one.

20. **a.** Prove that the set of all onto mappings from A to A is closed under composition of mappings.

   b. Prove that the set of all one-to-one mappings from A to A is closed under mapping composition.

1.5 MATRICES

The material in this section provides a rich source of examples for many of the concepts treated later in the text. The basic element under consideration here will be a *matrix* (plural *matrices*).

The word **matrix** is used in mathematics to denote a rectangular array of elements in rows or columns. The elements in the array are usually numbers, and brackets may be used to mark the beginning and the end of the array. Two illustrations of this type of matrix are

$$\begin{bmatrix} 5 & -1 & 0 & 3 \\ 2 & 1 & -2 & 7 \\ 4 & -6 & 4 & 3 \end{bmatrix} \text{ and } \begin{bmatrix} 9 & 1 \\ -1 & 0 \\ 6 & -3 \end{bmatrix}.$$

The formal notation for a matrix is introduced in the following definition. We shall soon see that this notation is extremely useful in proving certain facts about matrices.

Definition 1.26 MATRIX

An **m by n matrix** over a set S is a rectangular array of elements of S, arranged in m rows and n columns. It is customary to write an m by n matrix using notation such as

$$A = \begin{bmatrix} a_{11} & a_{12} & \cdots & a_{1n} \\ a_{21} & a_{22} & \cdots & a_{2n} \\ \vdots & \vdots & & \vdots \\ a_{m1} & a_{m2} & \cdots & a_{mn} \end{bmatrix},$$

where the uppercase letter A denotes the matrix and the lowercase a_{ij} denotes the element in row i and column j of the matrix A. The rows are numbered from the top down, and the columns are numbered from left to right. The matrix A is referred to as a matrix of **dimension $m \times n$** (read "m by n").

The $m \times n$ matrix A in Definition 1.26 can be written compactly as $A = [a_{ij}]_{m \times n}$ or simply as $A = [a_{ij}]$ if the dimension is known from the context.

Example 1   In compact notation, $B = [b_{ij}]_{2 \times 4}$ is shorthand for the matrix

$$B = \begin{bmatrix} b_{11} & b_{12} & b_{13} & b_{14} \\ b_{21} & b_{22} & b_{23} & b_{24} \end{bmatrix}.$$

As a more concrete example, the matrix A defined by $A = [a_{ij}]_{3 \times 3}$ with $a_{ij} = (-1)^{i+j}$ would appear written out as

$$A = \begin{bmatrix} 1 & -1 & 1 \\ -1 & 1 & -1 \\ 1 & -1 & 1 \end{bmatrix}.$$

(This matrix describes the sign pattern in the cofactor expansion of third-order determinants that is used with Cramer's Rule for solving systems of linear equations in intermediate algebra.)   ∎

An $n \times n$ matrix is called a **square matrix of order** n; and a square matrix $A = [a_{ij}]_{n \times n}$ with $a_{ij} = 0$ whenever $i \neq j$ is a **diagonal matrix**. The matrices

$$\begin{bmatrix} 5 & 0 & 0 \\ 0 & 7 & 0 \\ 0 & 0 & -2 \end{bmatrix} \quad \text{and} \quad \begin{bmatrix} 8 & 0 & 0 \\ 0 & 0 & 0 \\ 0 & 0 & 8 \end{bmatrix}$$

are diagonal matrices.

Definition 1.27   MATRIX EQUALITY

> Two matrices $A = [a_{ij}]_{m \times n}$ and $B = [b_{ij}]_{p \times q}$ over a set S are **equal** if and only if $m = p, n = q$, and $a_{ij} = b_{ij}$ for all pairs i, j.

The set of all $m \times n$ matrices over S will be denoted in this book by $M_{m \times n}(S)$. When $m = n$, we simply write $M_n(S)$ instead of $M_{n \times n}(S)$. For the remainder of this section, we will restrict our attention to the sets $M_{m \times n}(\mathbf{R})$, where $\mathbf{R}$ is the set of all real numbers. Our goal is to define binary operations of addition and multiplication on certain sets of matrices and to investigate the basic properties of these operations.

Definition 1.28   MATRIX ADDITION

> **Addition** in $M_{m \times n}(\mathbf{R})$ is defined by
>
> $$[a_{ij}]_{m \times n} + [b_{ij}]_{m \times n} = [c_{ij}]_{m \times n}$$
>
> where $c_{ij} = a_{ij} + b_{ij}$.

To form the sum of two elements in $M_{m \times n}(\mathbf{R})$, we simply add the elements that are placed in corresponding positions.

Example 2   In $M_{2 \times 3}(\mathbf{R})$, an example of addition is

$$\begin{bmatrix} 3 & -1 & 1 \\ 2 & -7 & -4 \end{bmatrix} + \begin{bmatrix} 2 & 1 & 0 \\ 1 & 3 & -1 \end{bmatrix} = \begin{bmatrix} 5 & 0 & 1 \\ 3 & -4 & -5 \end{bmatrix}.$$

We note that a sum of two matrices with *different* dimensions is *not defined*. For instance, the sum

$$\begin{bmatrix} 1 & 2 & 0 \\ 3 & 4 & 0 \end{bmatrix} + \begin{bmatrix} 5 & 6 \\ 7 & 8 \end{bmatrix}$$

is undefined because the dimensions of the two matrices involved are not equal.   ∎

Definition 1.28 can be written in shorter form as

$$[a_{ij}]_{m \times n} + [b_{ij}]_{m \times n} = [a_{ij} + b_{ij}]_{m \times n}$$

and this shorter form is efficient to use in proving the basic properties of addition in $M_{m \times n}(\mathbf{R})$. These basic properties are stated in the next theorem.

Theorem 1.29   **PROPERTIES OF MATRIX ADDITION**

Addition in $M_{m \times n}(\mathbf{R})$ has the following properties.

a. Addition as defined in Definition 1.28 is a binary operation on $M_{m \times n}(\mathbf{R})$.
b. Addition is associative in $M_{m \times n}(\mathbf{R})$.
c. $M_{m \times n}(\mathbf{R})$ contains an identity element for addition.
d. Each element of $M_{m \times n}(\mathbf{R})$ has an additive inverse in $M_{m \times n}(\mathbf{R})$.
e. Addition is commutative in $M_{m \times n}(\mathbf{R})$.

Proof   Let $A = [a_{ij}]_{m \times n}$, $B = [b_{ij}]_{m \times n}$, and $C = [c_{ij}]_{m \times n}$ be arbitrary elements of $M_{m \times n}(\mathbf{R})$.

a. The addition defined in Definition 1.28 is a binary operation on $M_{m \times n}(\mathbf{R})$ because the rule

$$[a_{ij}] + [b_{ij}] = [a_{ij} + b_{ij}]$$

yields a result that is both unique and an element of $M_{m \times n}(\mathbf{R})$.

b. The following equalities establish the associative property for addition.

$$\begin{aligned}
A + (B + C) &= [a_{ij}] + [b_{ij} + c_{ij}] & \text{by Definition 1.28} \\
&= [a_{ij} + (b_{ij} + c_{ij})] & \text{by Definition 1.28} \\
&= [(a_{ij} + b_{ij}) + c_{ij}] & \text{since addition in } \mathbf{R} \text{ is associative} \\
&= [a_{ij} + b_{ij}] + [c_{ij}] & \text{by Definition 1.28} \\
&= (A + B) + C & \text{by Definition 1.28}
\end{aligned}$$

c. Let $O_{m \times n}$ denote the $m \times n$ matrix that has all elements zero. Then

$$\begin{aligned}
A + O_{m \times n} &= [a_{ij}]_{m \times n} + [0]_{m \times n} \\
&= [a_{ij} + 0]_{m \times n} & \text{by Definition 1.28} \\
&= [a_{ij}]_{m \times n} & \text{since 0 is the additive identity in } \mathbf{R} \\
&= A.
\end{aligned}$$

A similar computation shows that $O_{m \times n} + A = A$, and therefore $O_{m \times n}$ is the additive identity for $M_{m \times n}(\mathbf{R})$, called the **zero matrix** of dimension $m \times n$.

d. It is left as an exercise to verify that the matrix $-A$ defined by

$$-A = [-a_{ij}]_{m \times n}$$

is the additive inverse for A in $M_{m \times n}(\mathbf{R})$.

e. The proof that addition in $M_{m \times n}(\mathbf{R})$ is commutative is also left as an exercise.

Part **d** of Theorem 1.29 leads to the definition of **subtraction** in $M_{m \times n}(\mathbf{R})$: for A and B in $M_{m \times n}(\mathbf{R})$,

$$A - B = A + (-B),$$

where $-B = [-b_{ij}]$ is the additive inverse of $B = [b_{ij}]$.

The definition of **multiplication** that we present is a standard definition universally used in linear algebra, operations research, and other branches of mathematics.

Its widespread acceptance is due to its usefulness in a great variety of important applications and not due to its simplicity, for the definition of multiplication is much more complicated and unnatural than the definition of addition. We first state the definition and then illustrate it with an example.

Definition 1.30 **MATRIX MULTIPLICATION**

The **product** of an $m \times n$ matrix A over **R** and an $n \times p$ matrix B over **R** is an $m \times p$ matrix $C = AB$, where the element c_{ij} in row i and column j of AB is found by using the elements in row i of A and the elements in column j of B in the following manner:

$$\text{row } i \text{ of } A \begin{bmatrix} \vdots & \vdots & \vdots & & \vdots \\ a_{i1} & a_{i2} & a_{i3} & \cdots & a_{in} \\ \vdots & \vdots & \vdots & & \vdots \end{bmatrix} \cdot \begin{matrix} \text{column } j \\ \text{of } B \end{matrix} \begin{bmatrix} \cdots & b_{1j} & \cdots \\ \cdots & b_{2j} & \cdots \\ \cdots & b_{3j} & \cdots \\ & \vdots & \\ \cdots & b_{nj} & \cdots \end{bmatrix} = \begin{matrix} \text{column } j \\ \text{of } C \end{matrix} \begin{bmatrix} & \vdots & \\ \cdots & c_{ij} & \cdots \\ & \vdots & \end{bmatrix} \text{row } i \text{ of } C$$

where

$$c_{ij} = a_{i1}b_{1j} + a_{i2}b_{2j} + a_{i3}b_{3j} + \cdots + a_{in}b_{nj}.$$

That is, the element

$$c_{ij} = a_{i1}b_{1j} + a_{i2}b_{2j} + a_{i3}b_{3j} + \cdots + a_{in}b_{nj}$$

in row i and column j of AB is found by adding the products formed from corresponding elements of row i in A and column j in B (first times first, second times second, and so on). Note that the elements of C are real numbers.

Notice that the number of columns in A *must* equal the number of rows in B in order to form the product AB. If this is the case, then A and B are said to be **conformable for multiplication**. A simple diagram illustrates this fact.

$$A_{m \times n} \quad \cdot \quad B_{n \times p} \quad = \quad C_{m \times p}$$

must be equal

dimension of product matrix

Example 3 Consider the products that can be formed using the matrices

$$A = \begin{bmatrix} 3 & -2 \\ 0 & 4 \\ 1 & -3 \\ 5 & 1 \end{bmatrix} \quad \text{and} \quad B = \begin{bmatrix} 2 & 1 & 0 \\ 4 & -3 & 7 \end{bmatrix}.$$

Since the number of columns in A is equal to the number of rows in B, the product AB is defined. Performing the multiplication, we obtain

$$
\begin{bmatrix} 3 & -2 \\ 0 & 4 \\ 1 & -3 \\ 5 & 1 \end{bmatrix} \begin{bmatrix} 2 & 1 & 0 \\ 4 & -3 & 7 \end{bmatrix}
$$

$$
= \begin{bmatrix} 3(2) + (-2)(4) & 3(1) + (-2)(-3) & 3(0) + (-2)(7) \\ 0(2) + 4(4) & 0(1) + 4(-3) & 0(0) + 4(7) \\ 1(2) + (-3)(4) & 1(1) + (-3)(-3) & 1(0) + (-3)(7) \\ 5(2) + 1(4) & 5(1) + 1(-3) & 5(0) + 1(7) \end{bmatrix}.
$$

Thus, AB is the 4×3 matrix given by

$$
AB = \begin{bmatrix} -2 & 9 & -14 \\ 16 & -12 & 28 \\ -10 & 10 & -21 \\ 14 & 2 & 7 \end{bmatrix}.
$$

Since the number of columns in B is not equal to the number of rows in A, the product BA is not defined. Similarly, the products $A \cdot A$ and $B \cdot B$ are also not defined.   ∎

The work in Example 3 shows that multiplication of matrices does not have the commutative property. Some of the computations in the exercises for this section illustrate cases where $AB \neq BA$, even when both products are defined and have the same dimension.

It should also be noted in connection with Example 3 that the product of matrices we are working with is not a true binary operation as defined in Section 1.4. With a binary operation on a set A, it must always be possible to combine any two elements of A and obtain a unique result of the operation. Multiplication of matrices does not have this feature, since the product of two matrices may not be defined. If consideration is restricted to the set $M_n(\mathbf{R})$ of all $n \times n$ matrices of a fixed order n, this difficulty disappears, and multiplication is a true binary operation on $M_n(\mathbf{R})$.

Although matrix multiplication is not commutative, it does have several properties that are analogous to corresponding properties in the set $\mathbf{R}$ of all real numbers. The *sigma notation* is useful in writing out proofs of these properties.

In the **sigma notation**, the capital Greek letter Σ (sigma) is used to indicate a sum:

$$
\sum_{i=1}^{n} a_i = a_1 + a_2 + \cdots + a_n.
$$

The variable i is called the **index of summation**, and the notations below and above the sigma indicate the value of i where the sum starts and the value of i where it ends. For example,

$$
\sum_{i=3}^{5} b_i = b_3 + b_4 + b_5.
$$

The index of summation is sometimes called a "dummy variable" because the value of the sum is unaffected if the index is changed to a different letter:

$$\sum_{i=0}^{3} a_i = \sum_{j=0}^{3} a_j = \sum_{k=0}^{3} a_k = a_0 + a_1 + a_2 + a_3.$$

Using the distributive properties in **R**, we can write

$$a\left(\sum_{k=1}^{n} b_k\right) = a(b_1 + b_2 + \cdots + b_n)$$
$$= ab_1 + ab_2 + \cdots + ab_n$$
$$= \sum_{k=1}^{n} ab_k.$$

Similarly,

$$\left(\sum_{k=1}^{n} b_k\right)a = \sum_{k=1}^{n} b_k a.$$

In the definition of the matrix product AB, the element

$$c_{ij} = a_{i1}b_{1j} + a_{i2}b_{2j} + \cdots + a_{in}b_{nj}$$

can be written compactly by use of the sigma notation as

$$c_{ij} = \sum_{k=1}^{n} a_{ik}b_{kj}.$$

If all necessary comformability is assumed, the following theorem asserts that matrix multiplication is associative.

Theorem 1.31   **ASSOCIATIVE PROPERTY OF MULTIPLICATION**

Let $A = [a_{ij}]_{m \times n}$, $B = [b_{ij}]_{n \times p}$, and $C = [c_{ij}]_{p \times q}$ be matrices over **R**. Then $A(BC) = (AB)C$.

Proof   From Definition 1.30, $BC = [d_{ij}]_{n \times q}$, where $d_{ij} = \sum_{k=1}^{p} b_{ik}c_{kj}$, and $A(BC) = \left[\sum_{r=1}^{n} a_{ir}d_{rj}\right]_{m \times q}$ where

$$\sum_{r=1}^{n} a_{ir}d_{rj} = \sum_{r=1}^{n} a_{ir}\left(\sum_{k=1}^{p} b_{rk}c_{kj}\right)$$
$$= \sum_{r=1}^{n}\left(\sum_{k=1}^{p} a_{ir}(b_{rk}c_{kj})\right).$$

Also, $AB = [f_{ij}]_{m \times p}$ where $f_{ij} = \sum_{r=1}^{n} a_{ir}b_{rj}$, and $(AB)C = \left[\sum_{k=1}^{p} f_{ik}c_{kj}\right]_{m \times q}$ where

$$\sum_{k=1}^{p} f_{ik}c_{kj} = \sum_{k=1}^{p}\left(\sum_{r=1}^{n} a_{ir}b_{rk}\right)c_{kj}$$
$$= \sum_{k=1}^{p}\left(\sum_{r=1}^{n} (a_{ir}b_{rk})c_{kj}\right)$$
$$= \sum_{k=1}^{p}\left(\sum_{r=1}^{n} a_{ir}(b_{rk}c_{kj})\right).$$

The last equality follows from the associative property

$$(a_{ir}b_{rk})c_{kj} = a_{ir}(b_{rk}c_{kj})$$

of multiplication of real numbers. Comparing the elements in row i, column j, of $A(BC)$ and $(AB)C$, we see that

$$\sum_{r=1}^{n} \left(\sum_{k=1}^{p} a_{ir}(b_{rk}c_{kj}) \right) = \sum_{k=1}^{p} \left(\sum_{r=1}^{n} a_{ir}(b_{rk}c_{kj}) \right),$$

since each of these double sums consists of all the np terms that can be made by using a product of the form $a_{ir}(b_{rk}c_{kj})$ with $1 \leq r \leq n$ and $1 \leq k \leq p$. Thus, $A(BC) = (AB)C$.

Similar but simpler use of the sigma notation can be made to prove the distributive properties stated in the following theorem. Proofs are requested in the exercises.

Theorem 1.32   DISTRIBUTIVE PROPERTIES

Let A be an $m \times n$ matrix over $\mathbf{R}$, let B and C be $n \times p$ matrices over $\mathbf{R}$, and let D be a $p \times q$ matrix over $\mathbf{R}$. Then

a. $A(B + C) = AB + AC$, and
b. $(B + C)D = BD + CD$.

For each positive integer n, we define a special matrix I_n by

$$I_n = [\delta_{ij}]_{n \times n} \qquad \text{where} \qquad \delta_{ij} = \begin{cases} 1 & \text{if } i = j \\ 0 & \text{if } i \neq j. \end{cases}$$

(The symbol δ_{ij} used in defining I_n is called the **Kronecker delta**.) For $n = 2$ and $n = 3$, these special matrices are given by

$$I_2 = \begin{bmatrix} 1 & 0 \\ 0 & 1 \end{bmatrix} \qquad \text{and} \qquad I_3 = \begin{bmatrix} 1 & 0 & 0 \\ 0 & 1 & 0 \\ 0 & 0 & 1 \end{bmatrix}.$$

The matrices I_n have special properties in matrix multiplication, as stated in Theorem 1.33.

Theorem 1.33   SPECIAL PROPERTIES OF I_n

Let A be an arbitrary $m \times n$ matrix over $\mathbf{R}$. With I_n as defined in the preceding paragraph,

a. $I_m A = A$, and
b. $A I_n = A$.

Proof   To prove part a, let $A = [a_{ij}]_{m \times n}$ and consider $I_m A$. By Definition 1.30,

$$I_m A = [c_{ij}]_{m \times n}$$

where

$$c_{ij} = \sum_{k=1}^{m} \delta_{ik} a_{kj}.$$

Since $\delta_{ik} = 0$ for $k \neq i$ and $\delta_{ii} = 1$, the expression for c_{ij} simplifies to

$$c_{ij} = \delta_{ii}a_{ij} = 1 \cdot a_{ij} = a_{ij}.$$

Thus, $c_{ij} = a_{ij}$ for all pairs i, j and $I_m A = A$.

The proof that $AI_n = A$ is left as an exercise.

Because the equations $I_m A = A$ and $AI_n = A$ hold for any $m \times n$ matrix A, the matrix I_n is called the **identity matrix of order n**. In a more general context, the terms *left identity* and *right identity* are defined as follows.

Definition 1.34 **LEFT IDENTITY, RIGHT IDENTITY**

Let $*$ be a binary operation on the nonempty set A. If an element e in A has the property that

$$e * x = x \text{ for all } x \in A,$$

then e is called a **left identity element** with respect to $*$. Similarly, if

$$x * e = x \text{ for all } x \in A,$$

then e is a **right identity element** with respect to $*$.

If the same element e is both a left identity and a right identity with respect to $*$, then e is an *identity element* as defined in Definition 1.20. An identity element is sometimes called a **two-sided identity**, to emphasize that both of the required equations hold.

Even though matrix multiplication is not a binary operation on $M_{m \times n}(\mathbf{R})$ when $m \neq n$, we call I_m a *left identity* and I_n a *right identity* for multiplication with elements of $M_{m \times n}(\mathbf{R})$. In the set $M_n(\mathbf{R})$ of all square matrices of order n over $\mathbf{R}$, I_n is a two-sided identity element with respect to multiplication.

The fact that I_n is a multiplicative identity for $M_n(\mathbf{R})$ leads immediately to the question: Does every nonzero element A of $M_n(\mathbf{R})$ have a multiplicative inverse? The answer is not what one might expect, because some nonzero square matrices do not have multiplicative inverses. This fact is illustrated in the next example.

Example 4 Let $A = \begin{bmatrix} 1 & 3 \\ 2 & 6 \end{bmatrix}$, and consider the problem of finding a matrix $B = \begin{bmatrix} x & z \\ y & w \end{bmatrix}$ such that $AB = I_2$. Computation of AB leads at once to

$$\begin{bmatrix} x + 3y & z + 3w \\ 2x + 6y & 2z + 6w \end{bmatrix} = \begin{bmatrix} 1 & 0 \\ 0 & 1 \end{bmatrix},$$

or

$$\begin{bmatrix} x + 3y & z + 3w \\ 2(x + 3y) & 2(z + 3w) \end{bmatrix} = \begin{bmatrix} 1 & 0 \\ 0 & 1 \end{bmatrix}.$$

This matrix equality is equivalent to the following system of four linear equations.

$$\begin{array}{ll} x + 3y = 1 & z + 3w = 0 \\ 2(x + 3y) = 0 & 2(z + 3w) = 1 \end{array}$$

Since $x + 3y = 1$ requires $2(x + 3y) = 2$, and this contradicts $2(x + 3y) = 0$, there is no solution to the system of equations and therefore no matrix B such that $AB = I_2$. That is, A does not have a multiplicative inverse.   ∎

When working with matrices, the convention is to use the term *inverse* to mean "multiplicative inverse." If the matrix A has an inverse, Exercise 9 of Section 1.4 assures us that the inverse is unique. In this case, A is **invertible**, and its inverse is denoted by A^{-1}. A few properties of inverses are included in the exercises for this section, but an in-depth investigation of inverses is more appropriate for a linear algebra course.

■ EXERCISES 1.5

1. Write out the matrix that matches the given description.
   a. $A = [a_{ij}]_{3\times2}$ with $a_{ij} = 2i - j$
   b. $A = [a_{ij}]_{4\times2}$ with $a_{ij} = (-1)^i j$
   c. $B = [b_{ij}]_{2\times4}$ with $b_{ij} = (-1)^{i+j}$
   d. $B = [b_{ij}]_{3\times4}$ with $b_{ij} = 1$ if $i < j$ and $b_{ij} = 0$ if $i \geq j$
   e. $C = [c_{ij}]_{4\times3}$ with $c_{ij} = i + j$ if $i \geq j$ and $c_{ij} = 0$ if $i < j$
   f. $C = [c_{ij}]_{4\times3}$ with $c_{ij} = 0$ if $i \neq j$ and $c_{ij} = 1$ if $i = j$

2. Perform the indicated operations, if possible.
   a. $\begin{bmatrix} -1 & 2 & 5 \\ 0 & -3 & 7 \end{bmatrix} + \begin{bmatrix} 4 & -2 & -9 \\ 8 & -5 & -1 \end{bmatrix}$
   b. $\begin{bmatrix} 8 & 9 \\ 3 & 7 \end{bmatrix} - \begin{bmatrix} 7 & 0 \\ 6 & 5 \end{bmatrix}$
   c. $\begin{bmatrix} 1 & 2 & 3 \\ 0 & 4 & 5 \end{bmatrix} + \begin{bmatrix} 4 & 9 \\ -5 & -8 \\ 6 & 7 \end{bmatrix}$
   d. $\begin{bmatrix} 3 & 0 \\ 8 & 0 \end{bmatrix} + \begin{bmatrix} -1 \\ 4 \end{bmatrix}$

3. Perform the following multiplications, if possible.
   a. $\begin{bmatrix} 2 & 0 & -3 \\ -4 & 1 & -1 \end{bmatrix}\begin{bmatrix} -1 & 2 \\ 5 & 6 \\ 1 & -1 \end{bmatrix}$
   b. $\begin{bmatrix} -1 & 2 \\ 5 & 6 \\ 1 & -1 \end{bmatrix}\begin{bmatrix} 2 & 0 & -3 \\ -4 & 1 & -1 \end{bmatrix}$
   c. $\begin{bmatrix} 2 & 0 \\ 0 & -3 \\ -1 & 5 \end{bmatrix}\begin{bmatrix} 3 & 2 & -1 \\ 6 & -2 & 0 \\ 1 & 0 & 4 \end{bmatrix}$
   d. $\begin{bmatrix} 3 & 2 & -1 \\ 6 & -2 & 0 \\ 1 & 0 & 4 \end{bmatrix}\begin{bmatrix} 2 & 0 \\ 0 & -3 \\ -1 & 5 \end{bmatrix}$
   e. $\begin{bmatrix} -6 & 4 \\ 1 & 3 \end{bmatrix}\begin{bmatrix} 0 & 1 \\ 1 & 2 \end{bmatrix}$
   f. $\begin{bmatrix} 0 & 1 \\ 1 & 2 \end{bmatrix}\begin{bmatrix} -6 & 4 \\ 1 & 3 \end{bmatrix}$
   g. $\begin{bmatrix} 5 \\ -3 \\ 2 \end{bmatrix}\begin{bmatrix} -1 \\ 4 \\ 1 \end{bmatrix}$
   h. $[-4 \ 6 \ 2][-1 \ 0 \ 5]$
   i. $[3 \ -2 \ 1]\begin{bmatrix} -4 \\ -5 \\ 6 \end{bmatrix}$
   j. $\begin{bmatrix} -4 \\ -5 \\ 6 \end{bmatrix}[3 \ -2 \ 1]$

4. Let $A = [a_{ij}]_{2\times3}$ where $a_{ij} = i + j$, and let $B = [b_{ij}]_{3\times4}$ where $b_{ij} = 2i - j$. If $AB = [c_{ij}]_{2\times4}$, write a formula for c_{ij} in terms of i and j.

5. Show that the matrix equation

$$\begin{bmatrix} 1 & -2 & 7 \\ 5 & -1 & 6 \\ 3 & 4 & -8 \end{bmatrix} \begin{bmatrix} x \\ y \\ z \end{bmatrix} = \begin{bmatrix} 9 \\ -4 \\ 2 \end{bmatrix}$$

is equivalent to a system of linear equations in x, y, and z.

6. Write a single matrix equation of the form $AX = B$ that is equivalent to the following system of equations.

$$w + 6x - 3y + 2z = 9$$
$$4w - 7x + y + 5z = 0$$

7. Let δ_{ij} denote the Kronecker delta: $\delta_{ij} = 1$ if $i = j$, and $\delta_{ij} = 0$ if $i \neq j$. Find the value of the following expressions.

a. $\displaystyle\sum_{i=1}^{n}\left(\sum_{j=1}^{n}\delta_{ij}\right)$

b. $\displaystyle\sum_{i=1}^{n}\left(\sum_{j=1}^{n}(1 - \delta_{ij})\right)$

c. $\displaystyle\sum_{i=1}^{5}\left(\sum_{j=1}^{4}(-1)^{\delta_{ij}}\right)$

d. $\displaystyle\sum_{j=1}^{n}\delta_{ij}\delta_{jk}$

Sec. 1.4, #3 ≫ **8.** Let S be the set of four matrices $S = \{I, A, B, C\}$ where

$$I = \begin{bmatrix} 1 & 0 \\ 0 & 1 \end{bmatrix}, \quad A = \begin{bmatrix} 0 & -1 \\ 1 & 0 \end{bmatrix}, \quad B = \begin{bmatrix} -1 & 0 \\ 0 & -1 \end{bmatrix}, \quad C = \begin{bmatrix} 0 & 1 \\ -1 & 0 \end{bmatrix}.$$

Follow the procedure described in Exercise 3 of Section 1.4 to complete the following multiplication table for S. (In this case, the product $BC = A$ is entered as shown in the row with B at the left end and in the column with C at the top.) Is S closed under multiplication?

·	I	A	B	C
I				
A				
B	B	C	I	A
C				

9. Find two square matrices A and B such that $AB \neq BA$.

10. Find two nonzero matrices A and B such that $AB = BA$.

11. Find two nonzero matrices A and B such that $AB = O_{n\times n}$.

12. Give an example of 2×2 matrices A, B, and C over $\mathbf{R}$ such that $AB = AC$, but $B \neq C$, and A is not a zero matrix.

13. Positive integral powers of a square matrix are defined by $A^1 = A$ and $A^{n+1} = A^n \cdot A$ for every positive integer n. Evaluate $(A - B)(A + B)$ and $A^2 - B^2$ and compare the results for

$$A = \begin{bmatrix} 1 & 2 \\ 4 & 0 \end{bmatrix} \quad \text{and} \quad B = \begin{bmatrix} 3 & -1 \\ 2 & 1 \end{bmatrix}.$$

14. For the matrices in Exercise 13, evaluate $(A + B)^2$ and $A^2 + 2AB + B^2$, and compare the results.

15. Assume that A^{-1} exists and find a solution X to $AX = B$ where A and B are in $M_n(\mathbf{R})$.

16. Assume that A, B, C, and X are in $M_n(\mathbf{R})$, and $AXC = B$ with A and C invertible. Solve for X.

17. a. Prove part d of Theorem 1.29.
   b. Prove part e of Theorem 1.29.

18. a. Prove part a of Theorem 1.32.
   b. Prove part b of Theorem 1.32.

19. Prove part b of Theorem 1.33.

20. Let G be the set of all elements of $M_2(\mathbf{R})$ that have one row that consists of zeros, and one row of the form $[a \quad a]$, with $a \neq 0$.
   a. Show that G is closed under multiplication.
   b. Show that for each x in G, there is an element y in G such that $xy = yx = x$.
   c. Show that G does not have an identity element with respect to multiplication.

21. Prove that if $A \in M_{m \times n}(\mathbf{R})$, then $A \cdot O_{n \times p} = O_{m \times p}$.

22. Suppose that A is an invertible matrix over $\mathbf{R}$ and O is a zero matrix. Prove that if $AX = O$, then $X = O$.

23. Let a, b, c, and d be real numbers. If $ad - bc \neq 0$, show that the multiplicative inverse of $\begin{bmatrix} a & b \\ c & d \end{bmatrix}$ is given by

Sec. 3.5, #3 ≪

$$\begin{bmatrix} \dfrac{d}{ad - bc} & \dfrac{-b}{ad - bc} \\ \dfrac{-c}{ad - bc} & \dfrac{a}{ad - bc} \end{bmatrix}.$$

Sec. 3.5, #3 ≪ **24.** Let $A = \begin{bmatrix} a & b \\ c & d \end{bmatrix}$ over $\mathbf{R}$. Prove that if $ad - bc = 0$, then A does not have an inverse.

25. Let A, B, and C be square matrices of order n over $\mathbf{R}$. Prove that if A is invertible and $AB = AC$, then $B = C$.

26. Let A and B be $n \times n$ matrices over $\mathbf{R}$ such that A^{-1} and B^{-1} exist. Prove that $(AB)^{-1}$ exists and that $(AB)^{-1} = B^{-1}A^{-1}$.

▬▬▬ 1.6 RELATIONS

In the study of mathematics, we deal with many examples of relations between elements of various sets. In working with the integers, we encounter relations such as "x is less than y," or "x is a factor of y." In the calculus, one function may be the derivative of some other function, or perhaps an integral of another function. The property that these examples of relations have in common is that there is an association of some sort between two elements of a set, and the ordering of the elements is important. These relations can all be described by the following definition.

Definition 1.35   RELATION

> A **relation** (or a **binary relation**) on a nonempty set A is a nonempty set R of ordered pairs (x, y) of elements x and y of A.

That is, a relation R is a subset of the Cartesian product $A \times A$. If the pair (a, b) is in R, we write aRb, and say that a has the relation R to b. If $(a, b) \notin R$, we write $a \not{R} b$. This notation agrees with the customary notations for relations, such as $a = b$ and $a < b$.

Example 1 Let $A = \{-2, -5, 2, 5\}$ and $R = \{(5, -2), (5, 2), (-5, -2), (-5, 2)\}$. Then $5R2$, $-5R2$, $5R(-2)$, and $(-5)R(-2)$, but $2\not{R}5$, $5\not{R}5$, and so on. As is frequently the case, this relation can be described by a simple rule: xRy if and only if the absolute value of x is the same as $y^2 + 1$; that is, if $|x| = y^2 + 1$. ∎

Any mapping from A to A is an example of a relation, but not all relations are mappings, as Example 1 illustrates. We have $(5, 2) \in R$ and $(5, -2) \in R$, and for a mapping from A to A, the second element y in $(5, y)$ would have to be unique.

Our main concern is with relations that have additional special properties. More precisely, we are interested for the most part in *equivalence relations*.

Definition 1.36 **EQUIVALENCE RELATION**

> A relation R on a nonempty set A is an **equivalence relation** if the following conditions are satisfied for arbitrary x, y, z in A:
>
> **1.** xRx for all $x \in A$.
> **2.** If xRy, then yRx.
> **3.** If xRy and yRz, then xRz.
>
> Properties 1, 2, and 3 are called the **reflexive**, **symmetric**, and **transitive properties**, respectively. They are the familiar basic properties of equality.

Example 2 The relation R defined on the set of integers $\mathbf{Z}$ by

$$xRy \quad \text{if and only if} \quad |x| = |y|$$

is reflexive, symmetric, and transitive. For arbitrary x, y, and z in $\mathbf{Z}$,

1. xRx, since $|x| = |x|$.
2. $xRy \Rightarrow |x| = |y|$
$\qquad \Rightarrow |y| = |x|$
$\qquad \Rightarrow yRx$.
3. xRy and $yRz \Rightarrow |x| = |y| \quad \text{and} \quad |y| = |z|$
$\qquad\qquad\qquad\quad \Rightarrow |x| = |z|$
$\qquad\qquad\qquad\quad \Rightarrow xRz$. ∎

Example 3 The relation R defined on the set of integers $\mathbf{Z}$ by

$$xRy \quad \text{if and only if} \quad x > y$$

is not an equivalence relation, since it is neither reflexive nor symmetric.

1. $x \not> x$ for all $x \in \mathbf{Z}$.
2. $x > y \not\Rightarrow y > x$.

Note that R is transitive:

3. $x > y \quad \text{and} \quad y > z \Rightarrow x > z$. ∎

The following example is a special case of an equivalence relation on the integers that will be extremely important in later work.

Example 4 The relation "congruence modulo 4" is defined on the set $\mathbf{Z}$ of all integers as follows: x is congruent to y modulo 4 if and only if $x - y$ is a multiple of 4. We write $x \equiv y \pmod 4$ as shorthand for " x is congruent to y modulo 4." Thus, $x \equiv y \pmod 4$ if and only if $x - y = 4k$ for some integer k. We demonstrate that this is an equivalence relation. For arbitrary x, y, z in $\mathbf{Z}$,

1. $x \equiv x \pmod 4$, since $x - x = (4)(0)$.
2. $x \equiv y \pmod 4 \Rightarrow x - y = 4k$ for some $k \in \mathbf{Z}$
$\Rightarrow y - x = 4(-k)$ and $-k \in \mathbf{Z}$
$\Rightarrow y \equiv x \pmod 4$.
3. $x \equiv y \pmod 4$ and $y \equiv z \pmod 4$
$\Rightarrow x - y = 4k$ and $y - z = 4m$ for some $k, m \in \mathbf{Z}$
$\Rightarrow x - z = x - y + y - z$
$= 4(k + m), \quad$ and $k + m \in \mathbf{Z}$
$\Rightarrow x \equiv z \pmod 4$.

Thus, congruence modulo 4 has the reflexive, symmetric, and transitive properties, and is an equivalence relation on $\mathbf{Z}$. ∎

Definition 1.37 EQUIVALENCE CLASS

> Let R be an equivalence relation on the nonempty set A. For each $a \in A$, the set
>
> $$[a] = \{x \in A \mid xRa\}$$
>
> is called the **equivalence class** containing a.

Example 5 The relation R in Example 2 defined on $\mathbf{Z}$ by $xRy \Leftrightarrow |x| = |y|$ is an equivalence relation. The equivalence class containing 0 is

$$[0] = \{0\}$$

since 0 is the only element $x \in \mathbf{Z}$ such that $|x| = 0$. Some other equivalence classes are given by

$$[1] = \{1, -1\} \quad \text{and} \quad [-3] = \{-3, 3\}.$$

For $a \neq 0$, the equivalence class $[a]$ is given by

$$[a] = \{-a, a\}$$

since a and $-a$ are the only elements in $\mathbf{Z}$ with absolute value equal to $|a|$. ∎

Example 6 The relation "congruence modulo 4" that is defined in Example 4 is also an equivalence relation. Since $x \equiv y \pmod 4$ if and only if $x - y$ is a multiple of 4, the equivalence class $[a]$ consists of all those integers that differ from a by a multiple of 4. Thus, $[0]$ consists of all multiples of 4:

$$[0] = \{\ldots, -8, -4, 0, 4, 8, \ldots\}.$$

Similarly, the other equivalence classes are given by

$$[1] = \{\ldots, -7, -3, 1, 5, 9, \ldots\},$$
$$[2] = \{\ldots, -6, -2, 2, 6, 10, \ldots\},$$
$$[3] = \{\ldots, -5, -1, 3, 7, 11, \ldots\}.$$

∎

In both Examples 5 and 6, the equivalence classes separate the set $\mathbf{Z}$ into mutually disjoint nonempty subsets. Recall from Section 1.1 that a separation of the elements of a nonempty set A into mutually disjoint nonempty subsets is called a *partition* of A. It is not difficult to show that if R is an equivalence relation on A, then the distinct equivalence classes of R form a partition of A. Conversely, if a partition of A is given, then we can find an equivalence relation R on A that has the given subsets as its equivalence classes. We simply define R by xRy if and only if x and y are in the same subset. The proofs of these statements are requested in the exercises for this section.

The discussion in the last paragraph illustrates a situation where we are dealing with a collection of sets about which very little is explicit. For example, the collection may be finite, or it may be infinite. In such situations, it is sometimes desirable to use the notational convenience known as indexing. We assume that the sets in the collection are **labeled**, or **indexed**, by a set $\mathcal{L}$ of symbols λ. That is, a typical set in the collection is denoted by a symbol such as A_λ, and the index λ takes on values from the set $\mathcal{L}$. For such a collection $\{A_\lambda\}$, we write $\bigcup_{\lambda \in \mathcal{L}} A_\lambda$ for the union of the collection of sets and $\bigcap_{\lambda \in \mathcal{L}} A_\lambda$ for the intersection. That is,

$$\bigcup_{\lambda \in \mathcal{L}} A_\lambda = \{x \mid x \in A_\lambda \text{ for at least one } \lambda \in \mathcal{L}\}$$

and

$$\bigcap_{\lambda \in \mathcal{L}} A_\lambda = \{x \mid x \in A_\lambda \text{ for every } \lambda \in \mathcal{L}\}.$$

If the indexing set $\mathcal{L}$ is given by $\mathcal{L} = \{1, 2, \ldots, n\}$, then the union of the collection of sets $\{A_i\}$ might be written in any one of the following three ways.

$$A_1 \cup A_2 \cup \cdots \cup A_n = \bigcup_{i \in \mathcal{L}} A_i = \bigcup_{i=1}^{n} A_i$$

The index notation is useful in describing a partition of a set. An alternate definition can be made in the following manner.

Definition 1.38 PARTITION

Let $\{A_\lambda\}$, $\lambda \in \mathcal{L}$, be a collection of subsets of the nonempty set A. Then $\{A_\lambda\}$ is a **partition** of A if these conditions are satisfied:

1. each A_λ is nonempty

2. $A = \bigcup_{\lambda \in \mathcal{L}} A_\lambda$

3. if $A_\alpha \cap A_\beta \neq \varnothing$, then $A_\alpha = A_\beta$.

EXERCISES 1.6

1. For $A = \{1, 3, 5\}$, determine which of the following relations on A are mappings from A to A, and justify your answer.

   a. $\{(1, 3), (3, 5), (5, 1)\}$ **b.** $\{(1, 1), (3, 1), (5, 1)\}$
   c. $\{(1, 1), (1, 3), (1, 5)\}$ **d.** $\{(1, 3), (3, 1), (5, 5)\}$
   e. $\{(1, 5), (3, 3), (5, 3)\}$ **f.** $\{(5, 1), (5, 3), (5, 5)\}$

2. In each of the following parts, a relation R is defined on the set $\mathbf{Z}$ of all integers. Determine in each case whether or not R is reflexive, symmetric, or transitive. Justify your answers.

   a. xRy if and only if $x = 2y$.
   b. xRy if and only if $x = -y$.
   c. xRy if and only if $x < y$.
   d. xRy if and only if $x \geq y$.
   e. xRy if and only if $x - y = 5k$ for some k in $\mathbf{Z}$.
   f. xRy if and only if the greatest common divisor of x and y is 1.
   g. xRy if and only if $|x| \leq |y + 1|$.
   h. xRy if and only if $xy \geq 0$.
   i. xRy if and only if $y = xk$ for some k in $\mathbf{Z}$.
   j. xRy if and only if $|x - y| = 1$.

3. **a.** Let R be the equivalence relation defined on $\mathbf{Z}$ in Example 2, and write out the elements of the equivalence class $[3]$.

   b. Let R be the equivalence relation "congruence modulo 4" that is defined on $\mathbf{Z}$ in Example 4. For this R, list five members of the equivalence class $[3]$.

4. Consider the set $\mathscr{P}(A) - \{\varnothing\}$ of all nonempty subsets of $A = \{1, 2, 3, 4, 5\}$. Determine whether or not the given relation R on $\mathscr{P}(A) - \{\varnothing\}$ is reflexive, symmetric, or transitive. Justify your answers.

   a. xRy if and only if x is a subset of y.
   b. xRy if and only if x is a proper subset of y.
   c. xRy if and only if x and y have the same number of elements.

5. In each of the following parts, a relation is defined on the set of all human beings. Determine whether or not the relation is reflexive, symmetric, or transitive, and justify your answers.

   a. xRy if and only if x lives within 400 miles of y.
   b. xRy if and only if x is the father of y.
   c. xRy if and only if x is a first cousin of y.
   d. xRy if and only if x and y were born in the same year.
   e. xRy if and only if x and y have the same mother.

6. Let $A = \mathbf{R} - \{0\}$, the set of all nonzero real numbers, and consider the following relations on $A \times A$. Decide in each case whether R is an equivalence relation, and justify your answers.

   a. $(a, b)R(c, d)$ if and only if $ad = bc$.
   b. $(a, b)R(c, d)$ if and only if $ab = cd$.
   c. $(a, b)R(c, d)$ if and only if $a^2 + b^2 = c^2 + d^2$.
   d. $(a, b)R(c, d)$ if and only if $a - b = c - d$.

7. Let $A = \{1, 2, 3, 4\}$ and define R on $\mathscr{P}(A) - \{\varnothing\}$ by xRy if and only if $x \cap y \neq \varnothing$. Determine whether or not R is reflexive, symmetric, or transitive.

8. In each of the following parts, a relation R is defined on the power set $\mathscr{P}(A)$ of the nonempty set A. Determine in each case whether or not R is reflexive, symmetric, or transitive. Justify your answers.

   a. xRy if and only if $x \cap y \neq \varnothing$.
   b. xRy if and only if $x \subseteq y$.

9. Let $\mathcal{P}(A)$ be the power set of the nonempty set A, and let C denote a fixed subset of A. Define R on $\mathcal{P}(A)$ by xRy if and only if $x \cap C = y \cap C$. Prove that R is an equivalence relation on $\mathcal{P}(A)$.

10. For each of the following relations R defined on the set A of all triangles in a plane, determine whether or not R is reflexive, symmetric, or transitive. Justify your answers.
    a. aRb if and only if a is similar to b.
    b. aRb if and only if a is congruent to b.

11. A relation R on a nonempty set A is called **irreflexive** if $x \not\!\!R\, x$ for all $x \in A$. Which of the relations in Exercise 2 are irreflexive?

12. A relation R on a nonempty set A is called **asymmetric** if, for x and y in A, xRy implies $y \not\!\!R\, x$. Which of the relations in Exercise 2 are asymmetric?

13. A relation R on a nonempty set A is called **antisymmetric** if, for x and y in A, xRy and yRx together imply $x = y$. (That is, R is antisymmetric if $x \neq y$ implies that either $x \not\!\!R\, y$ or $y \not\!\!R\, x$.) Which of the relations in Exercise 2 are antisymmetric?

14. For any relation R on the nonempty set A, the **inverse** of R is the relation R^{-1} defined by $xR^{-1}y$ if and only if yRx. Prove the following statements.
    a. R is symmetric if and only if $R = R^{-1}$.
    b. R is antisymmetric if and only if $R \cap R^{-1}$ is a subset of $\{(a, a) | a \in A\}$.
    c. R is asymmetric if and only if $R \cap R^{-1} = \varnothing$.

15. Let $\mathcal{L} = \{1, 2, 3\}$, $A_1 = \{a, b, c, d\}$, $A_2 = \{c, d, e, f\}$, and $A_3 = \{a, c, f, g\}$. Write out $\bigcup_{\lambda \in \mathcal{L}} A_\lambda$ and $\bigcap_{\lambda \in \mathcal{L}} A_\lambda$.

16. Let $\mathcal{L} = \{\alpha, \beta, \gamma\}$, $A_\alpha = \{1, 2, 3, \ldots\}$, $A_\beta = \{-1, -2, -3, \ldots\}$, and $A_\gamma = \{0\}$. Write out $\bigcup_{\lambda \in \mathcal{L}} A_\lambda$ and $\bigcap_{\lambda \in \mathcal{L}} A_\lambda$.

17. Suppose R is an equivalence relation on the nonempty set A. Prove that the distinct equivalence classes of R separate the elements of A into mutually disjoint subsets.

18. Let $A = \{1, 2, 3\}$, $B_1 = \{1, 2\}$, and $B_2 = \{2, 3\}$. Define the relation R on A by aRb if and only if there is a set B_i ($i = 1$ or 2) such that $a \in B_i$ and $b \in B_i$. Determine which of the properties of an equivalence relation hold for R, and give an example for each property that fails to hold.

19. Suppose $\{A_\lambda\}$, $\lambda \in \mathcal{L}$, represents a partition of the nonempty set A. Define R on A by xRy if and only if there is a subset A_λ such that $x \in A_\lambda$ and $y \in A_\lambda$. Prove that R is an equivalence relation on A and that the equivalence classes of R are the subsets A_λ.

20. Suppose that f is an onto mapping from A to B. Prove that if $\{B_\lambda\}$, $\lambda \in \mathcal{L}$, is a partition of B, then $\{f^{-1}(B_\lambda)\}$, $\lambda \in \mathcal{L}$, is a partition of A.

KEY WORDS AND PHRASES

A PIONEER IN MATHEMATICS ||||||||||||||||||||||||||||||||||||||||||||||||||||||||||||||||||||||||

Arthur Cayley (1821–1895)

The English mathematician Arthur Cayley, one of the three most prolific writers in mathematics, authored more than 200 mathematical papers. He founded the theory of matrices and was one of the first writers to describe abstract groups. According to mathematical historian Howard Eves, Cayley was one of the 19th-century algebraists who "opened the floodgates of modern abstract algebra."

Cayley displayed superior mathematical talent early in his life. At the age of 17 he studied at Trinity College of Cambridge University. Upon graduation, he accepted a position as assistant tutor at the college. At the end of his third year as tutor, his appointment was not renewed because he declined to take the holy orders to become a parson. Cayley then turned to law and spent the next 15 years as a practicing lawyer. It was during this period that he wrote most of his mathematical papers, many of which are now classics.

Mathematics was not Cayley's only love, though. He was also an avid novel reader, a talented watercolor artist, an ardent mountain climber, and a passionate nature lover. However, even on his mountaineering trips, he spent a few hours each day on mathematics.

Cayley spent the last 32 years of his life as a professor of mathematics at Cambridge University. During this period, he campaigned successfully for the admission of women to the university.

2

THE INTEGERS

INTRODUCTION

It is unusual for a chapter to begin with an optional section, but there is an explanation for doing so here. Whether Section 2.1 is to be included or skipped is a matter of attitude or emphasis. If the approach is to emphasize the development of the basic properties of addition, multiplication, and ordering of integers from an initial list of postulates for the integers, then Section 2.1 should be included. As an alternative approach, these properties can be taken as known material from earlier experience, and Section 2.1 can be skipped. Whichever approach is taken, Section 2.1 summarizes the knowledge that is needed for the subsequent material in the chapter, and it separates "what we know" from "what we must prove."

Although Section 2.2 on mathematical induction is not labeled as optional, this material may be familiar from calculus or previous algebra courses, and it might also be skipped.

The set $\mathbf{Z}_n$ of congruence classes modulo n makes its first appearance in Section 2.5 as a set of equivalence classes. Binary operations of addition and multiplication are defined on $\mathbf{Z}_n$ in Section 2.6. Both the additive and the multiplicative structures are drawn upon for examples in Chapters 3 and 4.

Sections 2.7 and 2.8 present optional introductions to coding theory and cryptography. The primary purpose of these sections is to demonstrate that the material in this text has usefulness other than as a foundation for mathematics courses at a higher level.

2.1   POSTULATES FOR THE INTEGERS (OPTIONAL)

The material in this chapter is concerned exclusively with integers. For this reason, we make a notational agreement that *all variables represent integers*. As our starting point, we shall take the system of integers as given and assume that the system of integers satisfies a certain list of basic axioms, or postulates. More precisely, we assume that there is a set **Z** of elements, called the **integers**, that satisfies the following conditions.

Postulates for the Set Z of Integers

1. **Addition postulates**. There is a binary operation defined in **Z** that is called **addition** and denoted by $+$, and that has the following properties:
   a. **Z** is **closed** under addition.
   b. Addition is **associative**.
   c. **Z** contains an element 0 that is an **identity element for addition**.
   d. For each $x \in \mathbf{Z}$, there is an **additive inverse** of x in **Z**, denoted by $-x$, such that
      $x + (-x) = 0 = (-x) + x$.
   e. Addition is **commutative**.

2. **Multiplication postulates**. There is a binary operation defined in **Z** that is called **multiplication** and denoted by $\cdot$, and that has the following properties:
   a. **Z** is **closed** under multiplication.
   b. Multiplication is **associative**.
   c. **Z** contains an element 1 that is different from 0 and that is an **identity element for multiplication**.
   d. Multiplication is **commutative**.

3. The **distributive law**,

$$x \cdot (y + z) = x \cdot y + x \cdot z,$$

   holds for all elements $x, y, z \in \mathbf{Z}$.

4. **Z** contains a subset $\mathbf{Z}^+$, called the **positive integers**, that has the following properties:
   a. $\mathbf{Z}^+$ is **closed** under addition.
   b. $\mathbf{Z}^+$ is **closed** under multiplication.
   c. For each x in **Z**, one and only one of the following statements is true
      i. $x \in \mathbf{Z}^+$
      ii. $x = 0$
      iii. $-x \in \mathbf{Z}^+$.

5. **Induction postulate**. If S is a subset of $\mathbf{Z}^+$ such that
   a. $1 \in S$, and
   b. $x \in S$ always implies $x + 1 \in S$,
   then $S = \mathbf{Z}^+$.

Note that we are taking the entire list of postulates as *assumptions* concerning **Z**. This list is our set of basic properties. In this section we shall investigate briefly some of the consequences of this set of properties.

After the term *group* has been defined in Chapter 3, we shall see that the addition postulates state that **Z** is a commutative group with respect to addition. Note that there is a major difference between the multiplication and the addition postulates, in that no inverses are required with respect to multiplication.

Postulate 3, the distributive law, is sometimes known as the **left distributive law**. The requirement that

$$(y + z) \cdot x = y \cdot x + z \cdot x$$

is known as the **right distributive law**. This property can be deduced from those in our list, as can all the familiar properties of addition and multiplication of integers.

Postulate 4c is referred to as the **law of trichotomy** because of its assertion that *exactly one of three possibilities* must hold. In case iii, where $-x \in \mathbf{Z}^+$, we say that x is a **negative integer**, and the set $\{x \mid -x \in \mathbf{Z}^+\}$ is the **set of all negative integers**.

The induction postulate is so named because it provides a basis for proofs by mathematical induction. Section 2.2 is devoted to the method of proof by induction, and the method is used from time to time throughout this book.

The right distributive law can be shown to follow from the set of postulates for the integers. We do this formally in the following theorem.

Theorem 2.1   **RIGHT DISTRIBUTIVE LAW**

The equality

$$(y + z) \cdot x = y \cdot x + z \cdot x$$

holds for all x, y, z in **Z**.

Proof   For arbitrary x, y, z in **Z**, we have

$$
\begin{aligned}
(y + z) \cdot x &= x \cdot (y + z) &&\text{by postulate 2d} \\
&= x \cdot y + x \cdot z &&\text{by postulate 3} \\
&= y \cdot x + z \cdot x &&\text{by postulate 2d.}
\end{aligned}
$$

The foregoing proof is admittedly trivial, but the point is that the usual manipulations involving integers are indeed consequences of our basic set of postulates. As another example, consider the statement[†] that $(-x)y = -(xy)$. In this equation, $-(xy)$ denotes the additive inverse of xy, just as $-x$ denotes the additive inverse of x.

Theorem 2.2   **ADDITIVE INVERSE OF A PRODUCT**

For arbitrary x and y in **Z**,
$$(-x)y = -(xy).$$

Instead of attempting to prove this statement directly, we shall first prove a lemma.

[†]We adopt the usual convention that the juxtaposition of x and y in xy indicates the operation of multiplication.

Lemma 2.3 **CANCELLATION LAW FOR ADDITION**

If a, b, and c are integers and $a + b = a + c$, then $b = c$.

$p \Rightarrow q$ **Proof of the Lemma**   Suppose $a + b = a + c$. Now $-a$ is in $\mathbf{Z}$, and hence

$$
\begin{aligned}
a + b = a + c &\Rightarrow (-a) + (a + b) = (-a) + (a + c) \\
&\Rightarrow [(-a) + a] + b = [(-a) + a] + c && \text{by postulate 1b} \\
&\Rightarrow \qquad\quad 0 + b = 0 + c && \text{by postulate 1d} \\
&\Rightarrow \qquad\qquad\quad b = c && \text{by postulate 1c.}
\end{aligned}
$$

This completes the proof of the lemma.

Proof of the Theorem   Returning to the theorem, we see that we only need to show that $xy + (-x)y = xy + [-(xy)]$. That is, we need only show that $xy + (-x)y = 0$. We have

$$
\begin{aligned}
xy + (-x)y &= [x + (-x)]y && \text{by Theorem 2.1} \\
&= 0 \cdot y && \text{by postulate 1d} \\
&= 0 \cdot y + 0 && \text{by postulate 1c} \\
&= 0 \cdot y + \{0 \cdot y + [-(0 \cdot y)]\} && \text{by postulate 1d} \\
&= (0 \cdot y + 0 \cdot y) + [-(0 \cdot y)] && \text{by postulate 1b} \\
&= (0 + 0)y + [-(0 \cdot y)] && \text{by Theorem 2.1} \\
&= 0 \cdot y + [-(0 \cdot y)] && \text{by postulate 1c} \\
&= 0 && \text{by postulate 1d.}
\end{aligned}
$$

We have shown that $xy + (-x)y = 0$, and the theorem is proven.

The proof of Theorem 2.2 would have been shorter if the fact that $0 \cdot y = 0$ had been available. However, it is our attitude at present to use in a proof only the basic postulates for $\mathbf{Z}$ and those facts previously proven. Several statements similar to the last two theorems are given to be proved in the exercises at the end of this section. After this section, we assume the usual properties of addition and multiplication in $\mathbf{Z}$.

Postulate 4, which asserts the existence of the set $\mathbf{Z}^+$ of positive integers, can be used to introduce the order relation "less than" on the set of integers. We make the following definition.

Definition 2.4   **THE ORDER RELATION LESS THAN**

For integers x and y,

$$x < y \quad \text{if and only if} \quad y - x \in \mathbf{Z}^+$$

where $y - x = y + (-x)$.

The symbol $<$ is read "less than," as usual. Here we have defined the relation, but we have not assumed any of its usual properties. However, they can be obtained by use of this definition and the properties of $\mathbf{Z}^+$. Before illustrating this with an example, we note that $0 < y$ if and only if $y \in \mathbf{Z}^+$.

For an arbitrary $x \in \mathbf{Z}$ and a positive integer n, we define x^n as follows:

$$x^1 = x$$
$$x^{k+1} = x^k \cdot x \quad \text{for any positive integer } k.$$

Similarly, positive multiples nx of x are defined by

$$1x = x$$
$$(k + 1)x = kx + x \quad \text{for any positive integer } k.$$

STRATEGY ▶

> Some proofs must be divided into different cases because the same argument does not apply to all elements under consideration. The proof of the next theorem separates naturally into two cases, based on the law of trichotomy (postulate 4c).

Theorem 2.5   **SQUARES OF NONZERO INTEGERS**

For any $x \neq 0$ in $\mathbf{Z}$, $x^2 \in \mathbf{Z}^+$.

$p \Rightarrow q$ **Proof**   Let $x \neq 0$ in $\mathbf{Z}$. By postulate 4, either $x \in \mathbf{Z}^+$ or $-x \in \mathbf{Z}^+$. If $x \in \mathbf{Z}^+$, then $x^2 = x \cdot x$ is in $\mathbf{Z}^+$ by postulate 4b. And if $-x \in \mathbf{Z}^+$, then $(-x)^2 = (-x) \cdot (-x)$ is in $\mathbf{Z}^+$, by the same postulate. But

$$x^2 = x \cdot x$$
$$= (-x) \cdot (-x) \quad \text{by Exercise 5 in this section,}$$

so x^2 is in $\mathbf{Z}^+$ if $-x \in \mathbf{Z}^+$. In each possible case, x^2 is in $\mathbf{Z}^+$, and this completes the proof.

As a particular case of this theorem, $1 \in \mathbf{Z}^+$, since $1 = (1)^2$. That is, 1 must be a positive integer, a fact that may not be immediately evident in postulate 4. This in turn implies that $2 = 1 + 1$ is in $\mathbf{Z}^+$, by postulate 4a. Repeated application of 4a gives $3 = 2 + 1 \in \mathbf{Z}^+$, $4 = 3 + 1 \in \mathbf{Z}^+$, $5 = 4 + 1 \in \mathbf{Z}^+$, and so on. It turns out that $\mathbf{Z}^+$ must necessarily be the set

$$\mathbf{Z}^+ = \{1, 2, 3, \ldots, n, n + 1, \ldots\}.$$

Although our discussion of order has been in terms of *less than*, the relations *greater than*, *less than or equal to*, and *greater than or equal to* can be introduced in $\mathbf{Z}$ and similarly treated. We consider this treatment to be trivial and do not bother with it. At the same time, we accept terms such as *nonnegative* and *nonpositive* with their usual meanings and without formal definitions.

EXERCISES 2.1

Prove that the qualities in Exercises 1–10 hold for all x, y, and z in $\mathbf{Z}$. Assume only the basic postulates for $\mathbf{Z}$ and those properties proved in this section. **Subtraction** is defined by $x - y = x + (-y)$.

1. $x \cdot 0 = 0$
2. $-x = (-1) \cdot x$
3. $-(-x) = x$
4. $(-1)(-1) = 1$
5. $(-x)(-y) = xy$
6. $x - 0 = x$
7. $x(y - z) = xy - xz$
8. $(y - z)x = yx - zx$
9. $-(x + y) = (-x) + (-y)$
10. $(x - y) + (y - z) = x - z$

In Exercises 11–18, prove the statements concerning the relation $<$ on the set $\mathbf{Z}$ of all integers.

11. If $x < y$, then $x + z < y + z$.
12. If $x < y$ and $z < w$, then $x + z < y + w$.
13. If $x < y$ and $y < z$, then $x < z$.
14. If $x < y$ and $0 < z$, then $xz < yz$.
15. If $x < y$ and $z < 0$, then $yz < xz$.
16. If $0 < x < y$, then $x^2 < y^2$.
17. If $0 < x < y$ and $0 < z < w$, then $xz < yw$.
18. If $0 < z$ and $xz < yz$, then $x < y$.
19. Prove that if x and y are integers and $xy = 0$, then either $x = 0$ or $y = 0$. (*Hint:* If $x \neq 0$, then either $x \in \mathbf{Z}^+$ or $-x \in \mathbf{Z}^+$, and similarly for y. Consider xy for the various cases.)
20. Prove that the cancellation law for multiplication holds in $\mathbf{Z}$. That is, if $xy = xz$ and $x \neq 0$, then $y = z$.

For an integer x, the **absolute value** of x is denoted by $|x|$ and is defined by

$$|x| = \begin{cases} x & \text{if } 0 \leq x \\ -x & \text{if } x < 0. \end{cases}$$

Use this definition for the proofs in Exercises 21–23.

21. Prove that $-|x| \leq x \leq |x|$ for any integer x.
22. Prove that $|xy| = |x| \cdot |y|$ for all x and y in $\mathbf{Z}$.

Sec. 2.2, #35 ◄ 23. Prove that $|x + y| \leq |x| + |y|$ for all x and y in $\mathbf{Z}$.
24. Prove that if a is positive and b is negative, then ab is negative.
25. Prove that if a is positive and ab is positive, then b is positive.
26. Prove that if a is positive and ab is negative, then b is negative.
27. Consider the set $\{0\}$ consisting of 0 alone, with $0 + 0 = 0$ and $0 \cdot 0 = 0$. Which of the postulates for $\mathbf{Z}$ are satisfied?

2.2 MATHEMATICAL INDUCTION

From this point on, full knowledge of the properties of addition, subtraction, and multiplication of integers is assumed. A study of divisibility begins in Section 2.3.

As mentioned in the last section, the induction postulate forms a basis for the method of proof known as mathematical induction. Some students may have encountered this method of proof in the calculus or in other previous courses. In this case, it is possible to skip this section and continue with Section 2.3.

STRATEGY ▶

Proof by Mathematical Induction   In a typical proof by induction, there is a statement P_n to be proved true for every positive integer n. The proof consists of three steps:

1. the statement is verified for $n = 1$;
2. the statement is assumed true for $n = k$;
3. with this assumption made, the statement is then proved to be true for $n = k + 1$.

The assumption that is made in step 2 is called the **inductive assumption** or **induction hypothesis**.

Principle of Mathematical Induction

The logic of the method is that

a. if P_n is true for $n = 1$, and
b. if the truth of P_k always implies that P_{k+1} is true,

then the statement P_n is true for all positive integers n. This logic fits the induction postulate perfectly if we let S be the set of all positive integers n for which P_n is true. When the induction postulate is used in this form, it is frequently called the **Principle of Mathematical Induction**.

Example I   We shall prove that

$$\frac{1}{1 \cdot 3} + \frac{1}{3 \cdot 5} + \frac{1}{5 \cdot 7} + \cdots + \frac{1}{(2n - 1)(2n + 1)} = \frac{n}{2n + 1}$$

for every positive integer n.

For each positive integer n, let P_n be the statement

$$\frac{1}{1 \cdot 3} + \frac{1}{3 \cdot 5} + \frac{1}{5 \cdot 7} + \cdots + \frac{1}{(2n - 1)(2n + 1)} = \frac{n}{2n + 1}.$$

In an equation of this type, it is understood that $1/[(2n - 1)(2n + 1)]$ is the last term on the left side. When $n = 1$, there is only one term, and no addition is actually performed.

When $n = 1$, the value of the left side is

$$\frac{1}{[2(1) - 1][2(1) + 1]} = \frac{1}{1 \cdot 3} = \frac{1}{3}$$

and the value of the right side is

$$\frac{1}{2(1) + 1} = \frac{1}{3}.$$

Thus, P_1 is true.

Assume now that P_k is true. That is, assume that the equation

$$\frac{1}{1 \cdot 3} + \frac{1}{3 \cdot 5} + \frac{1}{5 \cdot 7} + \cdots + \frac{1}{(2k-1)(2k+1)} = \frac{k}{2k+1}$$

is true. With this assumption made, we need to prove that P_{k+1} is true. By adding

$$\frac{1}{[2(k+1)-1][2(k+1)+1]} = \frac{1}{(2k+1)(2k+3)}$$

to both sides of the assumed equality, we obtain

$$\frac{1}{1 \cdot 3} + \frac{1}{3 \cdot 5} + \cdots + \frac{1}{(2k-1)(2k+1)} + \frac{1}{(2k+1)(2k+3)}$$

$$= \frac{k}{2k+1} + \frac{1}{(2k+1)(2k+3)}$$

$$= \frac{k(2k+3)+1}{(2k+1)(2k+3)}$$

$$= \frac{2k^2 + 3k + 1}{(2k+1)(2k+3)}$$

$$= \frac{(2k+1)(k+1)}{(2k+1)(2k+3)} = \frac{k+1}{2(k+1)+1}.$$

The last expression matches exactly the fraction

$$\frac{n}{2n+1}$$

when n is replaced by $k+1$. Thus, P_{k+1} is true whenever P_k is true.

It follows from the induction postulate that P_n is true for all positive integers n. ∎

Example 2 We shall prove that any even positive power of a nonzero integer is positive. That is, if $x \neq 0$ in **Z**, then x^{2n} is positive for every positive integer n.

The second formulation of the statement is suitable for a proof by induction on n. For $n = 1$, $x^{2n} = x^2$, and x^2 is positive by Theorem 2.5. Assume the statement is true for $n = k$; that is, x^{2k} is positive. For $n = k + 1$, we have

$$x^{2n} = x^{2(k+1)}$$
$$= x^{2k+2}$$
$$= x^{2k} \cdot x^2.$$

Since x^{2k} and x^2 are positive, the product is positive by postulate 4b. Thus, the statement is true for $n = k + 1$. It follows from the induction postulate that the statement is true for all positive integers. ∎

In Section 2.3 and in some of the exercises at the end of this section, we need to use the fact that 1 is the least positive integer. It might seem a bit strange to prove something so obvious, but the proof does reveal how this familiar fact is a consequence of the induction postulate.

Theorem 2.6   LEAST POSITIVE INTEGER

The integer 1 is the least positive integer. That is, $1 \leq x$ for all $x \in \mathbf{Z}^+$.

Induction   **Proof**   Let S be the set of all positive integers x such that $1 \leq x$. Then $1 \in S$. Suppose $k \in S$. Now $0 < 1$ implies $k = k + 0 < k + 1$, by Exercise 11 of Section 2.1, so we have $1 \leq k < k + 1$. Thus, $k \in S$ implies $k + 1 \in S$, and $S = \mathbf{Z}^+$ by postulate 5.

Mathematical induction can sometimes be used in more complicated situations involving integers. Some statements that involve positive integers n are false for some values of the positive integer n but are true for all positive integers that are sufficiently large. Statements of this type can be proved by a modified form of mathematical induction. If a is a positive integer, and we wish to prove that a statement P_n is true for all positive integers $n \geq a$, we alter the three steps described in the strategy box of this section to the following form.

STRATEGY ▶

Proof by Generalized Induction

1. The statement is verified for $n = a$.
2. The statement is assumed true for $n = k$, where $k \geq a$.
3. With this assumption made, the statement is then proved to be true for $n = k + 1$.

A proof of this type with $a = 4$ is given in Example 3.

Example 3   We shall prove that

$$1 + 3n < n^2$$

for every positive integer $n \geq 4$.

Notice that the statement is actually false for $n = 1, 2$, and 3. For $n = 4$,

$$1 + 3n = 1 + 12 = 13 \quad \text{and} \quad n^2 = 4^2 = 16.$$

Since $13 < 16$, the statement is true for $n = 4$.

Assume now that the inequality is true for k where $k \geq 4$:

$$1 + 3k < k^2.$$

When $n = k + 1$, the left side of the inequality is $1 + 3(k + 1)$, and

$$
\begin{aligned}
1 + 3(k + 1) &= 1 + 3k + 3 \\
&< k^2 + 3 && \text{since } 1 + 3k < k^2 \\
&= k^2 + 2 + 1 \\
&< k^2 + 2k + 1 && \text{since } 1 < k \text{ implies } 2 < 2k \\
&= (k + 1)^2.
\end{aligned}
$$

(In the steps involving $<$, we have used Exercises 11 and 14 of Section 2.1.) Since $(k + 1)^2$ is the right side of the inequality when $n = k + 1$, we have proved that

$$1 + 3n < n^2$$

is true when $n = k + 1$. Therefore, the inequality is true for all positive integers $n \geq 4$. ∎

The modification of mathematical induction that is described just before Example 3 can be extended even more by allowing a to be 0 or a negative integer and using the same three steps listed in the strategy box to prove that a statement P_n is true for all integers $n \geq a$. This type of proof is requested in Exercise 22 of this section.

In some cases, it is more convenient to use yet another form of the induction postulate. This form is known by three different titles: It is called the **Second Principle of Finite Induction**, the method of proof by **Complete Induction**, and the method of **Strong Mathematical Induction**. In this form, a proof that a statement P_n is true for all integers $n \geq a$ consists of the following three steps.

STRATEGY ▶

Proof by Complete Induction

1. The statement is proved true for $n = a$, where $a \in \mathbf{Z}$.
2. For an integer k, the statement is assumed true for all integers m such that $a \leq m < k$.
3. Under this assumption, the statement is proved to be true for $m = k$.

Our next example presents a proof by complete induction, and another example is provided by the proof of Theorem 2.18 in Section 2.4.

The fact stated in Example 4 is that every positive integer can be written as a sum of nonnegative powers of 2. This fact is a point of departure for developing the **binary representation** of real numbers, a representation that uses 2 as the number base instead of 10 as used in our familiar decimal system. Binary representations are used extensively in computer science.

Example 4 We shall use complete induction to prove the statement that every positive integer n can be expressed in the form

$$n = c_0 + c_1 \cdot 2 + c_2 \cdot 2^2 + \cdots + c_{j-1} \cdot 2^{j-1} + c_j \cdot 2^j,$$

where j is a nonnegative integer, $c_i \in \{0, 1\}$ for all i, and $c_j = 1$.
For $n = 1$, let $j = 0$ and $c_0 = 1$. Then

$$c_0 \cdot 2^0 = (1)(1) = 1,$$

and the statement is true for $n = 1$.
Assume now that $k > 1$ and the statement is true for all positive integers m such that $m < k$. We consider two cases: where k is even, and where k is odd.

If k is even, then $k = 2p$ for some $p \in \mathbf{Z}^+$ with $p < k$. Since $p < k$, the induction hypothesis applies to p, and p can be written in the form

$$p = c_0 + c_1 \cdot 2 + c_2 \cdot 2^2 + \cdots + c_{j-1} \cdot 2^{j-1} + c_j \cdot 2^j,$$

where j is a nonnegative integer, $c_i \in \{0, 1\}$ for all i, and $c_j = 1$. Multiplying both sides of the equation for p by 2 gives

$$k = 2p = c_0 \cdot 2 + c_1 \cdot 2^2 + c_2 \cdot 2^3 + \cdots + c_{j-1} \cdot 2^j + c_j \cdot 2^{j+1},$$

and this is an equation for k that has the required form (when k is even).

Suppose now that k is odd, say $k = 2p + 1$ for some $p \in \mathbf{Z}^+$. Since $k > 1$, this means that $k - 1 = 2p$ is in $\mathbf{Z}^+$ and

$$0 < p = \frac{k-1}{2} < \frac{k+k}{2} = k.$$

But $p < k$ implies that p can be written in the form

$$p = c_0 + c_1 \cdot 2 + c_2 \cdot 2^2 + \cdots + c_{j-1} \cdot 2^{j-1} + c_j \cdot 2^j$$

where $c_i \in \{0, 1\}$, and $c_j = 1$. Therefore

$$2p = c_0 \cdot 2 + c_1 \cdot 2^2 + c_2 \cdot 2^3 + \cdots + c_{j-1} \cdot 2^j + c_j \cdot 2^{j+1}$$

and

$$
\begin{aligned}
k &= 2p + 1 \\
&= 1 + c_0 \cdot 2 + c_1 \cdot 2^2 + \cdots + c_{j-1} \cdot 2^j + c_j \cdot 2^{j+1},
\end{aligned}
$$

which is an equation for k of the required form (when k is odd).

Combining the arguments for k even and k odd, we have proved that if $k > 1$ and the statement is true for all positive integers less than k, then it is also true for $n = k$. By the Second Principle of Finite Induction, the statement is true for all positive integers n. ∎

EXERCISES 2.2

Prove that the statements in Exercises 1–14 are true for every positive integer n.

1. $1 + 2 + 3 + \cdots + n = \dfrac{n(n+1)}{2}$

2. $1 + 3 + 5 + \cdots + (2n - 1) = n^2$

3. $1^2 + 2^2 + 3^2 + \cdots + n^2 = \dfrac{n(n+1)(2n+1)}{6}$

4. $1^2 + 3^2 + 5^2 + \cdots + (2n - 1)^2 = \dfrac{n(2n-1)(2n+1)}{3}$

5. $2 + 2^2 + 2^3 + \cdots + 2^n = 2(2^n - 1)$

6. $1^3 + 2^3 + 3^3 + \cdots + n^3 = \dfrac{n^2(n+1)^2}{4}$

7. $1^3 + 3^3 + 5^3 + \cdots + (2n - 1)^3 = n^2(2n^2 - 1)$

8. $1 \cdot 2 + 2 \cdot 3 + 3 \cdot 4 + \cdots + n(n+1) = \dfrac{n(n+1)(n+2)}{3}$

9. $1 \cdot 2 + 2 \cdot 2^2 + 3 \cdot 2^3 + \cdots + n \cdot 2^n = (n-1)2^{n+1} + 2$

10. $\dfrac{1}{1 \cdot 2} + \dfrac{1}{2 \cdot 3} + \dfrac{1}{3 \cdot 4} + \cdots + \dfrac{1}{n(n+1)} = \dfrac{n}{n+1}$

11. $\dfrac{1}{1 \cdot 4} + \dfrac{1}{4 \cdot 7} + \dfrac{1}{7 \cdot 10} + \cdots + \dfrac{1}{(3n-2)(3n+1)} = \dfrac{n}{3n+1}$

12. $\dfrac{1}{1 \cdot 2 \cdot 3} + \dfrac{1}{2 \cdot 3 \cdot 4} + \dfrac{1}{3 \cdot 4 \cdot 5} + \cdots + \dfrac{1}{n(n+1)(n+2)} = \dfrac{n(n+3)}{4(n+1)(n+2)}$

13. $a + (a+d) + (a+2d) + \cdots + [a + (n-1)d] = \dfrac{n}{2}[2a + (n-1)d]$

14. $a + ar + ar^2 + \cdots + ar^{n-1} = a\dfrac{1-r^n}{1-r}$ if $r \neq 1$

Let x and y be integers, and let m and n be positive integers. Use mathematical induction to prove the statements in Exercises 15–20. (The definitions of x^n and nx are given before Theorem 2.5 in Section 2.1.)

15. $(xy)^n = x^n y^n$

16. $x^m \cdot x^n = x^{m+n}$

17. $(x^m)^n = x^{mn}$

18. $n(x + y) = nx + ny$

19. $(m + n)x = mx + nx$

20. $m(nx) = (mn)x$

21. Let a and b be real numbers, and let n and r be integers with $0 \leq r \leq n$. The **binomial theorem** states that

$$(a+b)^n = \binom{n}{0}a^n + \binom{n}{1}a^{n-1}b + \binom{n}{2}a^{n-2}b^2 + \cdots + \binom{n}{r}a^{n-r}b^r + \cdots$$

$$+ \binom{n}{n-2}a^2 b^{n-2} + \binom{n}{n-1}ab^{n-1} + \binom{n}{n}b^n,$$

where the **binomial coefficients** $\binom{n}{r}$ are defined by

$$\binom{n}{r} = \dfrac{n!}{(n-r)!r!},$$

with $r! = r(r-1)\cdots(2)(1)$ for $r \geq 1$ and $0! = 1$. Prove that the binomial coefficients satisfy the equation

$$\binom{n}{r-1} + \binom{n}{r} = \binom{n+1}{r}$$ for $1 \leq r \leq n$.

22. Use Exercise 21 and mathematical induction to prove that $\binom{n}{r}$ is an integer for all integers n and r with $0 \leq r \leq n$.

Sec. 2.3, #34 ≪

23. Use the equation

$$\binom{n}{r-1} + \binom{n}{r} = \binom{n+1}{r}$$ for $1 \leq r \leq n$

and mathematical induction on n to prove the binomial theorem as it is stated in Exercise 21.

If B_1, B_2, and B_3 are matrices in $M_{n \times p}(\mathbf{R})$, part b of Theorem 1.29 implies that $B_1 + (B_2 + B_3) = (B_1 + B_2) + B_3$. For each positive integer $j \geq 3$, this associative property can be extended to the following generalized statement: Regardless of how symbols of grouping are introduced in the sum $B_1 + B_2 + \cdots + B_j$, the resulting value is the same matrix, and this justifies writing the sum without symbols of grouping. (This generalized statement is proved in Theorem 3.7.)

24. Let A be an $m \times n$ matrix over $\mathbf{R}$, and let $B_1, B_2, \ldots, B_j$ be $n \times p$ matrices over $\mathbf{R}$. Use Theorem 1.32 and mathematical induction to prove that

$$A(B_1 + B_2 + \cdots + B_j) = AB_1 + AB_2 + \cdots + AB_j$$

for every positive integer j.

25. Let C be a $p \times q$ matrix over $\mathbf{R}$, and let $B_1, B_2, \ldots, B_j$ be $n \times p$ matrices over $\mathbf{R}$. Use Theorem 1.32 and mathematical induction to prove that

$$(B_1 + B_2 + \cdots + B_j)C = B_1C + B_2C + \cdots + B_jC$$

for every positive integer j.

In Exercises 26 and 27, use mathematical induction to prove that the given statement is true for all positive integers n.

26. $n < 2^n$

27. $1 + 2n \leq 3^n$

28. Prove that if a set S has n distinct elements, then S has 2^n distinct subsets.

In Exercises 29–32, use mathematical induction to prove the given statement.

29. $1 + n < n^2$ for all integers $n \geq 3$

30. $1 + 2n < n^3$ for all integers $n \geq 2$

31. $1 + 2n < 2^n$ for all integers $n \geq 3$

32. $2^n < n!$ for all integers $n \geq 4$

33. Use mathematical induction and Exercise 29 to prove that $n^2 < n!$ for all integers $n \geq 4$.

34. Use mathematical induction and Exercise 31 to prove that $n^2 < 2^n$ for every positive integer $n \geq 5$. (In connection with this result, see the discussion of *counterexamples* in the appendix.)

Sec. 2.1, #23 ≫   **35.** Assume the statement from Exercise 23 in Section 2.1 that $|x + y| \leq |x| + |y|$ for all x and y in $\mathbf{Z}$. Use this assumption and mathematical induction to prove that

$$|a_1 + a_2 + \cdots + a_n| \leq |a_1| + |a_2| + \cdots + |a_n|$$

for all integers $n \geq 2$ and arbitrary integers $a_1, a_2, \ldots, a_n$.

36. Show that if the statement

$$1 + 2 + 2^2 + \cdots + 2^{n-1} = 2^n$$

is assumed to be true for $n = k$, then it can be proved to be true for $n = k + 1$. Is the statement true for all positive integers n? Why?

37. Show that if the statement

$$1 + 2 + 3 + \cdots + n = \frac{n(n + 1)}{2} + 2$$

is assumed to be true for $n = k$, the same equation can be proved to be true for $n = k + 1$. Explain why this does not prove that the statement is true for all positive integers. Is the statement true for all positive integers? Why?

38. The **Fibonacci sequence** $\{f_n\} = 1, 1, 2, 3, 5, 8, 13, 21, \ldots$ is defined by

$$f_1 = 1, \quad f_2 = 1 \quad f_{n+2} = f_{n+1} + f_n \quad \text{for } n = 1, 2, 3, \ldots.$$

Prove that $f_n < 2^n$ for all positive integers n.

39. Let f_n denote the nth term of the Fibonacci sequence that is defined in Exercise 38. Use mathematical induction to prove that f_n is given by the explicit formula

$$f_n = \frac{(1 + \sqrt{5})^n - (1 - \sqrt{5})^n}{2^n \sqrt{5}}.$$

(This equation is known as **Binet's formula**, named after the 19th-century French mathematician Jacques Binet.)

2.3 DIVISIBILITY

We turn now to a study of divisibility in the set of integers. Our main goal in this section is to obtain the **Division Algorithm** (Theorem 2.10). To achieve this, we need an important consequence of the induction postulate, known as the **Well-Ordering Theorem.**

Theorem 2.7 THE WELL-ORDERING THEOREM

Every nonempty set S of positive integers contains a least element. That is, there is an element $m \in S$ such that $m \le x$ for all $x \in S$.

$p \Rightarrow q$ **Proof**   Let S be a nonempty set of positive integers. If $1 \in S$, then $1 \le x$ for all $x \in S$, by Theorem 2.6. In this case, $m = 1$ is the least element in S.

Consider now the case where $1 \notin S$, and let L be the set of all positive integers p such that $p < x$ for all $x \in S$. That is,

$$L = \{p \in \mathbf{Z}^+ \mid p < x \text{ for all } x \in S\}.$$

Since $1 \notin S$, Theorem 2.6 assures us that $1 \in L$. We shall show that there is a positive integer p_0 such that $p_0 \in L$ and $p_0 + 1 \notin L$. Suppose this is not the case. Then we have that $p \in L$ implies $p + 1 \in L$, and $L = \mathbf{Z}^+$ by the induction postulate. This contradicts the fact that S is nonempty (note that $L \cap S = \varnothing$). Therefore, there is a p_0 such that $p_0 \in L$ and $p_0 + 1 \notin L$.

We must show that $p_0 + 1 \in S$. We have $p_0 < x$ for all $x \in S$, so $p_0 + 1 \le x$ for all $x \in S$ (see Exercise 18 at the end of this section). If $p_0 + 1 < x$ were always true, then $p_0 + 1$ would be in L. Hence, $p_0 + 1 = x$ for some $x \in S$, and $m = p_0 + 1$ is the required least element in S.

Definition 2.8 DIVISOR, MULTIPLE

Let a and b be integers. We say that a **divides** b if there is an integer c such that $b = ac$.

If a divides b, we write $a \mid b$. Also, we say that b is a **multiple** of a, or that a is a **factor** of b, or that a is a **divisor** of b. If a does not divide b, we write $a \nmid b$.

It may come as a surprise that we can use our previous results to prove the following simple theorem.

Theorem 2.9 DIVISORS OF 1

The only divisors of 1 are 1 and -1.

$p \Rightarrow (q \lor r)$

Proof Suppose a is a divisor of 1. Then $1 = ac$ for some integer c. The equation $1 = ac$ requires $a \neq 0$, so either $a \in \mathbf{Z}^+$ or $-a \in \mathbf{Z}^+$.

Consider first the case where $a \in \mathbf{Z}^+$. This requires $c \in \mathbf{Z}^+$ (see Exercise 25 of Section 2.1), so we have $1 \leq a$ and $1 \leq c$, by Theorem 2.6. Now

$$1 < a \quad \Rightarrow \quad 1 \cdot c < a \cdot c \quad \text{by Exercise 14 of Section 2.1}$$
$$\Rightarrow \quad c < 1 \quad \quad \text{since } ac = 1,$$

and this is a contradiction of $1 \leq c$. Thus, $1 = a$ is the only possibility when $a \in \mathbf{Z}^+$.

Consider[†] now the case where $-a \in \mathbf{Z}^+$. By Exercise 5 of Section 2.1, we have

$$(-a)(-c) = ac = 1,$$

and $-a \in \mathbf{Z}^+$ implies that $-c \in \mathbf{Z}^+$ by Exercise 25 of Section 2.1. Therefore, $1 \leq -a$ and $1 \leq -c$ by Theorem 2.6. Now

$$1 < -a \quad \Rightarrow (1)(-c) < (-a)(-c) \quad \text{by Exercise 14 of Section 2.1}$$
$$\Rightarrow \quad -c < 1 \quad \quad \text{since } (-a)(-c) = 1,$$

and $-c < 1$ is a contradiction to $1 \leq -c$. Therefore, $1 = -a$ is the only possibility when $-a \in \mathbf{Z}^+$, and we have

$$a = -(-a) \quad \text{by Exercise 3 of Section 2.1}$$
$$= -1 \quad \quad \text{since } -a = 1.$$

Combining the cases where $a \in \mathbf{Z}^+$ and where $-a \in \mathbf{Z}^+$, we have shown that either $a = 1$ or $a = -1$ if a is a divisor of 1.

Our next result is the basic theorem on divisibility.

Theorem 2.10 THE DIVISION ALGORITHM

Let a and b be integers with $b > 0$. Then there exist unique integers q and r such that

$$a = bq + r \quad \text{with} \quad 0 \leq r < b.$$

Existence

Proof Let S be the set of all integers x that can be written in the form $x = a - bn$ for $n \in \mathbf{Z}$, and let S' denote the set of all nonnegative integers in S. The set S' is nonempty. (See Exercise 19 at the end of this section.) If $0 \in S'$, we have $a - bq = 0$ for some q, and $a = bq + 0$. If $0 \notin S'$, then S' contains a least element $r = a - bq$ by the Well-Ordering Theorem, and

$$a = bq + r$$

[†]The proof for this case is similar to that where $a \in \mathbf{Z}^+$, but we include it here because it illustrates several uses of results from Section 2.1.

where r is positive. Now

$$r - b = a - bq - b = a - b(q + 1),$$

so $r - b \in S$. Since r is the least element in S' and $r - b < r$, it must be true that $r - b$ is negative. That is, $r - b < 0$, and $r < b$. Combining the two cases (where $0 \in S'$ and where $0 \notin S'$), we have

$$a = bq + r \quad \text{with} \quad 0 \leq r < b.$$

Uniqueness   To show that q and r are unique, suppose $a = bq_1 + r_1$ and $a = bq_2 + r_2$, where $0 \leq r_1 < b$ and $0 \leq r_2 < b$. We may assume that $r_1 \leq r_2$ without loss of generality. This means that

$$0 \leq r_2 - r_1 \leq r_2 < b.$$

However, we also have

$$0 \leq r_2 - r_1 = (a - bq_2) - (a - bq_1) = b(q_1 - q_2).$$

That is, $r_2 - r_1$ is a nonnegative multiple of b that is less than b. For any positive integer n, $1 \leq n$ implies $b \leq bn$. Therefore, $r_2 - r_1 = 0$ and $r_1 = r_2$. It follows that $bq_1 = bq_2$, where $b \neq 0$. This implies that $q_1 = q_2$ (see Exercise 20 of Section 2.1). We have shown that $r_1 = r_2$ and $q_1 = q_2$, and this proves that q and r are unique.

The word *algorithm* in the heading of Theorem 2.10 may seem strange at first glance, since an algorithm is usually a repetitive procedure for obtaining a result. The use of the word here is derived from the fact that the elements $a - bn$ of S' in the proof may be found by repeated subtraction of b:

$$a - b, a - 2b, a - 3b,$$

and so on.

In the Division Algorithm, the integer q is called the **quotient** and r is called the **remainder** in the division of a by b. The conclusion of the theorem may be more familiar in the form

$$\frac{a}{b} = q + \frac{r}{b},$$

but we are restricting our work here so that only integers are involved.

Example 1   When a and b are both positive integers, the quotient q and remainder r can be found by the familiar routine of long division. For instance, if $a = 357$ and $b = 13$, long division gives

$$
\begin{array}{r}
27 \\
13\overline{)357} \\
\underline{26} \\
97 \\
\underline{91} \\
6
\end{array}
$$

so $q = 27$ and $r = 6$ in $a = bq + r$, with $0 \leq r < b$:

$$357 = (13)(27) + 6.$$

If a is negative, a minor adjustment can be made to obtain the expression in the Division Algorithm. With $a = -357$ and $b = 13$, the preceding equation can be multiplied by -1 to obtain

$$-357 = (13)(-27) + (-6).$$

To obtain an expression with a positive remainder, we need only subtract and add 13 in the right member of the equation:

$$-357 = (13)(-27) + (13)(-1) + (-6) + 13$$
$$= (13)(-28) + 7.$$

Thus, $q = -28$ and $r = 7$ in the Division Algorithm, with $a = -357$ and $b = 13$.   ∎

EXERCISES 2.3

1. List all divisors of the following integers.
   a. 30   b. 42   c. 28   d. 45
   e. 24   f. 40   g. 32   h. 210

With a and b as given in Exercises 2–15, find the q and r that satisfy the conditions in the Division Algorithm.

2. $a = 796, b = 26$ 3. $a = 512, b = 15$
4. $a = 1149, b = 52$ 5. $a = 1205, b = 37$
6. $a = -12, b = 5$ 7. $a = -27, b = 7$
8. $a = -863, b = 17$ 9. $a = -921, b = 18$
10. $a = 26, b = 796$ 11. $a = 15, b = 512$
12. $a = -4317, b = 12$ 13. $a = -5316, b = 171$
14. $a = 0, b = 3$ 15. $a = 0, b = 5$

16. Prove that if $a, b,$ and c are integers such that $a|b$ and $a|c$, then $a|(b + c)$.
17. Let $a, b,$ and c be integers such that $a|b$ and $b|c$. Prove that $a|c$.
18. Let m be an arbitrary integer. Prove that there is no integer n such that $m < n < m + 1$.
19. Let S be as described in the proof of Theorem 2.10. Give a specific example of a positive element of S.
20. Let $a, b, c, m,$ and n be integers such that $a|b$ and $a|c$. Prove that $a|(mb + nc)$.
21. Prove that if a and b are integers such that $a|b$ and $b|a$, then either $a = b$ or $a = -b$.
22. Prove that if a and b are integers such that $b \neq 0$ and $a|b$, then $|a| \leq |b|$.
23. Use the Division Algorithm to prove that if a and b are integers, with $b \neq 0$, then there exist unique integers q and r such that $a = bq + r$, with $0 \leq r < |b|$.
24. Prove that the Well-Ordering Theorem implies the finite induction postulate 5.
25. Assume that the Well-Ordering Theorem holds, and prove the second principle of finite induction.

In Exercises 26–31, use mathematical induction to prove that the given statement is true for all positive integers n.

26. 3 is a factor of $4^n - 1$.   27. 8 is a factor of $9^n - 1$.

28. 3 is a factor of $n^3 + 2n$.

29. 3 is a factor of $n^3 - 7n$.

30. 6 is a factor of $n^3 - n$.

31. 6 is a factor of $n^3 + 5n$.

32. For all a and b in $\mathbf{Z}$, $a - b$ is a factor of $a^n - b^n$. [*Hint:* $a^{k+1} - b^{k+1} = a^k(a - b) + (a^k - b^k)b$.]

33. For all a and b in $\mathbf{Z}$, $a + b$ is a factor of $a^{2n} - b^{2n}$.

Sec. 2.2, #22 ≫ 34. The binomial coefficients $\binom{n}{r}$ are defined in Exercise 21 of Exercises 2.2. Use induction on r to prove that if p is a prime integer, then p is a factor of $\binom{p}{r}$ for $r = 1, 2, \ldots, p - 1$. (From Exercise 22 of Exercises 2.2, it is known that $\binom{p}{r}$ is an integer.)

2.4   PRIME FACTORS AND GREATEST COMMON DIVISOR

In this section we establish the existence of the greatest common divisor of two integers when at least one of them is nonzero. The **Unique Factorization Thoerem**, also known as the **Fundamental Theorem of Arithmetic**, is obtained.

Definition 2.11   **GREATEST COMMON DIVISOR**

> An integer d is a **greatest common divisor** of a and b if these conditions are satisfied:
> 1. d is a positive integer
> 2. $d|a$ and $d|b$
> 3. $c|a$ and $c|b$ imply $c|d$.

The next theorem shows that the greatest common divisor d of a and b exists when at least one of them is not zero. Our proof also shows that d is a **linear combination** of a and b; that is, $d = ma + nb$ for integers m and n.

STRATEGY ▶   The technique of proof by use of the Well-Ordering Theorem in Theorem 2.12 should be compared to that used in the proof of the Division Algorithm (Theorem 2.10).

Theorem 2.12   **GREATEST COMMON DIVISOR**

Let a and b be integers, at least one of them not 0. Then there exists a greatest common divisor d of a and b. Moreover, d can be written as

$$d = am + bn$$

for integers m and n, and d is the smallest positive integer that can be written in this form.

Existence   **Proof**   Let a and b be integers, at least one of them not 0. If $b = 0$, then $a \neq 0$, so $|a| > 0$. It is easy to see that $d = |a|$ is a greatest common divisor of a and b in this case, and either $d = a \cdot (1) + b \cdot (0)$ or $d = a \cdot (-1) + b \cdot (0)$.

Suppose now that $b \neq 0$. Consider the set S of all integers that can be written in the form $ax + by$ for some integers x and y, and let S^+ be the set of all positive integers in S. The set S contains $b = a \cdot (0) + b \cdot (1)$ and $-b = a \cdot (0) + b \cdot (-1)$, so S^+ is not empty. By the Well-Ordering Theorem, S^+ has a least element d,

$$d = am + bn.$$

We have d positive, and we shall show that d is a greatest common divisor of a and b.

By the Division Algorithm, there are integers q and r such that

$$a = dq + r \quad \text{with} \quad 0 \leq r < d.$$

From this equation

$$
\begin{aligned}
r &= a - dq \\
&= a - (am + bn)q \\
&= a(1 - mq) + b(-nq).
\end{aligned}
$$

Thus, r is in $S = \{ax + by\}$, and $0 \leq r < d$. By choice of d as the least element in S^+, it must be true that $r = 0$, and $d \mid a$. Similarly, it can be shown that $d \mid b$.

If $c \mid a$ and $c \mid b$, then $a = ch$ and $b = ck$ for integers h and k. Therefore,

$$
\begin{aligned}
d &= am + bn \\
&= chm + ckn \\
&= c(hm + kn),
\end{aligned}
$$

and this shows that $c \mid d$. By Definition 2.11, $d = am + bn$ is a greatest common divisor of a and b. It follows from the choice of d as least element of S^+ that d is the smallest positive integer that can be written in this form.

Whenever the greatest common divisor of a and b exists, it is unique. For if d_1 and d_2 are both greatest common divisors of a and b, it must be true that $d_1 \mid d_2$ and $d_2 \mid d_1$. Since d_1 and d_2 are positive integers, this means that $d_1 = d_2$ (see Exercise 21 of Section 2.3). We shall write (a, b) or $\gcd(a, b)$ to indicate the unique greatest common divisor of a and b.

When at least one of a and b is not 0, the proof of the last theorem establishes the existence of (a, b), but looking for a smallest positive integer in $S = \{ax + by\}$ is not a very satisfactory method for finding this greatest common divisor. A procedure known as the **Euclidean Algorithm** furnishes a systematic method for finding (a, b) where $b > 0$. It can also be used to find integers m and n such that $(a, b) = am + bn$. This procedure consists of repeated applications of the Division Algorithm according to the following pattern where a and b are integers with $b > 0$.

The Euclidean Algorithm

$$
\begin{array}{ll}
a = bq_0 + r_1, & 0 \leq r_1 < b \\
b = r_1 q_1 + r_2, & 0 \leq r_2 < r_1 \\
r_1 = r_2 q_2 + r_3, & 0 \leq r_3 < r_2 \\
\quad \vdots & \quad \vdots \\
r_k = r_{k+1} q_{k+1} + r_{k+2}, & 0 \leq r_{k+2} < r_{k+1}.
\end{array}
$$

Since the integers $r_1, r_2, \ldots, r_{k+2}$ are decreasing and are all nonnegative, there is a smallest integer n such that $r_{n+1} = 0$:

$$r_{n-1} = r_n q_n + r_{n+1}, \qquad 0 = r_{n+1}.$$

If we put $r_0 = b$, this last nonzero remainder r_n is always the greatest common divisor of a and b. The proof of this statement is left as an exercise.

As an example, we shall find the greatest common divisor of 1492 and 1776.

Example 1   Performing the arithmetic for the Euclidean Algorithm, we have

$$
\begin{array}{ll}
1776 = (1)(1492) + \mathbf{284} & (q_0 = 1, r_1 = 284) \\
1492 = (5)(\mathbf{284}) + \mathbf{72} & (q_1 = 5, r_2 = 72) \\
\mathbf{284} = (3)(\mathbf{72}) + \mathbf{68} & (q_2 = 3, r_3 = 68) \\
\mathbf{72} = (1)(\mathbf{68}) + \mathbf{4} & (q_3 = 1, r_4 = 4) \\
\mathbf{68} = (\mathbf{4})(17) & (q_4 = 17, r_5 = 0).
\end{array}
$$

Thus, the last nonzero remainder is $r_n = r_4 = 4$, and $(1776, 1492) = 4$.   ∎

As mentioned earlier, the Euclidean Algorithm can also be used to find integers m and n such that

$$(a, b) = am + bn.$$

We can obtain these integers by solving for the last nonzero remainder and substituting the remainders from the preceding equations successively until a and b are present in the equation. For example, the remainders in Example 1 can be expressed as

$$
\begin{array}{l}
\mathbf{284} = (1776)(1) + (1492)(-1) \\
\mathbf{72} = (1492)(1) + (\mathbf{284})(-5) \\
\mathbf{68} = (\mathbf{284})(1) + (\mathbf{72})(-3) \\
\mathbf{4} = (\mathbf{72})(1) + (\mathbf{68})(-1).
\end{array}
$$

Substituting the remainders from the preceding equations successively, we have

$$
\begin{aligned}
4 &= (\mathbf{72})(1) + [(\mathbf{284})(1) + (\mathbf{72})(-3)](-1) \\
&= (\mathbf{72})(1) + (\mathbf{284})(-1) + (\mathbf{72})(3) \\
&= (\mathbf{72})(4) + (\mathbf{284})(-1) \quad \text{after the first substitution} \\
&= [(1492)(1) + (\mathbf{284})(-5)](4) + (\mathbf{284})(-1) \\
&= (1492)(4) + (\mathbf{284})(-20) + (\mathbf{284})(-1) \\
&= (1492)(4) + (\mathbf{284})(-21) \quad \text{after the second substitution} \\
&= (1492)(4) + [(1776)(1) + (1492)(-1)](-21) \\
&= (1492)(4) + (1776)(-21) + (1492)(21) \\
&= (1776)(-21) + (1492)(25) \quad \text{after the third substitution.}
\end{aligned}
$$

Thus, $m = -21$ and $n = 25$ are integers such that

$$4 = 1776m + 1492n.$$

The remainders are printed in bold type in each of the preceding steps, and we carefully avoided performing a multiplication that involved a remainder.

The m and n are not unique in the equation

$$(a, b) = am + bn.$$

To see this, simply add and subtract the product ab:

$$(a, b) = am + ab + bn - ab$$
$$= a(m + b) + b(n - a).$$

Thus, $m' = m + b$ and $n' = n - a$ are another pair of integers such that

$$(a, b) = am' + bn'.$$

Definition 2.13   **RELATIVELY PRIME INTEGERS**

> Two integers a and b are **relatively prime** if their greatest common divisor is 1.

In the next two sections of this chapter, we prove some interesting results concerning those integers that are relatively prime to a given integer n. Theorem 2.14 is useful in the proofs of those results.

Theorem 2.14

If a and b are relatively prime and $a|bc$, then $a|c$.

$(p \wedge q) \Rightarrow r$ **Proof**   Assume that $(a, b) = 1$ and $a|bc$. Since $(a, b) = 1$, there are integers m and n such that $1 = am + bn$, by Theorem 2.12. Since $a|bc$, there exists an integer q such that $bc = aq$. Now,

$$1 = am + bn \Rightarrow c = acm + bcn$$
$$\Rightarrow c = acm + aqn \quad \text{since } bc = aq$$
$$\Rightarrow c = a(cm + qn)$$
$$\Rightarrow a|c.$$

Thus, the theorem is proven.

Among the integers, there are those that have the fewest number of factors possible. Some of these are the *prime integers*.

Definition 2.15   **PRIME INTEGER**

> An integer p is a **prime integer** if $p > 1$ and the only divisors of p are ± 1 and $\pm p$.

Notice that the condition $p > 1$ makes p positive and ensures that $p \neq 1$. The exclusion of 1 from the set of primes makes possible the statement of the Unique Factorization Theorem. Before delving into that, we prove the important property of primes in Theorem 2.16.

STRATEGY ▶

The conclusion in the next theorem has the form " r or s." One technique that can be used to prove an "or" statement such as this is to assume that one part (such as r) does not hold, and use this assumption to help prove that the other part must then hold.

Theorem 2.16   EUCLID'S LEMMA

If p is a prime and $p|ab$, then either $p|a$ or $p|b$.

$(p \wedge q) \Rightarrow (r \vee s)$ **Proof**   Assume p is a prime and $p|ab$. If $p|a$, the conclusion of the theorem is satisfied.

Suppose, then, that p does not divide a. This implies that $1 = (p, a)$, since the only positive divisors of p are 1 and p. Then Theorem 2.14 implies that $p|b$. Thus, $p|b$ if p does not divide a, and the theorem is true in any case.

The following corollary generalizes Theorem 2.16 to products with more than two factors. Its proof is requested in the exercises.

Corollary 2.17

If p is a prime and $p|(a_1 a_2 \cdots a_n)$, then p divides some a_j.

This brings us to the **Unique Factorization Theorem**, a result of such importance that it is frequently called the **Fundamental Theorem of Arithmetic**.

STRATEGY ▶   Note the proof of the uniqueness part of Theorem 2.18: Two factorizations are assumed, and then it is proved that the two are equal.

Theorem 2.18   UNIQUE FACTORIZATION THEOREM

Every positive integer n is either 1 or can be expressed as a product of prime integers, and this factorization is unique except for the order of the factors.

Induction   **Proof**   In the statement of the theorem, the word "product" is used in an extended sense: The "product" may have just one factor.

Let P_n be the statement that either $n = 1$ or n can be expressed as a product of primes. We shall prove that P_n is true for all $n \in \mathbf{Z}^+$ by the second principle of finite induction.

Now P_1 is trivially true. Assume that P_m is true for all positive integers $m < k$. If k is a prime, then k is a product with one prime factor, and P_k is true. Suppose k is not a prime. Then $k = ab$, where neither a nor b is 1. Therefore, $1 < a < k$ and $1 < b < k$. By the induction hypothesis, P_a is true and P_b is true. That is,

$$a = p_1 p_2 \cdots p_r \quad \text{and} \quad b = q_1 q_2 \cdots q_s$$

for primes p_i and q_j. These factorizations give

$$k = ab = p_1 p_2 \cdots p_r q_1 q_2 \cdots q_s,$$

and k is thereby expressed as a product of primes. Thus, P_k is true, and therefore, P_n is true for all positive integers n.

Uniqueness   To prove that the factorization is unique, suppose that

$$n = p_1 p_2 \cdots p_t \quad \text{and} \quad n = q_1 q_2 \cdots q_v$$

are factorizations of n as products of prime factors p_i and q_j. Then

$$p_1p_2 \cdots p_t = q_1q_2 \cdots q_v,$$

so $p_1|(q_1q_2 \cdots q_v)$. By Corollary 2.17, $p_1|q_j$ for some j, and there is no loss of generality if we assume $j = 1$. However, p_1 and q_1 are primes, so $p_1|q_1$ implies $q_1 = p_1$. This gives

$$p_1p_2 \cdots p_t = p_1q_2 \cdots q_v,$$

and therefore

$$p_2 \cdots p_t = q_2 \cdots q_v$$

by the cancellation law. This argument can be repeated, removing one factor p_i with each application of the cancellation law, until we obtain

$$p_t = q_t \cdots q_v.$$

Since the only positive factors of p_t are 1 and p_t, and since each q_j is a prime, this means that there must be only one q_j on the right in this equation, and it is q_t. That is, $v = t$ and $q_t = p_t$. This completes the proof.

The Unique Factorization Theorem can be used to describe a standard form of a positive integer n. Suppose $p_1, p_2, \ldots, p_r$ are the *distinct* prime factors of n, arranged in order of magnitude so that

$$p_1 < p_2 < \cdots < p_r.$$

Then all repeated factors may be collected together and expressed by use of exponents to yield

$$n = p_1^{m_1} p_2^{m_2} \cdots p_r^{m_r}$$

where each m_i is a positive integer. Each m_i is called the **multiplicity** of p_i, and this factorization is known as the **standard form** for n.

Example 2 The standard forms for two positive integers a and b can be used to find their greatest common divisor (a,b) and their least common multiple (see Exercises 23 and 24 at the end of this section). For instance, if

$$a = 31{,}752 = 2^3 \cdot 3^4 \cdot 7^2 \quad \text{and} \quad b = 126{,}000 = 2^4 \cdot 3^2 \cdot 5^3 \cdot 7,$$

then (a,b) can be found by forming the product of all the common prime factors, with each common factor raised to the least power to which it appears in either factorization:

$$(a,b) = 2^3 \cdot 3^2 \cdot 7 = 504. \qquad \blacksquare$$

From one point of view, the Unique Factorization Theorem says that the prime integers are building blocks for the integers, where the "building" is done by using multiplication and forming products. A natural question, then, is: How many blocks? Our next theorem states the answer given by the ancient Greek mathematician Euclid, that the number of primes is infinite. The proof is also credited to Euclid.

Theorem 2.19   **Euclid's Theorem on Primes**

The number of primes is infinite.

Contradiction   **Proof**   Assume there are only a finite number, n, of primes. Let these n primes be denoted by $p_1, p_2, \ldots, p_n$, and consider the integer

$$m = p_1 p_2 \cdots p_n + 1.$$

It is clear that the remainder in the division of m by any prime p_i is 1, so each p_i is not a factor of m. Thus, there are two possibilities: either m is itself a prime, or it has a prime factor that is different from every one of the p_i. In either case, we have a prime integer that was not in the list $p_1, p_2, \ldots, p_n$. Therefore, there are more than n primes, and this contradiction establishes the theorem.

EXERCISES 2.4

In this set of exercises, all variables represent integers.

1. List all the primes less than 100.

2. For each of the following pairs, write a and b in standard form and use these factorizations to find (a, b).
   a. $a = 1400, b = 980$
   b. $a = 4950, b = 10{,}500$
   c. $a = 3780, b = 16{,}200$
   d. $a = 52{,}920, b = 25{,}200$

3. In each part, find the greatest common divisor (a, b) and integers m and n such that $(a, b) = am + bn$.
   a. $a = 0, b = -3$
   b. $a = 65, b = -91$
   c. $a = 102, b = 66$
   d. $a = 52, b = 124$
   e. $a = 414, b = -33$
   f. $a = 252, b = -180$
   g. $a = 414, b = 693$
   h. $a = 382, b = 26$
   i. $a = 1197, b = 312$
   j. $a = 3780, b = 1200$
   k. $a = 6420, b = 132$
   l. $a = 602, b = 252$
   m. $a = 5088, b = -156$
   n. $a = 8767, b = 252$

4. Find the smallest integer in the given set.
   a. $\{x \in \mathbf{Z} \mid x > 0 \text{ and } x = 4s + 6t \text{ for some } s, t \text{ in } \mathbf{Z}\}$
   b. $\{x \in \mathbf{Z} \mid x > 0 \text{ and } x = 6s + 15t \text{ for some } s, t \text{ in } \mathbf{Z}\}$

5. If c is a divisor of a and b, prove that c is a divisor of $ax + by$ for all $x, y \in \mathbf{Z}$.

6. Prove that if p and q are distinct primes, then there exist integers m and n such that $pm + qn = 1$.

7. Give an example where $a \mid (bc)$, but $a \nmid b$ and $a \nmid c$.

8. Show that $n^2 - n + 5$ is a prime integer when $n = 1, 2, 3, 4$ but that it is not true that $n^2 - n + 5$ is always a prime integer. Write out a similar set of statements for the polynomial $n^2 - n + 11$.

9. If $a \mid c$ and $b \mid c$, and $(a, b) = 1$, prove that ab divides c.

10. If $b > 0$ and $a = bq + r$, prove that $(a, b) = (b, r)$.

11. Let $r_0 = b > 0$. With the notation used in the description of the Euclidean Algorithm, use the result in Exercise 10 to prove that $(a, b) = r_n$, the last nonzero remainder.

12. Prove that every remainder r_j in the Euclidean Algorithm is a "linear combination" of a and b: $r_j = s_j a + t_j b$, for integers s_j and t_j.

13. Let a and b be integers, at least one of them not 0. Prove that an integer c can be expressed as a linear combination of a and b if and only if $(a, b)|c$.

14. Prove Corollary 2.17: If p is a prime and $p|(a_1 a_2 \cdots a_n)$, then p divides some a_j.

15. Prove that if n is a positive integer greater than 1 such that n is not a prime, then n has a divisor d such that $1 < d \leq \sqrt{n}$.

16. Prove that $(ab, c) = 1$ if and only if $(a, c) = 1$ and $(b, c) = 1$.

17. Prove that if $m > 0$ and (a, b) exists, then $(ma, mb) = m \cdot (a, b)$.

Sec. 2.5, #22 ⩤   18. Prove that if $d = (a, b)$, $a = a_0 d$, and $b = b_0 d$, then $(a_0, b_0) = 1$.

19. Prove that if $d = (a, b)$, $a|c$, and $b|c$, then $ab|cd$.

20. A *least common multiple* of two nonzero integers a and b is an integer m that satisfies the conditions
    1. m is a positive integer,
    2. $a|m$ and $b|m$,
    3. $a|c$ and $b|c$ imply $m|c$.

    Prove that the least common multiple of two nonzero integers exists and is unique.

21. Let a and b be positive integers. If $d = (a, b)$ and m is the least common multiple of a and b, prove that $dm = ab$.

22. Let a and b be positive integers. Prove that if $d = (a, b)$, $a = a_0 d$, and $b = b_0 d$, then the least common multiple of a and b is $a_0 b_0 d$.

23. Describe a procedure for using the standard forms of two positive integers to find their least common multiple.

24. For each pair of integers a, b in Exercise 2, find the least common multiple of a and b by using their standard forms.

25. Let $a, b,$ and c be three nonzero integers.
    a. Use Definition 2.11 as a pattern to define a greatest common divisor of $a, b,$ and c.
    b. Use Theorem 2.12 and its proof as a pattern to prove the existence of a greatest common divisor of $a, b,$ and c.
    c. If d is the greatest common divisor of $a, b,$ and c, show that $d = ((a, b), c)$.

26. Use the Second Principle of Finite Induction to prove that every positive integer n can be expressed in the form
$$n = c_0 + c_1 \cdot 3 + c_2 \cdot 3^2 + \cdots + c_{j-1} \cdot 3^{j-1} + c_j \cdot 3^j,$$
where j is a nonnegative integer, $c_i \in \{0, 1, 2\}$ for all i, and $c_j \in \{1, 2\}$.

27. Use the fact that 2 is a prime to prove that there do not exist nonzero integers a and b such that $a^2 = 2b^2$. Explain how this proves that $\sqrt{2}$ is not a rational number.

28. Use the fact that 3 is a prime to prove that there do not exist nonzero integers a and b such that $a^2 = 3b^2$. Explain how this proves that $\sqrt{3}$ is not a rational number.

2.5   CONGRUENCE OF INTEGERS

In Example 4 of Section 1.6, we defined the relation "congruence modulo 4" on the set $\mathbf{Z}$ of all integers, and we proved this relation to be an equivalence relation on $\mathbf{Z}$. That example is a special case of **congruence modulo n**, as defined next.

Definition 2.20 **CONGRUENCE MODULO n**

Let n be a positive integer, $n > 1$. For integers x and y, x is **congruent** to y **modulo n** if and only if $x - y$ is a multiple of n. We write

$$x \equiv y \;(\mathrm{mod}\; n)$$

to indicate that x is congruent to y modulo n.

Thus, $x \equiv y(\mathrm{mod}\; n)$ if and only if n divides $x - y$, and this is equivalent to $x - y = nq$, or $x = y + nq$. Another way to describe this relation is to say that x and y yield the same remainder when each is divided by n. To see that this is true, let

$$x = nq_1 + r_1 \quad \text{with} \quad 0 \le r_1 < n$$

and

$$y = nq_2 + r_2 \quad \text{with} \quad 0 \le r_2 < n.$$

Then

$$x - y = n(q_1 - q_2) + (r_1 - r_2) \quad \text{with} \quad 0 \le |r_1 - r_2| < n.$$

Thus, $x - y$ is a multiple of n if and only if $r_1 - r_2 = 0$; that is, if and only if $r_1 = r_2$. In particular, any integer x is congruent to its remainder when divided by n. This means that any x is congruent to one of

$$0, 1, 2, \ldots, n - 1.$$

Congruence modulo n is an equivalence relation on **Z**, and this fact is important enough to be stated as a theorem.

Theorem 2.21 **EQUIVALENCE RELATION**

The relation of congruence modulo n is an equivalence relation on **Z**.

Proof We shall show that congruence modulo n is (1) reflexive, (2) symmetric, and (3) transitive. Let $n > 1$, and let x, y, and z be arbitrary in **Z**.

Reflexive   **1.** $x \equiv x \;(\mathrm{mod}\; n)$ since $x - x = (n)(0)$.

Symmetric   **2.** $x \equiv y \;(\mathrm{mod}\; n) \Rightarrow x - y = nq$ for some $q \in \mathbf{Z}$
$$\Rightarrow y - x = n(-q) \quad \text{and} \quad -q \in \mathbf{Z}$$
$$\Rightarrow y \equiv x \;(\mathrm{mod}\; n).$$

Transitive   **3.** $x \equiv y \;(\mathrm{mod}\; n)$ and $y \equiv z \;(\mathrm{mod}\; n)$
$$\Rightarrow x - y = nq \quad \text{and} \quad y - z = nk \quad \text{and} \quad q, k \in \mathbf{Z}$$
$$\Rightarrow x - z = x - y + y - z$$
$$= n(q + k), \quad \text{and} \quad q + k \in \mathbf{Z}$$
$$\Rightarrow x \equiv z \;(\mathrm{mod}\; n).$$

As with any equivalence relation, the equivalence classes for congruence modulo n form a *partition* of **Z**; that is, they separate **Z** into mutually disjoint subsets. These subsets are called **congruence classes** or **residue classes**. Referring to our discussion concerning remainders, we see that there are n distinct congruence classes modulo n, given by

$$[0] = \{\ldots, -2n, -n, 0, n, 2n, \ldots\}$$
$$[1] = \{\ldots, -2n + 1, -n + 1, 1, n + 1, 2n + 1, \ldots\}$$
$$[2] = \{\ldots, -2n + 2, -n + 2, 2, n + 2, 2n + 2, \ldots\}$$
$$\vdots$$
$$[n - 1] = \{\ldots, -n - 1, -1, n - 1, 2n - 1, 3n - 1, \ldots\}.$$

When $n = 4$, these classes appear as

$$[0] = \{\ldots, -8, -4, 0, 4, 8, \ldots\}$$
$$[1] = \{\ldots, -7, -3, 1, 5, 9, \ldots\}$$
$$[2] = \{\ldots, -6, -2, 2, 6, 10, \ldots\}$$
$$[3] = \{\ldots, -5, -1, 3, 7, 11, \ldots\}.$$

Congruence classes are useful in connection with numerous examples, and we shall see more of them later.

Although $x \equiv y \pmod{n}$ is certainly not an equation, in many ways congruences can be handled in the same fashion as equations. The next theorem asserts that the same integer can be added to both members, and that both members can be multiplied by the same integer.

Theorem 2.22 **ADDITION AND MULTIPLICATION PROPERTIES**

If $a \equiv b \pmod{n}$ and x is any integer, then

$$a + x \equiv b + x \pmod{n} \quad \text{and} \quad ax \equiv bx \pmod{n}.$$

$p \Rightarrow q$ **Proof**   Let $a \equiv b \pmod{n}$, and $x \in \mathbf{Z}$. We shall prove that $ax \equiv bx \pmod{n}$, and leave the other part as an exercise. We have

$$\begin{aligned}
a \equiv b \pmod{n} &\Rightarrow a - b = nq && \text{for } q \in \mathbf{Z} \\
&\Rightarrow (a - b)x = (nq)x && \text{for } q, x \in \mathbf{Z} \\
&\Rightarrow ax - bx = n(qx) && \text{for } qx \in \mathbf{Z} \\
&\Rightarrow ax \equiv bx \pmod{n}.
\end{aligned}$$

Congruence modulo n also has substitution properties that are analogous to those possessed by equality.

Theorem 2.23 **SUBSTITUTION PROPERTIES**

Suppose $a \equiv b \pmod{n}$ and $c \equiv d \pmod{n}$. Then

$$a + c \equiv b + d \pmod{n} \quad \text{and} \quad ac \equiv bd \pmod{n}.$$

$(p \wedge q) \Rightarrow r$ **Proof**   Let $a \equiv b \pmod{n}$ and $c \equiv d \pmod{n}$. By Theorem 2.22,

$$a \equiv b \pmod{n} \Rightarrow ac \equiv bc \pmod{n}$$

and

$$c \equiv d \pmod{n} \Rightarrow bc \equiv bd \pmod{n}.$$

But $ac \equiv bc \pmod{n}$ and $bc \equiv bd \pmod{n}$ imply $ac \equiv bd \pmod{n}$, by the transitive property.

The proof that $a + c \equiv b + d \pmod{n}$ is left as an exercise.

It is easy to show that there is a "cancellation law" for addition that holds for congruences: $a + x \equiv a + y \pmod{n}$ implies $x \equiv y \pmod{n}$. This is not the case, however, with multiplication:

$$ax \equiv ay \pmod{n} \quad \text{and} \quad a \not\equiv 0 \pmod{n} \quad \text{do not imply} \quad x \equiv y \pmod{n}.$$

As an example,

$$(4)(6) \equiv (4)(21) \pmod{30} \quad \text{but} \quad 6 \not\equiv 21 \pmod{30}.$$

It is important to notice here that $a = 4$ and $n = 30$ are not relatively prime. When the condition that a and n be relatively prime is imposed, we can obtain a cancellation law for multiplication.

Theorem 2.24   CANCELLATION LAW

If $ax \equiv ay \pmod{n}$ and $(a, n) = 1$, then

$$x \equiv y \pmod{n}.$$

$(p \wedge q) \Rightarrow r$ **Proof**   Assume that $ax \equiv ay \pmod{n}$ and that a and n are relatively prime.

$$\begin{aligned} ax \equiv ay \pmod{n} &\Rightarrow n \mid (ax - ay) \\ &\Rightarrow n \mid a(x - y) \\ &\Rightarrow n \mid (x - y) \quad \text{by Theorem 2.14} \\ &\Rightarrow x \equiv y \pmod{n} \end{aligned}$$

This completes the proof.

We have seen that there are analogues for many of the manipulations that may be performed with equations. There are also techniques for obtaining solutions to congruence equations of certain types. The basic technique makes use of Theorem 2.23 and the Euclidean Algorithm. It is illustrated following the proof of the next theorem.

Theorem 2.25   LINEAR CONGRUENCES

If a and n are relatively prime, the congruence $ax \equiv b \pmod{n}$ has a solution x in the integers, and any two solutions in $\mathbf{Z}$ are congruent modulo n.

$p \Rightarrow q$ **Proof**   Since a and n are relatively prime, there exist integers s and t such that

$$\begin{aligned} & 1 = as + nt \\ \Rightarrow \quad & b = asb + ntb \\ \Rightarrow \quad & a(sb) - b = n(-tb) \\ \Rightarrow \quad & n \mid [a(sb) - b] \\ \Rightarrow \quad & a(sb) \equiv b \pmod{n}. \end{aligned}$$

Thus, $x = sb$ is a solution to $ax \equiv b \pmod{n}$.

To complete the proof, suppose that both x and y are integers that are solutions to $ax \equiv b \pmod{n}$. Then we have

$$ax \equiv b \pmod{n} \quad \text{and} \quad ay \equiv b \pmod{n}.$$

Using the symmetric and transitive properties of congruence modulo n, these relations imply that

$$ax \equiv ay \pmod{n}.$$

Since $(a, n) = 1$, this requires that $x \equiv y \pmod{n}$, by Theorem 2.24. Hence any two solutions in **Z** are congruent modulo n.

Example 1   The Euclidean Algorithm can be used to find a solution x to $ax \equiv b \pmod{n}$ when $(a, n) = 1$. Consider the congruence

$$20x \equiv 14 \pmod{63}.$$

We first obtain s and t such that

$$1 = 20s + 63t.$$

Applying the Euclidean Algorithm, we have

$$63 = (20)(3) + \mathbf{3}$$
$$20 = (\mathbf{3})(6) + \mathbf{2}$$
$$\mathbf{3} = (\mathbf{2})(1) + \mathbf{1}$$
$$\mathbf{2} = (\mathbf{1})(2).$$

Solving for the nonzero remainders,

$$\mathbf{3} = 63 - (20)(3)$$
$$\mathbf{2} = 20 - (\mathbf{3})(6)$$
$$\mathbf{1} = \mathbf{3} - (\mathbf{2})(1).$$

Substituting the remainders in turn, we obtain

$$\begin{aligned}
1 = \mathbf{3} - (\mathbf{2})(1) &= \mathbf{3} - [20 - (\mathbf{3})(6)](1) \\
&= (\mathbf{3})(7) + (20)(-1) \\
&= [63 - (20)(3)][7] + (20)(-1) \\
&= (20)(-22) + (63)(7).
\end{aligned}$$

Multiplying this equation by $b = 14$, we have

$$14 = (20)(-308) + (63)(98)$$
$$\Rightarrow 14 \equiv (20)(-308) \pmod{63}.$$

Thus, $x = -308$ is a solution. However, any number is congruent modulo 63 to its remainder when divided by 63, and

$$-308 = (63)(-5) + 7.$$

By Theorem 2.23, $x = 7$ is also a solution, one that is in the range $0 \le x < 63$.   ■

The preceding example illustrates the basic technique for obtaining a solution to $ax \equiv b \pmod{n}$ when a and n are relatively prime, but other methods are also very useful. Some of them make use of Theorems 2.23 and 2.24. Theorem 2.24 can be used to remove a factor c from both sides of the congruence, provided c and n are relatively prime. That is, c may be canceled from $crx \equiv ct \pmod{n}$ to obtain the equivalent congruence $rx \equiv t \pmod{n}$.

Example 2   Since 2 and 63 are relatively prime, the factor 2 in both sides of

$$20x \equiv 14 \pmod{63}$$

can be removed to obtain

$$10x \equiv 7 \pmod{63}.$$

Theorem 2.21 allows us to replace an integer by any other integer that is congruent to it modulo n. Now $7 \equiv 70 \pmod{63}$, and this substitution yields

$$10x \equiv 70 \pmod{63}.$$

Removing the factor 10 from both sides, we have

$$x \equiv 7 \pmod{63}.$$

Thus, we have obtained the solution $x = 7$ much more easily than by the method of Example 1. However, this method is less systematic, and it requires more ingenuity.   ■

EXERCISES 2.5

In this exercise set, all variables are integers.

1. List the distinct congruence classes modulo 5, exhibiting at least three elements in each class.

2. Follow the instructions in Exercise 1 for the congruence classes modulo 6.

Find a solution $x \in \mathbf{Z}, 0 \le x < n$, for each of the congruences $ax \equiv b \pmod{n}$ in Exercises 3–18. Note that in each case, a and n are relatively prime.

3. $2x \equiv 3 \pmod 7$ 4. $2x \equiv 3 \pmod 5$

5. $3x \equiv 7 \pmod{13}$ 6. $3x \equiv 4 \pmod{13}$

7. $8x \equiv 1 \pmod{21}$ 8. $14x \equiv 8 \pmod{15}$

9. $11x \equiv 1 \pmod{317}$ 10. $11x \equiv 3 \pmod{138}$

11. $8x \equiv 66 \pmod{79}$ 12. $6x \equiv 14 \pmod{55}$

13. $8x + 3 \equiv 5 \pmod 9$ 14. $19x + 7 \equiv 27 \pmod{18}$

15. $25x \equiv 31 \pmod 7$ 16. $45x \equiv 17 \pmod{313}$

17. $35x + 14 \equiv 3 \pmod{27}$ 18. $57x + 7 \equiv 78 \pmod{53}$

19. Complete the proof of Theorem 2.22: If $a \equiv b \pmod{n}$ and x is any integer, then $a + x \equiv b + x \pmod{n}$.

20. Complete the proof of Theorem 2.23: If $a \equiv b \pmod{n}$ and $c \equiv d \pmod{n}$, then $a + c \equiv b + d \pmod{n}$.

21. Prove that if $a + x \equiv a + y \pmod{n}$, then $x \equiv y \pmod{n}$.

Sec. 2.4, #18 ≫ 22. If $ca \equiv cb \pmod{n}$ and $d = (c,n)$ where $n = dm$, prove that $a \equiv b \pmod{m}$.

23. Find the least positive integer that is congruent modulo 7 to the given product.
    a. $(4)(9)(15)(59)$ b. $(5)(11)(17)(65)$

24. If $a \equiv b \pmod{n}$, prove that $a^m \equiv b^m \pmod{n}$ for every positive integer m.

25. Prove that if m is an integer, then either $m^2 \equiv 0 \pmod 4$ or $m^2 \equiv 1 \pmod 4$. (*Hint:* Consider the cases where m is even and where m is odd.)

26. If m is an integer, show that m^2 is congruent modulo 8 to one of the integers $0, 1,$ or 4. (*Hint:* Use the Division Algorithm, and consider the possible remainders in $m = 4q + r$.)

27. Let x and y be integers. Prove that if there is an equivalence class $[a]$ modulo n such that $x \in [a]$ and $y \in [a]$, then $(x, n) = (y, n)$.

28. Prove that if p is a prime and $c \not\equiv 0 \pmod p$, then $cx \equiv b \pmod p$ has a unique solution modulo p. That is, a solution exists, and any two solutions are congruent modulo p.

29. Let $d = (a, n)$ where $n > 1$. Prove that if there is a solution to $ax \equiv b \pmod n$, then d divides b.

30. (See Exercise 29.) Suppose that $n > 1$ and that $d = (a, n)$ is a divisor of b. Let $a = a_0 d$, $n = n_0 d$, and $b = b_0 d$, where $a_0, n_0,$ and b_0 are integers. The following statements a–e lead to a proof that the congruence $ax \equiv b \pmod n$ has exactly d incongruent solutions modulo n, and they show how such a set of solutions can be found.
   a. Prove that $ax \equiv b \pmod n$ if and only if $a_0 x \equiv b_0 \pmod{n_0}$.
   b. Prove that if x_1 and x_2 are any two solutions to $a_0 x \equiv b_0 \pmod{n_0}$, then $x_1 \equiv x_2 \pmod{n_0}$.
   c. Let x_1 be a fixed solution to $a_0 x \equiv b_0 \pmod{n_0}$ and prove that each of the d integers in the list
$$x_1, x_1 + n_0, x_1 + 2n_0, \ldots, x_1 + (d - 1)n_0$$
   is a solution to $ax \equiv b \pmod n$.
   d. Prove that no two of the solutions listed in Part c are congruent modulo n.
   e. Prove that any solution to $ax \equiv b \pmod n$ is congruent to one of the numbers listed in Part **c**.

In the congruences $ax \equiv b \pmod n$ in Exercises 31–40, a and n may not be relatively prime. Use the results in Exercises 29 and 30 to determine if there are solutions. If there are, find d incongruent solutions modulo n.

31. $6x \equiv 33 \pmod{27}$ **32.** $18x \equiv 33 \pmod{15}$

33. $8x \equiv 66 \pmod{78}$ **34.** $35x \equiv 10 \pmod{20}$

35. $34x \equiv 18 \pmod{20}$ **36.** $21x \equiv 18 \pmod{30}$

37. $24x + 5 \equiv 50 \pmod{348}$ **38.** $36x + 1 \equiv 49 \pmod{270}$

39. $42x + 67 \equiv 23 \pmod{74}$ **40.** $38x + 54 \equiv 20 \pmod{60}$

Sec. 4.4, #43 ≪
41. Let p be a prime integer. Prove **Fermat's Little Theorem**: For any positive integer a, $a^p \equiv a \pmod p$. (*Hint:* Use induction on a, with p held fixed.)

42. Assume that m and n are relatively prime, and let a and b be integers. Prove that there exists an integer x such that $x \equiv a \pmod m$ and $x \equiv b \pmod n$.

43. In Exercise 42, prove that any two solutions are congruent modulo mn.

44. Solve the following simultaneous congruences. (See Exercise 42.)
   a. $x \equiv 2 \pmod 5$ and $x \equiv 3 \pmod 8$
   b. $x \equiv 4 \pmod 5$ and $x \equiv 2 \pmod 3$

2.6 CONGRUENCE CLASSES

In connection with the relation of congruence modulo n, we have observed that there are n distinct congruence classes. Let $\mathbf{Z}_n$ denote this set of classes:

$$\mathbf{Z}_n = \{[0], [1], [2], \ldots, [n - 1]\}.$$

When addition and multiplication are defined in a natural and appropriate manner in $\mathbf{Z}_n$, these sets provide useful examples for our work in later chapters.

Theorem 2.26   ADDITION IN $\mathbf{Z}_n$

Consider the rule given by

$$[a] + [b] = [a + b].$$

a. This rule defines an addition that is a binary operation on $\mathbf{Z}_n$.
b. Addition is associative in $\mathbf{Z}_n$:

$$[a] + ([b] + [c]) = ([a] + [b]) + [c].$$

c. $\mathbf{Z}_n$ has the additive identity $[0]$.
d. Each $[a]$ in $\mathbf{Z}_n$ has $[-a]$ as its additive inverse in $\mathbf{Z}_n$.
e. Addition is commutative in $\mathbf{Z}_n$:

$$[a] + [b] = [b] + [a].$$

Proof

a. It is clear that the rule $[a] + [b] = [a + b]$ yields an element of $\mathbf{Z}_n$, but the uniqueness of this result needs to be verified. In other words, closure is obvious, but we need to show that the operation is well-defined. To do this, suppose that $[a] = [x]$ and $[b] = [y]$. Then

$$[a] = [x] \Rightarrow a \equiv x \pmod{n}$$

and

$$[b] = [y] \Rightarrow b \equiv y \pmod{n}.$$

By Theorem 2.23,

$$a + b \equiv x + y \pmod{n},$$

and therefore $[a + b] = [x + y]$.

b. The associative property follows from

$$\begin{aligned}
[a] + ([b] + [c]) &= [a] + [b + c] \\
&= [a + (b + c)] \\
&= [(a + b) + c] \\
&= [a + b] + [c] \\
&= ([a] + [b]) + [c].
\end{aligned}$$

Notice that the key step here is the fact that addition is associative in $\mathbf{Z}$:

$$a + (b + c) = (a + b) + c.$$

c. $[0]$ is the additive identity, since

$$[a] + [0] = [a + 0] = [a]$$

and

$$[0] + [a] = [0 + a] = [a].$$

d. $[-a] = [n - a]$ is the additive inverse of $[a]$, since

$$[a] + [-a] = [a + (-a)] = [0]$$

and

$$[-a] + [a] = [-a + a] = [0].$$

e. The commutative property follows from

$$[a] + [b] = [a + b]$$
$$= [b + a]$$
$$= [b] + [a].$$

Example 1   Following the procedure described in Exercise 3 of Section 1.4, we can construct an addition table for $\mathbf{Z}_4 = \{[0], [1], [2], [3]\}$. In computing the entries for this table, $[a] + [b]$ is entered in the row with $[a]$ at the left and in the column with $[b]$ at the top. For instance,

$$[3] + [2] = [5] = [1]$$

is entered in the row with $[3]$ at the left and in the column with $[2]$ at the top. The complete addition table is shown in Figure 2-1.

Figure 2-1

+	[0]	[1]	[2]	[3]
[0]	[0]	[1]	[2]	[3]
[1]	[1]	[2]	[3]	[0]
[2]	[2]	[3]	[0]	[1]
[3]	[3]	[0]	[1]	[2]

■

In the following theorem, multiplication in $\mathbf{Z}_n$ is defined in a natural way, and the basic properties for this operation are stated. The proofs of the various parts of the theorem are quite similar to those for the corresponding parts of Theorem 2.26, and are left as exercises.

Theorem 2.27   **MULTIPLICATION IN $\mathbf{Z}_n$**

Consider the rule for multiplication in $\mathbf{Z}_n$ given by

$$[a][b] = [ab].$$

a. Multiplication as defined by this rule is a binary operation on $\mathbf{Z}_n$.
b. Multiplication is associative in $\mathbf{Z}_n$:

$$[a]([b][c]) = ([a][b])[c].$$

c. $\mathbf{Z}_n$ has the multiplicative identity $[1]$.
d. Multiplication is commutative in $\mathbf{Z}_n$:

$$[a][b] = [b][a].$$

When we compare the properties listed in Theorems 2.26 and 2.27, we see that the existence of multiplicative inverses, even for the nonzero elements, is conspicuously missing. The following example shows that this is appropriate because it illustrates a case where some of the nonzero elements of $\mathbf{Z}_n$ do not have multiplicative inverses.

Example 2   A multiplication table for $\mathbf{Z}_4$ is shown in Figure 2-2.

Figure 2-2

$\times$	[0]	[1]	[2]	[3]
[0]	[0]	[0]	[0]	[0]
[1]	[0]	[1]	[2]	[3]
[2]	[0]	[2]	[0]	[2]
[3]	[0]	[3]	[2]	[1]

The third row of the table shows that [2] is a nonzero element of $\mathbf{Z}_4$ that has no multiplicative inverse; there is no $[x]$ in $\mathbf{Z}_4$ such that $[2][x] = [1]$. Another interesting point in connection with this table is that the equality $[2][2] = [0]$ shows that in $\mathbf{Z}_n$, the product of nonzero factors may be zero.   ■

The next theorem characterizes those elements of $\mathbf{Z}_n$ that have multiplicative inverses.

Theorem 2.28   **MULTIPLICATIVE INVERSES IN $\mathbf{Z}_n$**

An element $[a]$ of $\mathbf{Z}_n$ has a multiplicative inverse in $\mathbf{Z}_n$ if and only if a and n are relatively prime.

$p \Rightarrow q$ **Proof**   Suppose first that $[a]$ has a multiplicative inverse $[b]$ in $\mathbf{Z}_n$. Then

$$[a][b] = [1].$$

This means that

$$[ab] = [1] \quad \text{and} \quad ab \equiv 1 \ (\text{mod } n).$$

Therefore,

$$ab - 1 = nq$$

for some integer q, and

$$a(b) + n(-q) = 1.$$

By Theorem 2.12, we have $(a, n) = 1$.

$p \Leftarrow q$ Conversely, if $(a, n) = 1$, then there exist integers s and t such that

$$
\begin{aligned}
as + nt &= 1 \\
\Rightarrow \quad as - 1 &= n(-t) \\
\Rightarrow \quad as &\equiv 1 \ (\text{mod } n) \\
\Rightarrow \quad [a][s] &= [1].
\end{aligned}
$$

Thus, $[a]$ has a multiplicative inverse $[s]$ in $\mathbf{Z}_n$.

Corollary 2.29

Every nonzero element of $\mathbf{Z}_n$ has a multiplicative inverse if and only if n is a prime.

$p \Leftrightarrow q$ **Proof**   The corollary follows from the fact that n is a prime if and only if every integer a such that $1 \leq a < n$ is relatively prime to n.

Example 3   The elements of $\mathbf{Z}_{15}$ that have multiplicative inverses can be listed by writing down those $[a]$ that are such that $(a, 15) = 1$. These elements are

$$[1], [2], [4], [7], [8], [11], [13], [14].$$ ∎

Example 4   Suppose we wish to find the multiplicative inverse of $[13]$ in $\mathbf{Z}_{191}$. The modulus $n = 191$ is so large that it is not practical to test all of the elements in $\mathbf{Z}_{191}$, so we utilize the Euclidean Algorithm and proceed according to the last part of the proof of Theorem 2.28:

$$191 = (13)(14) + \mathbf{9}$$
$$13 = (\mathbf{9})(1) + \mathbf{4}$$
$$\mathbf{9} = (\mathbf{4})(2) + \mathbf{1}.$$

Substituting the remainders in turn, we have

$$\mathbf{1} = \mathbf{9} - (\mathbf{4})(2)$$
$$= \mathbf{9} - [13 - (\mathbf{9})(1)](2)$$
$$= (\mathbf{9})(3) - (13)(2)$$
$$= [191 - (13)(14)](3) - (13)(2)$$
$$= (191)(3) + (13)(-44).$$

Thus,

$$(13)(-44) \equiv 1 \ (\text{mod } 191)$$

or

$$[13][-44] = [1].$$

The desired inverse is

$$[13]^{-1} = [-44] = [147].$$ ∎

Since every element in $\mathbf{Z}_n$ has an additive inverse, **subtraction** can be defined in $\mathbf{Z}_n$ by the equation

$$[a] - [b] \quad = [a] + (-[b])$$
$$= [a] + [-b]$$
$$= [a - b].$$

We now have at hand the basic knowledge about addition, subtraction, multiplication, and multiplicative inverses in $\mathbf{Z}_n$. Utilizing this knowledge, many of the techniques that we use to solve equations in real numbers can be successfully imitated to solve equations involving elements of $\mathbf{Z}_n$. For example, Exercise 9 of this section states that $[x] = [a]^{-1}[b]$ is the unique solution to $[a][x] = [b]$ in $\mathbf{Z}_n$ whenever $[a]^{-1}$ exists. In Exercise 19, some quadratic equations are to be solved by factoring. The next example shows how we can solve a simple system of linear equations in $\mathbf{Z}_n$ by using the same kinds of steps that we use when working with numbers.

Example 5   We shall solve the following system of linear equations in $\mathbf{Z}_{26}$.

$$[4][x] + [y] = [22]$$
$$[19][x] + [y] = [15]$$

We can eliminate $[y]$ by subtracting the top equation from the bottom one:

$$[19][x] - [4][x] = [15] - [22].$$

This simplifies to

$$[15][x] = [-7]$$

or

$$[15][x] = [19].$$

Using the Euclidean Algorithm as we did in Example 4, we find that $[15]$ in $\mathbf{Z}_{26}$ has the multiplicative inverse given by $[15]^{-1} = [7]$. Using the result in Exercise 9 of this section, the solution $[x]$ to $[15][x] = [19]$ is

$$
\begin{aligned}
[x] &= [15]^{-1}[19] \\
&= [7][19] \\
&= [133] \\
&= [3].
\end{aligned}
$$

Solving for $[y]$ in the equation $[4][x] + [y] = [22]$ yields

$$
\begin{aligned}
[y] &= [22] - [4][x] \\
&= [22] - [4][3] \\
&= [22] - [12] \\
&= [10].
\end{aligned}
$$

It is easy to check that $[x] = [3], [y] = [10]$ is indeed a solution to the system.   ∎

EXERCISES 2.6

1. Perform the following computations in $\mathbf{Z}_{12}$.
   a. $[8] + [7]$ b. $[10] + [9]$
   c. $[8][11]$ d. $[6][9]$
   e. $[6]([9] + [7])$ f. $[5]([8] + [11])$
   g. $[6][9] + [6][7]$ h. $[5][8] + [5][11]$

2. a. Verify that $[1][2][3][4] = [4]$ in $\mathbf{Z}_5$.
   b. Verify that $[1][2][3][4][5][6] = [6]$ in $\mathbf{Z}_7$.
   c. Evaluate $[1][2][3]$ in $\mathbf{Z}_4$.
   d. Evaluate $[1][2][3][4][5]$ in $\mathbf{Z}_6$.

3. Make addition tables for each of the following.
   a. $\mathbf{Z}_2$ b. $\mathbf{Z}_3$ c. $\mathbf{Z}_5$
   d. $\mathbf{Z}_6$ e. $\mathbf{Z}_7$ f. $\mathbf{Z}_8$

4. Make multiplication tables for each of the following.
   a. $\mathbf{Z}_2$ **b.** $\mathbf{Z}_3$ **c.** $\mathbf{Z}_5$
   d. $\mathbf{Z}_6$ **e.** $\mathbf{Z}_7$ **f.** $\mathbf{Z}_8$

5. Find the multiplicative inverse of each given element.
   a. $[3]$ in $\mathbf{Z}_{13}$ **b.** $[7]$ in $\mathbf{Z}_{11}$
   c. $[17]$ in $\mathbf{Z}_{20}$ **d.** $[16]$ in $\mathbf{Z}_{27}$
   e. $[11]$ in $\mathbf{Z}_{317}$ **f.** $[9]$ in $\mathbf{Z}_{128}$

6. For each of the following $\mathbf{Z}_n$, list all the elements in $\mathbf{Z}_n$ that have multiplicative inverses in $\mathbf{Z}_n$.
   a. $\mathbf{Z}_6$ **b.** $\mathbf{Z}_8$ **c.** $\mathbf{Z}_{10}$
   d. $\mathbf{Z}_{12}$ **e.** $\mathbf{Z}_{18}$ **f.** $\mathbf{Z}_{20}$

7. For each of the following $\mathbf{Z}_n$, find all the nonzero elements $[a]$ in $\mathbf{Z}_n$, for which the equation $[a][x] = [0]$ has a nonzero solution $[x] \neq [0]$ in $\mathbf{Z}_n$.
   a. $\mathbf{Z}_6$ **b.** $\mathbf{Z}_8$ **c.** $\mathbf{Z}_{10}$
   d. $\mathbf{Z}_{12}$ **e.** $\mathbf{Z}_{18}$ **f.** $\mathbf{Z}_{20}$

8. Whenever possible, find a solution for each of the following equations in the given $\mathbf{Z}_n$.
   a. $[4][x] = [2]$ in $\mathbf{Z}_6$ **b.** $[6][x] = [4]$ in $\mathbf{Z}_{12}$
   c. $[6][x] = [4]$ in $\mathbf{Z}_8$ **d.** $[10][x] = [6]$ in $\mathbf{Z}_{12}$
   e. $[8][x] = [6]$ in $\mathbf{Z}_{12}$ **f.** $[4][x] = [6]$ in $\mathbf{Z}_8$
   g. $[10][x] = [4]$ in $\mathbf{Z}_{12}$ **h.** $[9][x] = [3]$ in $\mathbf{Z}_{12}$

9. Let $[a]$ be an element of $\mathbf{Z}_n$ that has a multiplicative inverse $[a]^{-1}$ in $\mathbf{Z}_n$. Prove that $[x] = [a]^{-1}[b]$ is the unique solution in $\mathbf{Z}_n$ to the equation $[a][x] = [b]$.

10. Solve each of the following equations by finding $[a]^{-1}$ and using the result in Exercise 9.
    a. $[4][x] = [5]$ in $\mathbf{Z}_{13}$ **b.** $[8][x] = [7]$ in $\mathbf{Z}_{11}$
    c. $[7][x] = [11]$ in $\mathbf{Z}_{12}$ **d.** $[8][x] = [11]$ in $\mathbf{Z}_{15}$
    e. $[9][x] = [14]$ in $\mathbf{Z}_{20}$ **f.** $[8][x] = [15]$ in $\mathbf{Z}_{27}$
    g. $[6][x] = [5]$ in $\mathbf{Z}_{319}$ **h.** $[9][x] = [8]$ in $\mathbf{Z}_{242}$

In Exercises 11–14, solve the systems of equations in $\mathbf{Z}_7$.

11. $[2][x] + [y] = [4]$
    $[2][x] + [4][y] = [5]$

12. $[4][x] + [2][y] = [1]$
    $[3][x] + [2][y] = [5]$

13. $[3][x] + [2][y] = [1]$
    $[5][x] + [6][y] = [5]$

14. $[2][x] + [5][y] = [6]$
    $[4][x] + [6][y] = [6]$

15. Prove Theorem 2.27.

16. Prove the following distributive property in $\mathbf{Z}_n$:
$$[a]([b] + [c]) = [a][b] + [a][c].$$

17. Prove the following equality in $\mathbf{Z}_n$:
$$([a] + [b])([c] + [d]) = [a][c] + [a][d] + [b][c] + [b][d].$$

18. Let p by a prime integer. Prove that if $[a][b] = [0]$ in $\mathbf{Z}_p$, then either $[a] = [0]$ or $[b] = [0]$.

19. Use the results in Exercises 16–18 and find all solutions $[x]$ to the following quadratic equations by the factoring method.
    a. $[x]^2 + [5][x] + [6] = [0]$ in $\mathbf{Z}_7$ **b.** $[x]^2 + [4][x] + [3] = [0]$ in $\mathbf{Z}_5$
    c. $[x]^2 + [x] + [5] = [0]$ in $\mathbf{Z}_7$ **d.** $[x]^2 + [x] + [3] = [0]$ in $\mathbf{Z}_5$

20. Let p be a prime integer. Prove that $[1]$ and $[p - 1]$ are the only elements in $\mathbf{Z}_p$ that are their own multiplicative inverses.

21. Show that if n is not a prime, then there exist $[a]$ and $[b]$ in $\mathbf{Z}_p$ such that $[a] \neq [0]$ and $[b] \neq [0]$, but $[a][b] = [0]$.

22. Let p be a prime integer. Prove the following cancellation law in $\mathbf{Z}_p$: If $[a][x] = [a][y]$ and $[a] \neq [0]$, then $[x] = [y]$.

23. Show that if n is not a prime, the cancellation law stated in Exercise 22 does not hold in $\mathbf{Z}_n$.

24. Suppose that x and y are integers such that $x \in [a]$ and $y \in [a]$ for some $[a] \in \mathbf{Z}_n$. Prove that $(x, n) = (y, n)$.

2.7   INTRODUCTION TO CODING THEORY (OPTIONAL)

In this section we present some applications of congruence modulo n found in basic coding theory. When information is transmitted from one satellite to another or stored and retrieved in a computer or on a compact disc, the information is usually expressed in some sort of code. The ASCII code (American Standard Code for Information Interchange) of 256 characters used in computers is one example. However, errors can occur during the transmission or retrieval processes. The detection and correction of such errors are the fundamental goals of coding theory.

In binary coding theory, we omit the brackets on the elements in $\mathbf{Z}_2$ and call $\{0, 1\}$ the **binary alphabet**. A **bit**[†] is an element of the binary alphabet. A **word** (or **block**) is a sequence of bits, where all words in a message have the same **length**; that is, they contain the same number of bits. Thus a 2-bit word is an element of $\mathbf{Z}_2 \times \mathbf{Z}_2$. For notational convenience, we omit the comma and parentheses in the 2-bit word (a, b) and write ab, where $a \in \{0, 1\}$ and $b \in \{0, 1\}$. Thus,

$$
\begin{array}{cccc}
000 & 010 & 001 & 011 \\
100 & 110 & 101 & 111
\end{array}
$$

are all eight possible 3-bit words using the binary alphabet. Since there are thirty-two 5-bit words, then 5-bit words are frequently used to represent the 26 letters of our alphabet, along with six punctuation marks.

During the process of sending a message using k-bit words, one or more bits may be received incorrectly. It is essential that errors be detected and, if possible, corrected. The general idea is to generate a **code**, send the coded message, and then decode the coded message, as illustrated in the following diagram.

$$\text{message} \xrightarrow{\textit{encode}} \text{coded message} \xrightarrow{\textit{send}} \text{received message} \xrightarrow{\textit{decode}} \text{message}$$

Ideally, the code is devised in such a way as to detect and/or correct any errors in the received message. Most codes require appending extra bits to each k-bit word, forming an n-bit code word. The next example illustrates an **error-detecting** scheme.

[†]*Bit* is an abbreviation for *binary digit*.

Example 1 Parity Check Consider 3-bit words of the form *abc*. One coding scheme maps *abc* onto *abcd*, where

$$d \equiv a + b + c (\text{mod } 2)$$

is called the **parity check digit**. If $d = 0$, we say that the word *abc* has **even parity**. If $d = 1$, we say *abc* has **odd parity**. Thus, the eight possible 3-bit words are mapped onto the eight 4-bit code words as follows:

$$\text{word} \xrightarrow{\text{encode}} \text{codeword}$$
$$000 \xrightarrow{\text{encode}} 0000$$
$$010 \xrightarrow{\text{encode}} 0101$$
$$001 \xrightarrow{\text{encode}} 0011$$
$$011 \xrightarrow{\text{encode}} 0110$$
$$100 \xrightarrow{\text{encode}} 1001$$
$$110 \xrightarrow{\text{encode}} 1100$$
$$101 \xrightarrow{\text{encode}} 1010$$
$$111 \xrightarrow{\text{encode}} 1111$$

Notice that each 4-bit code word has even parity. Therefore, a simple parity check on the code word will detect any single-bit error. For example, suppose that the coded message of five 4-bit code words

$$1101 \qquad 1011 \qquad 0000 \qquad 0110 \qquad 0011$$

is received. It is obvious that the first two code words 1101 and 1011 each contain at least one error. This parity check scheme does not correct single-bit errors, nor will it detect which bit is in error. It also will not detect 2-bit errors. In this situation, the safest action is to request retransmission of the message, if retransmission is feasible.
∎

Example 2 Repetition Codes Multiple errors can be detected (but not corrected) in a scheme in which a *k*-bit word is mapped onto a 2*k*-bit code word according to the following:

$$x_1 x_2 \cdots x_k \xrightarrow{\text{encode}} x_1 x_2 \cdots x_k x_1 x_2 \cdots x_k.$$

In the coded message with $k = 3$,

$$110110 \qquad 010011 \qquad 011011 \qquad 101000,$$

errors occur in the second code word 010011 and in the last code word 101000. All other code words seem to be correct. If, upon retransmission, the coded message is received as

$$110110 \qquad 011011 \qquad 011011 \qquad 100100,$$

it would be decoded as

$$110 \qquad 011 \qquad 011 \qquad 100.$$
∎

Example 3 Maximum Likelihood Decoding Multiple errors can be detected and *corrected* if each k-bit word is mapped onto a $3k$-bit code word according to the following scheme (called a **triple repetition code**).

$$x_1 x_2 \cdots x_k \xrightarrow{\text{encode}} x_1 x_2 \cdots x_k x_1 x_2 \cdots x_k x_1 x_2 \cdots x_k$$

For example, if the 6-bit code word (for a 2-bit word)

$$010111$$

is received, then an error is detected. By separating the code word into three equal parts

$$01 \quad 01 \quad 11$$

and comparing bit by bit, we note that the first bits in each part do not agree. We correct the error by choosing the digit that occurs most often, in this case a 0. Thus, the corrected code word would be

$$010101,$$

and more than likely the correct message is 01. The main disadvantage to this type of coding is that each message requires three times as many bits as the decoded message, whereas with the parity check scheme, only one extra bit is needed for each word. ∎

A combination of a parity check and a repetition code allows detection and correction of coded messages without requiring quite as many bits as in the maximum likelihood scheme. We illustrate this in the next example.

Example 4 Error Detection and Correction Suppose 4-bit words are mapped onto 9-bit code words using the scheme

$$x_1 x_2 x_3 x_4 \xrightarrow{\text{encode}} x_1 x_2 x_3 x_4 x_1 x_2 x_3 x_4 p,$$

where p is the parity check digit

$$p \equiv x_1 + x_2 + x_3 + x_4 (\text{mod } 2).$$

For example, the 4-bit word 0110 is encoded as 011001100. Suppose, upon transmission, a code word 101011100 is received. Breaking 101011100 into three parts,

$$1010 \quad 1110 \quad 0,$$

indicates that an error occurs in the second bit. To have parity 0, the correct word must be 1010.

Errors might also occur in the parity digit. For example, if 001100111 is received, an error is detected, and more than likely the error has been made in the parity check digit. Thus, the correct word is 0011. ∎

The last two examples bring up the question of probability of errors occurring in any one or more bits of a n-bit code word. We make the following assumptions:

1. The probability of any single bit being transmitted incorrectly is P.

2. The probability of any single bit being transmitted correctly or incorrectly is independent of the probability of any other single bit being transmitted correctly or incorrectly.

Thus, the probability of transmitting a 5-bit code word with only one incorrect bit is $\binom{5}{1}P(1 - P)^4$. If it happens that $P = 0.01$ (approximately 1 of every 100 bits are transmitted incorrectly), then the probability of transmitting a 5-bit code word with only one incorrect bit is $\binom{5}{1}0.01(0.99)^4 = 0.04803$, and the probability of transmitting a 5-bit code word with no errors is $\binom{5}{0}(0.01)^0(0.99)^5 = 0.95099$. Hence, the probability of transmitting a 5-bit code word with at most one error is $\binom{5}{1}0.01(0.99)^4 + \binom{5}{0}(0.01)^0(0.99)^5 = 0.99902$.

Up to this point, $\mathbf{Z}_2$ has been used in all of our examples. We next look at some instances in which other congruence classes play a role.

Example 5 Using Check Digits Many companies use **check digits** for security purposes or for error detection. For example, an 11th digit may be appended to a 10-bit identification number to obtain the 11-digit invoice number of the form

$$x_1 x_2 x_3 x_4 x_5 x_6 x_7 x_8 x_9 x_{10} c,$$

where the 11th bit, c, is the check digit, computed as

$$x_1 x_2 x_3 x_4 x_5 x_6 x_7 x_8 x_9 x_{10} \equiv c (\bmod\ n).$$

If congruence modulo 9 is used, then the check digit for an identification number 3254782201 is 7, since $3254782201 \equiv 7 (\bmod\ 9)$. Thus, the complete correct invoice number would appear as 32547822017. If the invoice number 31547822017 were used instead and checked, an error would be detected, since $3154782201 \not\equiv 7 (\bmod\ 9)$. [$3154782201 \equiv 6 (\bmod\ 9)$.]

This particular scheme is not infallible in detecting errors. For example, if a transposition error (a common keyboarding error) occurred and the invoice number were erroneously entered as 32548722017, an error would not be detected, since $3254872201 \equiv 7 (\bmod\ 9)$. It can be shown that transposition errors will never be detected with this scheme (using congruence modulo 9) unless one of the digits is the check digit. (See Exercise 12.) ■

Even more sophisticated schemes of using check digits appear in such places as the ISBN numbers assigned to all books, the UPC (Universal Product Codes) assigned to products in the marketplace, passport numbers, and the driver's licenses and license plate numbers in some states. Some of the schemes are very good at detecting errors, and others are surprisingly faulty. In these schemes, a *weighting vector* is used in conjunction with arithmetic on congruence classes modulo n (modular arithmetic). The **dot product** notation is useful in describing the situation. We define the dot product $(x_1, x_2, \ldots, x_n) \cdot (y_1, y_2, \ldots, y_n)$ of two ordered n-tuples (*vectors*) $(x_1, x_2, \ldots, x_n)$ and $(y_1, y_2, \ldots, y_n)$ by

$$(x_1, x_2, \ldots, x_n) \cdot (y_1, y_2, \ldots, y_n) = x_1 y_1 + x_2 y_2 + \cdots + x_n y_n.$$

For example, $(1, 2, 3) \cdot (-3, 7, -1) = -3 + 14 - 3 = 8$. The next example describes the use of the dot product and weighting vector in bank identification numbers.

Example 6   Bank Identification Numbers   Identification numbers for banks have eight digits, $x_1 x_2, \ldots, x_8$, and a check digit x_9, given by

$$(x_1, x_2, \ldots, x_8) \cdot (7, 3, 9, 7, 3, 9, 7, 3) \equiv x_9 (\text{mod } 10).$$

The weighting vector is $(7, 3, 9, 7, 3, 9, 7, 3)$. Thus, a bank with identification number 05320044 has check digit

$$(0, 5, 3, 2, 0, 0, 4, 4) \cdot (7, 3, 9, 7, 3, 9, 7, 3) = 0 + 15 + 27 + 14 + 0 + 0 + 28 + 12$$
$$= 96$$
$$\equiv 6 \ (\text{mod } 10)$$

and appears as 053200446 at the bottom of the check. This particular scheme detects all one-digit errors. However, suppose that this same bank identification number is coded in as 503200446, with a transposition of the first and second digits. The check digit 6 does not detect the error:

$$(5, 0, 3, 2, 0, 0, 4, 4) \cdot (7, 3, 9, 7, 3, 9, 7, 3) = 35 + 0 + 27 + 14 + 0 + 0 + 28 + 12$$
$$= 116$$
$$\equiv 6 \ (\text{mod } 10).$$

Transposition errors of adjacent digits x_i and x_{i+1} will be detected by this scheme except when $|x_i - x_{i+1}| = 5$. (See Exercise 13.)   ∎

The next example illustrates the use of another weighting vector in Universal Product Codes.

Example 7   UPC Symbols   UPC symbols consist of 12 digits $x_1 x_2 \cdots x_{12}$, with the last, x_{12}, being the check digit. The weighting vector used for the UPC symbols is the 11-tuple $(3, 1, 3, 1, 3, 1, 3, 1, 3, 1, 3)$. The check digit x_{12} can be computed as

$$-(x_1, x_2, \ldots, x_{11}) \cdot (3, 1, 3, 1, 3, 1, 3, 1, 3, 1, 3) \equiv x_{12} \ (\text{mod } 10) .$$

The computation

$$-(0, 2, 1, 2, 0, 0, 6, 9, 1, 1, 3) \cdot (3, 1, 3, 1, 3, 1, 3, 1, 3, 1, 3) = -47$$

$$\equiv 3 \ (\text{mod } 10)$$

verifies the check digit 3 shown in the UPC symbol in Figure 2-3. As in the bank identification scheme, some transposition errors may go undetected.

Figure 2-3
UPC Symbol

0   21200 69113   3   ∎

In this section, we have attempted to introduce only the basic concepts of coding theory; more sophisticated coding schemes are constantly being developed. Much research is being done in this branch of mathematics, based not only on group and field theory but also on linear algebra and probability theory.

EXERCISES 2.7

1. Suppose 4-bit words *abcd* are mapped onto 5-bit code words *abcde*, where *e* is the parity check digit. Detect any errors in the following six-word coded message.

    11101   00101   00010   11100   00011   10100

2. Suppose 3-bit words *abc* are mapped onto 6-bit code words *abcabc* under a repetition scheme. Detect any errors in the following five-word coded message.

    111011   101101   011110   001000   011011

3. Use maximum likelihood decoding to correct the following six-word coded message generated by a triple repetition code. Then decode the message.

    101101101   110110101   110100101   101000111   110010011   011011011

4. Suppose 2-bit words *ab* are mapped onto 5-bit code words *ababc*, where *c* is the parity check digit. Correct the following seven-word coded message. Then decode the message.

    11100   01011   01010   10101   00011   10111   11111

5. Suppose a coding scheme is devised that maps *k*-bit words onto *n*-bit code words. The **efficiency** of the code is the ratio k/n. Compute the efficiency of the coding scheme described in each of the following examples.
    a. Example 1
    b. Example 2
    c. Example 3
    d. Example 4

6. Suppose the probability of erroneously transmitting a single digit is $P = 0.03$. Compute the probability of transmitting a 4-bit code word with **(a)** at most one error, and **(b)** exactly four errors.

7. Suppose the probability of erroneously transmitting a single digit is $P = 0.0001$. Compute the probability of transmitting an 8-bit code word with **(a)** no errors, **(b)** exactly one error, **(c)** at most one error, **(d)** exactly two errors, and **(e)** at most two errors.

8. Suppose the probability of incorrectly transmitting a single bit is $P = 0.001$. Compute the probability of correctly receiving a 100-word coded message made up of 4-bit words.

9. Compute the check digit for the 8-digit identification number 41126450 if the check digit is computed using congruence modulo 7.

10. Is the identification number 11257402 correct if the last digit is the check digit computed using congruence modulo 7?

11. Show that the check digit x_9 in bank identification numbers satisfies the congruence equation:

    $$(x_1, x_2, \ldots, x_8, x_9) \cdot (7, 3, 9, 7, 3, 9, 7, 3, 9) \equiv 0 \ (\text{mod } 10).$$

12. Suppose that the check digit is computed as described in Example 5. Prove that transposition errors of adjacent digits will not be detected unless one of the digits is the check digit.

13. Verify that transposition errors of adjacent digits x_i and x_{i+1} will be detected in a bank identification number except when $|x_i - x_{i+1}| = 5$.

14. Compute the check digit for the UPC symbols whose first 11 digits are given.

    a.
  0 7599-24511-4

    b.
  0 21200 00339

15. Verify that the check digit x_{12} in a UPC symbol satisfies the following congruence equation:

$$(x_1, x_2, \ldots, x_{12}) \cdot (3, 1, 3, 1, 3, 1, 3, 1, 3, 1, 3, 1) \equiv 0 \ (\text{mod } 10).$$

16. Show that transposition errors of the type

$$x_1 \ldots x_{i-1} x_i x_{i+1} \ldots x_{12} \rightarrow x_1 \ldots x_{i+1} x_i x_{i-1} \ldots x_{12}$$

    $(i = 2, 3, \ldots, 11)$ in a UPC symbol will not be detected by the check digit.

17. Passports contain identification codes of the form:

passport number	check digit	birth date	check digit	date of expiry	check digit	final check
012345678	4	USA 480517	7	F 020721	2 <<<<<<<<<<<<<<<	8

    Each of the first three check digits is computed on the preceding identification numbers by using a weighting vector of the form

    $$(7, 3, 1, 7, 3, 1, \ldots)$$

    in conjunction with congruence modulo 10. For example, in this passport identification code, the check digit 4 checks the *passport number*, the check digit 7 checks the *birth date*, and the check digit 2 checks the *date of expiry*. The final check digit is then computed by using the same type of weighting vector with all the digits (including check digits, excluding letters). Verify that this passport identification code is valid. Then check the validity of the following passport identification codes.
    a. 0987654326USA1512269F9901018 <<<<<<<<<<<<<<< 4
    b. 0444555331USA4609205M0409131 <<<<<<<<<<<<<<< 8
    c. 0123987457USA7803012M9711219 <<<<<<<<<<<<<<< 3
    d. 0246813570USA8301047F0312203 <<<<<<<<<<<<<<< 6

18. ISBN numbers are ten-digit numbers that identify books, where x_{10} is the check digit and $(x_1, x_2, \ldots, x_{10}) \cdot (10, 9, 8, 7, 6, 5, 4, 3, 2, 1) \equiv 0 \ (\text{mod } 11)$. Only digits 0 through 9 are used for the first nine digits, and if the check digit is required to be 10, then an X is used in place of the 10. If possible, detect any errors in the following ISBN numbers.
    a. ISBN 0-534-92888-9
    b. ISBN 0-543-91568-X
    c. ISBN 0-87150-334-X
    d. ISBN 0-87150-063-4

19. In the ISBN scheme, write the check digit x_{10} in the form

    $$(x_1, x_2, \ldots, x_9) \cdot \mathbf{y} \equiv x_{10}(\text{mod } 11),$$

    where $\mathbf{y}$ is obtained from the weighting vector $(10, 9, 8, 7, 6, 5, 4, 3, 2, 1)$.

20. Suppose $\mathbf{x} = x_1 x_2 \ldots x_k$ and $\mathbf{y} = y_1 y_2 \ldots y_k$ are k-bit words. The **Hamming**[†] **distance** $d(\mathbf{x}, \mathbf{y})$ between $\mathbf{x}$ and $\mathbf{y}$ is defined to be the number of bits in which $\mathbf{x}$ and $\mathbf{y}$ differ. More precisely,

---

[†]This distance function is named in honor of Richard Hamming (1915–), who pioneered the development of error-correcting codes.

$d(\mathbf{x}, \mathbf{y})$ is the number of indices in which $x_i \neq y_i$. Find the Hamming distance between the following pairs of words.

  a. 0011010 and 1011001

  b. 01000 and 10100

  c. 11110011 and 00110001

  d. 011000 and 110111

21. Let $\mathbf{x}, \mathbf{y}$, and $\mathbf{z}$ be k-bit words. Prove the following properties of the Hamming distance.

  a. $d(\mathbf{x}, \mathbf{y}) = d(\mathbf{y}, \mathbf{x})$

  b. $d(\mathbf{x}, \mathbf{y}) = 0$ if and only if $\mathbf{x} = \mathbf{y}$

  c. $d(\mathbf{x}, \mathbf{z}) \leq d(\mathbf{x}, \mathbf{y}) + d(\mathbf{y}, \mathbf{z})$

22. The **Hamming weight** $wt(\mathbf{x})$ of a k-bit word is defined to be $wt(\mathbf{x}) = d(\mathbf{x}, \mathbf{0})$, where $\mathbf{0}$ is the k-bit word in which every bit is 0. Find the Hamming weight of each of the following words.

  a. 0011100

  b. 11110

  c. 10100001

  d. 000110001

23. Let $\mathbf{x}$ and $\mathbf{y}$ be k-bit words. Prove that $d(\mathbf{x}, \mathbf{y}) = wt(\mathbf{x} - \mathbf{y})$, where $d(\mathbf{x}, \mathbf{y})$ is the Hamming distance between $\mathbf{x}$ and $\mathbf{y}$ and $wt(\mathbf{x})$ is the Hamming weight of $\mathbf{x}$.

24. The **minimum distance** of a code is defined to be the smallest distance between any pair of distinct code words in a code. Suppose a code consists of the following code words. This is the *repetition code on 2-bit words*.

$$0000 \qquad 0101 \qquad 1010 \qquad 1111$$

Find the minimum distance of this code.

25. Repeat Exercise 24 for the code consisting of the following code words. This code is a *repetition code on 3-bit words with a parity check digit*.

0000000	0100101	0010011	0110110
1001001	1101100	1011010	1111111

26. Repeat Exercise 24 for the code consisting of the following code words:

0000000	0001011	0010111	0011100
0100101	0101110	0110010	0111001
1000110	1001101	1010001	1011010
1100011	1101000	1110100	1111111.

This code is called the **Hamming (7,4) code**. Each code word $x_1x_2x_3x_4x_5x_6x_7$, with $x_i \in \{0, 1\}$, can be decoded by using the first four digits $x_1x_2x_3x_4$. The last three digits are parity check digits, where

$$x_5 \equiv x_1 + x_2 + x_3 \ (\text{mod } 2)$$
$$x_6 \equiv x_1 + x_3 + x_4 \ (\text{mod } 2)$$
$$x_7 \equiv x_2 + x_3 + x_4 \ (\text{mod } 2).$$

27. Write out the eight code words in the $(5, 3)$ code where each code word $x_1x_2x_3x_4x_5$ is generated in the following way:

$$x_i \in \{0, 1\},$$
$$x_4 \equiv x_1 + x_2 \ (\text{mod } 2),$$
$$x_5 \equiv x_1 + x_3 \ (\text{mod } 2).$$

■■■■■ 2.8   INTRODUCTION TO CRYPTOGRAPHY (OPTIONAL)

An additional application of congruence modulo n is found in **cryptography**, the designing of secret codes. **Cryptanalysis** is the process of breaking the secret codes and **cryptology** encompasses both cryptography and cryptanalysis. Cryptography differs from code theory in that code theory concentrates on the detection and correction of errors in messages whereas cryptography concentrates on concealing a message from an unauthorized person.

History is rich with examples of secret writings, dating back as far as 1900 B.C. when an Egyptian master scribe altered hieroglyphic writing, thus forming "secret messages" in the tomb of the nobleman Khnumhotep II. Later in 400 B.C., the Spartans used a device, called a *skytale*, to conceal messages. A ribbon was wound around a cylinder (the skytale), then a message was written on the ribbon. When the ribbon was removed, the message appeared scrabbled. However, the recipient of the ribbon had a similar skytale upon which he wound the ribbon and then easily read the message. An early cryptological system, called the *Caesar cipher*, was employed by Julius Caesar in the Gallic wars. In this system, Caesar simply replaced (*substituted*) each letter of the alphabet (the *plaintext*) by the letter three positions to the right (the *ciphertext*). The complete substitution for our alphabet[†] would thus appear as:

Plaintext:	a	b	c	d	e	f	g	$\cdots$	t	u	v	w	x	y	z
Ciphertext:	D	E	F	G	H	I	J	$\cdots$	W	X	Y	Z	A	B	C

and the plaintext message "attack at dawn" could easily be enciphered and deciphered using the substitution alphabet.

Plaintext:	a	t	t	a	c	k	a	t	d	a	w	n
Ciphertext:	D	W	W	D	F	N	D	W	G	D	Z	Q

The Caesar cipher is an example of an **additive cipher** or a **translation cipher**. All such translation ciphers can be illustrated in a cipher wheel made up of two concentric circles each containing the entire alphabet. One such cipher wheel is shown in Figure 2-4. The inner alphabet, representing the plaintext, is fixed while the outer alphabet, representing the ciphertext, spins. One pair of plaintext/ciphertext letters determines the entire scheme. This **key** is all that is needed to decipher any message. Caesar's plaintext/ciphertext key would appear as a/D.

A translation cipher, as used by Caesar, and other more sophisticated ciphers can be described mathematically. We first accept the following notational convention:

$$a \bmod n \text{ is the remainder when } a \text{ is divided by } n,$$

or symbolically,

$$r = a \bmod n \Leftrightarrow a = nq + r \text{ where } q \text{ and } r \text{ are integers with } 0 \le r < n.$$

Although this notation closely resembles the congruence notation defined in Section 2.5, the meaning is quite different and the distinction must be kept in mind. For a fixed y, the notation

$$x \equiv y \pmod{n}$$

[†]The letters j, u, and w were not in the Roman alphabet.

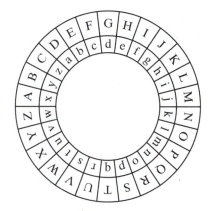

Figure 2-4
Cipher wheel

allows x to be *any integer* such that $x - y$ is a multiple of n, but the notation

$$x = y \bmod n$$

requires x to be *the unique integer* in the range $0 \le x < n$ such that $x - y$ is a multiple of n. All of the statements

$$27 \equiv 19 \ (\text{mod } 8), \quad 11 \equiv 19 \ (\text{mod } 8), \quad \text{and} \quad 3 \equiv 19 \ (\text{mod } 8)$$

are true, but the statement

$$x = 19 \bmod 8$$

is true if and only if $x = 3$.

Example I

a. $3 = 23 \bmod 5$ since $23 = 5(4) + 3$.
b. $1 = 37 \bmod 4$ since $37 = 4(9) + 1$.
c. $21 = 47 \bmod 26$ since $47 = 26(1) + 21$.
d. $19 = -7 \bmod 26$ since $-7 = 26(-1) + 19$. ■

Next we describe a translation cipher in terms of congruence modulo n.

Example 2 Translation Cipher   Associate the n letters of the "alphabet" with the integers $0, 1, 2, 3, \ldots, n - 1$. Let $A = \{0, 1, 2, 3, \ldots, n - 1\}$ and define the mapping $f : A \to A$ by

$$f(x) = x + k \bmod n$$

where k is the **key**, the number of positions from the plaintext to the ciphertext. If our alphabet consists of a through z, in natural order, followed by a blank, then we have 27 "letters" that we associate with the integers $0, 1, 2, \ldots, 26$ as follows:

Alphabet:	a	b	c	d	e	f	...	v	w	x	y	z	"blank"
A:	0	1	2	3	4	5	...	21	22	23	24	25	26

Now if our key is $k = 12$, then the plaintext message "send money" translates into the ciphertext message "DQZPLY ZQJ" as follows:

	translate to A $\longrightarrow$	18	4	13	3	26	12	14	13	4	24
send money	$f(x) = x + 12 \bmod 27$ $\longrightarrow$	3	16	25	15	11	24	26	25	16	9
	translate from A $\longrightarrow$	D	Q	Z	P	L	Y		Z	Q	J

The mapping f, given by

$$f(x) = x + k \bmod n$$

can be shown to be one-to-one and onto, so the inverse exists and is given by

$$f^{-1}(x) = x - k \bmod n.$$

The mapping f^{-1} can then be used to decipher the ciphertext.

	translate to A $\longrightarrow$	3	16	25	15	11	24	26	25	16	9
DQZPLY ZQJ	$f(x) = x - 12 \bmod 27$ $\longrightarrow$	18	4	13	3	26	12	14	13	4	24
	translate from A $\longrightarrow$	s	e	n	d		m	o	n	e	y ∎

A natural extension of the translation (or shift) cipher is found in a mapping of the form

$$f(x) = ax + b \bmod n$$

where a and b are fixed integers. This type of mapping is called an **affine mapping**. The ordered pair a, b of integers forms the key for this type of cipher. If $a = 1$, we simply have a translation cipher; whereas if $b = 0$, we have what's called a **multiplicative cipher**. It follows from Theorem 2.25 that an affine mapping $f : A \to A$ has an inverse $f^{-1} : A \to A$, if a and n are relatively prime. When $(a, n) = 1$, it can be shown that the inverse f^{-1} is given

$$f^{-1}(x) = a'x + b' \bmod n$$

where a' is defined by

$$1 = a'a \bmod n, \text{ with } 0 < a' < n$$

and

$$b' = -a'b \bmod n.$$

Example 3 Affine Mapping   We shall use an affine mapping with $a = 5$ and $b = 7$ as the key in our 27-letter alphabet. The mapping $f : A \to A$, where $A = \{0, 1, 2, \ldots, 26\}$, is given by

$$f(x) = 5x + 7 \bmod 27.$$

The plaintext message "hi mom" is translated into the ciphertext "PUCNXN" as follows:

	translate to A $\longrightarrow$	7	8	26	12	14	12
hi mom	$f(x) = 5x + 7 \bmod 27$ $\longrightarrow$	15	20	2	13	23	13
	translate from A $\longrightarrow$	P	U	C	N	X	N

Note that $(5, 27) = 1$, so the mapping f has an inverse given by:

$$
\begin{aligned}
f^{-1}(x) &= 11x - 11(7) \bmod 27 \quad \text{since } 1 = 11 \cdot 5 \bmod 27 \\
&= 11x + 16(7) \bmod 27 \quad \text{since } 16 = -11 \bmod 27 \\
&= 11x + 112 \bmod 27 \\
&= 11x + 4 \bmod 27,
\end{aligned}
$$

which can then be used to decipher the ciphertext.

PUCNXN	$\xrightarrow{\text{translate to } A}$	15	20	2	13	23	13
	$\xrightarrow{f(x) = 11x + 4 \bmod 27}$	7	8	26	12	14	12
	$\xrightarrow{\text{translate from } A}$	h	i		m	o	m	∎

Example 4 Affine Mapping with Unknown Key If a ciphertext message is relatively long, a frequency analysis of letters in a ciphertext can be used to "break the code" when the key to the affine mapping $f(x) = ax + b \bmod n$ is not known. Suppose we associate the letters a through z, in natural order, with the integers 0 through 25, respectively to form the 26-"letter" alphabet $A = \{0, 1, 2, \ldots, 25\}$. In the English language, with this alphabet the letter e occurs most often in a lengthy message, with the letters t, a, and o being the next most common. With this in mind, suppose in a ciphertext message the letter W occurred most frequently followed in frequency by P. It seems reasonable that the ciphertext letters W and P correspond to the plaintext letters e and t, respectively. Translating these into the set A, we have

		CIPHERTEXT			PLAINTEXT	
most frequent:	W	$\xrightarrow{\text{translate to } A}$	22	e	$\xrightarrow{\text{translate to } A}$	4
next most frequent:	P	$\xrightarrow{\text{translate to } A}$	15	t	$\xrightarrow{\text{translate to } A}$	19

Therefore, we can determine the key from the solution of the following system of equations for a and b:

$$
\begin{aligned}
22 &= a(4) + b \bmod 26 \\
15 &= a(19) + b \bmod 26.
\end{aligned}
$$

From Example 5 in Section 2.6, this solution is given by $a = 3, b = 10$. Thus, we find the affine mapping $f: A \rightarrow A$ to be given by

$$
f(x) = 3x + 10 \bmod 26,
$$

with inverse $f^{-1}: A \rightarrow A$ defined by

$$
f^{-1}(x) = 9x + 14 \bmod 26. \qquad \blacksquare
$$

In each of the preceding examples, once the mapping f was known, finding the inverse mapping f^{-1} was not difficult. In other words, once the key is known, a message can easily be deciphered. If security is an important issue (which is usually the case in sending **secret** messages), then it would certainly be advantageous to devise a system that would be difficult to break even if the key were known. Such systems are called **Public Key Cryptosystems**. We examine the RSA[†] cryptosystem next. The RSA system is based on the difficulty of factoring large numbers.

We begin by first choosing two distinct prime numbers, which we label as p and q and form the product

$$m = pq.$$

The value of m can be made known to the public. However, the factorization of m as pq shall be kept secret. The larger the value of m the more secure this system will be, since breaking the code relies on knowing the prime factors p and q of m. Next we choose e to be relatively prime to the product $(p - 1)(q - 1)$, that is, e is defined by

$$(e, (p - 1)(q - 1)) = 1.$$

Finally, solve for d in the equation

$$1 = ed \bmod (p - 1)(q - 1).$$

The public keys (the keys to be made known) are e and m whereas, the secret keys are p, q, and d.

Theorem 2.30 **RSA PUBLIC KEY CRYPTOSYSTEM**

Suppose $A = \{0, 1, 2, \ldots, m - 1\}$ is an alphabet, consisting of m "letters." With m, p, q, e, and d as described in the preceding paragraph, let the mapping $f : A \to A$ be defined by

$$f(x) = x^e \bmod m.$$

Then f has the inverse mapping $g : A \to A$ given by

$$g(x) = x^d \bmod m.$$

$p \Rightarrow q$ **Proof** Let $y = x^e \bmod m$. Then

$$y^d \equiv (x^e)^d \pmod{m}$$
$$\equiv x^{ed} \pmod{m}.$$

Since

$$1 = ed \bmod (p - 1)(q - 1),$$

then

$$ed = k(p - 1)(q - 1) + 1$$

for some integer k.

If $x \not\equiv 0 \pmod{p}$, then

$$x^{ed} \equiv x^{k(p-1)(q-1)+1} \pmod{p}$$
$$\equiv x^{k(p-1)(q-1)}x \pmod{p}$$
$$\equiv (x^{p-1})^{k(q-1)}x \pmod{p}$$
$$\equiv (1)^{k(q-1)}x \pmod{p}$$
$$\equiv x \pmod{p}$$

[†]RSA comes from the initials of the last names of Ronald Rivest, Adi Shamir, and Len Asleman who devised this system in 1977.

since $x^{p-1} \equiv 1 \pmod{p}$, from Exercise 41 and Theorem 2.24 in Section 2.5.
If $x \equiv 0 \pmod{p}$, it is clear that $x^{ed} \equiv 0^{ed} \pmod{p} \equiv 0 \pmod{p}$. Thus, $x^{ed} \equiv x \pmod{p}$ in all cases.

Similarly,

$$x^{ed} \equiv x \pmod{q}.$$

Hence

$$p \mid (x^{ed} - x) \quad \text{and} \quad q \mid (x^{ed} - x).$$

By Exercise 8 in Section 2.4, this implies that

$$pq \mid (x^{ed} - x),$$

and since $m = pq$, we have

$$x^{ed} \equiv x \pmod{m}.$$

Thus, $y^d \equiv x^{ed} \pmod{m} \equiv x \pmod{m}$, and it follows that $y^d \bmod m = x \bmod m$.
We have shown that $g(f(x)) = x$, and analogous steps can be used to verify that $f(g(x)) = x$. Therefore, g is the inverse mapping of f.

We illustrate the RSA cryptosystem with relatively small primes p and q. For the RSA system to be secure, it is recommended that the primes p and q be chosen so as to contain more than 100 digits.

Example 5 RSA Public Key Cryptosystem   We first choose two primes (which are to be kept **secret**):

$$p = 17, \text{ and } q = 43,$$

and compute m (which is to be made public):

$$m = pq = 17 \cdot 43 = 731.$$

Next we choose e (which is to be made public), where e must be relatively prime to $(p-1)(q-1) = 16 \cdot 42 = 672$. Suppose we take $e = 205$. The Euclidean Algorithm can be used to verify that $(205, 672) = 1$. Then d is determined by the equation

$$1 = 205d \bmod 672.$$

Using the Euclidean Algorithm, we find $d = 613$ (which is kept **secret**). The mapping $f: A \to A$, where $A = \{0, 1, 2, \ldots, 730\}$, defined by

$$f(x) = x^{205} \bmod 731$$

is used to encrypt a message and the inverse mapping $g: A \to A$, defined by

$$g(x) = x^{613} \bmod 731$$

can be used to recover the original message.
Using the 27-letter alphabet as in Examples 2 and 3, the plaintext message "no problem" is translated into the message as follows:

plaintext:	n	o		p	r	o	b	l	e	m
message:	13	14	26	15	17	14	01	11	04	12

and the message becomes

$$13142615171401110412$$

This message must be broken into blocks m_i, each of which is contained in A. If we choose three-digit blocks, each block $m_i < m = 731$.

$$m_i: \quad 131 \quad 426 \quad 151 \quad 714 \quad 011 \quad 104 \quad 12$$
$$f(m_i) = m_i^{205} \bmod 731 = c_i: \quad 082 \quad 715 \quad 376 \quad 459 \quad 551 \quad 593 \quad 320$$

and the enciphered message becomes

$$082 \qquad 715 \qquad 376 \qquad 459 \qquad 551 \qquad 593 \qquad 320$$

where we choose to report each c_i with three digits by appending any leading zeros as necessary.

To decipher the message, one must know the secret key $d = 613$, and apply the inverse mapping g to each enciphered message block $c_i = f(m_i)$:

$$c_i: \quad 082 \quad 715 \quad 376 \quad 459 \quad 551 \quad 593 \quad 320$$
$$g(c_i) = c_i^{613} \bmod 731: \quad 131 \quad 426 \quad 151 \quad 714 \quad 011 \quad 104 \quad 12$$

Finally, rebreaking the "message" back into two-digit blocks, it can be translated back into plaintext.

three-digit block message:	131	426	151	714	011	104	12
two-digit block message:	13 14	26	15	17	14	01 11	04 12
plaintext:	n o	p	r	o	b	l e	m ∎

The RSA Public Key Cipher is an example of an **exponentiation cipher**. As in coding theory, we have only barely touched on the basics of cryptography. It is our hope that this short introduction might spark further interest into a topic whose basis lies in modern algebra.

EXERCISES 2.8

1. In the 27-letter alphabet A described in Example 2, use the translation cipher with key $k = 8$ to encipher the following message.

   the check is in the mail

   What is the inverse mapping that will decipher the ciphertext?

2. Suppose the alphabet consists of a through z, in natural order, followed by a blank, a comma, a period, an apostrophe, and a question mark in that order. Associate these "letters" with the numbers $0, 1, 2, \ldots, 30$, respectively, thus forming a 31-letter alphabet B. Use the translation cipher with key $k = 21$ to encipher the following message.

   what's up, doc?

   What is the inverse mapping that will decipher the ciphertext?

3. In the 31-letter alphabet B as in Exercise 2, use the translation cipher with key $k = 11$ to decipher the following message.

   ?TRP.HGOZGEZAG.PLOGXPK

   What is the inverse mapping that deciphers this ciphertext?

4. In the 27-letter alphabet A described in Example 2, use the translation cipher with key $k = 15$ to decipher the following message.

FXGTOPBSOGWXBT

What is the inverse mapping that deciphers this ciphertext?

5. In the 27-letter alphabet A described in Example 2, use the affine cipher with key $a = 7$ and $b = 5$ to encipher the following message.

all systems go

What is the inverse mapping that will decipher the ciphertext?

6. In the 31-letter alphabet B described in Exercise 2, use the affine cipher with key $a = 15$ and $b = 22$ to encipher the following message.

Houston, we have a problem.

What is the inverse mapping that will decipher the ciphertext?

7. Suppose the alphabet consists of a through z, in natural order, followed by a blank and then a period. Associate these "letters" with the numbers $0, 1, 2, \ldots, 27$, respectively, thus forming a 28-letter alphabet, C. Use the affine cipher with key $a = 3$ and $b = 22$ to decipher the message

EEETZRIIYUAI.GTAIC

and state the inverse mapping that deciphers this ciphertext.

8. Use the alphabet C from the preceding problem and the affine cipher with key $a = 11$ and $b = 7$ to decipher the message

ZZZYDJBJYXMD

and state the inverse mapping that deciphers this ciphertext.

9. Suppose in a long ciphertext message the letter x occurred most frequently followed in frequency by c. Using the fact that in the 26-letter alphabet A, described in Example 4, e occurs most frequently followed in frequency by t, read the portion of the message

RNCYXRNCHFT

enciphered using an affine mapping on A. Write out the affine mapping f and its inverse.

10. Suppose in a long ciphertext message the letter D occurred most frequently followed in frequency by N. Using the fact that in the 27-letter alphabet A, described in Example 2, "blank" occurs most frequently followed in frequency by e, read the portion of the message

GENDOCFAADOQNIDPGMDCFE

enciphered using an affine mapping on A. Write out the affine mapping f and its inverse.

11. Suppose the alphabet consists of a through z, in natural order, followed by a blank and then the digits 0 through 9, in natural order. Associate these "letters" with the numbers $0, 1, 2, \ldots, 36$, respectively, thus forming a 37-letter alphabet, D. Use the affine cipher to decipher the message

X01916R916546M9CN1L6B1LL6X0RZ6UII

if you know that the plaintext message begins with "th". Write out the affine mapping f and its inverse.

12. Suppose the alphabet consists of a through z, in natural order, followed by a blank, a comma, and a period, in that order. Associate these "letters" with the numbers $0, 1, 2, \ldots, 28$, respectively, thus forming a 29-letter alphabet, E. Use the affine cipher to decipher the message

BZZK,AUZNZG,RSKZ,AUWAO

if you know that the plaintext message begins with "b" and ends with ".". Write out the affine mapping f and its inverse.

13. Let $f: A \to A$ be defined by $f(x) = ax + b \bmod n$. Show that $f^{-1}: A \to A$ exists if $(a, n) = 1$, and is given by $f^{-1}(x) = a'x + b' \bmod n$ where a' is defined by

$$1 = a'a \bmod n, \text{ with } 0 < a' < n$$

and

$$b' = -a'b \bmod n.$$

14. Suppose we encipher a plaintext message M using the mapping $f_1: A \to A$ resulting in the ciphertext C. Next we treat this ciphertext as plaintext and encipher it using the mapping $f_2: A \to A$ resulting in the ciphertext D. The composition mapping $f: A \to A$, where $f = f_2 \circ f_1$ could be used to encipher the plaintext message M resulting in the ciphertext D.
    a. Prove that if f_1 and f_2 are translation ciphers, then $f = f_2 \circ f_1$ is a translation cipher.
    b. Prove that if f_1 and f_2 are affine ciphers, then $f = f_2 \circ f_1$ is an affine cipher.

15. **a.** Excluding the identity cipher, how many different translation ciphers are there using an alphabet of n "letters"?
    b. Excluding the identity cipher, how many different affine ciphers are there using an alphabet of n "letters", where n is a prime?

16. Rework Example 5 by breaking the message into two-digit blocks instead of three-digit blocks. What is the enciphered message using the two-digit blocks?

17. Suppose in a RSA Public Key Cryptosystem, the public key is $e = 13, m = 77$. Encrypt the message "go for it" using two-digit blocks and the 27-letter alphabet A from Example 2. What is the **secret** key d?

18. Suppose in a RSA Public Key Cryptosystem, the public key is $e = 35, m = 64$. Encrypt the message "pay me later" using two-digit blocks and the 27-letter alphabet A from Example 2. What is the **secret** key d?

19. Suppose in a RSA Public Key Cryptosystem, $p = 11, q = 13$, and $e = 7$. Encrypt the message "algebra" using the 26-letter alphabet from Example 4.
    a. Use two-digit blocks.
    b. Use three-digit blocks.
    c. What is the **secret** key d?

20. Suppose in a RSA Public Key Cryptosystem, $p = 17, q = 19$, and $e = 19$. Encrypt the message "pascal" using the 26-letter alphabet from Example 4.
    a. Use two-digit blocks.
    b. Use three-digit blocks.
    c. What is the **secret** key d?

21. Suppose in a RSA Public Key Cryptosystem, the public key is $e = 23, m = 55$. The ciphertext message

26	25	00	39	09	18	52	17	49	52	02

    was intercepted; what was the message that was sent? Use the 27-letter alphabet from Example 2.

22. Suppose in a RSA Public Key Cryptosystem, the public key is $e = 5, m = 51$. The ciphertext message

04	05	32	44	26	39	04	00	13	08	00	44	24	29	17	26	49	28	03

    was intercepted; what was the message that was sent? Use the 27-letter alphabet from Example 2.

23. The **Euler phi-function** is defined for positive integers n as follows: $\phi(n)$ is the number of positive integers m such that $1 \leq m \leq n$ and $(m, n) = 1$. Evaluate each of the following and list each of the integers m relatively prime to the given n.

   a. $\phi(5)$ **b.** $\phi(19)$

   c. $\phi(15)$ **d.** $\phi(27)$

   e. $\phi(12)$ **f.** $\phi(36)$

24. If p is a prime, then $\phi(p) = p - 1$, since all positive integers less than p are relatively prime to p. Prove that if p and q are distinct primes, then $\phi(pq) = (p - 1)(q - 1)$.

25. If p is a prime and j is a positive integer, prove $\phi(p^j) = p^{j-1}(p - 1)$.

KEY WORDS AND PHRASES

A Pioneer in Mathematics ||||||||||||||||||||||||||||||||||||||||||||||||||||||||||||||||||||||||||||||||||
Blaise Pascal (1623–1662)

Blaise Pascal is most commonly associated with *Pascal's triangle,* a triangular-shaped pattern in which the binomial coefficients are generated. Although Pascal was not the first to discover this pattern, it was through his study of the pattern that he became the first writer to precisely describe the process of mathematical induction.

As a child, Pascal was frequently ill. His father, a mathematician himself, used to hide all his own mathematics books because he felt that his son's study of mathematics would be too strenuous. But when he was 12, Pascal was found in his playroom folding pieces of paper, doing an experiment by which he discovered that the sum of the angles in any triangle is equal to 180°. Pascal's father was so impressed that he gave his son Euclid's *Elements* to study, and Pascal soon discovered, on his own, many of the propositions of geometry.

At the age of 14, Pascal was allowed to actively participate in the gatherings of a group of French mathematicians. At 16, he had established significant results in projective geometry. Also at this time, he began developing a calculator to assist his father's work of auditing chaotic government tax records. Pascal perfected the machine over a period of ten years by building 50 various models, but ultimately it was too expensive to be practical.

Pascal made many contributions in the fields of mechanics and physics as well. The one-wheeled wheelbarrow is another of his inventions. Through his correspondence with the French mathematician Pierre de Fermat, he and Fermat laid the foundations of probability theory.

Pascal died in 1662 at the age of 39. His contributions to 17th-century mathematics were large compared to his short life. Scholars wonder how much more mathematics would have issued from his gifted mind had he lived longer.

GROUPS

INTRODUCTION

Some of the standard topics in elementary group theory are treated in this chapter: subgroups, cyclic groups, isomorphisms, and homomorphisms.

In the development here, the topic of isomorphism appears before homomorphism. Some instructors prefer a different order and teach Section 3.5 (Homomorphisms) before Section 3.4 (Isomorphisms). Logic can be used to support either approach. Isomorphism is a special case of homomorphism, while homomorphism is a generalization of isomorphism. Isomorphisms were placed first in this book with the thought that "same structure" is the simpler idea.

Both the additive and multiplicative structures in $\mathbf{Z}_n$ serve as a basis for some of the examples in this chapter.

3.1   DEFINITION OF A GROUP

The fundamental notions of set, mapping, binary operation, and binary relation were presented in Chapter 1. These notions are essential for the study of an algebraic system. An **algebraic structure**, or **algebraic system**, is a nonempty set in which at least one equivalence relation (equality) and one or more binary operations are defined. The simplest structures occur when there is only one binary operation, as is the case with the algebraic system known as a *group*.

An introduction to the theory of groups is presented in this chapter, and it is appropriate to point out that this is only an introduction. Entire books have been devoted to the theory of groups; the group concept is extremely useful in both pure and applied mathematics.

A group may be defined as follows.

Definition 3.1   GROUP

Suppose the binary operation $*$ is defined for elements of the set G. Then G is a **group** with respect to $*$ provided these conditions hold:

1. G is **closed** under $*$. That is, $x \in G$ and $y \in G$ imply that $x * y$ is in G.
2. $*$ is **associative**. For all x, y, z in G, $x * (y * z) = (x * y) * z$.
3. G has an **identity element** e. There is an e in G such that $x * e = e * x = x$ for all $x \in G$.
4. G contains **inverses**. For each $a \in G$, there exists $b \in G$ such that $a * b = b * a = e$.

The phrase "with respect to $*$" should be noted. For example, the set **Z** of all integers is a group with respect to addition, but not with respect to multiplication (it has no inverses for elements other than ± 1). Similarly, the set $G = \{1, -1\}$ is a group with respect to multiplication but not with respect to addition. In most instances, however, only one binary operation is under consideration, and we say simply that "G is a group." If the binary operation is unspecified, we adopt the multiplicative notation and use the juxtaposition xy to indicate the result of combining x and y. It should be kept in mind, though, that the binary operation is not necessarily multiplication.

Definition 3.2   ABELIAN GROUP

Let G be a group with respect to $*$. Then G is called a **commutative group**, or an **abelian**[†] **group**, if $*$ is commutative. That is, $x * y = y * x$ for all x, y in G.

Example 1   We can obtain some simple examples of groups by considering appropriate subsets of the familiar number systems.

a. The set of all *complex numbers* is an abelian group with respect to addition.
b. The set of all *nonzero rational numbers* is an abelian group with respect to multiplication.

[†]The term **abelian** is used in honor of Niels Henrik Abel (1802–1829). A biographical sketch of Abel appears on the last page of this chapter.

c. The set of all *positive real numbers* is an abelian group with respect to multiplication, but it is not a group with respect to addition (it has no additive identity and no additive inverses).   ∎

The following examples give some indication of the great variety there is in groups.

Example 2   Recall from Chapter 1 that a permutation on a set A is a one-to-one mapping from A onto A, and that $S(A)$ denotes the set of all permutations on A. We have seen that $S(A)$ is closed with respect to the binary operation $\circ$ of mapping composition, and that the operation $\circ$ is associative. The identity mapping I_A is an identity element:

$$f \circ I_A = f = I_A \circ f$$

for all $f \in S(A)$, and each $f \in S(A)$ has an inverse in $S(A)$. Thus, we may conclude from results in Chapter 1 that $S(A)$ is a group with respect to composition of mappings.   ∎

Example 3   We shall take $A = \{1, 2, 3\}$ and obtain an explicit example of $S(A)$. In order to define an element f of $S(A)$, we need to specify $f(1)$, $f(2)$, and $f(3)$. There are three possible choices for $f(1)$. Since f is to be bijective, there are two choices for $f(2)$ after $f(1)$ has been designated, and then only one choice for $f(3)$. Hence, there are $3! = 3 \cdot 2 \cdot 1$ different mappings f in $S(A)$. These are given by

$$e = I_A : \begin{cases} e(1) = 1 \\ e(2) = 2 \\ e(3) = 3 \end{cases} \qquad \sigma : \begin{cases} \sigma(1) = 2 \\ \sigma(2) = 1 \\ \sigma(3) = 3 \end{cases}$$

$$\rho : \begin{cases} \rho(1) = 2 \\ \rho(2) = 3 \\ \rho(3) = 1 \end{cases} \qquad \gamma : \begin{cases} \gamma(1) = 3 \\ \gamma(2) = 2 \\ \gamma(3) = 1 \end{cases}$$

$$\tau : \begin{cases} \tau(1) = 3 \\ \tau(2) = 1 \\ \tau(3) = 2 \end{cases} \qquad \delta : \begin{cases} \delta(1) = 1 \\ \delta(2) = 3 \\ \delta(3) = 2. \end{cases}$$

Thus, $S(A) = \{e, \rho, \tau, \sigma, \gamma, \delta\}$. Following the same convention as in Exercise 3 of Section 1.4, we shall construct a "multiplication" table for $S(A)$. As shown in Figure 3-1, the result of $f \circ g$ is entered in the row with f at the left and in the column with g at the top.

Figure 3-1

In constructing the table for $\mathcal{S}(A)$, the elements of $\mathcal{S}(A)$ are listed in a column at the left and in a row at the top, as shown in Figure 3-2. When the product $\rho^2 = \rho \circ \rho$ is computed, we have

$$\rho^2(1) = \rho(\rho(1)) = \rho(2) = 3$$
$$\rho^2(2) = \rho(\rho(2)) = \rho(3) = 1$$
$$\rho^2(3) = \rho(\rho(3)) = \rho(1) = 2,$$

so $\rho^2 = \tau$. Similarly, $\rho \circ \sigma = \gamma$, $\sigma \circ \rho = \delta$, and so on.

Figure 3-2

$\circ$	e	ρ	ρ^2	σ	γ	δ
e	e	ρ	ρ^2	σ	γ	δ
ρ	ρ	ρ^2	e	γ	δ	σ
ρ^2	ρ^2	e	ρ	δ	σ	γ
σ	σ	δ	γ	e	ρ^2	ρ
γ	γ	σ	δ	ρ	e	ρ^2
δ	δ	γ	σ	ρ^2	ρ	e

■

A table such as the one in Figure 3-2 is referred to in various texts as a **multiplication table**, a **group table**, or a **Cayley table**.[†] With such a table, it is easy to locate the identity and inverses of elements. An element e is a left identity if and only if the row headed by e at the left end reads exactly the same as the column headings in the table. Similarly, e is a right identity if and only if the column headed by e at the top reads exactly the same as the row headings in the table. If it exists, the inverse of a certain element a can be found by searching for the identity e in the row headed by a and again in the column headed by a.

If the elements in the row headings are listed in the same order from top to bottom as the elements in the column headings are listed from left to right, it is also possible to use the table to check for commutativity. The operation is commutative if and only if equal elements appear in all positions that are symmetrically placed relative to the diagonal from upper left to lower right. In Example 3, the group is not abelian since the table in Figure 3-2 is not symmetric. For example $\gamma \circ \rho^2 = \delta$ is in row 5, column 3, and $\rho^2 \circ \gamma = \sigma$ is in row 3, column 5.

Example 4   Let G be the set of complex numbers given by $G = \{1, -1, i, -i\}$, where $i = \sqrt{-1}$, and consider the operation of multiplication of complex numbers in G. The table in Figure 3-3 shows that G is closed with respect to multiplication.

Multiplication in G is associative and commutative, since multiplication has these properties in the set of all complex numbers. We can observe from Figure 3-3 that 1 is the identity element, and that all elements have inverses. Each of 1 and -1 is its own inverse, and i and $-i$ are inverses of each other. Thus, G is a group with respect to multiplication.

[†]The term **Cayley table** is in honor of Arthur Cayley (1821–1895). A biographical sketch of Cayley appears on the last page of Chapter 1.

Figure 3-3

×	1	−1	i	$-i$
1	1	−1	i	$-i$
−1	−1	1	$-i$	i
i	i	$-i$	−1	1
$-i$	$-i$	i	1	−1

■

Example 5 It is an immediate corollary of Theorem 2.26 that the set

$$\mathbf{Z}_n = \{[0], [1], [2], \ldots, [n-1]\}$$

of congruence classes modulo n forms a group with respect to addition. ■

Example 6 Let $G = \{e, a, b, c\}$ with multiplication as defined by the table in Figure 3-4.

Figure 3-4

·	e	a	b	c
e	e	a	b	c
a	a	b	c	e
b	b	c	e	a
c	c	e	a	b

From the table, we observe that

1. G is closed under this multiplication.
2. e is the identity element.
3. each of e and b is its own inverse, and c and a are inverses of each other.

This multiplication is associative, but we shall not verify it here because it is a laborious task. It follows that G is a group. ■

Example 7 The table in Figure 3-5 defines a binary operation $*$ on the set $S = \{A, B, C, D\}$.

Figure 3-5

$*$	A	B	C	D
A	B	C	A	B
B	C	D	B	A
C	A	B	C	D
D	A	B	D	D

From the table, we see that

1. G is closed under $*$.
2. C is an identity element.
3. D does not have an inverse since $DX = C$ has no solution.

Thus, S is not a group with respect to $*$. ■

Definition 3.3   **FINITE GROUP, INFINITE GROUP, ORDER OF A GROUP**

> If a group G has a finite number of elements, G is called a **finite group**, or a **group of finite order**. The number of elements in G is called the **order** of G, and is denoted by either $o(G)$ or $|G|$. If G does not have a finite number of elements, G is called an **infinite group**, or a **group of infinite order**.

Example 8   In Example 3, the group

$$G = \{e, \rho, \rho^2, \sigma, \gamma, \delta\}$$

has order $o(G) = 6$. In Example 5, $o(\mathbf{Z}_n) = n$. The set $\mathbf{Z}$ of all integers is a group under addition, and this is an example of an infinite group. If A is an infinite set, then $\mathcal{S}(A)$ furnishes an example of an infinite group. ∎

Several consequences of the definition of a group are recorded in Theorem 3.4.

STRATEGY ▶   Parts a and b of the next theorem are statements about uniqueness, and they can be proved by the standard type of uniqueness proof: Assume that two such quantities exist, and then prove the two to be equal.

Theorem 3.4   **PROPERTIES OF GROUP ELEMENTS**

Let G be a group with respect to a binary operation that is written as multiplication.

a. The identity element e in G is unique.
b. For each $x \in G$, the inverse x^{-1} in G is unique.
c. For each $x \in G$, $(x^{-1})^{-1} = x$.
d. **Reverse order law.** For any x and y in G, $(xy)^{-1} = y^{-1}x^{-1}$.
e. **Cancellation laws.** If a, x, and y are in G, then either of the equations $ax = ay$ or $xa = ya$ implies that $x = y$.

Uniqueness   **Proof**   We prove parts b and d and leave the others as exercises. To prove part b, let $x \in G$, and suppose that each of y and z are inverses of x. That is,

$$xy = e = yx \quad \text{and} \quad xz = e = zx.$$

Then

$$
\begin{aligned}
y &= ey & &\text{since } e \text{ is an identity} \\
  &= (zx)y & &\text{since } zx = e \\
  &= z(xy) & &\text{by associativity} \\
  &= z(e) & &\text{since } xy = e \\
  &= z & &\text{since } e \text{ is an identity.}
\end{aligned}
$$

Thus, $y = z$, and this justifies the notation x^{-1} as the unique inverse of x in G.

$(p \wedge q) \Rightarrow r$ We shall use part b in the proof of part d. Specifically, we shall use the fact that the inverse $(xy)^{-1}$ is unique. This means that in order to show that $y^{-1}x^{-1} = (xy)^{-1}$, we only need to verify that $(xy)(y^{-1}x^{-1}) = e = (y^{-1}x^{-1})(xy)$. These calculations are straightforward:

$$(y^{-1}x^{-1})(xy) = y^{-1}(x^{-1}x)y = y^{-1}ey = y^{-1}y = e$$

and

$$(xy)(y^{-1}x^{-1}) = x(yy^{-1})x^{-1} = xex^{-1} = xx^{-1} = e.$$

The order of the factors y^{-1} and x^{-1} in the reverse order law $(xy)^{-1} = y^{-1}x^{-1}$ is crucial in a nonabelian group. An example where $(xy)^{-1} \neq x^{-1}y^{-1}$ is requested in Exercise 40 at the end of this section.

Part e of Theorem 3.4 implies that in the table for a finite group G, no element of G appears twice in the same row, and no element of G appears twice in the same column. These results can be extended to the statement in the following strategy box. The proof of this fact is requested in Exercise 44.

STRATEGY ▶ : In the multiplication table for a group G, each element of G appears exactly once in each row and also appears exactly once in each column.

Although our definition of a group is a standard one, alternate forms can be made. One of these is given in the next theorem.

Theorem 3.5 **EQUIVALENT CONDITIONS FOR A GROUP**

Let G be a nonempty set that is closed under an associative binary operation called multiplication. Then G is a group if and only if the equations $ax = b$ and $ya = b$ have solutions x and y in G for all choices of a and b in G.

$p \Rightarrow (q \wedge r)$ **Proof** Assume first that G is a group, and let a and b represent arbitrary elements of G. Now a^{-1} is in G, and so are $x = a^{-1}b$ and $y = ba^{-1}$. With these choices for x and y, we have

$$ax = a(a^{-1}b) = (aa^{-1})b = eb = b$$

and

$$ya = (ba^{-1})a = b(a^{-1}a) = be = b.$$

Thus, G contains solutions x and y to $ax = b$ and $ya = b$.

$(q \wedge r) \Rightarrow p$ Suppose now that the equations always have solutions in G. We first show that G has an identity element. Let a represent an arbitrary but fixed element in G. The equation $ax = a$ has a solution $x = u$ in G. We shall show that u is a right identity for every element in G. To do this, let b be arbitrary in G. With z a solution to $ya = b$, we have $za = b$ and

$$bu = (za)u = z(au) = za = b.$$

Thus, u is a right identity for every element in G. In a similar fashion, there exists an element v in G such that $vb = b$ for all b in G. Then $vu = v$, since u is a right identity, and $vu = u$, since v is a left identity. That is, the element $e = u = v$ is an identity element for G.

Now for any a in G, let x be a solution to $ax = e$, and let y be a solution to $ya = e$. Combining these equations, we have

$$\begin{aligned} x &= ex \\ &= yax \\ &= ye \\ &= y, \end{aligned}$$

and $x = y$ is an inverse for a. This proves that G is a group.

In a group G, the associative property can be extended to products involving more than three factors. For example, if a_1, a_2, a_3, and a_4 are elements of G, then applications of condition 2 in Definition 3.1 yield

$$[a_1(a_2 a_3)]a_4 = [(a_1 a_2)a_3]a_4$$

and

$$(a_1 a_2)(a_3 a_4) = [(a_1 a_2)a_3]a_4.$$

These equalities suggest (but do not completely prove) that, regardless of how symbols of grouping are introduced in a product $a_1 a_2 a_3 a_4$, the resulting expression can be reduced to

$$[(a_1 a_2)a_3]a_4.$$

With these observations in mind, we make the following definition.

Definition 3.6   PRODUCT NOTATION

> Let n be a positive integer, $n \geq 2$. For elements a_1, a_2, $\ldots$, a_n in a group G, the expression $a_1 a_2 \cdots a_n$ is defined recursively by
>
> $$a_1 a_2 \cdots a_k a_{k+1} = (a_1 a_2 \cdots a_k)a_{k+1} \quad \text{for} \quad k \geq 1.$$

We can now prove the following generalization of the associative property.

Theorem 3.7   GENERALIZED ASSOCIATIVE LAW

Let $n \geq 2$ be a positive integer, and let a_1, a_2, $\ldots$, a_n denote elements of a group G. For any positive integer m such that $1 \leq m < n$,

$$(a_1 a_2 \cdots a_m)(a_{m+1} \cdots a_n) = a_1 a_2 \cdots a_n.$$

Induction   **Proof**   For $n \geq 2$, let P_n denote the statement of the theorem. With $n = 2$, the only possible value for m is $m = 1$, and P_2 asserts the trivial equality

$$(a_1)(a_2) = a_1 a_2.$$

Assume now that P_k is true: for any positive integer m such that $1 \leq m < k$,

$$(a_1 a_2 \cdots a_m)(a_{m+1} \cdots a_k) = a_1 a_2 \cdots a_k.$$

Consider the statement P_{k+1}, and let m be a positive integer such that $1 \le m < k + 1$. We treat separately the cases where $m = k$ and where $1 \le m < k$. If $m = k$, the desired equality is true at once from Definition 3.6, as follows:

$$(a_1 a_2 \cdots a_m)(a_{m+1} \cdots a_{k+1}) = (a_1 a_2 \cdots a_k)a_{k+1}.$$

If $1 \le m < k$, then

$$a_{m+1} \cdots a_k a_{k+1} = (a_{m+1} \cdots a_k)a_{k+1}$$

by Definition 3.6, and consequently,

$$
\begin{aligned}
(a_1 a_2 \cdots a_m)&(a_{m+1} \cdots a_k a_{k+1}) \\
&= (a_1 a_2 \cdots a_m)[(a_{m+1} \cdots a_k)a_{k+1}] \\
&= [(a_1 a_2 \cdots a_m)(a_{m+1} \cdots a_k)]a_{k+1} && \text{by the associative property} \\
&= [a_1 a_2 \cdots a_k]a_{k+1} && \text{by } P_k \\
&= a_1 a_2 \cdots a_{k+1} && \text{by Definition 3.6.}
\end{aligned}
$$

Thus, P_{k+1} is true whenever P_k is true, and the proof of the theorem is complete.

The material in Section 1.5 on matrices leads to some interesting examples of groups, both finite and infinite. This is pursued now in Examples 9 and 10.

Example 9 Theorem 1.29 translates directly into the statement that $M_{m \times n}(\mathbf{R})$ is an abelian group with respect to addition. This is an example of another infinite group.

When the proof of each part of Theorem 1.29 is examined, it becomes clear that each group property in $M_{m \times n}(\mathbf{R})$ derives in a natural way from the corresponding property in $\mathbf{R}$. If the set $\mathbf{R}$ is replaced by the set $\mathbf{Z}$ of all integers, the steps in the proof of each part of Theorem 1.29 can be paralleled to prove the same group property for $M_{m \times n}(\mathbf{Z})$. Thus, $M_{m \times n}(\mathbf{Z})$ is also a group under addition. The same reasoning is valid if $\mathbf{R}$ is replaced by the set $\mathbf{Q}$ of all rational numbers, by the set $\mathbf{C}$ of all complex numbers, or by the set $\mathbf{Z}_k$ of all congruence classes modulo k. That is, each of $M_{m \times n}(\mathbf{Q})$, $M_{m \times n}(\mathbf{C})$, and $M_{m \times n}(\mathbf{Z}_k)$ is a group with respect to addition.

We thus have a family of groups, with $M_{m \times n}(\mathbf{Z}_k)$ finite and all the others infinite. Some aspects of computation in $M_{m \times n}(\mathbf{Z}_k)$ may appear strange at first. For instance,

$$B = \begin{bmatrix} [1] & [3] & [0] \\ [2] & [4] & [2] \end{bmatrix}$$

is the additive inverse of

$$A = \begin{bmatrix} [4] & [2] & [0] \\ [3] & [1] & [3] \end{bmatrix}$$

in $M_{2 \times 3}(\mathbf{Z}_5)$, since

$$A + B = \begin{bmatrix} [0] & [0] & [0] \\ [0] & [0] & [0] \end{bmatrix} = B + A. \qquad \blacksquare$$

In Example 4 of Section 1.5, it was shown that the matrix

$$A = \begin{bmatrix} 1 & 3 \\ 2 & 6 \end{bmatrix}$$

in $M_2(\mathbf{R})$ does not have an inverse, so the nonzero elements of $M_2(\mathbf{R})$ do not form a group with respect to multiplication. This result generalizes to arbitrary $M_n(\mathbf{R})$ with $n > 1$; that is, the nonzero elements of $M_n(\mathbf{R})$ do not form a group with respect to multiplication. However, the next example shows that the invertible elements[†] of $M_n(\mathbf{R})$ form a group under multiplication.

Example 10   We shall show that the invertible elements of $M_n(\mathbf{R})$ form a group G with respect to matrix multiplication.

We have seen in Section 1.5 that matrix multiplication is a binary operation on $M_n(\mathbf{R})$, that this operation is associative (Theorem 1.31), and that $I_n = [\delta_{ij}]_{n \times n}$ is an identity element (Theorem 1.33). These properties remain valid when attention is restricted to the set G of invertible elements of $M_n(\mathbf{R})$, so we need only show that G is closed under multiplication. To this end, suppose that A and B are elements of $M_n(\mathbf{R})$ such that A^{-1} and B^{-1} exist. Using the associative property of matrix multiplication, we can write

$$
\begin{aligned}
(AB)(B^{-1}A^{-1}) &= A(BB^{-1})A^{-1} \\
&= AI_nA^{-1} \\
&= AA^{-1} \\
&= I_n.
\end{aligned}
$$

Although matrix multiplication is not commutative, a similar simplification shows that

$$ (B^{-1}A^{-1})(AB) = I_n, $$

and it follows that $(AB)^{-1}$ exists and $(AB)^{-1} = B^{-1}A^{-1}$. Thus, G is a group.

As in Example 9, the discussion in the preceding paragraph can be extended by replacing $\mathbf{R}$ with one of the systems $\mathbf{Z}$, $\mathbf{Q}$, $\mathbf{C}$, or $\mathbf{Z}_k$. Once again, the computations in $M_n(\mathbf{Z}_k)$ may seem strange. As an illustration, it can be verified by multiplication that

$$
\begin{bmatrix} [3] & [1] \\ [5] & [2] \end{bmatrix} \quad \text{is the inverse of} \quad \begin{bmatrix} [2] & [6] \\ [2] & [3] \end{bmatrix}
$$

in $M_2(\mathbf{Z}_7)$.   ∎

EXERCISES 3.1

In Exercises 1–10, decide whether each of the given sets is a group with respect to the indicated operation. If it is not a group, state a condition in Definition 3.1 that fails to hold.

1. The set of all rational numbers with operation addition.

2. The set of all irrational numbers with operation addition.

3. The set of all positive irrational numbers with operation multiplication.

4. The set of all positive rational numbers with operation multiplication.

5. The set of all real numbers x such that $0 < x \leq 1$, with operation multiplication.

[†]Recall that a square matrix A is called *invertible* if its multiplicative inverse, A^{-1}, exists.

6. For a fixed positive integer n, the set of all complex numbers x such that $x^n = 1$ (that is, the set of all nth roots of 1), with operation multiplication.

7. The set of all complex numbers x that have absolute value 1, with operation multiplication.

8. The set in Exercise 7 with operation addition.

9. The set **E** all even integers with operation addition.

10. The set **E** of all even integers with operation multiplication.

In Exercises 11 and 12, the given table defines an operation of multiplication on the set $S = \{e, a, b, c\}$. In each case, find a condition in Definition 3.1 that fails to hold, and thereby show that S is not a group.

11. See Figure 3-6.

12. See Figure 3-7.

$\times$	e	a	b	c
e	e	a	b	c
a	a	b	a	b
b	b	c	b	c
c	c	e	c	e

Figure 3-6

$\times$	e	a	b	c
e	e	a	b	c
a	e	a	b	c
b	e	a	b	c
c	e	a	b	c

Figure 3-7

In Exercises 13 and 14, part of the multiplication table for the group $G = \{a, b, c, d\}$ is given. In each case, complete the table.

13. See Figure 3-8.

14. See Figure 3-9.

$\times$	a	b	c	d
a		d		
b				
c			c	
d				c

Figure 3-8

$\times$	a	b	c	d
a				
b		a		
c	a			
d				

Figure 3-9

In Exercises 15–20, let the binary operation $*$ be defined on **Z** by the given rule. Determine in each case whether **Z** is a group with respect to $*$, and whether it is an abelian group. State which, if any, conditions fail to hold.

15. $x * y = x + y + 1$

16. $x * y = x + y - 1$

17. $x * y = x + xy$

18. $x * y = xy + y$

19. $x * y = x + xy + y$

20. $x * y = x - y$

In Exercises 21–26, decide whether each of the given sets is a group with respect to the indicated operation. If it is not a group, state all of the conditions in Definition 3.1 that fail to hold.

21. The set $\{[1], [3]\} \subseteq \mathbf{Z}_8$ with operation multiplication.

22. The set $\{[1], [2], [3], [4]\} \subseteq \mathbf{Z}_5$ with operation multiplication.

23. The set $\{[0], [2], [4]\} \subseteq \mathbf{Z}_8$ with operation multiplication.

24. The set $\{[0], [2], [4], [6], [8]\} \subseteq \mathbf{Z}_{10}$ with operation multiplication.

25. The set $\{[0], [2], [4], [6], [8]\} \subseteq \mathbf{Z}_{10}$ with operation addition.

26. The set $\{[0], [2], [4], [6]\} \subseteq \mathbf{Z}_8$ with operation addition.

27. a. Let $G = \{[a] \mid [a] \neq [0]\} \subseteq \mathbf{Z}_n$. Show that G is a group with respect to multiplication in $\mathbf{Z}_n$ if and only if n is a prime.

Sec. 3.3, #9 ≪
Sec. 3.4, #15 ≪

   b. Construct a multiplication table for the group G of all nonzero elements in $\mathbf{Z}_7$, and identify the inverse of each element.

28. Let G be the set of eight elements $G = \{1, i, j, k, -1, -i, -j, -k\}$ with identity element 1 and noncommutative multiplication given by[†]

$$(-1)^2 = 1,$$
$$i^2 = j^2 = k^2 = -1,$$
$$ij = -ji = k,$$
$$jk = -kj = i,$$
$$ki = -ik = j,$$
$$-x = (-1)x = x(-1) \text{ for all } x \text{ in } G.$$

Figure 3-10

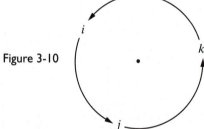

Sec 3.2, #18a ≪
Sec 3.3, #2 ≪
Sec. 4.4, #16, 19 ≪
Sec. 4.5, #3 ≪

(The circular order of multiplication is indicated by the diagram in Figure 3-10.) Given that G is a group, write out the multiplication table for G. This group is known as the **quaternion group**.

29. A **permutation matrix** is a matrix that can be obtained from an identity matrix I_n by interchanging the rows one or more times (that is, by *permuting* the rows). For $n = 3$, the permutation matrices are I_3 and the five matrices

$$P_1 = \begin{bmatrix} 1 & 0 & 0 \\ 0 & 0 & 1 \\ 0 & 1 & 0 \end{bmatrix} \quad P_2 = \begin{bmatrix} 0 & 1 & 0 \\ 1 & 0 & 0 \\ 0 & 0 & 1 \end{bmatrix} \quad P_3 = \begin{bmatrix} 0 & 1 & 0 \\ 0 & 0 & 1 \\ 1 & 0 & 0 \end{bmatrix}$$

$$P_4 = \begin{bmatrix} 0 & 0 & 1 \\ 0 & 1 & 0 \\ 1 & 0 & 0 \end{bmatrix} \quad P_5 = \begin{bmatrix} 0 & 0 & 1 \\ 1 & 0 & 0 \\ 0 & 1 & 0 \end{bmatrix}.$$

Sec. 3.2, #18c ≪
Sec. 3.3, #5 ≪
Sec. 4.2, #4 ≪

Given that $G = \{I_3, P_1, P_2, P_3, P_4, P_5\}$ is a group with respect to matrix multiplication, write out a multiplication table for G.

[†]In a multiplicative group, a^2 is defined by $a^2 = a \cdot a$.

30. Consider the matrices

$$R = \begin{bmatrix} 0 & -1 \\ 1 & 0 \end{bmatrix} \quad H = \begin{bmatrix} 1 & 0 \\ 0 & -1 \end{bmatrix} \quad V = \begin{bmatrix} -1 & 0 \\ 0 & 1 \end{bmatrix}$$

$$D = \begin{bmatrix} 0 & 1 \\ 1 & 0 \end{bmatrix} \quad T = \begin{bmatrix} 0 & -1 \\ -1 & 0 \end{bmatrix}$$

Sec. 3.2, #18b ≪
Sec. 4.1, #18 ≪
Sec. 4.5, #12 ≪

in $M_2(\mathbf{R})$, and let $G = \{I_2, R, R^2, R^3, H, D, V, T\}$. Given that G is a group with respect to multiplication, write out a multiplication table for G.

31. Prove that the set of all diagonal matrices in $M_n(\mathbf{R})$ forms a group with respect to addition.

32. Let G be the set of all matrices in $M_3(\mathbf{R})$ that have the form

$$\begin{bmatrix} a & 0 & 0 \\ 0 & b & 0 \\ 0 & 0 & c \end{bmatrix}$$

with all three numbers a, b, and c nonzero. Prove that G is a group with respect to multiplication.

33. Let G be the set of all matrices in $M_3(\mathbf{R})$ that have the form

$$\begin{bmatrix} 1 & a & b \\ 0 & 1 & c \\ 0 & 0 & 1 \end{bmatrix}$$

for arbitrary real numbers a, b, and c. Prove that G is a group with respect to multiplication.

34. Decide whether the set G in Exercise 32 is a group with respect to addition, and justify your answer.

35. Decide whether the set G in Exercise 33 is a group with respect to addition, and justify your answer.

36. Prove part a of Theorem 3.4.

37. Prove part c of Theorem 3.4.

38. Prove part e of Theorem 3.4.

39. An element x in a multiplicative group G is called **idempotent** if $x^2 = x$. Prove that the identity element e is the only idempotent element in a group G.

Sec. 5.1, #29 ≪

40. In Example 3, find elements a and b of $S(A)$ such that $(ab)^{-1} \neq a^{-1}b^{-1}$.

41. In Example 3, find elements a, b, and c of $S(A)$ such that $ab = bc$ but $a \neq c$.

42. In Example 3, find elements a and b of $S(A)$ such that $(ab)^2 \neq a^2b^2$.

43. Prove that in Theorem 3.5, the solutions to the equations $ax = b$ and $ya = b$ are actually unique.

44. Suppose that G is a finite group. Prove that each element of G appears in the multiplication table for G exactly once in each row and exactly once in each column.

45. Prove that if $x = x^{-1}$ for all x in the group G, then G is abelian.

46. Prove that if $(ab)^{-1} = a^{-1}b^{-1}$ for all a and b in the group G, then G is abelian.

47. Prove that if $(xy)^2 = x^2y^2$ for all x and y in the group G, then G is abelian.

48. Let a, b, c, and d be elements of a group G. Find an expression for $(abcd)^{-1}$ in terms of a^{-1}, b^{-1}, c^{-1}, and d^{-1}.

49. Use mathematical induction to prove that if $a_1, a_2, \ldots, a_n$ are elements of a group G, then $(a_1 a_2 \cdots a_n)^{-1} = a_n^{-1} a_{n-1}^{-1} \cdots a_2^{-1} a_1^{-1}$. (This is the general form of the **reverse order law** for inverses.)

50. For an arbitrary set A, the power set $\mathcal{P}(A)$ was defined in Section 1.1 by $\mathcal{P}(A) = \{X \mid X \subseteq A\}$, and addition in $\mathcal{P}(A)$ was defined by

$$X + Y = (X \cup Y) - (X \cap Y)$$
$$= (X - Y) \cup (Y - X).$$

Prove that $\mathcal{P}(A)$ is a group with respect to this operation of addition.

Sec. 1.1, #7c ≫   **51.** Write out the elements of $\mathcal{P}(A)$ for the set $A = \{a, b, c\}$, and construct an addition table for $\mathcal{P}(A)$ using addition as defined in Exercise 50.

Sec. 1.1, #7c ≫   **52.** Let $A = \{a, b, c\}$. Show that $\mathcal{P}(A)$ is not a group with respect to the operation of union.

Sec. 1.1, #7c ≫   **53.** Let $A = \{a, b, c\}$. Show that $\mathcal{P}(A)$ is not a group with respect to the operation of intersection.

54. Suppose G is a finite set with n distinct elements given by $G = \{a_1, a_2, \ldots, a_n\}$. Assume that G is closed under an associate binary operation $*$, and that the following two cancellation laws hold for all a, x, and y in G:

$$a * x = a * y \quad \text{implies} \quad x = y;$$
$$x * a = y * a \quad \text{implies} \quad x = y.$$

Prove that G is a group with respect to $*$.

55. Reword Definition 3.6 for a group G with respect to addition.

56. State and prove Theorem 3.7 for an additive group.

<hr>

▬▬▬▬ 3.2 SUBGROUPS

Among the nonempty subsets of a group G, there are some that themselves form a group with respect to the binary operation $*$ in G. That is, a subset $H \subseteq G$ may be such that H is also a group with respect to $*$. Such a subset H is called a *subgroup* of G.

Definition 3.8   SUBGROUP

> Let G be a group with respect to the binary operation $*$. A subset H of G is called a **subgroup** of G if H forms a group with respect to the binary operation $*$ that is defined in G.

The subsets $H = \{e\}$ and $H = G$ are always subgroups of the group G. They are referred to as **trivial subgroups**, and all other subgroups of G are called **nontrivial**.

Example 1   The set $\mathbf{Z}$ of all integers is a group with respect to addition, and the set $\mathbf{E}$ of all even integers is a nontrivial subgroup of $\mathbf{Z}$. (See Exercise 9 of Section 3.1.)   ∎

Example 2   The set of all nonzero complex numbers is a group under multiplication, and $G = \{1, -1, i, -i\}$ is a nontrivial subgroup of this group. (See Example 4 of Section 3.1.)   ∎

Example 3   From the discussion in Example 9 of Section 3.1, it is clear that for fixed m and n, each of the additive groups in the list

$$M_{m \times n}(\mathbf{Z}) \subseteq M_{m \times n}(\mathbf{Q}) \subseteq M_{m \times n}(\mathbf{R}) \subseteq M_{m \times n}(\mathbf{C})$$

is a subgroup of every listed group in which it is contained.   ■

If G is a group with respect to $*$, then $*$ is an associative operation on any non-empty subset of G. A subset H of G is a subgroup, provided

1. H contains the identity;
2. H is closed under $*$; and
3. H contains an inverse for each of its elements.

In connection with condition 1, consider the possibility that H might contain an identity e' for its elements that could be different from the identity e of G. Such an element e' would have the property that $e' * e' = e'$, and Exercise 39 of Section 3.1 then implies that $e' = e$. In connection with condition 3, we might consider the possibility that an element $a \in H$ might have one inverse as an element of the subgroup H and a different inverse as an element of the group G. In fact, this cannot happen because part b of Theorem 3.4 guarantees that the solution y to $a * y = y * a = e$ is unique in G. The following theorem gives a set of conditions that is slightly different from 1, 2, and 3.

Theorem 3.9   **EQUIVALENT SET OF CONDITIONS FOR A SUBGROUP**

A subset H of the group G is a subgroup of G if and only if these conditions are satisfied:

a. H is nonempty;
b. $x \in H$ and $y \in H$ imply $xy \in H$; and
c. $x \in H$ implies $x^{-1} \in H$.

$p \Rightarrow q$ **Proof**   If H is a subgroup of G, the conditions follow at once from Definitions 3.8 and 3.1.

$p \Leftarrow q$ Suppose that H is a subset of G that satisfies the conditions. Since H is non-empty, there is at least one $a \in H$. By condition c, $a^{-1} \in H$. But $a \in H$ and $a^{-1} \in H$ imply $aa^{-1} = e \in H$, by condition b. Thus, H contains e, is closed, and contains inverses. Hence, H is a subgroup.

Example 4   It follows from Example 5 of Section 3.1 that

$$G = \mathbf{Z}_8 = \{[0], [1], [2], [3], [4], [5], [6], [7]\}$$

forms an abelian group with respect to addition $[a] + [b] = [a + b]$. Consider the subset

$$H = \{[0], [2], [4], [6]\}$$

of G. An addition table for H is given in Figure 3-11. The subset H is nonempty, and it is evident from the table that H is closed and contains the inverse of each of its elements. Hence, H is a nontrivial abelian subgroup of $\mathbf{Z}_8$ under addition.

Figure 3-11

+	[0]	[2]	[4]	[6]
[0]	[0]	[2]	[4]	[6]
[2]	[2]	[4]	[6]	[0]
[4]	[4]	[6]	[0]	[2]
[6]	[6]	[0]	[2]	[4]

∎

Example 5   In Exercise 27 of Section 3.1, it was shown that

$$G = \{[1], [2], [3], [4], [5], [6]\} \subseteq \mathbf{Z}_7$$

is a group with respect to multiplication in $\mathbf{Z}_7$. The multiplication table in Figure 3-12 shows that the nonempty subset

$$H = \{[1], [2], [4]\}$$

is closed and contains inverses and therefore is an abelian subgroup of G.

Figure 3-12

·	[1]	[2]	[4]
[1]	[1]	[2]	[4]
[2]	[2]	[4]	[1]
[4]	[4]	[1]	[2]

∎

An even shorter set of conditions for a subgroup is given in the next theorem.

Theorem 3.10   EQUIVALENT SET OF CONDITIONS FOR A SUBGROUP

A subset H of the group G is a subgroup of G if and only if

a. H is nonempty, and
b. $a \in H$ and $b \in H$ imply $ab^{-1} \in H$.

$p \Rightarrow q$ **Proof**   Assume H is a subgroup of G. Then H is nonempty since $e \in H$. Let $a \in H$ and $b \in H$. Then $b^{-1} \in H$ since H contains inverses. Since $a \in H$ and $b^{-1} \in H$, the product $ab^{-1} \in H$ because H is closed. Thus, conditions a and b are satisfied.

$p \Leftarrow q$ Suppose, conversely, that conditions a and b hold for H. There is at least one $a \in H$, and condition b implies that $aa^{-1} = e \in H$. For an arbitrary $x \in H$, we have $e \in H$ and $x \in H$, which implies that $ex^{-1} = x^{-1} \in H$. Thus, H contains inverses. To show closure, let $x \in H$ and $y \in H$. Since H contains inverses, $y^{-1} \in H$. But $x \in H$ and $y^{-1} \in H$ imply $x(y^{-1})^{-1} = xy \in H$, by condition b. Hence, H is closed; therefore, H is a subgroup of G.

When the phrase "H is a subgroup of G" is used, it indicates that H is a group with respect to the group operation in G. Consider the following example.

Example 6   The operation of multiplication is defined in $\mathbf{Z}_{10}$ by

$$[a][b] = [ab].$$

This rule defines a binary operation that is associative, and $\mathbf{Z}_{10}$ is closed under this multiplication. Also, [1] is an identity element. However, $\mathbf{Z}_{10}$ is *not* a group with respect to multiplication since some of its elements do not have inverses. For example, the products

$$[2][0] = [0] \qquad [2][5] = [0]$$
$$[2][1] = [2] \qquad [2][6] = [2]$$
$$[2][2] = [4] \qquad [2][7] = [4]$$
$$[2][3] = [6] \qquad [2][8] = [6]$$
$$[2][4] = [8] \qquad [2][9] = [8]$$

show that $[2][x] = [1]$ has no solution in $\mathbf{Z}_{10}$.

Now let us examine the multiplication table for the subset $H = \{[2],[4],[6],[8]\}$ of $\mathbf{Z}_{10}$ (see Figure 3-13). It is surprising, perhaps, but the table shows that $[6]$ is an identity element for H and that H actually forms a group with respect to multiplication. However, H is *not* a subgroup of $\mathbf{Z}_{10}$ since $\mathbf{Z}_{10}$ is not a group with respect to multiplication.

Figure 3-13

$\times$	$[2]$	$[4]$	$[6]$	$[8]$
$[2]$	$[4]$	$[8]$	$[2]$	$[6]$
$[4]$	$[8]$	$[6]$	$[4]$	$[2]$
$[6]$	$[2]$	$[4]$	$[6]$	$[8]$
$[8]$	$[6]$	$[2]$	$[8]$	$[4]$

■

Integral exponents can be defined for elements of a group as follows.

Definition 3.11 **INTEGRAL EXPONENTS**

> Let G be a group with the binary operation written as multiplication. For any $a \in G$, we define **nonnegative integral exponents** by
>
> $$a^0 = e, \qquad a^1 = a,$$
>
> and
>
> $$a^{k+1} = a^k \cdot a \quad \text{for any positive integer } k.$$
>
> **Negative integral exponents** are defined by
>
> $$a^{-k} = (a^{-1})^k \quad \text{for any positive integer } k.$$

It is common practice to write the binary operation as addition in the case of abelian groups. When the operation is addition, the corresponding **multiples** of a are defined in a similar fashion. The following list shows how the notations correspond, where k is a positive integer.

Multiplicative Notation	**Additive Notation**
$a^0 = e$	$0a = 0$
$a^1 = a$	$1a = a$
$a^{k+1} = a^k \cdot a$	$(k+1)a = ka + a$
$a^{-k} = (a^{-1})^k$	$(-k)a = k(-a)$

The notation ka in additive notation does not represent a *product* of k and a, but rather a *sum*

$$ka = a + a + \cdots + a$$

with k terms. In $0a = 0$, the zero on the left is the zero integer, and the zero on the right represents the additive identity in the group.

Considering the rich variety of operations and sets that have been involved in our examples, it might be surprising and reassuring to find in the next theorem that the familiar **laws of exponents** hold in a group.

Theorem 3.12   LAWS OF EXPONENTS

Let x and y be elements of the group G, and let m and n denote integers. Then

a. $x^n \cdot x^{-n} = e$
b. $x^m \cdot x^n = x^{m+n}$
c. $(x^m)^n = x^{mn}$
d. If G is abelian, $(xy)^n = x^n y^n$.

Induction   **Proof**   The proof of each statement involves the use of mathematical induction. It would be redundant, and even boring, to include a complete proof of the theorem, so we shall assume statement **a** and prove **b** for the case where m is a positive integer. Even then, the argument is lengthy. The proofs of the statements **a**, **c**, and **d** are left as exercises.

Let m be an arbitrary, but fixed, positive integer. There are three cases to consider for n:

i. $n = 0$
ii. n a positive integer
iii. n a negative integer.

First, let $n = 0$ for case i. Then

$$x^m \cdot x^n = x^m \cdot x^0 = x^m \cdot e = x^m \quad \text{and} \quad x^{m+n} = x^{m+0} = x^m.$$

Thus, $x^m \cdot x^n = x^{m+n}$ in the case where $n = 0$.

Second, we shall use induction on n for case ii where n is a positive integer. If $n = 1$, we have

$$x^m \cdot x^n = x^m \cdot x = x^{m+1} = x^{m+n},$$

and statement b of the theorem holds when $n = 1$. Assume that b is true for $n = k$. That is, assume that

$$x^m \cdot x^k = x^{m+k}.$$

Then, for $n = k + 1$, we have

$$
\begin{aligned}
x^m \cdot x^n &= x^m \cdot x^{k+1} \\
&= x^m \cdot (x^k \cdot x) && \text{by definition of } x^{k+1} \\
&= (x^m \cdot x^k) \cdot x && \text{by associativity} \\
&= x^{m+k} \cdot x && \text{by the induction hypothesis} \\
&= x^{m+k+1} && \text{by definition of } x^{(m+k)+1} \\
&= x^{m+n} && \text{since } n = k + 1.
\end{aligned}
$$

Thus, b is true for $n = k + 1$, and it follows that it is true for all positive integers n.

Third, consider case iii where n is a negative integer. This means that $n = -p$, where p is a positive integer. We consider three possibilities for p: $p = m$, $p < m$, and $m < p$.

If $p = m$, then $n = -p = -m$, and we have

$$x^m \cdot x^n = x^m \cdot x^{-m} = e$$

by statement a of the theorem, and

$$x^{m+n} = x^{m-m} = x^0 = e.$$

We have $x^m \cdot x^n = x^{m+n}$ when $p = m$.

If $p < m$, let $m - p = q$, so that $m = q + p$ where q and p are positive integers. We have already proved statement b when m and n are positive integers, so we may use $x^{q+p} = x^q \cdot x^p$. This gives

$$
\begin{aligned}
x^m \cdot x^n &= x^{q+p} \cdot x^{-p} \\
&= x^q \cdot x^p \cdot x^{-p} \\
&= x^q \cdot e \qquad \text{by statement } \mathbf{a} \\
&= x^q \\
&= x^{q+p-p} \\
&= x^{m+n}.
\end{aligned}
$$

That is, $x^m \cdot x^n = x^{m+n}$ for the case where $p < m$.

Finally, suppose that $m < p$. Let $r = p - m$, so that r is a positive integer and $p = m + r$. By the definition of x^{-p},

$$
\begin{aligned}
x^{-p} &= (x^{-1})^p \\
&= (x^{-1})^{m+r} \\
&= (x^{-1})^m \cdot (x^{-1})^r \quad \text{since } m \text{ and } r \text{ are positive integers} \\
&= x^{-m} \cdot x^{-r}.
\end{aligned}
$$

Substituting this value for x^{-p} in $x^m \cdot x^n = x^m \cdot x^{-p}$, we have

$$
\begin{aligned}
x^m \cdot x^n &= x^m \cdot (x^{-m} \cdot x^{-r}) \\
&= (x^m \cdot x^{-m}) \cdot x^{-r} \\
&= e \cdot x^{-r} \\
&= x^{-r}.
\end{aligned}
$$

We also have

$$
\begin{aligned}
x^{m+n} &= x^{m-p} \\
&= x^{m-(m+r)} \\
&= x^{-r},
\end{aligned}
$$

so $x^m \cdot x^n = x^{m+n}$ when $m < p$.

We have proved that $x^m \cdot x^n = x^{m+n}$ in the cases where m is a positive integer and n is any integer (zero, positive, or negative). Of course, this is not a complete proof of statement b of the theorem. A complete proof would require considering cases where $m = 0$ or where m is a negative integer. The proofs for these cases are similar to those given here, and we omit them entirely.

The laws of exponents in Theorem 3.12 translate into the following **laws of multiples** for an additive group G.

Laws of Multiples

a. $nx + (-n)x = 0$
b. $mx + nx = (m + n)x$
c. $n(mx) = (nm)x$
d. If G is abelian, $n(x + y) = nx + ny$.

Let G be a group, let a be an element of G, and let H be the set of all elements of the form a^n, where n is an integer. That is,

$$H = \{x \in G \,|\, x = a^n \text{ for } n \in \mathbf{Z}\}.$$

Then H is nonempty and actually forms a subgroup of G. For if $x = a^m \in H$ and $y = a^n \in H$, then $xy = a^{m+n} \in H$ and $x^{-1} = a^{-m} \in H$. It follows from Theorem 3.9 that H is a subgroup.

Definition 3.13   CYCLIC SUBGROUP

Let G be a group. For any $a \in G$, the subgroup

$$H = \{x \in G \,|\, x = a^n \text{ for } n \in \mathbf{Z}\}$$

is the **subgroup generated by** a and is denoted by $\langle a \rangle$. A given subgroup K of G is a **cyclic subgroup** if there exists an element b in G such that

$$K = \langle b \rangle = \{y \in G \,|\, y = b^n \text{ for some } n \in \mathbf{Z}\}.$$

In particular, G is a **cyclic group** if there is an element $a \in G$ such that $G = \langle a \rangle$.

Example 7

a. The set $\mathbf{Z}$ of integers is a cyclic group under addition. We have $\mathbf{Z} = \langle 1 \rangle$ and $\mathbf{Z} = \langle -1 \rangle$.
b. The subgroup $\mathbf{E} \subseteq \mathbf{Z}$ of all even integers is a cyclic subgroup of the additive group $\mathbf{Z}$, generated by 2. Hence, $\mathbf{E} = \langle 2 \rangle$.
c. In Example 6, we saw that

$$H = \{[2], [4], [6], [8]\} \subseteq \mathbf{Z}_{10}$$

is an abelian group with respect to multiplication. Since

$$[2]^2 = [4], \quad [2]^3 = [8], \quad [2]^4 = [6],$$

then

$$H = \langle [2] \rangle.$$

d. The group $S(A) = \{e, \rho, \rho^2, \sigma, \gamma, \delta\}$ of Example 3 in Section 3.1 is not a cyclic group. This can be verified by considering $\langle a \rangle$ for all possible choices of a in $S(A)$. ∎

EXERCISES 3.2

1. Let $S(A) = \{e, \rho, \rho^2, \sigma, \gamma, \delta\}$ be as in Example 3 in Section 3.1. Decide whether each of the following subsets is a subgroup of $S(A)$. If a set is not a subgroup, give a reason why it is not. (*Hint:* Construct a multiplication table for each subset.)

   a. $\{e, \sigma\}$ **b.** $\{e, \delta\}$
   c. $\{e, \rho\}$ **d.** $\{e, \rho^2\}$
   e. $\{e, \rho, \rho^2\}$ **f.** $\{e, \rho, \sigma\}$
   g. $\{e, \sigma, \gamma\}$ **h.** $\{e, \sigma, \gamma, \delta\}$

2. Decide whether each of the following sets is a subgroup of the group $G = \{1, -1, i, -i\}$ under multiplication. If a set is not a subgroup, give a reason why it is not.

   a. $\{1, -1\}$ **b.** $\{1, i\}$
   c. $\{i, -i\}$ **d.** $\{1, -i\}$

3. Consider the group $\mathbf{Z}_{16}$ under addition. List all the elements of the subgroup $\langle [6] \rangle$.

4. List all the elements of the subgroup $\langle [8] \rangle$ in the group $\mathbf{Z}_{18}$ under addition.

5. Assume that the nonzero elements of $\mathbf{Z}_{13}$ form a group G under multiplication $[a][b] = [ab]$.
   a. List the elements of the subgroup $\langle [4] \rangle$ of G.
   b. List the elements of the subgroup $\langle [8] \rangle$ of G.

6. Let G be the group of all invertible matrices in $M_2(\mathbf{R})$ under multiplication, and list the elements of the subgroup $\langle A \rangle$ of G for the given A.

   a. $A = \begin{bmatrix} 0 & -1 \\ 1 & 0 \end{bmatrix}$ **b.** $A = \begin{bmatrix} 0 & -1 \\ -1 & 0 \end{bmatrix}$

7. Let G be the group $M_2(\mathbf{Z}_5)$ under addition, and list the elements of the subgroup $\langle A \rangle$ of G for the given A.

   a. $A = \begin{bmatrix} [2] & [0] \\ [0] & [3] \end{bmatrix}$ **b.** $A = \begin{bmatrix} [0] & [1] \\ [2] & [4] \end{bmatrix}$

8. Find a subset of $\mathbf{Z}$ that is closed under addition but is not a subgroup of the additive group $\mathbf{Z}$.

9. Let G be the group of all nonzero real numbers under multiplication. Find a subset of G that is closed under multiplication but is not a subgroup of G.

10. Let $n > 1$ be an integer, and let a be a fixed integer. Prove that the set

$$H = \{x \in \mathbf{Z} \mid ax \equiv 0 \ (\text{mod } n)\}$$

is a subgroup of $\mathbf{Z}$ under addition.

11. Let H be a subgroup of G, let a be a fixed element of G, and let K be the set of all elements of the form aha^{-1}, where $h \in H$. That is,

$$K = \{x \in G \mid x = aha^{-1} \text{ for some } h \in H\}.$$

Prove that K is a subgroup of G.

12. Prove that $H = \{h \in G \mid h^{-1} = h\}$ is a subgroup of the group G if G is abelian.

13. Prove that each of the following subsets H of $M_2(\mathbf{Z})$ is a subgroup of the group $M_2(\mathbf{Z})$ under addition.

a. $H = \left\{ \begin{bmatrix} x & y \\ z & w \end{bmatrix} \,\middle|\, w = 0 \right\}$

b. $H = \left\{ \begin{bmatrix} x & y \\ z & w \end{bmatrix} \,\middle|\, z = w = 0 \right\}$

c. $H = \left\{ \begin{bmatrix} x & y \\ 0 & 0 \end{bmatrix} \,\middle|\, x = y \right\}$

d. $H = \left\{ \begin{bmatrix} x & y \\ z & w \end{bmatrix} \,\middle|\, x + y + z + w = 0 \right\}$

14. Prove that each of the following subsets H of $M_2(\mathbf{R})$ is a subgroup of the group G of all invertible matrices in $M_2(\mathbf{R})$ under multiplication.

a. $H = \left\{ \begin{bmatrix} 1 & a \\ 0 & 1 \end{bmatrix} \,\middle|\, a \in \mathbf{R} \right\}$

b. $H = \left\{ \begin{bmatrix} a & -b \\ b & a \end{bmatrix} \,\middle|\, a^2 + b^2 = 1 \right\}$

Sec. 3.4, #9 ≪ **c.** $H = \left\{ \begin{bmatrix} a & -b \\ b & a \end{bmatrix} \,\middle|\, a^2 + b^2 \neq 0 \right\}$

d. $H = \left\{ \begin{bmatrix} 1 & a \\ 0 & b \end{bmatrix} \,\middle|\, b \neq 0 \right\}$

15. Prove that each of the following sets H is a subgroup of the group G of all invertible matrices in $M_2(\mathbf{C})$ under multiplication.

Sec. 4.3, #27 ≪ **a.** $H = \left\{ \begin{bmatrix} 1 & 0 \\ 0 & 1 \end{bmatrix}, \begin{bmatrix} 1 & 0 \\ 0 & -1 \end{bmatrix}, \begin{bmatrix} -1 & 0 \\ 0 & 1 \end{bmatrix}, \begin{bmatrix} -1 & 0 \\ 0 & -1 \end{bmatrix} \right\}$

Sec. 3.4, #8 ≪ **b.** $H = \left\{ \begin{bmatrix} 1 & 0 \\ 0 & 1 \end{bmatrix}, \begin{bmatrix} i & 0 \\ 0 & -i \end{bmatrix}, \begin{bmatrix} -i & 0 \\ 0 & i \end{bmatrix}, \begin{bmatrix} -1 & 0 \\ 0 & -1 \end{bmatrix} \right\}$

16. Consider the set of matrices $H = \{I_2, M_1, M_2, M_3, M_4, M_5\}$ where

$$I_2 = \begin{bmatrix} 1 & 0 \\ 0 & 1 \end{bmatrix}, \quad M_1 = \begin{bmatrix} 1 & 0 \\ -1 & -1 \end{bmatrix}, \quad M_2 = \begin{bmatrix} 0 & 1 \\ -1 & -1 \end{bmatrix},$$

$$M_3 = \begin{bmatrix} -1 & -1 \\ 1 & 0 \end{bmatrix}, \quad M_4 = \begin{bmatrix} -1 & -1 \\ 0 & 1 \end{bmatrix}, \quad M_5 = \begin{bmatrix} 0 & 1 \\ 1 & 0 \end{bmatrix}.$$

Sec. 3.4, #5 ≪ Show that H is a subgroup of the multiplicative group of all invertible matrices in $M_2(\mathbf{R})$.

17. For any group G, the set of all elements that commute with every element of G is called the **center** of G and is denoted by $Z(G)$:

$$Z(G) = \{a \in G \,|\, ax = xa \text{ for every } x \in G\}.$$

Sec. 4.4, #32 ≪ Prove that $Z(G)$ is a subgroup of G.

18. (See Exercise 17.) Find the center $Z(G)$ for each of the following groups G.

Sec. 3.1, #28 ≫ **a.** $G = \{1, i, j, k, -1, -i, -j, -k\}$ in Exercise 28 of Section 3.1.

Sec. 3.1, #30 ≫ **b.** $G = \{I_2, R, R^2, R^3, H, D, V, T\}$ in Exercise 30 of Section 3.1.

Sec. 3.1, #29 ≫ **c.** $G = \{I_3, P_1, P_2, P_3, P_4, P_5\}$ in Exercise 29 of Section 3.1.

d. G is the group of all invertible matrices in $M_2(\mathbf{R})$ under multiplication.

19. Let A be a given nonempty set. As noted in Example 2 of Section 3.1, $S(A)$ is a group with respect to mapping composition. For a fixed element a in A, let H_a denote the set of all $f \in S(A)$ such that $f(a) = a$. Prove that H_a is a subgroup of $S(A)$.

20. (See Exercise 19.) Let A be an infinite set, and let H be the set of all $f \in S(A)$ such that $f(x) = x$ for all but a finite number of elements x of A. Prove that H is a subgroup of $S(A)$.

21. Let G be an abelian group. For a fixed positive integer n, let

$$G_n = \{a \in G \,|\, a = x^n \text{ for some } x \in G\}.$$

Prove that G_n is a subgroup of G.

22. For fixed integers a and b, let

$$S = \{ax + by \mid x \in \mathbf{Z} \text{ and } y \in \mathbf{Z}\}.$$

   Prove that S is a subgroup of $\mathbf{Z}$ under addition. (A special form of this S is used in proving the existence of a greatest common divisor in Theorem 2.12.)

23. For a fixed element a of a group G, the set $C_a = \{x \in G \mid ax = xa\}$ is the **centralizer** of a in G. Prove that for any $a \in G$, C_a is a subgroup of G.

24. Suppose that H_1 and H_2 are subgroups of the group G. Prove that $H_1 \cap H_2$ is a subgroup of G.

25. For an arbitrary n in $\mathbf{Z}$, the cyclic subgroup $\langle n \rangle$ of $\mathbf{Z}$, generated by n under addition, is the set of all multiples of n. Describe the subgroup $\langle m \rangle \cap \langle n \rangle$ for arbitrary m and n in $\mathbf{Z}$.

26. Let $\{H_\lambda\}$, $\lambda \in \mathscr{L}$, be an arbitrary nonempty collection of subgroups H_λ of the group G, and let $K = \cap_{\lambda \in \mathscr{L}} H_\lambda$. Prove that K is a subgroup of G.

27. Find subgroups H_1 and H_2 of the group $S(A)$ in Example 3 of Section 3.1 such that $H_1 \cup H_2$ is *not* a subgroup of $S(A)$.

28. Assume H_1 and H_2 are subgroups of the abelian group G. Prove that the set of products $H_1 H_2 = \{g \in G \mid g = h_1 h_2 \text{ for } h_1 \in H_1 \text{ and } h_2 \in H_2\}$ is a subgroup of G.

29. Find subgroups H_1 and H_2 of the group $S(A)$ in Example 3 of Section 3.1 such that the set $H_1 H_2$ defined in Exercise 28 is not a subgroup of $S(A)$.

30. Let G be a cyclic group, $G = \langle a \rangle$. Prove that G is abelian.

31. Prove statement **a** of Theorem 3.12: $x^n \cdot x^{-n} = e$ for all integers n.

32. Prove statement **c** of Theorem 3.12: $(x^m)^n = x^{mn}$ for all integers m and n.

33. Prove statement **d** of Theorem 3.12: if G is abelian, $(xy)^n = x^n y^n$ for all integers n.

34. Suppose H is a nonempty subset of a group G. Prove that H is a subgroup of G if and only if $a^{-1}b \in H$ for all $a \in H$ and $b \in H$.

35. Assume that G is a finite group, and let H be a nonempty subset of G. Prove that H is closed if and only if H is a subgroup of G.

3.3   CYCLIC GROUPS

In the last section a group G was defined to be *cyclic* if there exists an element $a \in G$ such that $G = \langle a \rangle$. It may happen that there is more than one element $a \in G$ such that $G = \langle a \rangle$. For the additive group $\mathbf{Z}$, we have $\mathbf{Z} = \langle 1 \rangle$ and also $\mathbf{Z} = \langle -1 \rangle$, since any $n \in \mathbf{Z}$ can be written as $(-n)(-1)$. [Here $(-n)(-1)$ does not indicate a product but rather a multiple of -1, as described in Section 3.2.]

Definition 3.14   GENERATOR

> Any element a of the group G such that $G = \langle a \rangle$ is a **generator** of G.

If a is a generator of G, then a^{-1} is also, since any element $x \in G$ can be written as

$$x = a^n = (a^{-1})^{-n}$$

for some integer n.

Example 1   The additive group

$$\mathbf{Z}_n = \{[0], [1], \ldots, [n-1]\}$$

is a cyclic group with generator $[1]$, since any $[k]$ in $\mathbf{Z}_n$ can be written as

$$[k] = k[1]$$

where $k[1]$ indicates a multiple of $[1]$ as described in Section 3.2. Elements other than $[1]$ may also be generators. To illustrate this, consider the particular case

$$\mathbf{Z}_6 = \{[0], [1], [2], [3], [4], [5]\}.$$

The element $[5]$ is also a generator of $\mathbf{Z}_6$ since $[5]$ is the additive inverse of $[1]$. The following list shows how $\mathbf{Z}_6$ is generated by $[5]$; that is, how $\mathbf{Z}_6$ consists of multiples of $[5]$.

$$1[5] = [5]$$
$$2[5] = [5] + [5] = 4$$
$$3[5] = [5] + [5] + [5] = [3]$$
$$4[5] = [2]$$
$$5[5] = [1]$$
$$6[5] = [0]$$

The cyclic subgroups generated by the other elements of $\mathbf{Z}_6$ under addition are as follows:

$$\langle [0] \rangle = \{[0]\}$$
$$\langle [2] \rangle = \{[2], [4], [0]\}$$
$$\langle [3] \rangle = \{[3], [0]\}$$
$$\langle [4] \rangle = \{[4], [2], [0]\} = \langle [2] \rangle.$$

Thus, $[1]$ and $[5]$ are the only elements that are generators of the entire group.   ■

Example 2   We saw in Example 7 of Section 3.2 that

$$H = \{[2], [4], [6], [8]\} \subseteq \mathbf{Z}_{10}$$

forms a cyclic group with respect to multiplication, and that $[2]$ is a generator of H. The element $[8] = [2]^{-1}$ is also a generator of H, as the following computations confirm:

$$[8]^2 = [4], \qquad [8]^3 = [2], \qquad [8]^4 = [6].$$ ■

Example 3   In the quaternion group $G = \{\pm 1, \pm i, \pm j, \pm k\}$, described in Exercise 28 of Section 3.1, we have

$$i^2 = -1$$
$$i^3 = i^2 \cdot i = -i$$
$$i^4 = i^3 \cdot i = -i^2 = 1.$$

Thus, i generates the cyclic subgroup of order 4 given by

$$\langle i \rangle = \{i, -1, -i, 1\},$$

although the group G itself is not cyclic.   ∎

Whether a group G is cyclic or not, each element a of G generates the cyclic subgroup $\langle a \rangle$, and

$$\langle a \rangle = \{x \in G \,|\, x = a^n \text{ for } n \in \mathbf{Z}\}.$$

We shall see that the structure of $\langle a \rangle$ depends entirely on whether or not $a^n = e$ for some positive integer n. The next two theorems state the possibilities for the structure of $\langle a \rangle$.

STRATEGY ▶

The method of proof of the next theorem is by contradiction. A statement $p \Rightarrow q$ may be proved by assuming that p is true and q is false and then proving that this assumption leads to a situation where some statement is both true and false—a contradiction.

Theorem 3.15 **INFINITE CYCLIC GROUP**

Let a be an element in the group G. If $a^n \neq e$ for every positive integer n, then $a^p \neq a^q$ whenever $p \neq q$ in $\mathbf{Z}$, and $\langle a \rangle$ is an infinite cyclic group.

Contradiction
$(p \wedge \sim q) \Rightarrow \sim p$

Proof Assume that a is an element of the group G such that $a^n \neq e$ for every positive integer n. Having made this assumption, suppose now that

$$a^p = a^q$$

where $p \neq q$ in $\mathbf{Z}$. We may assume that $p > q$. Then

$$a^p = a^q \Rightarrow a^p \cdot a^{-q} = a^q \cdot a^{-q}$$
$$\Rightarrow a^{p-q} = e.$$

Since $p - q$ is a positive integer, this result contradicts $a^n \neq e$ for every positive integer n. Therefore, it must be that $a^p \neq a^q$ whenever $p \neq q$. Thus, all powers of a are distinct, and therefore $\langle a \rangle$ is an infinite cyclic group.

Corollary 3.16

If G is a finite group and $a \in G$, then $a^n = e$ for some positive integer n.

$p \Rightarrow q$

Proof Suppose G is a finite group and $a \in G$. Since the cyclic subgroup

$$\langle a \rangle = \{x \in G \,|\, x = a^m \text{ for } m \in \mathbf{Z}\}$$

is a subset of G, $\langle a \rangle$ must also be finite. It must therefore happen that $a^p = a^q$ for some integers p and q with $p \neq q$. It follows from Theorem 3.15 that $a^n = e$ for some positive integer n.

If it happens that $a^n \neq e$ for every positive integer n, then Theorem 3.15 states that all the powers of a are distinct and that $\langle a \rangle$ is an infinite group. Of course, it may happen that $a^n = e$ for some positive integers n. In this case, Theorem 3.17 describes $\langle a \rangle$ completely.

Theorem 3.17   FINITE CYCLIC GROUP

Let a be an element in a group G, and suppose $a^n = e$ for some positive integer n. If m is the least positive integer such that $a^m = e$, then

a. $\langle a \rangle$ has order m, and $\langle a \rangle = \{a^0 = e = a^m, a^1, a^2, \ldots, a^{m-1}\}$
b. $a^s = a^t$ if and only if $s \equiv t \pmod{m}$.

$p \Rightarrow q$ **Proof**   Assume that m is the least positive integer such that $a^m = e$. We first show that the elements

$$a^0 = e, a, a^2, \ldots, a^{m-1}$$

are all distinct. Suppose

$$a^i = a^j \quad \text{where} \quad 0 \leq i < m \quad \text{and} \quad 0 \leq j < m.$$

There is no loss of generality in assuming $i \geq j$. Then $a^i = a^j$ implies

$$a^{i-j} = a^i \cdot a^{-j} = e \quad \text{where} \quad 0 \leq i - j < m.$$

Since m is the least positive integer such that $a^m = e$, and since $i - j < m$, it must be true that $i - j = 0$, and therefore $i = j$. Thus, $\langle a \rangle$ contains the m distinct elements $a^0 = e, a, a^2, \ldots, a^{m-1}$. The proof of part **a** will be complete if we can show that any power of a is equal to one of these elements. Consider an arbitrary a^k. By the Division Algorithm, there exist integers q and r such that

$$k = mq + r, \quad \text{with } 0 \leq r < m.$$

Thus,

$$
\begin{aligned}
a^k &= a^{mq+r} \\
&= a^{mq} \cdot a^r && \text{by part } \mathbf{b} \text{ of Theorem 3.12} \\
&= (a^m)^q \cdot a^r && \text{by part } \mathbf{c} \text{ of Theorem 3.12} \\
&= e^q \cdot a^r \\
&= a^r
\end{aligned}
$$

where r is in the set $\{0, 1, 2, \ldots, m - 1\}$. It follows that

$$\langle a \rangle = \{e, a, a^2, \ldots, a^{m-1}\}, \quad \text{and } \langle a \rangle \text{ has order } m.$$

$p \Rightarrow (q \Leftrightarrow r)$ To obtain part **b**, we first observe that if $k = mq + r$, with $0 \leq r < m$, then $a^k = a^r$, where r is in the set $\{0, 1, 2, \ldots, m - 1\}$. In particular, $a^k = e$ if and only if $r = 0$; that is, if and only if $k \equiv 0 \pmod{m}$. Thus,

$$a^s = a^t \Leftrightarrow a^{s-t} = e$$

$$\Leftrightarrow s - t \equiv 0 \ (\text{mod } m)$$
$$\Leftrightarrow s \equiv t \ (\text{mod } m),$$

and the proof is complete.

We have defined the order $o(G)$ of a group G to be the number of elements in the group.

Definition 3.18   **ORDER OF AN ELEMENT**

> The **order** $o(a)$ of an element a of the group G is the order of the subgroup generated by a. That is, $o(a) = o(\langle a \rangle)$.

Part a of Theorem 3.17 immediately translates into the following corollary.

Corollary 3.19   **FINITE ORDER OF AN ELEMENT**

If $o(a)$ is finite, then $m = o(a)$ is the least positive integer such that $a^m = e$.

As might be expected, every subgroup of a cyclic group is also a cyclic group. It is even possible to predict a generator of the subgroup, as stated in Theorem 3.20.

STRATEGY ▶   The conclusion of the next theorem has the form "either a or b." To prove this statement, we can assume that a is false and prove that b must then be true.

Theorem 3.20   **SUBGROUP OF A CYCLIC GROUP**

Let G be a cyclic group with $a \in G$ as a generator, and let H be a subgroup of G. Then either

a. $H = \{e\} = \langle e \rangle$, or
b. if $H \neq \{e\}$, then $H = \langle a^k \rangle$ where k is the least positive integer such that $a^k \in H$.

$(p \wedge q \wedge \sim r)$ **Proof**   Let $G = \langle a \rangle$, and suppose H is a subgroup and $H \neq \{e\}$. Then H contains an
$\Rightarrow s$ element of the form a^j with $j \neq 0$. Since H contains inverses and $(a^j)^{-1} = a^{-j}$, both a^j and a^{-j} are in H. Thus, H contains positive powers of a. Let k be the least positive integer such that $a^k \in H$.

Since H is closed and contains inverses, and since $a^k \in H$, all powers $(a^k)^t = a^{kt}$ are in H. We need to show that any element of H is a power of a^k. Let $a^n \in H$. There are integers q and r such that

$$n = kq + r \quad \text{with} \quad 0 \leq r < k.$$

Now $a^{-kq} = (a^k)^{-q} \in H$ and $a^n \in H$ imply that

$$a^n \cdot a^{-kq} = a^{kq+r} \cdot a^{-kq} = a^r$$

is in H. Since $0 \leq r < k$ and k is the least positive integer such that $a^k \in H$, r must be zero and $a^n = a^{kq}$. Thus, $H = \langle a^k \rangle$.

Corollary 3.21

Any subgroup of a cyclic group is cyclic.

Notice that Theorem 3.20 and Corollary 3.21 apply to infinite cyclic groups as well as finite ones. The next theorem, however, applies only to finite groups.

STRATEGY ▶
In the proof of Theorem 3.22, we use the standard technique to prove that two sets A and B are equal: We show that $A \subseteq B$ and then that $B \subseteq A$.

Theorem 3.22   GENERATORS OF SUBGROUPS

Let G be a finite cyclic group of order n with $a \in G$ as a generator. For any integer m, the subgroup generated by a^m is the same as the subgroup generated by a^d where $d = (m, n)$.

$p \Rightarrow q$ **Proof**   Let $d = (m, n)$, and let $m = dp$, $n = dq$. Since $a^m = a^{dp} = (a^d)^p$, a^m is in $\langle a^d \rangle$, and therefore $\langle a^m \rangle \subseteq \langle a^d \rangle$.

In order to show that $\langle a^d \rangle \subseteq \langle a^m \rangle$, it is sufficient to show that a^d is in $\langle a^m \rangle$. By Theorem 2.12, there exist integers x and y such that

$$d = mx + ny.$$

Since a is a generator of G and $o(G) = n$, $a^n = e$. Using this fact, we have

$$a^d = a^{mx+ny}$$
$$= a^{mx} \cdot a^{ny}$$
$$= (a^m)^x \cdot (a^n)^y$$
$$= (a^m)^x \cdot (e)^y$$
$$= (a^m)^x.$$

Thus, a^d is in $\langle a^m \rangle$, and the proof of the theorem is complete.

As an immediate corollary to Theorem 3.22, we have the following result.

Corollary 3.23   DISTINCT SUBGROUPS OF A FINITE CYCLIC GROUP

Let G be a finite cyclic group of order n with $a \in G$ as a generator. The distinct subgroups of G are those subgroups $\langle a^d \rangle$ where d is a positive divisor of n.

Corollary 3.23 provides a systematic way to obtain all the subgroups of a cyclic group of order n.

Example 4 Let $G = \langle a \rangle$ be a cyclic group of order 12. The divisors of 12 are 1, 2, 3, 4, 6, and 12, so the distinct subgroups of G are

$$\langle a \rangle = G$$
$$\langle a^2 \rangle = \{a^2, a^4, a^6, a^8, a^{10}, a^{12} = e\}$$
$$\langle a^3 \rangle = \{a^3, a^6, a^9, a^{12} = e\}$$
$$\langle a^4 \rangle = \{a^4, a^8, a^{12} = e\}$$
$$\langle a^6 \rangle = \{a^6, a^{12} = e\}$$
$$\langle a^{12} \rangle = \langle e \rangle = \{e\}.$$

Thus, Corollary 3.23 makes it easy to list all the distinct subgroups of a cyclic group. Theorem 3.22 itself makes it easy to determine which subgroup is generated by each element of the group. For our cyclic group of order 12,

$$\langle a^5 \rangle = \langle a \rangle = G \quad \text{since } (5, 12) = 1$$
$$\langle a^7 \rangle = \langle a \rangle \quad \text{since } (7, 12) = 1$$
$$\langle a^8 \rangle = \langle a^4 \rangle \quad \text{since } (8, 12) = 4$$
$$\langle a^9 \rangle = \langle a^3 \rangle \quad \text{since } (9, 12) = 3$$
$$\langle a^{10} \rangle = \langle a^2 \rangle \quad \text{since } (10, 12) = 2$$
$$\langle a^{11} \rangle = \langle a \rangle \quad \text{since } (11, 12) = 1.$$ ∎

The results in Example 4 lead us to a method for finding all generators of a finite cyclic group. This method is described in the next theorem.

Theorem 3.24 **GENERATORS OF A FINITE CYCLIC GROUP**

Let $G = \langle a \rangle$ be a cyclic group of order n. Then a^m is a generator of G if and only if m and n are relatively prime.

$p \Leftarrow q$ **Proof** On the one hand, if m is such that m and n are relatively prime, then $d = (m, n) = 1$, and a^m is a generator of G by Theorem 3.22.

$p \Rightarrow q$ On the other hand, if a^m is a generator of G, then $a = (a^m)^p$ for some integer p. By part b of Theorem 3.17, this implies that $1 \equiv mp(\bmod\ n)$. That is,

$$1 - mp = nq$$

for some integer q. This gives

$$1 = mp + nq$$

and it follows from Theorem 2.12 that $(m, n) = 1$.

Example 5 Let $G = \langle a \rangle$ be a cyclic group of order 10. The positive integers less than 10 and relatively prime to 10 are 1, 3, 7, and 9. Therefore, all generators of G are included in the list

$$a, \quad a^3, \quad a^7, \quad \text{and} \quad a^9.$$ ∎

Example 6 Some other explicit uses of Theorem 3.24 can be demonstrated by using $\mathbf{Z}_7$.

The generators of the additive group $\mathbf{Z}_7$ are those $[a]$ in $\mathbf{Z}_7$ such that a and 7 are relatively prime, and this includes all nonzero $[a]$. Thus, every element of $\mathbf{Z}_7$, except $[0]$, generates $\mathbf{Z}_7$ under addition.

The situation is quite different when we consider the group G of nonzero elements of $\mathbf{Z}_7$ under multiplication. It is easy to verify that $[3]$ is a generator:

$$[3]^2 = [2], \quad [3]^3 = [6], \quad [3]^4 = [4],$$
$$[3]^5 = [5], \quad [3]^6 = [1], \quad [3]^7 = [3].$$

According to Theorem 3.24, the only other generator of G is $[3]^5 = [5]$, since 2, 3, 4, and 6 are not relatively prime to 6. ∎

EXERCISES 3.3

1. List all cyclic subgroups of the group $S(A)$ in Example 3 of Section 3.1.

Sec. 3.1, #28 ≫   **2.** Let $G = \{ \pm 1, \pm i, \pm j, \pm k \}$ be the quaternion group given in Exercise 28 of Section 3.1. List all cyclic subgroups of G.

3. Find the order of each element of the group $S(A)$ in Example 3 of Section 3.1.

4. Find the order of each element of the group G in Exercise 2.

Sec. 3.1, #29 ≫   **5.** The elements of the multiplicative group G of 3×3 permutation matrices are given in Exercise 29 of Section 3.1. Find the order of each element of the group.

6. In the multiplicative group of invertible matrices in $M_4(\mathbf{R})$, find the order of the given element A.

a. $A = \begin{bmatrix} 0 & 0 & 0 & 1 \\ 0 & 1 & 0 & 0 \\ 0 & 0 & 1 & 0 \\ 1 & 0 & 0 & 0 \end{bmatrix}$ **b.** $A = \begin{bmatrix} 0 & 0 & 1 & 0 \\ 0 & 0 & 0 & 1 \\ 0 & 1 & 0 & 0 \\ 1 & 0 & 0 & 0 \end{bmatrix}$

7. For each of the following values of n, find all generators of the cyclic group $\mathbf{Z}_n$ under addition.

  a. $n = 8$ **b.** $n = 12$
  c. $n = 10$ **d.** $n = 15$
  e. $n = 16$ **f.** $n = 18$

8. For each of the following values of n, find all subgroups of the cyclic group $\mathbf{Z}_n$ under addition.

  a. $n = 12$ **b.** $n = 8$
  c. $n = 10$ **d.** $n = 15$
  e. $n = 16$ **f.** $n = 18$

Sec. 3.1, #27 ≫   **9.** According to Exercise 27 of Section 3.1, the nonzero elements of $\mathbf{Z}_n$ form a group G with respect to multiplication if n is a prime. For each of the following values of n, show that this group G is cyclic.

  a. $n = 7$ **b.** $n = 5$
  c. $n = 11$ **d.** $n = 13$
  e. $n = 17$ **f.** $n = 19$

10. For each of the following values of n, find all generators of the group G described in Exercise 9.

  a. $n = 7$ **b.** $n = 5$
  c. $n = 11$ **d.** $n = 13$
  e. $n = 17$ **f.** $n = 19$

11. For each of the following values of n, find all subgroups of the group G described in Exercise 9.

  a. $n = 7$ **b.** $n = 5$

c. $n = 11$ **d.** $n = 13$

e. $n = 17$ **f.** $n = 19$

12. Prove that the set

$$H = \left\{ \begin{bmatrix} 1 & n \\ 0 & 1 \end{bmatrix} \middle| n \in \mathbf{Z} \right\}$$

Sec. 3.4, #7 ≪   is a cyclic subgroup of the group of all invertible matrices in $M_2(\mathbf{R})$.

13. **a.** Use trigonometric identities and mathematical induction to prove that

$$\begin{bmatrix} \cos\theta & -\sin\theta \\ \sin\theta & \cos\theta \end{bmatrix}^n = \begin{bmatrix} \cos n\theta & -\sin n\theta \\ \sin n\theta & \cos n\theta \end{bmatrix}$$

for all integers n (positive, zero, or negative). Hence, conclude that for a constant θ, the set

$$H = \left\{ \begin{bmatrix} \cos n\theta & -\sin n\theta \\ \sin n\theta & \cos n\theta \end{bmatrix} \middle| n \in \mathbf{Z} \right\}$$

is a cyclic subgroup of the group of all invertible matrices in $M_2(\mathbf{R})$.

b. Evaluate each element of H for $\theta = 90°$.

c. Evaluate each element of H for $\theta = 120°$.

14. For an integer $n > 1$, let G be the set of all $[a]$ in $\mathbf{Z}_n$ that have multiplicative inverses. Prove that G is a group with respect to multiplication.

15. Let G be as described in Exercise 14. Prove that $[a] \in G$ if and only if a and n are relatively prime.

16. Let G be as described in Exercise 14. For each value of n, write out the elements of G and construct a multiplication table for G.

a. $n = 20$ **b.** $n = 8$

Sec. 4.5, #8 ≪   **c.** $n = 24$ **d.** $n = 30$

17. Which of the groups in Exercise 16 are cyclic?

18. Consider the set G of all $[a]$ in $\mathbf{Z}_9$ that have multiplicative inverses. Given that G is a cyclic group under multiplication, find all subgroups of G.

19. Suppose $G = \langle a \rangle$ is a cyclic group of order 24. List all generators of G.

20. List all distinct subgroups of the group in Exercise 19.

21. Describe all subgroups of the group $\mathbf{Z}$ under addition.

22. Find all generators of an infinite cyclic group $G = \langle a \rangle$.

23. Let a and b be elements of the group G. Prove that if $a \in \langle b \rangle$, then $\langle a \rangle \subseteq \langle b \rangle$.

24. Let a and b be elements of a finite group G.

a. Prove that a and bab^{-1} have the same order.

b. Prove that ab and ba have the same order.

25. In Exercise 17 of Section 3.2, the center $Z(G)$ is defined as

$$Z(G) = \{a \in G \mid ax = xa \text{ for every } x \in G\}.$$

Prove that if b is the only element of order 2 in G, then $b \in Z(G)$.

26. If a is an element of order m in a group G and $a^k = e$, prove that m divides k.

27. If G is a cyclic group, prove that the equation $x^2 = e$ has at most two distinct solutions in G.

28. Let G be a finite cyclic group of order n. If d is a positive divisor of n, prove that the equation $x^d = e$ has exactly d distinct solutions in G.

29. If G is a cyclic group of order p and p is a prime, how many elements in G are generators of G?

30. Suppose that a and b are elements of finite order in a group such that $ab = ba$ and $\langle a \rangle \cap \langle b \rangle = \{e\}$. Prove that $o(ab)$ is the least common multiple of $o(a)$ and $o(b)$.

31. Suppose a is an element of order m in a group G, and let k be an integer. If $d = (k, m)$, prove that a^k has order m/d.

32. Assume that $G = \langle a \rangle$ is a cyclic group of order n. Prove that if r divides n, then G has a subgroup of order r.

Sec. 4.1, #11 ≪
33. Suppose a is an element of order mn in a group G, where m and n are relatively prime. Prove that a is the product of an element of order m and an element of order n.

34. Let G be an abelian group. Prove that the set of all elements of finite order in G forms a subgroup of G. This subgroup is called the **torsion subgroup** of G.

35. The **Euler phi-function** is defined for positive integers n as follows: $\phi(n)$ is the number of positive integers m such that $1 \leq m \leq n$ and m is relatively prime to n. Use Corollary 3.23 and the additive groups $\mathbf{Z}_d$ to show that

$$n = \sum_{d \mid n} \phi(d)$$

where the sum has one term for each positive divisor d of n.

3.4 ISOMORPHISMS

It turns out that the permutation groups can serve as models for all groups. For this reason, we examine permutation groups in great detail in the next chapter. In order to describe their relation to groups in general, we need the concept of an *isomorphism*. Before formally introducing this concept, however, we consider some examples.

Example 1 Consider a cyclic group of order 4. If G is a cyclic group of order 4, it must contain an identity element e and a generator $a \neq e$ in G. The proof of Theorem 3.17 shows that

$$G = \{e, a, a^2, a^3\}$$

where $a^4 = e$. A multiplication table for G would have the form shown in Figure 3-14.

Figure 3-14

$\cdot$	e	a	a^2	a^3
e	e	a	a^2	a^3
a	a	a^2	a^3	e
a^2	a^2	a^3	e	a
a^3	a^3	e	a	a^2

In a very definite way, then, the structure of G is determined. The details as to what the element a might be and what the operation in G might be may vary, but the basic structure of G fits the pattern in the table. ■

Example 2 Let us consider a group related to geometry. We begin with an equilateral triangle T with center point O and vertices labeled V_1, V_2, and V_3 (see Figure 3-15).

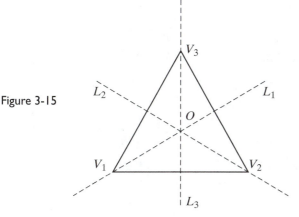

Figure 3-15

The equilateral triangle, of course, consists of the set of all points on the three sides of the triangle. By a **rigid motion** of the triangle, we mean a bijection of the set of points of the triangle onto itself that leaves the distance between any two points unchanged. In other words, a rigid motion of the triangle is a bijection that preserves distances. Such a rigid motion must map a vertex onto a vertex, and the entire mapping is determined by the images of the vertices V_1, V_2, and V_3. These rigid motions (or **symmetries** as they are often called) form a group with respect to mapping composition. (Verify this.) There are a total of six elements in the group, and they may be described as follows:

1. e, the identity mapping, that leaves all points unchanged;
2. r, a counterclockwise rotation through $120°$ about O in the plane of the triangle;
3. $r^2 = r \circ r$, a counterclockwise rotation through $240°$ about O in the plane of the triangle;
4. a reflection f about the line L_1 through V_1 and O;
5. a reflection g about the line L_2 through V_2 and O;
6. a reflection h about the line L_3 through V_3 and O.

These rigid motions can be described by indicating their values at the vertices as follows:

$$e: \begin{cases} e(V_1) = V_1 \\ e(V_2) = V_2 \\ e(V_3) = V_3 \end{cases} \qquad h: \begin{cases} h(V_1) = V_2 \\ h(V_2) = V_1 \\ h(V_3) = V_3 \end{cases}$$

$$r: \begin{cases} r(V_1) = V_2 \\ r(V_2) = V_3 \\ r(V_3) = V_1 \end{cases} \qquad g: \begin{cases} g(V_1) = V_3 \\ g(V_2) = V_2 \\ g(V_3) = V_1 \end{cases}$$

$$r^2: \begin{cases} r^2(V_1) = V_3 \\ r^2(V_2) = V_1 \\ r^2(V_3) = V_2 \end{cases} \qquad f: \begin{cases} f(V_1) = V_1 \\ f(V_2) = V_3 \\ f(V_3) = V_2 \end{cases}$$

We have a group

$$G = \{e, r, r^2, h, g, f\},$$

and G has the multiplication table shown in Figure 3-16.

Figure 3-16

$\circ$	e	r	r^2	h	g	f
e	e	r	r^2	h	g	f
r	r	r^2	e	g	f	h
r^2	r^2	e	r	f	h	g
h	h	f	g	e	r^2	r
g	g	h	f	r	e	r^2
f	f	g	h	r^2	r	e

We shall compare this group G with the group $S(A)$ from Example 3 of Section 3.1, and we shall see that they are the same except for notation. Let the elements of G correspond to those of $S(A)$ according to the mapping $\phi: G \to S(A)$ given by

$$\phi(e) = I_A \qquad \phi(h) = \sigma$$
$$\phi(r) = \rho \qquad \phi(g) = \gamma$$
$$\phi(r^2) = \rho^2 \qquad \phi(f) = \delta.$$

This mapping is a one-to-one correspondence from G to $S(A)$. Moreover, ϕ has the property that

$$\phi(xy) = \phi(x) \cdot \phi(y)$$

for all x and y in G. This statement can be verified by using the multiplication tables for G and $S(A)$ in the following manner: In the entire multiplication table for G, we replace each element $x \in G$ by its image $\phi(x)$ in $S(A)$. This yields the table in Figure 3-17 that has $\phi(xy)$ in the row with $\phi(x)$ at the left and in the column with $\phi(y)$ at the top.

Figure 3-17

	I_A	ρ	ρ^2	σ	γ	δ
I_A	I_A	ρ	ρ^2	σ	γ	δ
ρ	ρ	ρ^2	I_A	γ	δ	σ
ρ^2	ρ^2	I_A	ρ	δ	σ	γ
σ	σ	δ	γ	I_A	ρ^2	ρ
γ	γ	σ	δ	ρ	I_A	ρ^2
δ	δ	γ	σ	ρ^2	ρ	I_A

The multiplication table for $S(A)$ given in Example 3 of Section 3.1 furnishes a table of values for $\phi(x) \cdot \phi(y)$, and the two tables agree in every position.[†] This means that $\phi(xy) = \phi(x) \cdot \phi(y)$ for all x and y in G. Thus, G and $S(A)$ are the same except for notation.   ∎

A mapping such as ϕ in the preceding example is called an *isomorphism*.

[†]Note that the e in Example 3 of Section 3.1 stands for I_A.

Definition 3.25   ISOMORPHISM, AUTOMORPHISM

> Let G be a group with respect to $\circledast$, and let G' be a group with respect to $\boxdot$. A mapping $\phi\colon G \to G'$ is an **isomorphism** from G to G' if
>
> **1.** ϕ is a one-to-one correspondence from G to G', and
> **2.** $\phi(x \circledast y) = \phi(x) \boxdot \phi(y)$ for all x and y in G.
>
> If an isomorphism from G to G' exists, we say that G is **isomorphic** to G', and we use the notation $G \cong G'$ as shorthand for this phrase. An isomorphism from a group G to G itself is called an **automorphism** of G.

The use of $\circledast$ and $\boxdot$ in Definition 3.25 is intended to emphasize the fact that the group operations may be different. Now that this point has been made, we revert to our convention of using the multiplicative notation for the group operation. An isomorphism is said to "preserve the operation," since condition 2 of Definition 3.25 requires that the result be the same if the group operation is performed before or after the mapping.

The notation $\cong$ in Definition 3.25 is not standardized. The notations $\simeq$, $\approxeq$, and $\approx$ are used for the same purpose in some other texts.

Because an isomorphism preserves the group operation between two groups, it is not surprising that the identity elements always correspond under an isomorphism and that inverses are always mapped onto inverses. These results are stated more precisely in the next theorem.

Theorem 3.26   IMAGES OF IDENTITIES AND INVERSES

Suppose ϕ is an isomorphism from the group G to the group G'. If e denotes the identity in G and e' denotes the identity in G', then

a. $\phi(e) = e'$, and
b. $\phi(x^{-1}) = [\phi(x)]^{-1}$ for all x in G.

$p \Rightarrow q$ **Proof**   We have

$$\begin{aligned}
e \cdot e = e \Rightarrow\quad & \phi(e \cdot e) = \phi(e) \\
\Rightarrow\quad & \phi(e) \cdot \phi(e) = \phi(e) && \text{since } \phi \text{ is an isomorphism} \\
\Rightarrow\quad & \phi(e) \cdot \phi(e) = \phi(e) \cdot e' && \text{since } e' \text{ is an identity} \\
\Rightarrow\quad & \phi(e) = e' && \text{by Theorem 3.4e.}
\end{aligned}$$

$(p \wedge q) \Rightarrow r$ For any x in G,

$$\begin{aligned}
x \cdot x^{-1} = e \Rightarrow\quad & \phi(x \cdot x^{-1}) = \phi(e) \\
\Rightarrow\quad & \phi(x \cdot x^{-1}) = e' && \text{by part a} \\
\Rightarrow\quad & \phi(x) \cdot \phi(x^{-1}) = e'.
\end{aligned}$$

Similarly, $x^{-1} \cdot x = e$ implies $\phi(x^{-1}) \cdot \phi(x) = e'$, and therefore $\phi(x^{-1}) = [\phi(x)]^{-1}$.

The concept of isomorphism introduces the relation of being isomorphic on a collection $\mathcal{G}$ of groups. This relation is an equivalence relation, as the following statements show.

1. Any group G in the collection $\mathcal{G}$ is isomorphic to itself. The identity mapping I_G is an automorphism of G.

2. If G and G' are in $\mathcal{G}$ and G is isomorphic to G', then G' is isomorphic to G. In fact, if ϕ is an isomorphism from G to G', then ϕ^{-1} is an isomorphism from G' to G. (See Exercise 1 at the end of this section.)

3. Suppose G_1, G_2, G_3 are in $\mathcal{G}$. If G_1 is isomorphic to G_2 and G_2 is isomorphic to G_3, then G_1 is isomorphic to G_3. It is left as an exercise to show that if ϕ_1 is an isomorphism from G_1 to G_2 and ϕ_2 is an isomorphism from G_2 to G_3, then $\phi_2\phi_1$ is an isomorphism from G_1 to G_3.

The fundamental idea behind isomorphisms is this: Groups that are isomorphic have the same structure relative to their respective group operation. They are algebraically the same, although details such as the appearance of the elements or the rule defining the operation may vary.

From our discussion at the beginning of this section, we see that any two cyclic groups of order 4 are isomorphic. In fact, any two cyclic groups of the same order are isomorphic (see Exercises 21 and 22 at the end of this section).

The next two examples emphasize the fact that the elements of two isomorphic groups and their group operations may be quite different from each other.

Example 3 Consider $G = \{1,\ i,\ -1,\ -i\}$ under multiplication and $G' = \mathbf{Z}_4 = \{[0], [1], [2], [3]\}$ under addition. Let $\phi\colon G \to G'$ be defined by

$$\phi(1) = [0], \qquad \phi(i) = [1], \qquad \phi(-1) = [2], \qquad \phi(-i) = [3].$$

This defines a one-to-one correspondence ϕ from G to G'. To see that ϕ is an isomorphism from G to G', we use the group tables for G and G' in the same way as in Example 2 of this section. Beginning with the multiplication table for G, we replace each x in the table with $\phi(x)$ (see Figures 3-18 and 3-19). Since the resulting table (Figure 3-19) agrees completely with the addition table for $\mathbf{Z}_4$, we conclude that

$$\phi(xy) = \phi(x) + \phi(y)$$

for all $x \in G$, $y \in G$, and therefore that ϕ is an isomorphism from G to G'.

Multiplication Table for G

·	1	i	-1	$-i$
1	1	i	-1	$-i$
i	i	-1	$-i$	1
-1	-1	$-i$	1	i
$-i$	$-i$	1	i	-1

Figure 3-18

$\xrightarrow{\phi}$

Table of $\phi(xy)$

	[0]	[1]	[2]	[3]
[0]	[0]	[1]	[2]	[3]
[1]	[1]	[2]	[3]	[0]
[2]	[2]	[3]	[0]	[1]
[3]	[3]	[0]	[1]	[2]

Figure 3-19

We conclude this section with an example involving matrices.

Example 4 The multiplicative group G of 3×3 permutation matrices was introduced in Exercise 29 of Section 3.1. This group G is given by $G = \{I_3, P_1, P_2, P_3, P_4, P_5\}$ where

$$P_1 = \begin{bmatrix} 1 & 0 & 0 \\ 0 & 0 & 1 \\ 0 & 1 & 0 \end{bmatrix}, \qquad P_2 = \begin{bmatrix} 0 & 1 & 0 \\ 1 & 0 & 0 \\ 0 & 0 & 1 \end{bmatrix}, \qquad P_3 = \begin{bmatrix} 0 & 1 & 0 \\ 0 & 0 & 1 \\ 1 & 0 & 0 \end{bmatrix},$$

$$P_4 = \begin{bmatrix} 0 & 0 & 1 \\ 0 & 1 & 0 \\ 1 & 0 & 0 \end{bmatrix}, \qquad P_5 = \begin{bmatrix} 0 & 0 & 1 \\ 1 & 0 & 0 \\ 0 & 1 & 0 \end{bmatrix}.$$

We shall show that this group is isomorphic to the group $S(A) = \{I_A, \rho, \rho^2, \sigma, \gamma, \delta\}$ that appears in Example 2 of this section.

A multiplication table for G is needed as a guide in defining an isomorphism from G to $S(A)$. In constructing this table, we find that

$$P_3^2 = P_5, \qquad P_3^3 = I_3, \qquad P_3 P_1 = P_4, \qquad \text{and} \qquad P_3 P_4 = P_2.$$

Using the group table for $S(A)$ in Figure 3-17 as a pattern, we list the elements of G across the table in the order

$$I_3, P_3, P_3^2, P_1, P_4, P_2$$

and evaluate all the products as shown in Figure 3-20. A comparison of the group tables for G and $S(A)$ suggests that the one-to-one correspondence $\phi: G \to S(A)$ given by

$$\phi(I_3) = I_A \quad \phi(P_3) = \rho \quad \phi(P_3^2) = \rho^2$$
$$\phi(P_1) = \sigma \quad \phi(P_4) = \gamma \quad \phi(P_2) = \delta$$

might be an isomorphism. To verify the property $\phi(xy) = \phi(x)\phi(y)$, we replace each x in the table for G with its image $\phi(x)$ in $S(A)$. The resulting table is shown in Figure 3-21, and it agrees in every position with the group table for $S(A)$ in Figure 3-17. Thus, ϕ is an isomorphism from G to $S(A)$.

Multiplication Table for G

$\cdot$	I_3	P_3	P_3^2	P_1	P_4	P_2
I_3	I_3	P_3	P_3^2	P_1	P_4	P_2
P_3	P_3	P_3^2	I_3	P_4	P_2	P_1
P_3^2	P_3^2	I_3	P_3	P_2	P_1	P_4
P_1	P_1	P_2	P_4	I_3	P_3^2	P_3
P_4	P_4	P_1	P_2	P_3	I_3	P_3^2
P_2	P_2	P_4	P_1	P_3^2	P_3	I_3

$\xrightarrow{\phi}$

Table of $\phi(xy)$

	I_A	ρ	ρ^2	σ	γ	δ
I_A	I_A	ρ	ρ^2	σ	γ	δ
ρ	ρ	ρ^2	I_A	γ	δ	σ
ρ^2	ρ^2	I_A	ρ	δ	σ	γ
σ	σ	δ	γ	I_A	ρ^2	ρ
γ	γ	σ	δ	ρ	I_A	ρ^2
δ	δ	γ	σ	ρ^2	ρ	I_A

Figure 3-20

Figure 3-21

EXERCISES 3.4

1. Prove that if ϕ is an isomorphism from the group G to the group G', then ϕ^{-1} is an isomorphism from G' to G.

2. Let G_1, G_2, and G_3 be groups. Prove that if ϕ_1 is an isomorphism from G_1 to G_2 and ϕ_2 is an isomorphism from G_2 to G_3, then $\phi_2\phi_1$ is an ismorphism from G_1 to G_3.

3. Find an isomorphism from the additive group† $\mathbf{Z}_4 = \{[0]_4, [1]_4, [2]_4, [3]_4\}$ to the multiplicative group $G = \{[1]_5, [2]_5, [3]_5, [4]_5\} \subseteq \mathbf{Z}_5$.

4. Let $G = \{1, i, -1, -i\}$ under multiplication, and let $G' = \mathbf{Z}_4 = \{[0], [1], [2], [3]\}$ under addition. Find an isomorphism from G to G' that is different from the one given in Example 3 of this section.

Sec. 3.2, #16 ≫   5. Let H be the group given in Exercise 16 of Section 3.2, and let $S(A)$ be as given in Example 4 of this section. Find an isomorphism from H to $S(A)$.

6. Find an isomorphism from the additive group $\mathbf{Z}_6 = \{[a]_6\}$ to the multiplicative group $G = \{[a]_7 \in \mathbf{Z}_7 | [a]_7 \neq [0]_7\}$.

Sec. 3.3, #12 ≫   7. (See Exercise 12 of Section 3.3.) Find an isomorphism ϕ from the additive group $\mathbf{Z}$ to the multiplicative group

$$H = \left\{ \begin{bmatrix} 1 & n \\ 0 & 1 \end{bmatrix} \middle| n \in \mathbf{Z} \right\}$$

and prove that $\phi(x + y) = \phi(x)\phi(y)$.

Sec. 3.2, #15b ≫   8. (See Exercise 15b of Section 3.2.) Find an isomorphism from the group $G = \{1, i, -1, -i\}$ in Example 3 of this section to the multiplicative group

$$H = \left\{ \begin{bmatrix} 1 & 0 \\ 0 & 1 \end{bmatrix}, \begin{bmatrix} i & 0 \\ 0 & -i \end{bmatrix}, \begin{bmatrix} -i & 0 \\ 0 & i \end{bmatrix}, \begin{bmatrix} -1 & 0 \\ 0 & -1 \end{bmatrix} \right\}.$$

Sec. 3.2, #14c ≫   9. (See Exercise 14c of Section 3.2) Find an isomorphism ϕ from the multiplicative group G of nonzero complex numbers to the multiplicative group

$$H = \left\{ \begin{bmatrix} a & -b \\ b & a \end{bmatrix} \middle| a, b \in \mathbf{R} \text{ and } a^2 + b^2 \neq 0 \right\}$$

and prove that $\phi(xy) = \phi(x)\phi(y)$.

10. Let G be the additive group of all real numbers, and let G' be the group of all positive real numbers under multiplication. Verify that the mapping $\phi: G \to G'$ defined by $\phi(x) = 10^x$ is an isomorphism from G to G'.

11. Let G and G' be as given in Exercise 10. Verify that the mapping $\theta: G' \to G$ defined by $\theta(x) = \log x$ is an isomorphism from G' to G.

12. Assume that the nonzero complex numbers form a group G with respect to multiplication. If a and b are real numbers and $i = \sqrt{-1}$, the **conjugate** of the complex number $a + bi$ is defined to be $a - bi$. With this notation, let $\phi: G \to G$ be defined by $\phi(a + bi) = a - bi$ for all $a + bi$ in G. Prove that ϕ is an automorphism of G.

13. Let G be an arbitrary group. Prove or disprove that the mapping $\phi(a) = a^{-1}$ is an isomorphism from G to G.

---

†For clarity, we are temporarily writing $[a]_n$ for $[a] \in \mathbf{Z}_n$.

Sec. 3.1, #27a ≫

Sec. 4.5, #22 ≪

14. Suppose $(m, n) = 1$ and let $\phi: \mathbf{Z}_n \to \mathbf{Z}_n$ be defined by $\phi([a]) = m[a]$. Prove or disprove that ϕ is an automorphism of the additive group $\mathbf{Z}_n$.

15. According to Exercise 27a of Section 3.1, the set of nonzero elements of $\mathbf{Z}_n$ forms a group G with respect to multiplication if n is prime. Prove or disprove that the mapping $\phi: G \to G$ defined by the rule in Exercise 14 is an automorphism of G.

16. For each a in the group G, define a mapping $t_a: G \to G$ by $t_a(x) = axa^{-1}$. Prove that t_a is an automorphism of G.

17. Assume G is a (not necessarily finite) cyclic group generated by a in G, and let ϕ be an automorphism of G. Prove that each element of G is equal to a power of $\phi(a)$; that is, $\phi(a)$ is a generator of G.

18. Let G be as in Exercise 17. Suppose also that a^r is a generator of G. Define f on G by $f(a) = a^r$, $f(a^i) = (a^r)^i = a^{ri}$. Prove that f is an automorphism of G.

19. Let G be the group of all nonzero elements of $\mathbf{Z}_7$ under multiplication. Use the results of Exercises 17 and 18 to list all the automorphisms of G. For each automorphism ϕ, write out the images $\phi(x)$ for all x in G.

20. Use the results of Exercises 17 and 18 to find the *number* of automorphisms of the additive group $\mathbf{Z}_n$ for the given value of n.
   a. $n = 3$ **b.** $n = 4$
   c. $n = 8$ **d.** $n = 6$

21. For an arbitrary positive integer n, prove that any two cyclic groups of order n are isomorphic.

22. Prove that any infinite cyclic group is isomorphic to $\mathbf{Z}$ under addition.

23. Prove that any cyclic group of finite order n is isomorphic to $\mathbf{Z}_n$ under addition.

24. Let G be the group that consists of the nonzero elements of $\mathbf{Z}_7$ under multiplication, and let H be the group $\mathbf{Z}_6$ under addition. Find all isomorphisms from G to H.

25. Suppose that G and G' are isomorphic groups. Prove that if G is abelian, then G' is abelian.

26. Prove that if G and G' are two groups that contain exactly two elements each, then G and G' are isomorphic.

27. Prove that any two groups of order 3 are isomorphic.

28. If G and G' are groups and $\phi: G \to G'$ is an isomorphism, prove that a and $\phi(a)$ have the same order, for any $a \in G$.

29. Suppose that ϕ is an isomorphism from the group G to the group G'.
   a. Prove that if H is any subgroup of G, then $\phi(H)$ is a subgroup of G'.
   b. Prove that if K is any subgroup of G', then $\phi^{-1}(K)$ is a subgroup of G.

3.5   HOMOMORPHISMS

We saw in the last section that an isomorphism between two groups provides a connection which shows that the two groups have the same structure relative to their group operations. It is for this reason that the concept of an isomorphism is extremely important in algebra.

The name *homomorphism* is given to another important type of mapping that is related to, but different from, the isomorphism. The basic differences are that a homomorphism is not required to be one-to-one and also not required to be onto. The formal definition is as follows.

Definition 3.27 **HOMOMORPHISM, EPIMORPHISM**

> Let G be a group with respect to $\circledast$, and let G' be a group with respect to $\boxdot$. A **homomorphism** from G to G' is a mapping $\phi: G \to G'$ such that
>
> $$\phi(x \circledast y) = \phi(x) \boxdot \phi(y)$$
>
> for all x and y in G. If ϕ is a homomorphism from G to G' that is onto, ϕ is called an **epimorphism**.

As we did with isomorphisms, we drop the special symbols $\circledast$ and $\boxdot$ and simply write $\phi(xy) = \phi(x)\phi(y)$ for the given condition.

As already noted, a homomorphism ϕ from G to G' need not be one-to-one or onto. If ϕ is both (that is, if ϕ is a bijection), then ϕ is an isomorphism as defined in Definition 3.25.

Our first example of a homomorphism has a natural connection with our work in Chapter 2.

Example 1 For a fixed integer $n > 1$, consider the mapping ϕ from the additive group $\mathbf{Z}$ to the additive group $\mathbf{Z}_n$ defined by

$$\phi(x) = [x],$$

where $[x]$ is the congruence class in $\mathbf{Z}_n$ that contains x. Using the properties of addition in $\mathbf{Z}_n$ from Section 2.6, it follows that

$$\begin{aligned}
\phi(x + y) &= [x + y] \\
&= [x] + [y] \\
&= \phi(x) + \phi(y).
\end{aligned}$$

Thus, ϕ is a homomorphism. It follows from the definition of $\mathbf{Z}_n$ that ϕ is onto, so ϕ is in fact an epimorphism from $\mathbf{Z}$ to $\mathbf{Z}_n$. ∎

Example 2 For two arbitrary groups G and G', let e' denote the identity element in G' and define $\phi: G \to G'$ by $\phi(x) = e'$ for all $x \in G$. Then

$$\begin{aligned}
\phi(x) \cdot \phi(y) &= e' \cdot e' \\
&= e' \\
&= \phi(xy),
\end{aligned}$$

and ϕ is a homomorphism from G to G'. ∎

The two previous examples show that, unlike the situation with isomorphisms, the existence of a homomorphism from G to G' does not imply that G and G' have the same structure. However, we shall see that the existence of a homomorphism can reveal important and interesting information relating their structures. As with isomorphisms, we say that a homomorphism "preserves the group operation." Two simple consequences of this condition are that identities must correspond and inverses must be mapped onto inverses. This is stated in our next theorem, with the proofs requested in the exercises.

Theorem 3.28   IMAGES OF IDENTITIES AND INVERSES

Let ϕ be a homomorphism from the group G to the group G'. If e denotes the identity in G and e' denotes the identity in G', then

a. $\phi(e) = e'$, and
b. $\phi(x^{-1}) = [\phi(x)]^{-1}$ for all x in G.

The following examples give some indication of the variety that is in homomorphisms. Other examples appear in the exercises for this section.

Example 3   Consider the group $\mathbf{R}$ of nonzero real numbers under multiplication and the additive group $\mathbf{Z}$. Define $\phi: \mathbf{Z} \to \mathbf{R}$ by

$$\phi(n) = \begin{cases} 1 & \text{if } n \text{ is even} \\ -1 & \text{if } n \text{ is odd.} \end{cases}$$

Since every integer is either even or odd and not both, $\phi(n)$ is well-defined. The following table systematically checks the equality $\phi(m + n) = \phi(m) \cdot \phi(n)$.

	$m + n$	$\phi(m) \cdot \phi(n)$	$\phi(m + n)$
m, n both even	even	$(1)(1)$	1
one even, one odd	odd	$(1)(-1)$	-1
m, n both odd	even	$(-1)(-1)$	1

A comparison of the last two columns shows that ϕ is indeed a homomorphism from $\mathbf{Z}$ to $\mathbf{R}$. ∎

Example 4   Consider the additive group $\mathbf{Z}$ and the mapping $\phi: \mathbf{Z} \to \mathbf{Z}$ defined by $\phi(x) = 5x$ for all $x \in \mathbf{Z}$. Since

$$\begin{aligned} \phi(x + y) &= 5(x + y) \\ &= 5x + 5y \\ &= \phi(x) + \phi(y), \end{aligned}$$

ϕ is a homomorphism. ∎

We saw in the last section that the relation of being isomorphic is an equivalence relation on a given collection $\mathcal{G}$ of groups. The concept of homomorphism leads to a corresponding, but different, relation. If there exists an epimorphism from the group G to the group G', then G' is called a **homomorphic image** of G. Example 1 in this section shows that the additive group $\mathbf{Z}_n$ is a homomorphic image of the additive group $\mathbf{Z}$.

On a given collection $\mathcal{G}$ of groups, the relation of being a homomorphic image is reflexive and transitive, but may not be symmetric. These facts are brought out in the exercises for this section.

The real importance of homomorphisms will be much clearer at the end of Section 4.5 in the next chapter. The kernel of a homomorphism is one of the key concepts in that section.

Definition 3.29   **KERNEL**

> Let ϕ be a homomorphism from the group G to the group G'. The **kernel** of ϕ is the set
>
> $$\ker \phi = \{x \in G \mid \phi(x) = e'\}$$
>
> where e' denotes the identity in G'.

Example 5   To illustrate Definition 3.29, we list the kernels of the homomorphisms from the preceding examples in this section.

The kernel of the homomorphism $\phi: \mathbf{Z} \to \mathbf{Z}_n$ defined by $\phi(x) = [x]$ in Example 1 is given by

$$\ker \phi = \{x \in \mathbf{Z} \mid x = kn \text{ for some } k \in \mathbf{Z}\},$$

since $\phi(x) = [x] = [0]$ if and only if x is a multiple of n.

The homomorphism $\phi: \mathbf{Z} \to \mathbf{R}$ in Example 3 defined by

$$\phi(n) = \begin{cases} 1 & \text{if } n \text{ is even} \\ -1 & \text{if } n \text{ is odd} \end{cases}$$

has the set $\mathbf{E}$ of all even integers as its kernel, since 1 is the identity in $\mathbf{R}$.

For $\phi: \mathbf{Z} \to \mathbf{Z}$ defined by $\phi(x) = 5x$ in Example 4, we have $\ker \phi = \{0\}$, since $5x = 0$ if and only if $x = 0$. This kernel is an extreme case since part a of Theorem 3.28 assures us that the identity is always an element of the kernel.

At the other extreme, the homomorphism $\phi: G \to G'$ defined in Example 2 by $\phi(x) = e'$ for all $x \in G$ has $\ker \phi = G$.   ■

EXERCISES 3.5

1. Each of the following rules determines a mapping $\phi: G \to G$, where G is the group of all nonzero real numbers under multiplication. Decide in each case whether or not ϕ is a homomorphism, and state the kernel for those that are homomorphisms.
   a. $\phi(x) = |x|$ **b.** $\phi(x) = 1/x$
   c. $\phi(x) = -x$ **d.** $\phi(x) = x^2$

2. Consider the additive group $\mathbf{Z}_{12}$ and define $\phi: \mathbf{Z}_{12} \to \mathbf{Z}_{12}$ by $\phi([x]) = [3x]$. Prove that ϕ is a homomorphism and find $\ker \phi$.

Sec. 1.5, #23, 24 ≫
3. (See Exercises 23 and 24 of Section 1.5.) Let G be the multiplicative group of invertible matrices in $M_2(\mathbf{R})$, and let G' be the group of nonzero real numbers under multiplication. Prove that the mapping $\phi: G \to G'$ defined by

$$\phi\left(\begin{bmatrix} a & b \\ c & d \end{bmatrix}\right) = ad - bc$$

Sec. 4.5, #12 ≪
   is a homomorphism. (The value of this mapping is called the **determinant** of the matrix.)

4. Find an example of G, G', and ϕ such that G is a nonabelian group, G' is an abelian group, and ϕ is an epimorphism from G to G'.

5. Let ϕ be a homomorphism from the group G to the group G'.
   a. Prove part a of Theorem 3.28: If e denotes the identity in G and e' denotes the identity in G', then $\phi(e) = e'$.
   b. Prove part b of Theorem 3.28: $\phi(x^{-1}) = [\phi(x)]^{-1}$ for all x in G.

6. Prove that, on a given collection $\mathcal{G}$ of groups, the relation of being a homomorphic image has the reflexive property.

7. Suppose that G, G', and G'' are groups. If G' is a homomorphic image of G and G'' is a homomorphic image of G', prove that G'' is a homomorphic image of G. (Thus, the relation in Exercise 6 has the transitive property.)

8. Find two groups G and G' such that G' is a homomorphic image of G but G is not a homomorphic image of G'. (Thus, the relation in Exercise 6 does not have the symmetric property.)

9. Suppose that ϕ is an epimorphism from the group G to the group G'. Prove that ϕ is an isomorphism if and only if ker $\phi = \{e\}$, where e denotes to the identity in G.

10. If G is an abelian group and the group G' is a homomorphic image of G, prove that G' is abelian.

11. Let a be a fixed element of the multiplicative group G. Define ϕ from the additive group $\mathbf{Z}$ to G by $\phi(n) = a^n$ for all $n \in \mathbf{Z}$. Prove that ϕ is a homomorphism.

12. With ϕ as in Exercise 11, show that $\phi(\mathbf{Z}) = \langle a \rangle$ and describe the kernel of ϕ.

13. Assume that ϕ is a homomorphism from the group G to the group G'.
    a. Prove that if H is any subgroup of G, then $\phi(H)$ is a subgroup of G'.
Sec. 4.5, #18 ≪   **b.** Prove that if K is any subgroup of G', then $\phi^{-1}(K)$ is a subgroup of G.

14. Assume that the group G' is a homomorphic image of the group G.
    a. Prove that G' is cyclic if G is cyclic.
    b. Prove that $o(G')$ divides $o(G)$, whether G is cyclic or not.

15. For a fixed group G, prove that the set of all automorphisms of G forms a group with re-
Sec. 4.5, #22 ≪   spect to mapping composition.

KEY WORDS AND PHRASES

A PIONEER IN MATHEMATICS ||||||||||||||||||||||||||||||||||||||||||||||||||||||||||||||||||||||||||||

Niels Henrik Abel (1802–1829)

Niels Henrik Abel was a leading 19th-century Norwegian mathematician. Although he died at the age of 27, his accomplishments were so great that he is considered Norway's most noted mathematician. His memory is honored in many ways. A monument to him was erected at Froland Church, his burial place, by his friend Baltazar Mathias Keilhau. History tells us that on his deathbed, Abel jokingly asked his friend to care for his fiancée after his death, perhaps by marrying her. (After Abel died, Keilhau did marry Abel's fiancée.) A statue of Abel stands in the Royal Park of Oslo, and Norway has issued five postage stamps in his honor. Many theorems of advanced mathematics bear his name. Probably the most lasting and significant recognition is in the term *abelian group*, coined around 1870.

Abel was one of seven children of a pastor. When he was 18 his father died, and supporting the family became his responsibility. In spite of this burden, Abel continued his study of mathematics and successfully solved a problem that had baffled mathematicians for more than 300 years: He proved that the general fifth-degree polynomial equation could not be solved using the four basic arithmetic operations and extraction of roots.

Although Abel never held an academic position, he continued to pursue his mathematical research, contributing not only to the groundwork for what later became known as abstract algebra, but also to the theory of infinite series, elliptic functions, elliptic integrals, and abelian integrals.

In Berlin, Abel became friends with August Leopold Crelle (1780–1856), a civil engineer and founder of the first journal devoted entirely to mathematical research. It was only through Crelle's friendship and respect for Abel's talent that many of Abel's papers were published. In fact, Crelle finally obtained a faculty position for Abel at the University of Berlin, but unfortunately the news reached Norway two days after Abel's death.

MORE ON GROUPS

INTRODUCTION

The first two sections of this chapter present the standard material on permutation groups, and the optional Section 4.3 contains some real-world applications of such groups. The next two sections are devoted to normal subgroups and quotient groups. The chapter then concludes with two optional sections that present some results on finite abelian groups and give a sample of more advanced work.

The set $\mathbf{Z}_n$ of congruence classes modulo n makes isolated appearances in this chapter.

4.1 FINITE PERMUTATION GROUPS

An appreciation of the importance of permutation groups must be based to some extent on a knowledge of their structures. The basic facts about finite permutation groups are presented in this section, and their importance is revealed in the next two sections.

Suppose A is a finite set of n elements—say,

$$A = \{a_1, a_2, \ldots, a_n\}.$$

Any permutation f on A is determined by the choices for the n values

$$f(a_1), f(a_2), \ldots, f(a_n).$$

In assigning these values, there are n choices for $f(a_1)$, then $n - 1$ choices of $f(a_2)$, then $n - 2$ choices of $f(a_3)$, and so on. Thus, there are $n(n - 1) \ldots (2)(1) = n!$ different ways in which f can be defined, and $S(A)$ has $n!$ elements. Each element f in $S(A)$ can be represented by a matrix (rectangular array) in which the image of a_i is written under a_i:

$$f = \begin{bmatrix} a_1 & a_2 & \cdots & a_n \\ f(a_1) & f(a_2) & \cdots & f(a_n) \end{bmatrix}.$$

Each permutation f on A can be made to correspond to a permutation f' on $B = \{1, 2, \ldots, n\}$ by replacing a_k with k for $k = 1, 2, \ldots n$:

$$f' = \begin{bmatrix} 1 & 2 & \cdots & n \\ f'(1) & f'(2) & \cdots & f'(n) \end{bmatrix}.$$

The mapping $f \to f'$ is an isomorphism from $S(A)$ to $S(B)$, and the groups are the same except for notation. For this reason, we will henceforth consider a permutation on a set of n elements as being written on the set $B = \{1, 2, \ldots, n\}$. The group $S(B)$ is known as the **symmetric group** on n elements, and it is denoted by S_n.

Example 1   As an illustration of the matrix representation, the notation

$$f = \begin{bmatrix} 1 & 2 & 3 & 4 & 5 \\ 3 & 5 & 1 & 4 & 2 \end{bmatrix}$$

indicates that f is an element of S_5, and that $f(1) = 3$, $f(2) = 5$, $f(3) = 1$, $f(4) = 4$, and $f(5) = 2$.   ∎

Definition 4.1   **CYCLE**

An element f of S_n is a **cycle** if there exists a set $\{i_1, i_2, \ldots, i_r\}$ of distinct integers such that

$$f(i_1) = i_2, \quad f(i_2) = i_3, \quad \ldots, \quad f(i_{r-1}) = i_r, \quad f(i_r) = i_1,$$

and f leaves all other elements fixed.

By this definition, f is a cycle if there are distinct integers $i_1, i_2, \ldots, i_r$ such that f maps these elements according to the cyclic pattern

$$i_1 \to i_2 \to i_3 \to \cdots \to i_{r-1} \to i_r,$$

and f leaves all other elements fixed. A cycle such as this can be written in the form

$$f = (i_1, i_2, \ldots, i_r),$$

where it is understood that $f(i_k) = i_{k+1}$ for $1 \le k < r$, and $f(i_r) = i_1$.

Example 2   The permutation

$$f = \begin{bmatrix} 1 & 2 & 3 & 4 & 5 & 6 & 7 \\ 1 & 6 & 3 & 7 & 5 & 4 & 2 \end{bmatrix}$$

can be written simply as

$$f = (2, 6, 4, 7).$$

This expression is not unique, for

$$\begin{aligned} f &= (2, 6, 4, 7) \\ &= (6, 4, 7, 2) \\ &= (4, 7, 2, 6) \\ &= (7, 2, 6, 4). \end{aligned}$$

■

Example 3   It is easy to write the inverse of a cycle. Since $f(i_k) = i_{k+1}$ implies $f^{-1}(i_{k+1}) = i_k$, we only need to reverse the order of the cyclic pattern. For

$$f = (1, 2, 3, 4, 5, 6, 7, 8, 9),$$

we have

$$f^{-1} = (1, 9, 8, 7, 6, 5, 4, 3, 2).$$

■

Not all elements of S_n are cycles, but every permutation can be written as a product of mutually disjoint cycles. As an example, consider the permutation

$$f = \begin{bmatrix} 1 & 2 & 3 & 4 & 5 & 6 & 7 & 8 & 9 \\ 3 & 8 & 2 & 6 & 7 & 4 & 9 & 1 & 5 \end{bmatrix}.$$

Using the same representation scheme with $f(k)$ written beneath k, the result of a re-arrangement of the columns in the matrix still represents f:

$$f = \begin{bmatrix} 1 & 3 & 2 & 8 & 4 & 6 & 5 & 7 & 9 \\ 3 & 2 & 8 & 1 & 6 & 4 & 7 & 9 & 5 \end{bmatrix}.$$

The columns have been arranged in a special way: If $f(p) = q$, the column with q at the top has been written next after the column with p at the top. This arranges the elements in the first row so that f maps them according to the following pattern:

$$1 \to 3 \to 2 \to 8 \to 1$$
$$4 \to 6 \to 4$$
$$5 \to 7 \to 9 \to 5.$$

Thus, 1, 3, 2, and 8 are mapped in a circular pattern, and so are 4 and 6, or 5, 7, and 9. This procedure has led to a separation of the elements of $\{1, 2, 3, 4, 5, 6, 7, 8, 9\}$ into disjoint subsets $\{1, 3, 2, 8\}$, $\{4, 6\}$, and $\{5, 7, 9\}$ according to the pattern determined by the following computations[†]:

$$
\begin{array}{lll}
f(1) = 3 & f(4) = 6 & f(5) = 7 \\
f^2(1) = f(3) = 2 & f^2(4) = f(6) = 4 & f^2(5) = f(7) = 9 \\
f^3(1) = f(2) = 8 & & f^3(5) = f(9) = 5. \\
f^4(1) = f(8) = 1 & &
\end{array}
$$

The disjoint subsets $\{1, 3, 2, 8\}$, $\{4, 6\}$, and $\{5, 7, 9\}$ are called the **orbits** of f.

For each orbit of f, we define a cycle that maps the elements in that orbit in the same way as does f:

$$g_1 = (1, 3, 2, 8)$$
$$g_2 = (4, 6)$$
$$g_3 = (5, 7, 9).$$

These cycles are automatically on disjoint sets of elements since the orbits are disjoint, and we see that their product is f:

$$
\begin{aligned}
f &= g_1 g_2 g_3 \\
&= (1, 3, 2, 8)(4, 6)(5, 7, 9).
\end{aligned}
$$

Note that these cycles commute with each other because they are on disjoint sets of elements.

Example 4 The positive integral powers of a cycle f are easy to compute since f^m will map each integer in the cycle onto the integer located m places farther along in the cycle. For instance, if

$$f = (1, 2, 3, 4, 5, 6, 7, 8, 9),$$

then f^2 maps each element onto the element two places farther along, according to the pattern

$$\overbrace{1, 2, 3, 4, 5, 6, 7}, \ldots$$
$$f^2 = (1, 3, 5, 7, 9, 2, 4, 6, 8).$$

Similarly, f^3 maps each element onto the element three places farther along, and so on for higher powers:

$$f^3 = (1, 4, 7)(2, 5, 8)(3, 6, 9)$$
$$f^4 = (1, 5, 9, 4, 8, 3, 7, 2, 6),$$

and so on. ∎

[†]$f^2 = f \circ f, f^3 = f \circ f^2 = f \circ f \circ f$ and so on.

In connection with Example 4, we note that the order of an r-cycle (a cycle with r elements) is r.

Ordinarily, cycles that are not on disjoint sets of elements will not commute, but their product is defined using mapping composition. For example, suppose $f = (1, 3, 2, 4)$ and $g = (1, 7, 6, 2)$. Then [†]

$$fg = (1, 3, 2, 4)(1, 7, 6, 2) = (1, 7, 6, 4)(2, 3),$$

since

$$
\begin{array}{l}
\overbrace{\hspace{3cm}}^{fg} \\
1 \xrightarrow{g} 7 \xrightarrow{f} 7 \\
7 \xrightarrow{g} 6 \xrightarrow{f} 6 \\
6 \xrightarrow{g} 2 \xrightarrow{f} 4 \\
4 \xrightarrow{g} 4 \xrightarrow{f} 1 \\
2 \xrightarrow{g} 1 \xrightarrow{f} 3 \\
3 \xrightarrow{g} 3 \xrightarrow{f} 2 \\
5 \xrightarrow{g} 5 \xrightarrow{f} 5.
\end{array}
$$

The computation of fg may be easier to see in the following diagram:

$$
fg \left(
\begin{array}{ccccccc}
1 & 2 & 3 & 4 & 5 & 6 & 7 \\
7 & 1 & 3 & 4 & 5 & 2 & 6 \\
7 & 3 & 2 & 1 & 5 & 4 & 6
\end{array}
\right)
$$

$$fg = (1, 3, 2, 4)(1, 7, 6, 2) = (1, 7, 6, 4)(2, 3).$$

A similar diagram for gf appears as follows:

$$
gf \left(
\begin{array}{ccccccc}
1 & 2 & 3 & 4 & 5 & 6 & 7 \\
3 & 4 & 2 & 1 & 5 & 6 & 7 \\
3 & 4 & 1 & 7 & 5 & 2 & 6
\end{array}
\right)
$$

$$gf = (1, 7, 6, 2)(1, 3, 2, 4) = (1, 3)(2, 4, 7, 6).$$

Thus, $gf \neq fg$. We adopt the notation that a 1-cycle such as (5) indicates that the element is left fixed. For example, gf could also be written as

$$gf = (1, 3)(2, 4, 7, 6)(5).$$

This allows expressions such as $e = (1)$ or $e = (1)(2)$ for the identity permutation.

Example 5 A product of cycles with any number of factors can be expressed as a product of disjoint cycles by the same procedure that was used in computing fg with $f = (1, 3, 2, 4)$ and $g = (1, 7, 6, 2)$. To illustrate, suppose we wish to express

$$(1, 4, 3, 2)(1, 6, 2, 5)(1, 5, 3, 6, 2)$$

[†]The product fg is computed from right to left, according to $f(g(x))$. Some texts multiply permutations from left to right.

as a product of disjoint cycles. Let

$$f = (1, 4, 3, 2),$$
$$g = (1, 6, 2, 5),$$

and

$$h = (1, 5, 3, 6, 2).$$

The following computations can be done *mentally* to obtain fgh as a product of disjoint cycles:

$$\overset{\xrightarrow{\hspace{4cm}} fgh}{}$$

$$1 \xrightarrow{h} 5 \xrightarrow{g} 1 \xrightarrow{f} 4$$
$$4 \xrightarrow{h} 4 \xrightarrow{g} 4 \xrightarrow{f} 3$$
$$3 \xrightarrow{h} 6 \xrightarrow{g} 2 \xrightarrow{f} 1$$
$$2 \xrightarrow{h} 1 \xrightarrow{g} 6 \xrightarrow{f} 6$$
$$6 \xrightarrow{h} 2 \xrightarrow{g} 5 \xrightarrow{f} 5$$
$$5 \xrightarrow{h} 3 \xrightarrow{g} 3 \xrightarrow{f} 2.$$

Thus,

$$(1, 4, 3, 2)(1, 6, 2, 5)(1, 5, 3, 6, 2) = (1, 4, 3)(2, 6, 5). \qquad \blacksquare$$

When a permutation is written as a product of disjoint cycles, it is easy to find the order of the permutation if we use the result in Exercise 30 of Section 3.3: The order of the product is simply the least common multiple of the orders of the cycles. For example, the product $(1, 2, 3, 4)(5, 6, 7, 8, 9, 10)$ has order 12, the least common multiple of 4 and 6.

Example 6 The expression of permutations as products of cycles enables us to write the elements of S_n in a very compact form. The elements of S_3 are given by

$$\begin{array}{ll} e = (1) & \sigma = (1, 2) \\ \rho = (1, 2, 3) & \gamma = (1, 3) \\ \rho^2 = (1, 3, 2) & \delta = (2, 3). \end{array} \qquad \blacksquare$$

A 2-cycle such as $(3, 7)$ is called a **transposition**. Every permutation can be written as a product of transpositions, for every permutation can be written as a product of cycles, and any cycle $(i_1, i_2, \ldots, i_r)$ can be written as

$$(i_1, i_2, \ldots, i_r) = (i_1, i_r)(i_1, i_{r-1}) \cdots (i_1, i_3)(i_1, i_2).$$

For example,

$$(1, 3, 2, 4) = (1, 4)(1, 2)(1, 3).$$

The factorization into a product of transpositions is not unique, as the next example shows.

Example 7   Consider the product fg where $f = (1, 3, 2, 4)$ and $g = (1, 7, 6, 2)$. This product can be written as

$$(1, 3, 2, 4)(1, 7, 6, 2) = (1, 4)(1, 2)(1, 3)(1, 2)(1, 6)(1, 7)$$

and also as

$$(1, 3, 2, 4)(1, 7, 6, 2) = (1, 7, 6, 4)(2, 3)$$
$$= (1, 4)(1, 6)(1, 7)(2, 3).$$ ∎

Although the expression of a permutation as a product of transpositions is not unique, the number of transpositions used for a certain permutation is either *always odd* or else *always even*. Our proof of this fact takes us somewhat astray from our main course in this chapter. It involves consideration of a polynomial P in n variables $x_1, x_2, \ldots, x_n$ that is the product of all factors of the form $(x_i - x_j)$ with $1 \leq i < j \leq n$:

$$P = \prod_{i<j}^{n} (x_i - x_j).$$

(The symbol $\prod$ indicates a product in the same way that $\sum$ is used to indicate sums.) For example, if $n = 3$, then

$$P = \prod_{i<j}^{3} (x_i - x_j)$$
$$= (x_1 - x_2)(x_1 - x_3)(x_2 - x_3).$$

For $n = 4$, P is given by

$$P = \prod_{i<j}^{4} (x_i - x_j)$$
$$= (x_1 - x_2)(x_1 - x_3)(x_1 - x_4)(x_2 - x_3)(x_2 - x_4)(x_3 - x_4),$$

and similarly for larger values of n.

If f is any permutation on $\{1, 2, \ldots, n\}$, then f is applied to P by the rule

$$f(P) = \prod_{i<j}^{n} (x_{f(i)} - x_{f(j)}).$$

As an illustration, let us apply the transposition $t = (2, 4)$ to the polynomial

$$P = \prod_{i<j}^{4} (x_i - x_j)$$
$$= (x_1 - x_2)(x_1 - x_3)(x_1 - x_4)(x_2 - x_3)(x_2 - x_4)(x_3 - x_4).$$

We have

$$t(P) = (x_1 - x_4)(x_1 - x_3)(x_1 - x_2)(x_4 - x_3)(x_4 - x_2)(x_3 - x_2),$$

since 2 and 4 are interchanged by t. Analyzing this result, we see that

1. The factor $(x_2 - x_4)$ in P is changed to $(x_4 - x_2)$ in $t(P)$, so this factor changes sign.
2. The factor $(x_1 - x_3)$ is unchanged.

3. The remaining factors in $t(P)$ may be grouped in pairs as

$$(x_1 - x_4)(x_1 - x_2) \text{ and } (x_4 - x_3)(x_3 - x_2) = -(x_3 - x_4)(x_3 - x_2).$$

The products of these pairs are unchanged by t.

Thus, $t(P) = (-1)P$ in this particular case. The sort of analysis we have used here can be used to prove the following lemma.

Lemma 4.2 _____

If $t = (r, s)$ is any transposition on $\{1, 2, \ldots, n\}$ and $P = \prod_{i<j}^{n} (x_i - x_j)$, then

$$t(P) = (-1)P.$$

$(u \wedge v) \Rightarrow w$ **Proof** Since $t = (r, s) = (s, r)$, we may assume that $r < s$. We have

$$t(P) = \prod_{i<j}^{n} (x_{t(i)} - x_{t(j)}).$$

The factors of $t(P)$ may be analyzed as follows:

1. The factor $(x_r - x_s)$ in P is changed to $(x_s - x_r)$ in $t(P)$, so this factor changes sign.
2. The factors $(x_i - x_j)$ in P with both subscripts different from r and s are unchanged by t.
3. The remaining factors in P have exactly one subscript different from r and s, and they may be grouped into pairs as

$$(x_k - x_r)(x_k - x_s) \text{ or } -(x_k - x_r)(x_k - x_s).$$

The products of these pairs are unchanged by the interchange of r and s.

Thus, $t(P) = (-1)P$ and the proof of the lemma is complete.

STRATEGY ▶ The conclusion in the next theorem has the form "r or s." In previous conclusions of this type, we have assumed that r was false and proved that s must then be true. It is interesting to note that this time our technique is different and uses no negative assumption.

Theorem 4.3   PRODUCTS OF TRANSPOSITIONS

If a certain permutation f is expressed as a product of p transpositions and also as a product of q transpositions, then either p and q are both even, or else p and q are both odd.

$(u \wedge v) \Rightarrow (r \vee s)$ **Proof** Suppose

$$f = t_1 t_2 \cdots t_p \text{ and } f = t_1' t_2' \cdots t_q'$$

where each t_i and each t'_j are transpositions. With the first factorization, the result of applying f to

$$P = \prod_{i<j}^{n} (x_i - x_j)$$

can be obtained by successive application of the transpositions $t_p, t_{p-1}, \ldots, t_2, t_1$. By Lemma 4.2, each t_i changes the sign of P, so

$$f(P) = (-1)^p P.$$

Repeating this same line of reasoning with the second factorization, we obtain

$$f(P) = (-1)^q P.$$

This means that

$$(-1)^p P = (-1)^q P,$$

and consequently,

$$(-1)^p = (-1)^q.$$

Therefore, either p or q are both even, or p and q are both odd.

Theorem 4.3 assures us that when a particular permutation is expressed in different ways as a product of transpositions, the number of transpositions used either will always be an even number, or else will always be an odd number. This fact allows us to make the following definition.

Definition 4.4   **EVEN, ODD PERMUTATIONS**

> A permutation that can be expressed as a product of an even number of transpositions is called an **even permutation**, and a permutation that can be expressed as a product of an odd number of transpositions is called an **odd permutation**.

The product fg in Example 7 was written as a product of six transpositions and then as a product of four transpositions, and fg is an even permutation.

The factorization of an r-cycle $(i_1, i_2, \ldots, i_r)$ as

$$(i_1, i_2, \ldots, i_r) = (i_1, i_r)(i_1, i_{r-1}) \cdots (i_1, i_3)(i_1, i_2)$$

uses $r - 1$ transpositions. This shows that *an r-cycle is an even permutation if r is odd and an odd permutation if r is even*. The identity is an even permutation since $e = (1, 2)(1, 2)$. The product of two even permutations is clearly an even permutation. Since any permutation can be written as a product of disjoint cycles, and since the inverse of an r-cycle is an r-cycle, the inverse of an even permutation is an even permutation. These remarks shows that the set A_n of all even permutations in S_n is a subgroup of S_n. It is called the *alternating group* of n elements.

Definition 4.5 **ALTERNATING GROUP**

> The **alternating group** A_n is the subgroup of S_n that consists of all even permutations in S_n.

Example 8 The elements of the group A_4 are as follows:

$$
\begin{array}{llll}
(1) & (1, 2, 4) & (1, 4, 2) & (1, 2)(3, 4) \\
(1, 2, 3) & (1, 4, 3) & (2, 3, 4) & (1, 3)(2, 4) \\
(1, 3, 2) & (1, 3, 4) & (2, 4, 3) & (1, 4)(2, 3).
\end{array}
$$
∎

The concept of conjugate elements in a group is basic to the next chapter, which is devoted to certain aspects of the structure of groups. This concept is defined as follows.

Definition 4.6 **CONJUGATE ELEMENTS**

> If a and b are elements of the group G, the **conjugate** of a by b is the element bab^{-1}. We say that $c \in G$ is a **conjugate** of a if and only if $c = bab^{-1}$ for some b in G.

We should point out that this concept is trivial in an abelian group G, because $bab^{-1} = bb^{-1}a = ea = a$ for all $b \in G$.

There is a procedure by which conjugates of elements in a permutation group may be computed with ease. To see how this works, suppose that f and g are permutations on $\{1, 2, \ldots, n\}$ that have been written as products of disjoint cycles, and consider gfg^{-1}. If i_1 and i_2 are integers such that $f(i_1) = i_2$, then gfg^{-1} maps $g(i_1)$ to $g(i_2)$, as the following diagram shows:

$$
g(i_1) \xrightarrow{\ g^{-1}\ } i_1 \xrightarrow{\ f\ } i_2 \xrightarrow{\ g\ } g(i_2).
$$

This means that if

$$
(i_1, i_2, \ldots, i_r)
$$

is one of the disjoint cycles in f, then

$$
(g(i_1), g(i_2), \ldots, g(i_r))
$$

is a corresponding cycle in gfg^{-1}. Thus, if

$$
f = (i_1, i_2, \ldots, i_r)(j_1, j_2, \ldots, j_s) \cdots (k_1, k_2, \ldots, k_t),
$$

then

$$
gfg^{-1} = (g(i_1), g(i_2), \ldots, g(i_r))(g(j_1), \ldots, g(j_s)) \cdots (g(k_1), \ldots, g(k_t)).
$$

Example 9 If

$$
f = (1, 3, 6, 9, 5)(2, 4, 7),
$$

and

$$
g = (1, 2, 8)(3, 6)(4, 5, 7),
$$

then gfg^{-1} may be obtained from f as follows:

$$f = (1, 3, 6, 9, 5)(2, 4, 7)$$
$$\downarrow \downarrow \downarrow \downarrow \downarrow \; \downarrow \downarrow \downarrow$$
$$gfg^{-1} = (2, 6, 3, 9, 7)(8, 5, 4)$$
$$= (2, 6, 3, 9, 7)(4, 8, 5),$$

where the arrows indicate replacement of i by $g(i)$. This result may be verified by direct computation of g^{-1} and the product gfg^{-1}. ∎

The procedure for computing conjugates described just before Example 9 shows that any conjugate of a given permutation f has the same type of factorization into disjoint cycles as f does. If suitable permutations f and h are given, the procedure also indicates how g may be found so that $gfg^{-1} = h$. This is illustrated in Example 10.

Example 10 Suppose $f = (1, 4, 2)(3, 5)$, $h = (6, 8, 9)(5, 7)$, and we wish to find g such that $gfg^{-1} = h$. Using arrows to indicate replacements in the same way as in Example 9, we wish to obtain $gfg^{-1} = h$ from f as follows:

$$f = (1, 4, 2)(3, 5)$$
$$\downarrow \downarrow \downarrow \; \downarrow \downarrow$$
$$gfg^{-1} = (6, 8, 9)(5, 7).$$

From this diagram it is easy to see that

$$g = (1, 6)(4, 8)(2, 9)(3, 5, 7)$$

is a solution to our problem. It is also easy to see that g is not unique. For example,

$$(1, 6, 4, 8, 2, 9)(3, 5, 7)$$

is another value of g that works just as well. ∎

In Example 2 of Section 3.4, we considered the group of all *rigid motions,* or *symmetries,* of an equilateral triangle. Every geometric figure has an associated group of rigid motions. (We are considering only rigid motions in space here. For a plane figure, one can similarly consider rigid motions of the figure in that plane.) For simple figures such as a square, a regular pentagon, or a cube, a rigid motion is completely determined by the images of the vertices. If the vertices are labeled $1, 2, 3, \ldots$ rather than $V_1, V_2, V_3, \ldots$, the rigid motions may be represented by permutation notation. In Example 2 of Section 3.4, the mappings

$$h : \begin{cases} h(V_1) = V_2 \\ h(V_2) = V_1 \\ h(V_3) = V_3 \end{cases} \quad \text{and} \quad r : \begin{cases} r(V_1) = V_2 \\ r(V_2) = V_3 \\ r(V_3) = V_1 \end{cases}$$

can be written simply as

$$h = (1, 2) \quad \text{and} \quad r = (1, 2, 3).$$

Example 11   Using the notational convention described in the preceding paragraph, we shall write out the (space) group G of rigid motions of a square (see Figure 4-1).

Figure 4-1

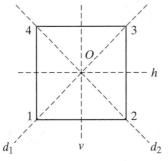

The elements of the group G are as follows:

1. the identity mapping $e = (1)$
2. the counterclockwise rotation $\alpha = (1, 2, 3, 4)$ through 90° about the center O
3. the counterclockwise rotation $\alpha^2 = (1, 3)(2, 4)$ through 180° about the center O
4. the counterclockwise rotation $\alpha^3 = (1, 4, 3, 2)$ through 270° about the center O
5. the reflection $\beta = (1, 4)(2, 3)$ about the horizontal line h
6. the reflection $\gamma = (2, 4)$ about the diagonal d_1
7. the reflection $\Delta = (1, 2)(3, 4)$ about the vertical line v
8. the reflection $\theta = (1, 3)$ about the diagonal d_2.

The group $G = \{e, \alpha, \alpha^2, \alpha^3, \beta, \gamma, \Delta, \theta\}$ of rigid motions of the square is known as the **octic group**. The multiplication table for G is requested in Exercise 16 of this section.   ∎

EXERCISES 4.1

1. Express each permutation as a product of disjoint cycles.

   a. $\begin{bmatrix} 1 & 2 & 3 & 4 & 5 \\ 4 & 5 & 3 & 1 & 2 \end{bmatrix}$

   b. $\begin{bmatrix} 1 & 2 & 3 & 4 & 5 \\ 1 & 3 & 2 & 5 & 4 \end{bmatrix}$

   c. $\begin{bmatrix} 1 & 2 & 3 & 4 & 5 \\ 4 & 1 & 3 & 5 & 2 \end{bmatrix}$

   d. $\begin{bmatrix} 1 & 2 & 3 & 4 & 5 \\ 3 & 5 & 2 & 4 & 1 \end{bmatrix}$

   e. $\begin{bmatrix} 1 & 2 & 3 & 4 & 5 & 6 & 7 \\ 3 & 4 & 5 & 6 & 1 & 2 & 7 \end{bmatrix}$

   f. $\begin{bmatrix} 1 & 2 & 3 & 4 & 5 & 6 & 7 \\ 5 & 1 & 3 & 7 & 2 & 6 & 4 \end{bmatrix}$

   g. $\begin{bmatrix} 1 & 2 & 3 & 4 & 5 \\ 1 & 3 & 4 & 5 & 2 \end{bmatrix}\begin{bmatrix} 1 & 2 & 3 & 4 & 5 \\ 3 & 2 & 4 & 1 & 5 \end{bmatrix}$

   h. $\begin{bmatrix} 1 & 2 & 3 & 4 & 5 \\ 2 & 3 & 4 & 1 & 5 \end{bmatrix}\begin{bmatrix} 1 & 2 & 3 & 4 & 5 \\ 1 & 3 & 5 & 4 & 2 \end{bmatrix}$

2. Express each permutation as a product of disjoint cycles.
   a. $(1, 9, 2, 3)(1, 9, 6, 5)(1, 4, 8, 7)$
   b. $(1, 2, 9)(3, 4)(5, 6, 7, 8, 9)(4, 9)$
   c. $(1, 4, 8, 7)(1, 9, 6, 5)(1, 5, 3, 2, 9)$
   d. $(1, 4, 2, 3, 5)(1, 3, 4, 5)$
   e. $(1, 3, 5, 4, 2)(1, 4, 3, 5)$

  f. $(1, 9, 2, 4)(1, 7, 6, 5, 9)(1, 2, 3, 8)$
  g. $(2, 3, 7)(1, 2)(3, 5, 7, 6, 4)(1, 4)$
  h. $(4, 9, 6, 7, 8)(2, 6, 4)(1, 8, 7)(3, 5)$

3. In each part of Exercise 1, decide whether the permutation is even or odd.

4. In each part of Exercise 2, decide whether the permutation is even or odd.

5. Find the order of each permutation in Exercise 1.

6. Find the order of each permutation in Exercise 2.

7. Express each permutation in Exercise 1 as a product of transpositions.

8. Express each permutation in Exercise 2 as a product of transpositions.

9. Compute gfg^{-1} for each pair f, g.
  a. $f = (1, 2, 4, 3);$ $g = (1, 3, 2)$
  b. $f = (1, 3, 5, 6);$ $g = (2, 5, 4, 6)$
  c. $f = (2, 3, 5, 4);$ $g = (1, 3, 2)(4, 5)$
  d. $f = (1, 4)(2, 3);$ $g = (1, 2, 3)$
  e. $f = (1, 3, 5)(2, 4);$ $g = (2, 5)(3, 4)$
  f. $f = (1, 3, 5, 2)(4, 6);$ $g = (1, 3, 6)(2, 4, 5)$

10. For the given permutations, f and h, find a permutation g such that $gfg^{-1} = h$.
  a. $f = (1, 5, 9);$ $h = (2, 6, 4)$
  b. $f = (1, 3, 5, 7);$ $h = (3, 4, 6, 8)$
  c. $f = (1, 3, 5)(2, 4);$ $h = (2, 4, 3)(1, 5)$
  d. $f = (1, 2, 3)(4, 5);$ $h = (2, 3, 4)(1, 6)$
  e. $f = (1, 4, 7)(2, 5, 8);$ $h = (1, 5, 4)(2, 3, 6)$
  f. $f = (1, 3, 5)(2, 4, 6);$ $h = (1, 2, 4)(3, 5, 6)$

Sec. 3.3, #33 ≫ **11.** Write the permutation $f = (1, 2, 3, 4, 5, 6)$ as a product of a permutation g of order 2 and a permutation h of order 3. (See Exercise 33 of Section 3.3.)

12. List all the elements of the alternating group A_3, written in cyclic notation.

13. List all the elements of S_4, written in cyclic notation.

14. Find all the distinct cyclic subgroups of A_4.

15. Find cyclic subgroups of S_4 that have three different orders.

16. Construct a multiplication table for the octic group described in Example 11 of this section.

17. Find all the distinct cyclic subgroups of the octic group in Exercise 16.

Sec. 3.1, #30 ≫ **18.** Find an isomorphism from the octic group G in Example 11 of this section to the group $G' = \{I_2, R, R^2, R^3, H, D, V, T\}$ in Exercise 30 of Section 3.1.

19. Prove that in any group, the relation "x is a conjugate of y" is an equivalence relation.

20. As stated in Exercise 23 of Section 3.2, the **centralizer** of an element a in the group G is the subgroup given by $C_a = \{x \in G \,|\, ax = xa\}$. Use the multiplication table constructed in Exercise 16 to find the centralizer C_a for each element a of the octic group.

21. A subgroup H of the group S_n is called **transitive** on $B = \{1, 2, \ldots, n\}$ if for each pair i, j of elements of B, there exists an element $h \in H$ such that $h(i) = j$. Show that there exists a cyclic subgroup H of S_n that is transitive on B.

4.2   CAYLEY'S THEOREM

At the opening of Section 3.4, we stated that permutation groups can serve as models for all groups. A more precise statement is that every group is isomorphic to a group of permutations; this is the reason for the fundamental importance of permutation groups in algebra.

Theorem 4.7   CAYLEY'S THEOREM

Every group is isomorphic to a group of permutations.

$p \Rightarrow q$ **Proof**   Let G be a given group. The permutations that we use in the proof will be mappings defined on the set of all elements in G.

For each element a in G, we define a mapping $f_a : G \rightarrow G$ by

$$f_a(x) = ax \text{ for all } x \text{ in } G.$$

That is, the image of each x in G is obtained by multiplying x on the left by a. Now f_a is one-to-one since

$$\begin{aligned} f_a(x) = f_a(y) &\Rightarrow ax = ay \\ &\Rightarrow x = y. \end{aligned}$$

To see that f_a is onto, let b be arbitrary in G. Then $x = a^{-1}b$ is in G, and for this particular x we have

$$\begin{aligned} f_a(x) &= ax \\ &= a(a^{-1}b) = b. \end{aligned}$$

Thus, f_a is a permutation on the set of elements of G.

We shall show that the set

$$G' = \{f_a \,|\, a \in G\}$$

actually forms a group of permutations. Since mapping composition is always associative, we only need to show that G' is closed, has an identity, and contains inverses.

For any f_a and f_b in G, we have

$$f_a f_b(x) = f_a(f_b(x)) = f_a(bx) = a(bx) = (ab)(x) = f_{ab}(x)$$

for all x in G. Thus, $f_a f_b = f_{ab}$, and G' is closed. Since

$$f_e(x) = ex = x$$

for all x in G, f_e is the identity permutation, $f_e = I_G$. Using the result $f_a f_b = f_{ab}$, we have

$$f_a f_{a^{-1}} = f_{aa^{-1}} = f_e$$

and

$$f_{a^{-1}} f_a = f_{a^{-1}a} = f_e.$$

Thus, $(f_a)^{-1} = f_{a^{-1}}$ is in G', and G' is a group of permutations.

All that remains is to show that G is isomorphic to G'. The mapping $\phi : G \rightarrow G'$ defined by

$$\phi(a) = f_a$$

is clearly onto. It is one-to-one since

$$\begin{aligned} \phi(a) = \phi(b) &\Rightarrow f_a = f_b \\ &\Rightarrow f_a(x) = f_b(x) \quad \text{for all } x \in G \\ &\Rightarrow ax = bx \qquad\quad \text{for all } x \in G \\ &\Rightarrow a = b. \end{aligned}$$

Finally, ϕ is an isomorphism since

$$\phi(a)\phi(b) = f_a f_b = f_{ab} = \phi(ab)$$

for all a, b in G.

Notice that the group $G' = \{f_a | a \in G\}$ is a subgroup of the group $S(G)$ of all permutations on G, and $G' \neq S(G)$ in most cases.

Example 1 We shall follow the proof of Cayley's Theorem with the group $G = \{1, i, -1, -i\}$ to obtain a group of permutations G' that is isomorphic to G and an isomorphism from G to G'.

With $f_a: G \rightarrow G$ defined by $f_a(x) = ax$ for each $a \in G$, we obtain the following permutations on the set of elements of G:

$$f_1: \begin{cases} f_1(1) = 1 \\ f_1(i) = i \\ f_1(-1) = -1 \\ f_1(-i) = -i \end{cases} \qquad f_i: \begin{cases} f_i(1) = i \\ f_i(i) = -1 \\ f_i(-1) = -i \\ f_i(-i) = 1 \end{cases}$$

$$f_{-1}: \begin{cases} f_{-1}(1) = -1 \\ f_{-1}(i) = -i \\ f_{-1}(-1) = 1 \\ f_{-1}(-i) = i \end{cases} \qquad f_{-i}: \begin{cases} f_{-i}(1) = -i \\ f_{-i}(i) = 1 \\ f_{-i}(-1) = i \\ f_{-i}(-i) = -1. \end{cases}$$

(In a more compact form, we could write $f_1(x) = x$, $f_i(x) = ix$, $f_{-1}(x) = -x$, and $f_{-i}(x) = -ix$ for all $x \in G$.) According to the proof of Cayley's Theorem, the set

$$G' = \{f_1, f_i, f_{-1}, f_{-i}\}$$

is a group of permutations, and the mapping $\phi: G \rightarrow G'$ defined by

$$\phi: \begin{cases} \phi(1) = f_1 \\ \phi(i) = f_i \\ \phi(-1) = f_{-1} \\ \phi(-i) = f_{-i} \end{cases}$$

is an isomorphism from G to G'. ∎

EXERCISES 4.2

Sec. 4.4, #13 ≪

1. Consider the group G of four elements $G = \{e, a, b, ab\}$ with the multiplication table in Figure 4-2. This group is known as the **Klein four group**. Write out the elements of a group of permutations that is isomorphic to G, and exhibit an isomorphism from G to this group.

Figure 4-2

·	e	a	b	ab
e	e	a	b	ab
a	a	e	ab	b
b	b	ab	e	a
ab	ab	b	a	e

2. Let G be the multiplicative group $G = \{[1], [2], [3], [4]\} \subseteq \mathbf{Z}_5$. Write out the elements of a group of permutations that is isomorphic to G, and exhibit an isomorphism from G to this group.

3. Let $G = \{[2], [4], [6], [8]\} \subseteq \mathbf{Z}_{10}$. It is given that G forms a group with respect to multiplication. Write out the elements of a group of permutations that is isomorphic to G, and exhibit an isomorphism from G to this group.

Sec. 3.1, #29 ≫   **4.** Consider the group of permutation matrices $G = \{I_3, P_1, P_2, P_3, P_4, P_5\}$ as given in Exercise 29 of Section 3.1. Write out the elements of a group of permutations that is isomorphic to G, and exhibit an isomorphism from G to this group.

5. For each a in the group G, define a mapping $h_a \colon G \to G$ by $h_a(x) = xa$ for all x in G.
   a. Prove that each h_a is a permutation on the set of elements in G.
   b. Prove that $H = \{h_a | a \in G\}$ is a group with respect to mapping composition.
   c. Define $\phi \colon G \to H$ by $\phi(a) = h_a$. Determine whether ϕ is always an isomorphism.

6. For each element a the group G, define a mapping $k_a \colon G \to G$ by $k_a(x) = xa^{-1}$ for all x in G.
   a. Prove that each k_a is a permutation on the set of elements of G.
   b. Prove that $K = \{k_a | a \in G\}$ is a group with respect to mapping composition.
   c. Define $\phi \colon G \to K$ by $\phi(a) = k_a$ for each a in G. Determine whether ϕ is always an isomorphism.

7. For each a in the group G, define a mapping $m_a \colon G \to G$ by $m_a(x) = a^{-1}x$ for all x in G.
   a. Prove that each m_a is a permutation on the set of elements of G.
   b. Prove that $M = \{m_a | a \in G\}$ is a group with respect to mapping composition.
   c. Define $\phi \colon G \to M$ by $\phi(a) = m_a$ for each a in G. Determine whether ϕ is always an isomorphism.

4.3   PERMUTATION GROUPS IN SCIENCE AND ART (OPTIONAL)

Often, the usefulness of certain knowledge in mathematics is neither obvious nor simple. So it is with permutation groups. Their applications in the real world come about through connections that are somewhat involved. Nevertheless, we shall indicate here some of their uses in both science and art.

Most of the scientific applications of permutation groups are in physics and chemistry. One of the most impressive applications occurred in 1962. In that year, physicists Murray Gell-Mann and Yuval Ne'eman used group theory to predict the existence of a new particle, which was designated the *omega minus particle*. It was not until 1964 that the existence of this particle was confirmed in laboratory experiments.

One of the most extensive uses made of permutation groups has been in the science of crystallography. As mentioned in Section 4.1, every geometric figure in two or three dimensions has its associated rigid motions, or *symmetries*. This association provides a natural connection between permutation groups and many objects in the real world. One of the most fruitful of these connections has been made in the study of the structure of crystals. Crystals are classified according to geometric symmetry based on a structure with a balanced arrangement of faces. One of the simplest and most common examples of such a structure is provided by the fact that a common table salt (NaCl) crystal is in the shape of a cube. (See photo, facing page.)

In this section we examine some groups related to the rigid motions of a plane figure. We have already seen two examples of this type of group. The first was the group of symmetries of an equilateral triangle in Example 2 of Section 3.4, and the other was the group of symmetries of a square in Example 11 of Section 4.1.

Salt crystals are in the form of cubes.

It is not hard to see that the symmetries of any plane figure F form a group under mapping composition. We already know that the permutations on the set F form a group $\mathcal{S}(F)$ with respect to mapping composition. The identity permutation I_F preserves distances and consequently is a symmetry of F. If two permutations on F preserve distances, their composition does also, and if a given permutation preserves distances, its inverse does also. Thus, the symmetries of F form a subgroup of $\mathcal{S}(F)$.

Before considering some other specific plane figures F, a discussion of the term *symmetry* is in order. In agreement with conventional terminology in algebra, we have used the word *symmetry* to refer to a rigid motion of a geometric figure. However, *symmetry* is commonly used in another way. For example, the pentagon shown in Figure 4-3 is said to have *symmetry* with respect to the vertical line ℓ through the center O and the vertex at the top, or to be *symmetric* with respect to ℓ. To make a distinction between the two uses, we shall use the phrase *geometric symmetry* for the latter type of symmetry.

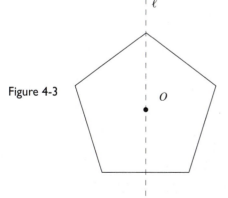

Figure 4-3

The groups of symmetries for regular polygons with three or four sides generalize to a regular polygon P with n sides, for any positive integer $n > 4$. Any symmetry f of P is determined by the images of the vertices of P. Let the vertices be numbered $1, 2, \ldots, n$, and consider the mapping that makes the symmetry f of P correspond to the permutation on $\{1, 2, \ldots, n\}$ that has the matrix form

$$\begin{bmatrix} 1 & 2 & \ldots & n \\ f(1) & f(2) & \ldots & f(n) \end{bmatrix}.$$

Since f is completely determined by the images of the vertices, this mapping is clearly a bijection between the rigid motions of P and a subset D_n of the symmetric group S_n of all permutations on $\{1, 2, \ldots, n\}$. This mapping is in fact an isomorphism, D_n is a subgroup of S_n, and we identify the rigid motions of P with the elements of D_n in the same way that we did in Example 11 of Section 4.1.

Regular polygons with $n = 5$ (a pentagon) and $n = 6$ (a hexagon) are shown in Figure 4-4. Bearing in mind that a symmetry is determined by the images of the vertices, it can be seen that D_n consists of n counterclockwise rotations and n reflections about a line through the center O of P. If n is odd, each reflection is about a line through a vertex and the midpoint of the opposite side. If n is even, half of the reflections are about lines through pairs of opposite vertices, and the other half are about lines through midpoints of opposite sides. Thus, D_n has order $2n$.

Figure 4-4

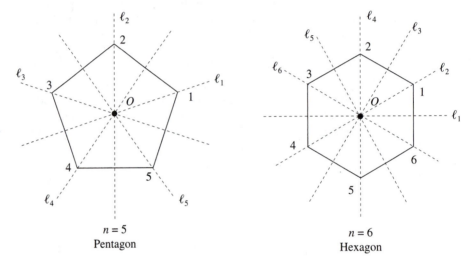

$n = 5$
Pentagon

$n = 6$
Hexagon

Example 1 Consider the pentagon in Figure 4-4. If we let R denote the rotation of $\frac{360°}{5} = 72°$ counterclockwise about the center O, then all possible rotations in D_5 are found in the following list.

$$R = (1, 2, 3, 4, 5), \quad R^2 = (1, 3, 5, 2, 4), \quad R^3 = (1, 4, 2, 5, 3),$$
$$R^4 = (1, 5, 4, 3, 2), \quad R^5 = (1)$$

If we let L_k denote the reflection about line ℓ_k for $k = 1, 2, 3, 4, 5$, then the reflections in D_5 appear as follows in cyclic notation.

$$L_1 = (2, 5)(3, 4), \quad L_2 = (1, 3)(4, 5), \quad L_3 = (1, 5)(2, 4),$$
$$L_4 = (1, 2)(3, 5), \quad L_5 = (1, 4)(2, 3)$$

Direct computations verify that

$$L_1 R = L_3, \quad L_1 R^2 = L_5, \quad L_1 R^3 = L_2, \text{ and } L_1 R^4 = L_4.$$

Thus, the elements of D_5 can be listed in the form

$$D_5 = \{I, R, R^2, R^3, R^4, L_1, L_1 R, L_1 R^2, L_1 R^3, L_1 R^4\}. \qquad \blacksquare$$

All the symmetries in our examples have been either rotations or reflections about a line. This is no accident because these are the only kinds of symmetries that exist for a bounded nonempty set. If the group of symmetries of a certain figure contains a rotation different from the identity mapping, then the figure is said to possess **rotational symmetry**. A figure with a group of symmetries that includes a reflection about a line is said to have **reflective symmetry**.

Example 2   Each part of Figure 4-5 has a group of symmetries that consists entirely of rotations, and each possesses only rotational symmetry. In contrast, the group of symmetries of the pentagon contains both reflections and rotations, and the pentagon has both reflective symmetry and rotational symmetry.

Figure 4-5

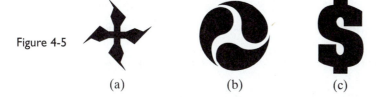

(a)   (b)   (c)   ∎

We have barely touched on the subject of symmetries in this section, concentrating primarily on bounded nonempty sets in the plane. When attention is extended to unbounded sets in the plane, there are two more types of symmetries that can be considered: the translations and the glide reflections.

A **translation** is simply a sliding (or glide) of the entire object through a certain distance in a fixed direction. A **glide reflection** consists of a translation (or glide) followed by a reflection about a line parallel to the direction of the translation. These types of symmetries are treated in detail in more advanced books than this one, and it can be shown that there are only four kinds of symmetries for plane figures: *rotations, reflections, translations,* and *glide reflections.*

As our final example in this section, we consider the group of symmetries of an unbounded set.

Example 3   The unbounded set shown in Figure 4-6 is composed of a horizontal string of copies of the letter **R**, equally spaced one unit from the beginning of one **R** to the beginning of the next **R**, and endless in both directions.

Figure 4-6

If t denotes a translation of the set in Figure 4-6 one unit to the right, then t^2 is a translation two units to the right and t^n is a translation n units to the right, for any positive integer n. Thus, all positive integral powers of t are symmetries on the set of **R**'s. The inverse mapping t^{-1} is a translation of the set one unit to the left, and t^{-n} is a translation n units to the left for any positive integer n. Thus, all integral powers of t are symmetries on the set of **R**'s, and the set

$$\{\cdots, t^{-2}, t^{-1}, t^0 = I, t, t^2, \cdots\}$$

is the (infinite) group of symmetries of this set.   ∎

Translations and glide reflections are common in the group of symmetries for wallpaper patterns, textile patterns, pottery, ribbons, and all sorts of decorative art. The interested reader can find an excellent exposition of the applications that we have touched on in Tannenbaum and Arnold's *Excursions in Modern Mathematics,* 2nd ed. (Englewood Cliffs, NJ: Prentice-Hall, 1995).

The outstanding connection between permutation groups and art is provided by the famous works of the great Dutch artist M. C. Escher. Concerning Escher, J. Taylor Hollist said, "Mathematicians continue to use his periodic patterns of animal figures as clever illustrations of translation, rotation and reflection symmetry. Psychologists use his optical illusions and distorted views of life as enchanting examples in the study of vision."[†]

EXERCISES 4.3

List all elements in the group of symmetries of the given set.

1. The letter **T**
2. The letter **M**
3. The letter **S**
4. The letter **H**

Determine if the given figure has rotational symmetry or reflective symmetry.

5.   6.   7.

8.   9.   10.

Describe the elements in the group of symmetries of the given bounded figure.

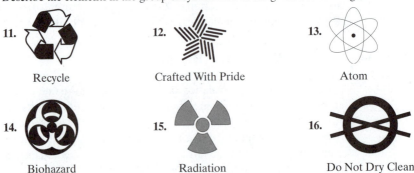

11.   12.   13.

Recycle   Crafted With Pride   Atom

14.   15.   16.

Biohazard   Radiation   Do Not Dry Clean

[†]J. Taylor Hollist, "Escher Correspondence in the Roosevelt Collection," *Leonardo,* Vol. 24, No. 3 (1991), p. 329.

Describe the elements in the groups of symmetries of the given unbounded figures.

17. $\xrightarrow{1 \text{ unit}}$
$\cdots$ E E E E E E E $\cdots$

18. $\xrightarrow{1 \text{ unit}}$
$\cdots$ ▷ ▷ ▷ ▷ ▷ ▷ ▷ $\cdots$

19. $\xrightarrow{1 \text{ unit}}$
$\cdots$ T T T T T T T $\cdots$

20. $\xrightarrow{1 \text{ unit}}$
$\cdots$ ★ ★ ★ ★ ★ ★ ★ $\cdots$

21. Show that the group of symmetries in Example 3 of this section is isomorphic to the group of integers under addition.

22. Construct a multiplication table for the group G of rigid motions of a rectangle with vertices 1, 2, 3, 4 if the rectangle is not a square.

23. Construct a multiplication table for the group G of rigid motions of a regular pentagon with vertices 1, 2, 3, 4, 5.

24. List the elements of the group G of rigid motions of a regular hexagon with vertices 1, 2, 3, 4, 5, 6.

25. Let G be the group of rigid motions of a cube. Find the order $o(G)$.

26. Let G be the group of rigid motions of a regular tetrahedron (see Figure 4-7). Find the order $o(G)$.

Figure 4-7

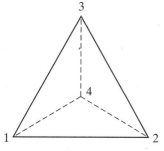

Sec. 3.2, #15a ≫ 27. (See Exercise 15a of Section 3.2.) Find an isomorphism from the group G in Exercise 22 of this exercise set to the multiplicative group

$$H = \left\{ \begin{bmatrix} 1 & 0 \\ 0 & 1 \end{bmatrix}, \begin{bmatrix} 1 & 0 \\ 0 & -1 \end{bmatrix}, \begin{bmatrix} -1 & 0 \\ 0 & 1 \end{bmatrix}, \begin{bmatrix} -1 & 0 \\ 0 & -1 \end{bmatrix} \right\}.$$

4.4   NORMAL SUBGROUPS

The binary operation in a given group can be used in a natural way to define a product between subsets of the group. The importance of this product is difficult to appreciate at this point in our development. It leads to the definition of *cosets:* Cosets in turn lead to *quotient groups*; and quotient groups provide a systematic description of all *homomorphic images* of a group in Section 4.5.

Definition 4.8 PRODUCT OF SUBSETS

> Let A and B be nonempty subsets of the group G. The **product** AB is defined by
>
> $$AB = \{x \in G \mid x = ab \text{ for some } a \in A, b \in B\}.$$

This product is formed by using the group operation in G. A more precise formulation would be

$$A * B = \{x \in G \mid x = a * b \text{ for some } a \in A, b \in B\},$$

where $*$ is the group operation in G.

Several properties of this product are worth mentioning. For nonempty subsets A, B, and C of the group G,

$$\begin{aligned} A(BC) &= \{a(bc) \mid a \in A, b \in B, c \in C\} \\ &= \{(ab)c \mid a \in A, b \in B, c \in C\} \\ &= (AB)C. \end{aligned}$$

It is obvious that

$$B = C \implies AB = AC \text{ and } BA = CA,$$

but we must be careful about the order because AB and BA may be different sets.

Example 1 Consider the subsets $A = \{(1, 2, 3), (1, 2)\}$, and $B = \{(1, 3), (2, 3)\}$ in $G = S_3$. We have

$$\begin{aligned} AB &= \{(1, 2, 3)(1, 3), (1, 2)(1, 3), (1, 2, 3)(2, 3), (1, 2)(2, 3)\} \\ &= \{(2, 3), (1, 3, 2), (1, 2), (1, 2, 3)\} \end{aligned}$$

and

$$\begin{aligned} BA &= \{(1, 3)(1, 2, 3), (2, 3)(1, 2, 3), (1, 3)(1, 2), (2, 3)(1, 2)\} \\ &= \{(1, 2), (1, 3), (1, 2, 3), (1, 3, 2)\}, \end{aligned}$$

so $AB \neq BA$. ∎

For a nonabelian group G, we would probably expect AB and BA to be different. A fact that is not quite so natural is that

$$AB = AC \not\Longrightarrow B = C.$$

Example 2 An example where $AB = AC$ but $B \neq C$ is provided by $A = \{(1, 2, 3), (1, 3, 2)\}$, $B = \{(1, 3), (2, 3)\}$, and $C = \{(1, 2), (1, 3)\}$ in $G = S_3$. Straightforward calculations show that

$$AB = \{(2, 3), (1, 2), (1, 3)\} = AC,$$

but $B \neq C$. ∎

If $B = \{g\}$ consists of a single element g of a group G, then AB is written simply as Ag instead of as $A\{g\}$:

$$Ag = \{x \in G \mid x = ag \text{ for some } a \in A\}.$$

Similarly,

$$gA = \{x \in G \,|\, x = ga \text{ for some } a \in A\}.$$

This is one instance in which a cancellation law does hold:

$$gA = gB \implies A = B.$$

This is true because

$$\begin{aligned}
gA = gB &\implies g^{-1}(gA) = g^{-1}(gB) \\
&\implies (g^{-1}g)A = (g^{-1}g)B \\
&\implies eA = eB \\
&\implies A = B.
\end{aligned}$$

For convenience of reference, we summarize these results in a theorem.

Theorem 4.9 PROPERTIES OF THE PRODUCT OF SUBSETS

Let $A, B,$ and C denote nonempty subsets of the group G, and let g denote an element of G. Then the following statements hold:

a. $A(BC) = (AB)C$.
b. $B = C$ implies $AB = AC$ and $BA = CA$.
c. The product AB is not commutative.
d. $AB = AC$ does not imply $B = C$.
e. $gA = gB$ implies $A = B$.

Statements d and e have obvious duals in which the common factor is on the right side.

We shall be concerned mainly with products of subsets in which one of the factors is a subgroup. The cosets of a subgroup are of special importance.

Definition 4.10 COSETS

> Let H be a subgroup of the group G. For any a in G,
>
> $$aH = \{x \in G \,|\, x = ah \text{ for some } h \in H\}$$
>
> is a **left coset** of H in G. Similarly, Ha is called a **right coset** of H in G.

The left coset aH and the right coset Ha are never disjoint, since $a = ae = ea$ is in both sets. In spite of this, aH and Ha may happen to be different sets, as the next example shows.

Example 3 Consider the subgroup

$$K = \{(1), (1, 2)\}$$

of

$$G = S_3 = \{(1), (1, 2, 3), (1, 3, 2), (1, 2), (1, 3), (2, 3)\}.$$

For $a = (1, 2, 3)$, we have

$$aK = \{(1, 2, 3), (1, 2, 3)(1, 2)\}$$
$$= \{(1, 2, 3), (1, 3)\}$$

and

$$Ka = \{(1, 2, 3), (1, 2)(1, 2, 3)\}$$
$$= \{(1, 2, 3), (2, 3)\}.$$

In this case, $aK \neq Ka$. ■

Although a left coset of H and a right coset of H may be neither equal nor disjoint, this cannot happen with two left cosets of H. This fact is fundamental to the proof of Lagrange's Theorem (Theorem 4.13), so we designate it as a lemma.

STRATEGY ▶ The proof of this lemma is by use of the *contrapositive*. The **contrapositive** of $p \Rightarrow q$ is $\sim q \Rightarrow \sim p$. As shown in the appendix of this book, any statement and its contrapositive are logically equivalent.

The following proof illustrates a case where it is easier to prove the contrapositive than the original statement.

Lemma 4.11   **LEFT COSET PARTITION**

Let H be a subgroup of the group G. The distinct left cosets of H in G form a partition of G; that is, they separate the elements of G into mutually disjoint subsets.

$\sim q \Rightarrow \sim p$ **Proof**   It is sufficient to show that any two left cosets of H that are not disjoint must be the same left coset.

Suppose aH and bH have at least one element in common—say, $z \in aH \cap bH$. Then $z = ah_1$ for some $h_1 \in H$, and $z = bh_2$ for some $h_2 \in H$. This means that $ah_1 = bh_2$ and $a = bh_2h_1^{-1}$. We have that $h_2h_1^{-1}$ is in H since H is a subgroup, so $a = bh_3$ where $h_3 = h_2h_1^{-1} \in H$. Now, for every $h \in H$,

$$ah = bh_3h$$
$$= bh_4$$

where $h_4 = h_3 \cdot h$ is in H. That is, $ah \in bH$ for all $h \in H$. This proves that $aH \subseteq bH$. A similar argument shows that $bH \subseteq aH$, and thus $aH = bH$.

Example 4   Consider again the subgroup

$$K = \{(1), (1, 2)\}$$

of

$$G = S_3 = \{(1), (1, 2, 3), (1, 3, 2), (1, 2), (1, 3), (2, 3)\}.$$

In Example 3 of this section, we saw that

$$(1, 2, 3)K = \{(1, 2, 3), (1, 3)\}.$$

Since $(1, 3)$ is in this left coset, it follows from Lemma 4.11 that

$$(1, 3)K = (1, 2, 3)K = \{(1, 2, 3), (1, 3)\}.$$

Straightforward computations show that

$$(1)K = (1, 2)K = \{(1), (1, 2)\}$$

and

$$(2, 3)K = (1, 3, 2)K = \{(1, 3, 2), (2, 3)\}.$$

Thus, the distinct left cosets of K in G are given by

$$K, \ (1, 2, 3)K, \ (1, 3, 2)K. \qquad\blacksquare$$

Definition 4.12   INDEX

> Let H be a subgroup of G. The number of distinct left cosets of H in G is called the **index** of H in G, and is denoted by $[G:H]$.

In the proof of the next theorem, we show that if $o(G)$ is finite, then the order of any subgroup of G must divide the order of the group G.

Theorem 4.13   LAGRANGE'S THEOREM

If G is a finite group and H is a subgroup of G, then

$$\text{order of } G = (\text{order of } H) \cdot (\text{index of } H \text{ in } G).$$

$(p \wedge q) \Rightarrow r$ **Proof**   Let G be a finite group of order n and let H be a subgroup of G with order k. We shall show that k divides n.

From Lemma 4.11, we know that the left cosets of H in G separate the elements of G into mutually disjoint subsets. Let m be the index of H in G; that is, there are m distinct left cosets of H in G. We shall show that each left coset has exactly k elements.

Let aH represent an arbitrary left coset of H. The mapping $\phi: H \to aH$ defined by

$$\phi(h) = ah$$

is one-to-one, because the left cancellation law holds in G. It is also onto, since any x in aH can be written as $x = ah$ for $h \in H$. Thus, ϕ is a one-to-one correspondence from H to aH, and this means that aH has the same number of elements as does H.

We have the n elements of G separated into m disjoint subsets, and each subset has k elements. Therefore, $n = km$, and

$$o(G) = o(H) \cdot [G:H].$$

Lagrange's Theorem is of great value if we are interested in finding all the subgroups of a finite group. In connection with this task, it is worthwhile to record this immediate corollary.

Corollary 4.14 $o(a) \mid o(G)$

The order of an element of a finite group must divide the order of the group.

Example 5 To illustrate the usefulness of the foregoing results, we shall exhibit all of the subgroups of S_3. Any subgroup of S_3 must be of order 1, 2, 3, or 6, since $o(S_3) = 6$. An element in a subgroup of order 3 must have order dividing 3, and therefore any subgroup of order 3 is cyclic. Similarly, any subgroup of order 2 is cyclic. The following list is thus a complete list of the subgroups of S_3:

$$H_1 = \{(1)\} \qquad H_4 = \{(1), (2, 3)\}$$
$$H_2 = \{(1), (1, 2)\} \qquad H_5 = \{(1), (1, 2, 3), (1, 3, 2)\}$$
$$H_3 = \{(1), (1, 3)\} \qquad H_6 = S_3.$$
■

It is easy to see that if p is a prime, any group G of order p must be cyclic (any a in G such that $a \neq e$ must be a generator). This means that, up to an isomorphism, there is only one group of order p, if p is a prime. In particular, the only groups of order 2, 3, or 5 are the cyclic groups.

By examination of the possible orders of the elements and the possible multiplication tables, it can be shown that a group of order 4 is either cyclic or is isomorphic to the Klein four group

$$G = \{e, a, b, ab = ba\}$$

of Exercise 1 in Section 4.2.

Among the subgroups of a group are those known as the *normal subgroups*. The significance of the normal subgroups is revealed in the next section.

Definition 4.15 **NORMAL SUBGROUP**

> Let H be a subgroup of G. Then H is a **normal** (or **invariant**) **subgroup** of G if $xH = Hx$ for all $x \in G$.

Note that the condition $xH = Hx$ is an equality of sets, and it does not require that $xh = hx$ for all h in H.

Example 6 Let

$$H = A_3 = \{(1), (1, 2, 3), (1, 3, 2)\} = \langle (1, 2, 3) \rangle$$

and

$$G = S_3 = \{(1), (1, 2, 3), (1, 3, 2), (1, 2), (1, 3), (2, 3)\}.$$

For $x = (1, 2)$ we have

$$xH = \{(1, 2)(1), (1, 2)(1, 2, 3), (1, 2)(1, 3, 2)\}$$
$$= \{(1, 2), (2, 3), (1, 3)\}$$

and

$$Hx = \{(1)(1, 2), (1, 2, 3)(1, 2), (1, 3, 2)(1, 2)\}$$
$$= \{(1, 2), (1, 3), (2, 3)\}.$$

We have $xH = Hx$, but $xh \neq hx$ when $h = (1, 2, 3) \in H$. Similar computations show that

$$(1)H = (1, 2, 3)H = (1, 3, 2)H = \{(1), (1, 2, 3), (1, 3, 2)\}$$
$$H(1) = H(1, 2, 3) = H(1, 3, 2) = \{(1), (1, 2, 3), (1, 3, 2)\}$$
$$(1, 2)H = (1, 3)H = (2, 3)H = \{(1, 2), (1, 3), (2, 3)\}$$
$$H(1, 2) = H(1, 3) = H(2, 3) = \{(1, 2), (1, 3), (2, 3)\}.$$

Thus, H is a normal subgroup of G. ∎

In Example 6, we have $hH = H = Hh$ for all $h \in H$. These equalities hold for all subgroups, as stated in the following theorem.

Theorem 4.16 A Special Coset H

If H is any subgroup of a group G, then $hH = H = Hh$ for all $h \in H$.

$p \Rightarrow q$ **Proof** Let h be an arbitrary element in the subgroup H of the group G.
   If $x \in hH$, then $x = hy$ for some $y \in H$. But $h \in H$ and $y \in H$ imply $hy = x$ is in H, since H is closed. Thus, $hH \subseteq H$.
   For any $x \in H$, the element $h^{-1}x$ is in H since H contains the inverse of h and H is closed. But

$$h^{-1}x \in H \Rightarrow h(h^{-1}x) = x \in hH,$$

and this proves that $H \subseteq hH$. It follows that $hH = H$.
   The proof of the equality $Hh = H$ is similar.

The proof of the following corollary is left as an exercise.

Corollary 4.17 The Square of a Subgroup

For any subgroup H of a group G, $H^2 = H$, where H^2 denotes the product HH as defined in Definition 4.8.

Example 7 As an example of a subgroup that is *not* normal, consider $K = \{(1), (1, 2)\}$ in S_3. With $x = (1, 2, 3)$, we have

$$xK = \{(1, 2, 3), (1, 2, 3)(1, 2)\}$$
$$= \{(1, 2, 3), (1, 3)\}$$

$$Kx = \{(1, 2, 3), (1, 2)(1, 2, 3)\}$$
$$= \{(1, 2, 3), (2, 3)\}.$$

Thus, $xK \neq Kx$, and K is not a normal subgroup of S_3. ∎

The definition of a normal subgroup can be formulated in several different ways. For instance, we can write

$$xH = Hx \quad \text{for all } x \in G \Leftrightarrow xHx^{-1} = H \quad \text{for all } x \in G$$
$$\Leftrightarrow x^{-1}Hx = H \quad \text{for all } x \in G.$$

Other formulations can be made. One that is frequently taken as the definition is given in Theorem 4.18.

Theorem 4.18 NORMAL SUBGROUPS AND CONJUGATES

Let H be a subgroup of G. Then H is a normal subgroup of G if and only if $xhx^{-1} \in H$ for every $h \in H$ and every $x \in G$.

$p \Rightarrow q$ **Proof** If H is a normal subgroup of G, then the condition follows easily, since H normal requires

$$xHx^{-1} = H \text{ for all } x \in G \Rightarrow xHx^{-1} \subseteq H \quad \text{for all } x \in G$$
$$\Rightarrow xhx^{-1} \in H \quad \text{for all } h \in H \text{ and all } x \in G.$$

$p \Leftarrow q$ Suppose now that the condition holds. For any $x \in G$, $xHx^{-1} \subseteq H$ follows immediately, and we only need to show that $H \subseteq xHx^{-1}$. Let h be arbitrary in H, and let $x \in G$. Now x^{-1} is an element in G, and the condition implies that

$$(x^{-1})(h)(x^{-1})^{-1} = x^{-1}hx$$

is in H; that is,

$$x^{-1}hx = h_1 \text{ for some } h_1 \in H \Rightarrow h = xh_1x^{-1} \text{ for some } h_1 \in H$$
$$\Rightarrow h \in xHx^{-1}.$$

Thus, $H \subseteq xHx^{-1}$, and we have $xHx^{-1} = H$ for all $x \in G$. It follows that H is a normal subgroup of G.

The concept of generators can be extended from cyclic subgroups $\langle a \rangle$ to more complicated situations where a subgroup is generated by more than one element. We only touch on this topic here, but it is a fundamental idea in more advanced study of groups.

Definition 4.19 SET GENERATED BY A

If A is a nonempty subset of the group G, then the **set generated by A**, denoted by $\langle A \rangle$, is the set defined by

$$\langle A \rangle = \{x \in G \mid x = a_1 a_2 \cdots a_n \text{ with either } a_i \in A \text{ or } a_i^{-1} \in A\}.$$

In other words, $\langle A \rangle$ is the set of all products that can be formed with a finite number of factors, each of which is either an element of A or has an inverse that is an element of A.

Theorem 4.20 SUBGROUP GENERATED BY A

For any nonempty subset A of a group G, the set $\langle A \rangle$ is a subgroup of G called the **subgroup of G generated by A**.

$p \Rightarrow q$ **Proof**   There exists at least one $a \in A$, since $A \neq \varnothing$. Then $e = aa^{-1} \in \langle A \rangle$, so $\langle A \rangle$ is nonempty.

If $x \in \langle A \rangle$ and $y \in \langle A \rangle$, then

$$x = x_1 x_2 \cdots x_n \text{ with either } x_i \in A \text{ or } x_i^{-1} \in A$$

and

$$y = y_1 y_2 \cdots y_k \text{ with either } y_j \in A \text{ or } y_j^{-1} \in A.$$

Thus,

$$xy = x_1 x_2 \cdots x_n y_1 y_2 \cdots y_k,$$

where each factor on the right is either in A or has an inverse that is an element of A. Also,

$$x^{-1} = x_n^{-1} \cdots x_2^{-1} x_1^{-1} \text{ with either } x_i^{-1} \in A \text{ or } x_i \in A.$$

The set $\langle A \rangle$ is closed and contains inverses, and therefore it is a subgroup of G.

In work with *finite groups,* the result in Exercise 35 of Section 3.2 is extremely helpful in finding $\langle A \rangle$, since it implies that $\langle A \rangle$ is the smallest subset of G that contains A and is closed under the operation. (This is true *only for finite groups.*) The subgroup $\langle A \rangle$ can be constructed systematically by starting a multiplication table using the elements of A and enlarging the table by adjoining additional elements until closure is obtained. A practical first step in this direction is to begin the table using all the elements of A and all their distinct powers. This is illustrated in the next example.

Example 8   Consider the problem of finding $\langle A \rangle$ in S_4 where $A = \{(1, 2, 3, 4), (1, 4)(2, 3)\}$. We begin by computing the distinct powers of the elements of A:

$$\alpha = (1, 2, 3, 4) \qquad \alpha^2 = (1, 3)(2, 4)$$
$$\alpha^3 = \alpha^{-1} = (1, 4, 3, 2) \qquad \alpha^4 = e = (1)$$
$$\beta = (1, 4)(2, 3) \qquad \beta^2 = e.$$

Starting a multiplication table using $e, \alpha, \alpha^2, \alpha^3, \beta$, we find the following new elements of $\langle A \rangle$:

$$\alpha\beta = (1, 2, 3, 4)(1, 4)(2, 3) = (2, 4) = \gamma$$
$$\alpha^2\beta = (1, 3)(2, 4)(1, 4)(2, 3) = (1, 2)(3, 4) = \Delta$$
$$\alpha^3\beta = (1, 4, 3, 2)(1, 4)(2, 3) = (1, 3) = \theta.$$

We then enlarge the table so as to use all eight elements

$$e, \alpha, \alpha^2, \alpha^3, \beta, \gamma, \Delta, \theta.$$

Proceeding to fill out the enlarged table, we obtain the table in Figure 4-8, which shows that the set

$$G = \{e, \alpha, \alpha^2, \alpha^3, \beta, \gamma, \Delta, \theta\}$$

is the subgroup of S_4 generated by $A = \{\alpha, \beta\}$. This group G is the **octic group** that was presented in Example 11 of Section 4.1.

Figure 4-8

$\circ$	e	α	α^2	α^3	β	γ	Δ	θ
e	e	α	α^2	α^3	β	γ	Δ	θ
α	α	α^2	α^3	e	γ	Δ	θ	β
α^2	α^2	α^3	e	α	Δ	θ	β	γ
α^3	α^3	e	α	α^2	θ	β	γ	Δ
β	β	θ	Δ	γ	e	α^3	α^2	α
γ	γ	β	θ	Δ	α	e	α^3	α^2
Δ	Δ	γ	β	θ	α^2	α	e	α^3
θ	θ	Δ	γ	β	α^3	α^2	α	e

EXERCISES 4.4

1. **a.** Let G be the octic group in Example 8 and let H be the subgroup $H = \{e, \beta\}$ of G. Find the distinct left cosets of H in G and write out their elements.

   b. With G and H as in part a, find the distinct right cosets of H in G and write out their elements.

2. **a.** With G as in Exercise 1 and $H = \{e, \Delta\}$, find the distinct left cosets of H in G, and write out their elements.

   b. Find the distinct right cosets of $H = \{e, \Delta\}$ in the octic group and write out their elements.

3. **a.** Let $G = \{I_3, P_3, P_3^2, P_1, P_4, P_2\}$ be the multiplicative group of permutation matrices in Example 4 of Section 3.4 and let H be the subgroup given by

$$H = \{I_3, P_4\} = \left\{ \begin{bmatrix} 1 & 0 & 0 \\ 0 & 1 & 0 \\ 0 & 0 & 1 \end{bmatrix}, \begin{bmatrix} 0 & 0 & 1 \\ 0 & 1 & 0 \\ 1 & 0 & 0 \end{bmatrix} \right\}.$$

   Find the distinct left cosets of H in G and write out their elements.

   b. With G and H as in part **a**, find the distinct right cosets of H in G and write out their elements.

4. With G as in Exercise 3, show that $H = \{I_3, P_3, P_3^2\}$ is a normal subgroup of G.

5. Show that

$$H = \left\{ \begin{bmatrix} 1 & 0 \\ 0 & 1 \end{bmatrix}, \begin{bmatrix} -1 & 0 \\ 0 & -1 \end{bmatrix} \right\}$$

   is a normal subgroup of the multiplicative group G of invertible matrices in $M_2(\mathbf{R})$.

6. Let H be a subgroup of the group G. Prove that if two right cosets Ha and Hb are not disjoint, then $Ha = Hb$.

7. For any subgroup H of the group G, let H^2 denote the product $H^2 = HH$ as defined in Definition 4.8. Prove Corollary 4.17: $H^2 = H$.

8. If H is a subgroup of the group G, prove that gHg^{-1} is a subgroup of G for any $g \in G$. We say that gHg^{-1} is a **conjugate** of H and that H and gHg^{-1} are **conjugate subgroups**.

9. Show that every subgroup of an abelian group is normal.

10. For an arbitrary subgroup H of the group G, define the mapping θ from the set of left cosets of H in G to the set of right cosets of H in G by $\theta(aH) = Ha^{-1}$. Prove that θ is a bijection.

11. Let H be a subgroup of the group G. Prove that the index of H in G is the number of distinct right cosets of H in G.

12. Consider the octic group G of Example 8.
    a. Find a subgroup of G that has order 2 and is a normal subgroup of G.
    b. Find a subgroup of G that has order 2 and is *not* a normal subgroup of G.

Sec. 4.2, #1 ≫ 13. Show that a group of order 4 is either cyclic or is isomorphic to the Klein four group $G = \{e, a, b, ab = ba\}$ of Exercise 1 in Section 4.2.

14. Find all subgroups of the octic group in Example 8.

15. Find all subgroups of the alternating group A_4.

Sec. 3.1, #28 ≫ 16. Find all subgroups of the quaternion group. (See Exercise 28 of Section 3.1.)

Sec. 4.5, #5 ≪ 17. Find all normal subgroups of the octic group in Example 8.

Sec. 4.5, #6 ≪ 18. Find all normal subgroups of the alternating group A_4.

Sec. 3.1, #28 ≫ 19. Find all normal subgroups of the quaternion group. (See Exercise 28 of Section 3.1.)

Sec. 4.5, #7 ≪

20. Find groups H and G such that $H \subseteq G \subseteq A_4$ and the following conditions are satisfied:
    a. H is a normal subgroup of G.
    b. G is a normal subgroup of A_4.
    c. H is not a normal subgroup of A_4.
    (Thus, the statement "A normal subgroup of a normal subgroup is a normal subgroup" is false.)

21. Find groups H and K such that the following conditions are satisfied:
    a. H is a normal subgroup of K.
    b. K is a normal subgroup of the octic group.
    c. H is not a normal subgroup of the octic group.

22. Find two groups of order 6 that are not isomorphic.

23. Let H be a subgroup of G and assume that every left coset aH of H in G is equal to a right coset Hb of H in G. Prove that H is a normal subgroup of G.

24. If $\{H_\lambda\}$, $\lambda \in \mathcal{L}$, is a collection of normal subgroups H_λ of G, prove that $\bigcap_{\lambda \in \mathcal{L}} H_\lambda$ is a normal subgroup of G.

Sec. 4.5, #26 ≪ 25. If H is a subgroup of G, and K is an normal subgroup of G, prove that $HK = KH$.

Sec. 4.5, #26 ≪ 26. With H and K as in Exercise 25, prove that HK is a subgroup of G.

Sec. 4.5, #26 ≪ 27. With H and K as in Exercise 25, prove that $H \cap K$ is a normal subgroup of H.

Sec. 4.5, #28 ≪ 28. With H and K as in Exercise 25, prove that K is a normal subgroup of HK.

29. If H and K are arbitrary subgroups of G, prove that $HK = KH$ if and only if HK is a subgroup of G.

30. If H and K are both normal subgroups of G, prove that HK is a normal subgroup of G.

31. Prove that if H and K are normal subgroups of G such that $H \cap K = \{e\}$, then $hk = kh$
Sec. 4.5, #24 ≪ for all $h \in H$, $k \in K$.

Sec. 3.2, #17 ≫ 32. The **center** $Z(G)$ of a group G is defined in Exercise 17 of Section 3.2 as

$$Z(G) = \{a \in G \,|\, ax = xa \text{ for all } x \in G\}.$$

Sec. 4.5, #20, 23 ≪ Prove that $Z(G)$ is a normal subgroup of G.

33. (See Exercise 32.) Find the center of the octic group. (See Example 8.)

34. (See Exercise 32.) Find the center of A_4.

35. For an arbitrary subgroup H of the group G, the **normalizer** of H in G is the set $\mathcal{N}(H) = \{x \in G \mid xHx^{-1} = H\}$.
   a. Prove that $\mathcal{N}(H)$ is a subgroup of G.
   b. Prove that H is a normal subgroup of $\mathcal{N}(H)$.
   c. Prove that if K is a subgroup of G that contains H as a normal subgroup, then $K \subseteq \mathcal{N}(H)$.

36. Find the normalizer of the subgroup $\{(1), (1,3)(2,4)\}$ of the octic group. (See Example 8.)

37. Find the normalizer of the subgroup $\{(1), (1,4)(2,3)\}$ of the octic group. (See Example 8.)

38. Let H be a subgroup of G. Define the relation "congruence modulo H" on G by

$$a \equiv b \pmod{H} \text{ if and only if } a^{-1}b \in H.$$

Prove that congruence modulo H is an equivalence relation on G.

39. Describe the equivalence classes in Exercise 38.

40. Let $n > 1$ in the group of integers under addition, and let $H = \langle n \rangle$. Prove that

$$a \equiv b \pmod{H} \text{ if and only if } a \equiv b \pmod{n}.$$

41. Prove that any subgroup H of G that has index 2 in G is a normal subgroup of G.

42. Show that A_n has index 2 in S_n, and thereby conclude that A_n is always a normal subgroup of S_n.

Sec. 2.5, #41 ≫ **43.** Let p be a prime and consider the multiplicative group G of all nonzero elements of $\mathbf{Z}_p$. Use Lagrange's Theorem in G to prove **Fermat's Little Theorem** in the form $[a]^p = [a]$ for any $a \in \mathbf{Z}$. (Compare with Exercise 41 in Section 2.5.)

44. Let G be a group of order pq, where p and q are distinct prime integers. If G has only one subgroup of order p and only one subgroup of order q, prove that G is cyclic.

45. Find the subgroup of S_4 that is generated by the given set.
   a. $\{(1,3), (1,2,3,4)\}$ **b.** $\{(1,2,4), (2,3,4)\}$ **c.** $\{(1,2), (1,3), (1,4)\}$

46. Let n be a positive integer, $n > 1$. Prove by induction that the set of transpositions $\{(1,2), (1,3), \ldots, (1,n)\}$ generates the entire group S_n.

47. A subgroup H of the group S_n is called **transitive** on $B = \{1, 2, \ldots, n\}$ if for each pair i, j of elements of B, there exists an element $h \in H$ such that $h(i) = j$. Suppose G is a group that is transitive on $\{1, 2, \ldots, n\}$, and let H_i be the subgroup of G that leaves i fixed:

$$H_i = \{g \in G \mid g(i) = i\}$$

for $i = 1, 2, \ldots, n$. Prove that $o(G) = n \cdot o(H_i)$.

48. (See Exercise 47.) Suppose G is a group that is transitive on $\{1, 2, \ldots, n\}$, and let K_i be the subgroup that leaves each of the elements $1, 2, \ldots, i$ fixed:

$$K_i = \{g \in G \mid g(k) = k \text{ for } k = 1, 2, \ldots, i\}$$

for $i = 1, 2, \ldots, n$. Prove that $G = S_n$ if and only if $H_i \neq H_j$ for all pairs i, j such that $i \neq j$ and $i < n - 1$.

4.5 QUOTIENT GROUPS

If H is a normal subgroup of G, then $xH = Hx$ for all x in G, so there is no distinction between left and right cosets of H in G. In this case, we refer simply to the cosets of H in G.

If H is any subgroup of G, then $hH = H = Hh$ for all h in H, according to Theorem 4.16. Corollary 4.17 states that $H^2 = H \cdot H = H$ for all subgroups H. We use this fact in proving the next theorem.

Theorem. 4.21   **GROUP OF COSETS**

Let H be a normal subgroup of G. Then the cosets of H in G form a group with respect to the product of subsets as given in Definition 4.8.

$p \Rightarrow q$ **Proof**   Let H be a normal subgroup of G. We shall denote the set of all distinct cosets of H in G by G/H. Multiplication in G/H is associative, by part a of Theorem 4.9.

We need to show that the cosets of H in G are closed under the given product. Let aH and bH be arbitrary cosets of H in G. Using the associative property freely, we have

$$
\begin{aligned}
(aH)(bH) &= a(Hb)H \\
&= a(bH)H && \text{since } H \text{ is normal} \\
&= (ab)H \cdot H \\
&= abH && \text{since } H^2 = H.
\end{aligned}
$$

Thus, G/H is closed and $(aH)(bH) = abH$.

The coset $H = eH$ is an identity element, since $(aH)(eH) = aeH = aH$ and $(eH)(aH) = eaH = aH$ for all aH in G/H.

The inverse of aH is $a^{-1}H$, since

$$
(aH)(a^{-1}H) = aa^{-1}H = eH = H
$$

and

$$
(a^{-1}H)(aH) = a^{-1}aH = eH = H.
$$

This completes the proof.

Definition 4.22   **QUOTIENT GROUP**

If H is a normal subgroup of G, the group G/H that consists of the cosets of H in G is called the **quotient group** or **factor group** of G by H.

Example 1   Let G be the octic group as given in Example 8 of Section 4.4:

$$
G = \{e, \alpha, \alpha^2, \alpha^3, \beta, \gamma, \Delta, \theta\}.
$$

It can be readily verified that $H = \{e, \gamma, \theta, \alpha^2\}$ is a normal subgroup of G. The distinct cosets of H in G are

$$
H = eH = \gamma H = \theta H = \alpha^2 H = \{e, \gamma, \theta, \alpha^2\}
$$

and

$$
\alpha H = \alpha^3 H = \beta H = \Delta H = \{\alpha, \alpha^3, \beta, \Delta\}.
$$

Thus, $G/H = \{H, \alpha H\}$, and a multiplication table for G/H is as follows.

$\cdot$	H	αH
H	H	αH
αH	αH	H

$\blacksquare$

There is a very important and natural relation between the quotient groups of a group G and the epimorphisms from G to another group G'. Our next theorem shows that every quotient group G/H is a homomorphic image of G.

Theorem 4.23   **QUOTIENT GROUP $\Rightarrow$ HOMOMORPHIC IMAGE**

Let G be a group and let H be a normal subgroup of G. The mapping $\phi: G \to G/H$ defined by

$$\phi(a) = aH$$

is an epimorphism from G to G/H.

$p \Rightarrow q$ **Proof**   The rule $\phi(a) = aH$ clearly defines a mapping from G to G/H. For any a and b in G,

$$\phi(a) \cdot \phi(b) = (aH)(bH)$$
$$= abH \qquad \text{since } H \text{ is normal in } G$$
$$= \phi(ab).$$

Thus, ϕ is a homomorphism. Every element of G/H is a coset of H in G that has the form aH for some a in G. For any such a, we have $\phi(a) = aH$. Therefore, ϕ is an epimorphism.

Example 2   Consider the octic group

$$G = \{e, \alpha, \alpha^2, \alpha^3, \beta, \gamma, \Delta, \theta\}$$

and its normal subgroup

$$H = \{e, \gamma, \theta, \alpha^2\}.$$

We saw in Example 1 that $G/H = \{H, \alpha H\}$. Theorem 4.23 assures us that the mapping $\phi: G \to G/H$ defined by

$$\phi(a) = aH$$

is an epimorphism. The values of ϕ are given in this case by

$$\phi(e) = \phi(\gamma) = \phi(\theta) = \phi(\alpha^2) = H$$
$$\phi(\alpha) = \phi(\alpha^3) = \phi(\beta) = \phi(\Delta) = \alpha H.$$

$\blacksquare$

Theorem 4.23 says that every quotient G/H is a homormophic image of G. We shall see that, up to an isomorphism, these quotient groups give all of the homomorphic images of G. In order to prove this, we need the following result about the kernel of a homomorphism.

Theorem 4.24   **KERNEL OF A HOMOMORPHISM**

For any homomorphism ϕ from the group G to the group G', ker ϕ is a normal subgroup of G.

$p \Rightarrow q$ **Proof**   The identity e is in ker ϕ since $\phi(e) = e'$, so ker ϕ is always nonempty. If $a \in$ ker ϕ and $b \in$ ker ϕ, then $\phi(a) = e'$ and $\phi(b) = e'$, so that

$$\phi(ab) = \phi(a) \cdot \phi(b) = e' \cdot e' = e',$$

and therefore $ab \in$ ker ϕ. By Theorem 3.28, $\phi(x) = e'$ implies $\phi(x^{-1}) = (e')^{-1} = e'$, so ker ϕ contains inverses and we have proved that ker ϕ is a subgroup of G.

To show that ker ϕ is normal, let $x \in G$ and $a \in$ ker ϕ. Then

$$
\begin{aligned}
\phi(xax^{-1}) &= \phi(x)\phi(a)\phi(x^{-1}) &&\text{since } \phi \text{ is a homomorphism} \\
&= \phi(x) \cdot e' \cdot \phi(x^{-1}) &&\text{since } a \in \text{ ker } \phi \\
&= \phi(x) \cdot \phi(x^{-1}) \\
&= e' &&\text{by part b of Theorem 3.28.}
\end{aligned}
$$

Thus, xax^{-1} is in ker ϕ and ker ϕ is a normal subgroup by Theorem 4.18.

The mapping ϕ in Theorem 4.23 has H as its kernel and this shows that every normal subgroup of G is the kernel of a homomorphism. Combining this fact with Theorem 4.24, we see that the normal subgroups of G and the kernels of the homomorphisms from G to another group are the same subgroups of G.

We can now prove that every homomorphic image of G is isomorphic to a quotient group of G.

Theorem 4.25   **HOMOMORPHIC IMAGE $\Rightarrow$ QUOTIENT GROUP**

Let G and G' be groups with G' a homomorphic image of G. Then G' is isomorphic to a quotient group of G.

$p \Rightarrow q$ **Proof**   Let ϕ be an epimorphism from G to G', and let $K = $ ker ϕ. For each aK in G/K, define $\theta(aK)$ by

$$\theta(aK) = \phi(a).$$

First we need to prove that this rule defines a mapping. For any aK and bK in G/K,

$$
\begin{aligned}
aK = bK &\Leftrightarrow b^{-1}aK = K \\
&\Leftrightarrow b^{-1}a \in K \\
&\Leftrightarrow \phi(b^{-1}a) = e' \\
&\Leftrightarrow \phi(b^{-1})\phi(a) = e' \\
&\Leftrightarrow [\phi(b)]^{-1}\phi(a) = e' \\
&\Leftrightarrow \phi(a) = \phi(b) \\
&\Leftrightarrow \theta(aK) = \theta(bK).
\end{aligned}
$$

Thus, θ is a well-defined mapping from G/K to G', and the $\Leftarrow$ parts of the $\Leftrightarrow$ statements show that θ is one-to-one as well.

We shall show that θ is an isomorphism from G/K to G'. Since

$$\begin{aligned}
\theta(aK \cdot bK) &= \theta(abK) \\
&= \phi(ab) \\
&= \phi(a) \cdot \phi(b) \\
&= \theta(aK) \cdot \theta(bK),
\end{aligned}$$

θ is a homomorphism. To show that θ is onto, let a' be arbitrary in G'. Since ϕ is an epimorphism, there exists an element a in G such that $\phi(a) = a'$. Then aK is in G/K, and

$$\theta(aK) = \phi(a) = a'.$$

Thus, every element in G' is an image under θ, and this proves that θ is an isomorphism.

Theorem 4.26 FUNDAMENTAL THEOREM OF HOMOMORPHISMS

If ϕ is an epimorphism from the group G to the group G', then G' is isomorphic to $G/\ker \phi$.

The Fundamental Theorem follows at once from the proof of Theorem 4.25.

In order to give nontrivial illustrations of Theorem 4.24 and 4.25, we need an example of a homomorphism that is somewhat involved. This homomorphism is presented in the next example.

Example 3 Consider the permutation group

$$G = S_3 = \{(1), (1, 2, 3), (1, 3, 2), (1, 2), (1, 3), (2, 3)\}$$

and the multiplicative group

$$G' = \{[1], [2]\} \subseteq \mathbf{Z}_3.$$

The mapping $\phi: G \to G'$ defined by

$$\begin{aligned}
\phi(1) = \phi(1, 2, 3) = \phi(1, 3, 2) = [1] \\
\phi(1, 2) = \phi(1, 3) = \phi(2, 3) = [2]
\end{aligned}$$

can be shown by direct computation to be an epimorphism from G to G', but it is tedious to verify $\phi(xy) = \phi(x)\phi(y)$ for all 36 choices of the pair of factors x, y in S_3. As an alternative to this chore, we shall obtain another description of ϕ. We first note that if $\alpha = (1, 2, 3)$ and $\beta = (1, 2)$, the elements of S_3 can be written as

$$\begin{aligned}
(1) = \alpha^0\beta^0 \quad (1, 2, 3) = \alpha\beta^0 \quad (1, 3, 2) = \alpha^2\beta^0 \\
(1, 2) = \alpha^0\beta \quad\quad (1, 3) = \alpha\beta \quad\quad (2, 3) = \alpha^2\beta.
\end{aligned}$$

We then make the following observations concerning S_3:

1. Any element of S_3 can be written in the form $\alpha^i\beta^k$, with $i \in \{0, 1, 2\}$ and $k \in \{0, 1\}$;
2. $\beta\alpha^i = \alpha^{-i}\beta$, and
3. Any $x \in S_3$ is either of the form $x = \alpha^i$ or of the form $x = \alpha^i\beta$.

Routine calculations will confirm that our mapping ϕ can be described by the rule

$$\phi(\alpha^r \beta^k) = [2]^k \text{ for any integer } r.$$

Having made these observations, we can now verify the equation $\phi(x)\phi(y) = \phi(xy)$ with a reasonable amount of work. For arbitrary x and y in S_3, we write either $x = \alpha^i$ or $x = \alpha^i\beta$, and $y = \alpha^m\beta^n$ where $m \in \{0, 1, 2\}$ and $n \in \{0, 1\}$.

If $x = \alpha^i$, we have

$$\phi(xy) = \phi(\alpha^i\alpha^m\beta^n) = \phi(\alpha^{i+m}\beta^n) = [2]^n$$

and

$$\phi(x)\phi(y) = \phi(\alpha^i)\phi(\alpha^m\beta^n) = [2]^0[2]^n = [2]^n.$$

If $x = \alpha^i\beta$, we have

$$\begin{aligned}
\phi(xy) &= \phi(\alpha^i\beta\alpha^m\beta^n) \\
&= \phi(\alpha^i\alpha^{-m}\beta\beta^n) \\
&= \phi(\alpha^{i-m}\beta^{n+1}) \\
&= [2]^{n+1}
\end{aligned}$$

and

$$\begin{aligned}
\phi(x)\phi(y) &= \phi(\alpha^i\beta)\phi(\alpha^m\beta^n) \\
&= [2][2]^n \\
&= [2]^{n+1}.
\end{aligned}$$

Thus, $\phi(xy) = \phi(x)\phi(y)$ in all cases, and ϕ is a homomorphism (an epimorphism, actually) from G to G'. ∎

Example 4   To illustrate Theorems 4.24 and 4.25, consider the groups $G = S_3$ and $G' = \{[1], [2]\}$ in the previous example. We see that the kernel of the epimorphism $\phi: G \rightarrow G'$ is the normal subgroup

$$\begin{aligned}
K &= \ker \phi \\
&= \{(1), (1, 2, 3), (1, 3, 2)\}
\end{aligned}$$

of G. The quotient group G/K is given by

$$G/K = \{K, (1, 2)K\}$$

where

$$(1, 2)K = \{(1, 2), (2, 3), (1, 3)\}.$$

The isomorphism $\theta: G/K \rightarrow G'$ has values

$$\begin{aligned}
\theta(K) &= \phi(1) = [1] \\
\theta((1, 2)K) &= \phi(1, 2) = [2].
\end{aligned}$$
∎

Using the results of this section, we can systematically find all of the homomorphic images of a group G. We now know that the homomorphic images of G are the same (in the sense of isomorphism) as the quotient groups of G.

Example 5   Let $G = S_3$, the symmetric group on three elements. In order to find all the homomorphic images of G, we only need to find all of the normal subgroups H of G and form all possible quotient groups G/H. As we saw in Section 4.4, a complete list of the subgroups of G is

$$H_1 = \{(1)\} \qquad\qquad H_4 = \{(1), (2, 3)\}$$
$$H_2 = \{(1), (1, 2)\} \qquad H_5 = \{(1), (1, 2, 3), (1, 3, 2)\}$$
$$H_3 = \{(1), (1, 3)\} \qquad H_6 = S_3.$$

Of these, H_1, H_5, and H_6 are the only normal subgroups. The possible homomorphic images of G, then, are

$$G/H_1 = \{H_1, (1, 2)H_1, (1, 3)H_1, (2, 3)H_1, (1, 2, 3)H_1, (1, 3, 2)H_1\}$$
$$G/H_5 = \{H_5, (1, 2)H_5\}$$
$$G/G = \{G\}.$$

Thus, any homomorphic image of S_3 is isomorphic either to S_3, to a cyclic group of order 2, or to a group with only the identity element.   ∎

EXERCISES 4.5

1. Let G be the octic group in Example 8 of Section 4.4, and let $H = \{e, \alpha^2\}$. Assume that H is a normal subgroup of G. Write out the distinct elements of G/H and construct a multiplication table for G/H.

2. Let $G = A_4$ and assume that the four group

$$H = \{(1), (1, 2)(3, 4), (1, 3)(2, 4), (1, 4)(2, 3)\}$$

is a normal subgroup of A_4. Write out the distinct elements of G/H and make a multiplication table for G/H.

Sec. 3.1, #28 ≫ **3.** Let G be the quaternion group in Exercise 28 of Section 3.1, and let $H = \{1, -1\}$. Assume that H is a normal subgroup of G. Write out the distinct elements of G/H and make a multiplication table for G/H.

4. Let G be the multiplicative group consisting of all $[a]$ in $\mathbf{Z}_{20}$ that have multiplicative inverses. Find a normal subgroup H of G that has order 2 and construct a multiplication table for G/H.

Sec. 4.4, #17 ≫ **5.** Find all homomorphic images of the octic group (see Exercise 17 of Section 4.4).

Sec. 4.4, #18 ≫ **6.** Find all homomorphic images of A_4 (see Exercise 18 of Section 4.4).

Sec. 4.4, #19 ≫ **7.** Find all homomorphic images of the quaternion group (see Exercise 19 of Section 4.4).

Sec. 3.3, #16 ≫ **8.** Find all homomorphic images of each group G in Exercise 16 of Section 3.3.

9. Let $G = S_3$. For each H that follows show that the set of all left cosets of H in G does *not* form a group with respect to a product defined by $(aH)(bH) = abH$.
a. $H = \{(1), (1, 2)\}$
b. $H = \{(1), (1, 3)\}$
c. $H = \{(1), (2, 3)\}$

10. Assume that the four group

$$H = \{(1), (1, 2)(3, 4), (1, 3)(2, 4), (1, 4)(2, 3)\}$$

is a normal subgroup of S_4. Write out the distinct cosets of H in S_4 and construct a multiplication table for S_4/H.

11. Consider the mapping from the symmetric group S_n to the additive group $\mathbf{Z}_2$ defined by

$$\phi(x) = \begin{cases} [0] & \text{if } x \text{ is an even permutation} \\ [1] & \text{if } x \text{ is an odd permutation.} \end{cases}$$

Decide whether or not ϕ is a homomorphism and state the kernel if it is a homomorphism.

Sec. 3.1, #30 ≫

Sec. 3.5, #3 ≫

12. Let $G = \{I_2, R, R^2, R^3, H, D, V, T\}$ be the multiplicative group of matrices in Exercise 30 of Section 3.1, let $G' = \{1, -1\}$ under multiplication, and define $\phi: G \to G'$ by

$$\phi\left(\begin{bmatrix} a & b \\ c & d \end{bmatrix}\right) = ad - bc.$$

   a. Assume that ϕ is an epimorphism and find the elements of $K = \ker \phi$.
   b. Write out the distinct elements of G/K.
   c. Let $\theta: G/K \to G'$ be the isomorphism described in the proof of Theorem 4.25 and write out the values of θ.

13. If H is a subgroup of the group G such that $(aH)(bH) = abH$ for all left cosets aH and bH of H in G, prove that H is normal in G.

14. Let H be a subgroup of the group G. Prove that H is a normal subgroup of G if and only if $(Ha)(Hb) = Hab$ for all right cosets Ha and Hb of H in G.

15. If H is a normal subgroup of the group G, prove that $(aH)^n = a^n H$ for every positive integer n.

16. Let G by a cyclic group. Prove that for every subgroup H of G, G/H is a cyclic group.

17. a. Show that a cyclic group of order 8 has a cyclic group of order 4 as a homomorphic image.
   b. Show that a cyclic group of order 6 has a cyclic group of order 2 as a homomorphic image.

Sec. 3.5, #13 ≫

18. (See Exercise 13 in Section 3.5.) Assume that ϕ is an epimorphism from the group G to the group G'.
   a. Prove that the mapping $H \to \phi(H)$ is a bijection from the set of all subgroups of G that contain $\ker \phi$ to the set of all subgroups of G'.
   b. Prove that if K is a normal subgroup of G', then $\phi^{-1}(K)$ is a normal subgroup of G.

19. Suppose ϕ is an epimorphism from the group G to the group G'. Let H be a normal subgroup of G containing $\ker \phi$, and let $H' = \phi(H)$.
   a. Prove that H' is a normal subgroup of G'.
   b. Prove that G/H is isomorphic to G'/H'.

Sec. 4.4, #32 ≫

20. (See Exercise 32 of Section 4.4.) Let G be a group with center $Z(G) = C$. Prove that if G/C is cyclic, then G is abelian.

21. (See Exercise 20.) Prove that if p and q are primes and G is a nonabelian group of order pq, then the center of G is the trivial subgroup $\{e\}$.

Sec. 3.5, #15 ≫

Sec. 3.4, #16 ≫

22. (See Exercise 15 of Section 3.5.) Let a be a fixed element of the group G. According to Exercise 16 of Section 3.4, the mapping $t_a: G \to G$ defined by $t_a(x) = axa^{-1}$ is an automorphism of G. Each of these automorphisms t_a is called an **inner automorphism** of G. Prove that the set $Inn(G) = \{t_a | a \in G\}$ forms a normal subgroup of the group of all automorphisms of G.

Sec. 4.4, #32 ≫

23. (See Exercise 22 here and Exercise 32 of Section 4.4.) Let G be a group with center $Z(G) = C$. Prove that $Inn(G)$ is isomorphic to G/C.

Sec. 4.4, #31 ≫

24. (See Exercise 31 of Section 4.4.) If H and K are normal subgroups of the group G such that $G = HK$ and $H \cap K = \{e\}$, then G is said to be the **internal direct product** of H and K, and we write $G = H \times K$ to denote this. If $G = H \times K$, prove that $\phi: H \to G/K$ defined by $\phi(h) = hK$ is an isomorphism from H to G/K.

25. (See Exercise 24.) If $G = H \times K$, prove that each element $g \in G$ can be written uniquely as $g = hk$ with $h \in H$ and $k \in K$.

Sec. 4.4, #25-28 ≫

26. (See Exercises 25–28 of Section 4.4) Let H be a subgroup of G and let K be a normal subgroup of G.

a. Prove that the mapping $\phi: H \to HK/K$ defined by $\phi(h) = hK$ is an epimorphism from H to HK/K.

b. Prove that $\ker \phi = H \cap K$.

Sec. 6.2, #20 ≪

c. Prove that $H/H \cap K$ is isomorphic to HK/K.

27. Let H and K be arbitrary groups and let $H \otimes K$ denote the Cartesian product of H and K:

$$H \otimes K = \{(h, k) | h \in H \text{ and } k \in K\}.$$

Equality in $H \otimes K$ is defined by $(h, k) = (h', k')$ if and only if $h = h'$ and $k = k'$. Multiplication in $H \otimes K$ is defined by

$$(h_1, k_1)(h_2, k_2) = (h_1 h_2, k_1 k_2).$$

a. Prove that $H \otimes K$ is a group. This group is called the **external direct product** of H and K.

b. Suppose that e_1 and e_2 are the identity elements of H and K, respectively. Show that $H' = \{(h, e_2) | h \in H\}$ is a normal subgroup of $H \otimes K$ that is isomorphic to H, and similarly that $K' = \{(e_1, k) | k \in K\}$ is a normal subgroup isomorphic to K.

c. Prove that $H \otimes K/H'$ is isomorphic to K and $H \otimes K/K'$ is isomorphic to H.

28. (See Exercise 27.) Let a and b be fixed elements of a group G, and let $\mathbf{Z} \otimes \mathbf{Z}$ be the external direct product of the additive group $\mathbf{Z}$ with itself. Prove that the mapping $\phi: \mathbf{Z} \otimes \mathbf{Z} \to G$ defined by $\phi(m, n) = a^m b^n$ is a homomorphism if and only if $ab = ba$ in G.

4.6 DIRECT SUMS (OPTIONAL)

The overall objective of this and the next section is to present some of the basic material on abelian groups. A tremendous amount of work has been done on the subject. One of the concepts fundamental to abelian groups is a *direct sum*, to be defined presently. Throughout this section we write all abelian groups in additive notation.

We begin by defining the sum of a finite number of subgroups in an abelian group and showing that this sum is a subgroup.

Definition 4.27 **SUM OF SUBGROUPS**

> If $H_1, H_2, \ldots, H_n$ are subgroups of the abelian group G, then the **sum** $H_1 + H_2 + \cdots + H_n$ of these subgroups is defined by
>
> $$H_1 + H_2 + \cdots + H_n = \{x \in G | x = h_1 + h_2 + \cdots + h_n \text{ with } h_i \in H_i\}.$$

Theorem 4.28 **SUM OF SUBGROUPS**

If $H_1, H_2, \ldots, H_n$ are subgroups of the abelian group G, then $H_1 + H_2 + \cdots + H_n$ is a subgroup of G.

$p \Rightarrow q$ **Proof** The sum $H_1 + H_2 + \cdots + H_n$ is clearly nonempty. For arbitrary

$$x = h_1 + h_2 + \cdots + h_n$$

with $h_i \in H_i$, the inverse

$$-x = (-h_1) + (-h_2) + \cdots + (-h_n)$$

is in the sum $H_1 + H_2 + \cdots + H_n$, since $-h_i \in H_i$ for each i. Also, if

$$y = h_1' + h_2' + \cdots + h_n'$$

with $h_i' \in H_i$, then

$$x + y = (h_1 + h_1') + (h_2 + h_2') + \cdots + (h_n + h_n')$$

is in the sum of the H_i, since $h_i + h_i' \in H_i$ for each i. Thus, $H_1 + H_2 + \cdots + H_n$ is a subgroup of G.

The contents of Definition 4.19 and Theorem 4.20 may be restated as follows, with addition as the binary operation:

If A is a nonempty subset of the group G, then the *subgroup of G generated by A* is the set

$$\langle A \rangle = \{x \in G \,|\, x = a_1 + a_2 + \cdots + a_n \text{ with } a_i \in A \text{ or } -a_i \in A\}.$$

It is left as an exercise to prove that if $H_1, H_2, \ldots, H_n$ are subgroups of an abelian group G, then $G = H_1 + H_2 + \cdots + H_n$ if and only if G is generated by $\bigcup_{i=1}^{n} H_i$.

Example I Let G be the group $G = \mathbf{Z}_{12}$ under addition, and consider the following sums of subgroups in G.

a. If

$$H_1 = \langle [3] \rangle = \{[3], [6], [9], [0]\}$$

and

$$H_2 = \langle [2] \rangle = \{[2], [4], [6], [8], [10], [0]\},$$

then

$$\begin{aligned} H_1 + H_2 &= \{r[3] + s[2] \mid r, s \in \mathbf{Z}\} \\ &= \{[3r + 2s] \mid r, s \in \mathbf{Z}\} \end{aligned}$$

is a subgroup. Since $[3(1) + 2(11)] = [25] = [1]$ in $\mathbf{Z}_{12}$ and $[1]$ generates $\mathbf{Z}_{12}$ under addition, we have

$$H_1 + H_2 = G.$$

b. Now let

$$\begin{aligned} K_1 &= H_1 = \langle [3] \rangle, \\ K_2 &= \langle [4] \rangle = \{[4], [8], [0]\}. \end{aligned}$$

The sum $K_1 + K_2$, is given by

$$\begin{aligned} K_1 + K_2 &= \{u[3] + v[4] \mid u, v \in \mathbf{Z}\} \\ &= \{[3u + 4v] \mid u, v \in \mathbf{Z}\}. \end{aligned}$$

Since $[3(-1) + 4(1)] = [1], [1] \in K_1 + K_2$, and hence

$$K_1 + K_2 = G.$$

c. With the same notation as in parts **a** and **b**,

$$H_2 + K_2 = H_2,$$

since $K_2 \subseteq H_2$.   ■

We now consider the definition of a direct sum.

Definition 4.29 **DIRECT SUM**

> If $H_1, H_2, \ldots, H_n$ are subgroups of the abelian group G, then $H_1 + H_2 + \cdots + H_n$ is a **direct sum** if and only if the expression for each x in the sum as
>
> $$x = h_1 + h_2 + \cdots + h_n$$
>
> with $h_i \in H_i$ is **unique**. We write
>
> $$H_1 \oplus H_2 \oplus \cdots \oplus H_n$$
>
> to indicate a direct sum.

The next theorem gives a simple fact about direct sums that can be very useful when we work with finite groups.

Theorem 4.30 **ORDER OF A DIRECT SUM**

If $H_1, H_2, \ldots, H_n$ are finite subgroups of the abelian group G such that their sum is direct, then the order of $H_1 \oplus H_2 \oplus \cdots \oplus H_n$ is the product of the orders of the subgroups H_i:

$$o(H_1 \oplus H_2 \oplus \cdots \oplus H_n) = o(H_1)o(H_2) \cdots o(H_n).$$

$p \Rightarrow q$ **Proof**   With $h_i \in H_i$ in the expression

$$x = h_1 + h_2 + \cdots + h_n,$$

there are $o(H_i)$ choices for each h_i. Any change in one of the h_i produces a different element x, by the uniqueness property stated in Definition 4.29. Hence, there are

$$o(H_1)o(H_2) \cdots o(H_n)$$

distinct elements x of the form $x = h_1 + h_2 + \cdots + h_n$, and the theorem follows.

There are several equivalent ways to formulate the definition of direct sum. One of these is presented in the following theorem.

Theorem 4.31 **EQUIVALENT CONDITION FOR A DIRECT SUM**

If each H_i is a subgroup of the abelian group G, the sum $H_1 + H_2 + \cdots + H_n$ is direct if and only if the following condition holds: Any equation of the form

$$h_1 + h_2 + \cdots + h_n = 0$$

with $h_i \in H_i$ implies that all $h_i = 0$.

$p \Leftarrow q$ **Proof**   Assume first that the condition holds. If an element x in the sum of the H_i is written as

$$x = h_1 + h_2 + \cdots + h_n$$

and also as

$$x = h'_1 + h'_2 + \cdots + h'_n$$

with h_i and $h'_i \in H_i$ for each i, then

$$h_1 + h_2 + \cdots + h_n = h'_1 + h'_2 + \cdots + h'_n$$

and

$$(h_1 - h'_1) + (h_2 - h'_2) + \cdots + (h_n - h'_n) = 0.$$

The condition implies that $h_i - h'_i = 0$, and hence $h_i = h'_i$ for each i. Thus, the sum $H_1 + H_2 + \cdots + H_n$ is direct.

$p \Rightarrow q$ Conversely, suppose the sum $H_1 + H_2 + \cdots + H_n$ is direct. Then the identity element 0 in the sum can be written *uniquely* as

$$0 = 0 + 0 + \cdots + 0$$

where the sum on the right indicates a choice of 0 as the term from each H_i. From the uniqueness property,

$$h_1 + h_2 + \cdots + h_n = 0$$

with $h_i \in H_i$ requires that all $h_i = 0$.

Some intuitive feeling for the concept of a direct sum is provided by considering the special case where the sum has only two terms.

Theorem 4.32   DIRECT SUM OF TWO SUBGROUPS

Let H_1 and H_2 be subgroups of the abelian group G. Then $G = H_1 \oplus H_2$ if and only if $G = H_1 + H_2$ and $H_1 \cap H_2 = \{0\}$.

$p \Rightarrow (q \wedge r)$ **Proof**   Assume first that $G = H_1 \oplus H_2$, and let $x \in H_1 \cap H_2$. Then $x = h_1$ for some $h_1 \in H_1$. Also, $x \in H_2$, and therefore $-x \in H_2$. Let $h_2 = -x$. Then

$$h_1 + h_2 = x + (-x)$$
$$= 0$$

where $h_i \in H_i$. This implies that $x = h_1 = h_2 = 0$, by Theorem 4.31.

$p \Leftarrow (q \wedge r)$ Assume now that $G = H_1 + H_2$ and $H_1 \cap H_2 = \{0\}$. If

$$h_1 + h_2 = 0$$

with $h_i \in H_i$, then $h_1 = -h_2 \in H_1 \cap H_2$. Therefore, $h_1 = 0$ and $h_2 = 0$. By Theorem 4.31, $G = H_1 \oplus H_2$.

Example 2   In Example 1, we saw that the equations $H_1 + H_2 = G$ and $K_1 + K_2 = G$ were both valid. Since $H_1 \cap H_2 = \{[0], [6]\}$, the sum $H_1 + H_2$ is not direct. However, $K_1 \cap K_2 = \{[0]\}$, so $G = K_1 \oplus K_2$ in Example 1.   ∎

Theorem 4.32 can be generalized to the results stated in the next theorem. A proof is requested in the exercises.

Theorem 4.33   DIRECT SUM OF n SUBGROUPS

Let $H_1, H_2, \ldots, H_n$ be subgroups of the abelian group G. The sum $H_1 + H_2 + \cdots + H_n$ is direct if and only if the intersection of each H_j with the subgroup generated by $\bigcup_{i=1, i \neq j}^{n} H_i$ is the identity subgroup $\{0\}$.

As a final result for this section, we prove the following theorem.

Theorem 4.34   DIRECT SUMS AND ISOMORPHISMS

Let H_1 and H_2 be subgroups of the abelian group G such that $G = H_1 \oplus H_2$. Then G/H_2 is isomorphic to H_1.

$p \Rightarrow q$ **Proof**   The rule $\phi(h_1) = h_1 + H_2$ defines a mapping ϕ from H_1 to G/H_2. This mapping is a homomorphism, since

$$\phi(h_1 + h_1') = (h_1 + h_1')H_2$$
$$= (h_1 + H_2) + (h_1' + H_2)$$
$$= \phi(h_1) + \phi(h_1').$$

Now

$$
\begin{aligned}
h_1 \in \ker \phi &\Leftrightarrow \phi(h_1) = H_2 \\
&\Leftrightarrow h_1 + H_2 = H_2 \\
&\Leftrightarrow h_1 \in H_2 \\
&\Leftrightarrow h_1 = 0 \quad \text{since } H_1 \cap H_2 = \{0\}.
\end{aligned}
$$

Thus, ϕ is one-to-one. Let $g + H_2$ be arbitrary in G/H_2. Since $G = H_1 \oplus H_2$, g can be written as $g = h_1 + h_2$ with $h_i \in H_i$.
Then

$$
\begin{aligned}
g + H_2 &= (h_1 + h_2) + H_2 \\
&= h_1 + H_2 \quad \text{since } h_2 + H_2 = H_2 \\
&= \phi(h_1),
\end{aligned}
$$

and this shows that ϕ is onto. Thus, ϕ is an isomorphism from H_1 to G/H_2.

EXERCISES 4.6

1. Suppose that H_1 and H_2 are subgroups of the abelian group G such that $H_1 \subseteq H_2$. Prove that $H_1 + H_2 = H_2$.

2. Suppose that H_1 and H_2 are subgroups of the abelian group G such that $G = H_1 \oplus H_2$. If K is a subgroup of G such that $K \supseteq H_1$, prove that $K = H_1 \oplus (K \cap H_2)$.

3. Assume that $H_1, H_2, \ldots, H_n$ are subgroups of the abelian group G such that the sum $H_1 + H_2 + \cdots + H_n$ is direct. If K_i is a subgroup of H_i for $i = 1, 2, \ldots, n$ prove that $K_1 + K_2 + \cdots + K_n$ is a direct sum.

4. Prove that if each H_i is a subgroup of the abelian group G, then $H_1 + H_2 + \cdots + H_n$ is the smallest subgroup of G that contains all the subgroups H_i.

5. If H_1, H_2, ..., H_n are subgroups of the abelian group G, prove that $G = H_1 + H_2 + \cdots + H_n$ if and only if G is generated by $\bigcup_{i=1}^{n} H_i$.

6. Write $\mathbf{Z}_{20}$ as the direct sum of two of its nontrivial subgroups.

7. Let G be an abelian group of order mn, where m and n are relatively prime. If $H_1 = \{x \in G \mid mx = 0\}$ and $H_2 = \{x \in G \mid nx = 0\}$, prove that $G = H_1 \oplus H_2$.

8. Let H_1 and H_2 be cyclic subgroups of the abelian group G, where $H_1 \cap H_2 = \{0\}$. Prove that $H_1 \oplus H_2$ is cyclic if and only if $o(H_1)$ and $o(H_2)$ are relatively prime.

9. Show that $\mathbf{Z}_{15}$ is isomorphic to $\mathbf{Z}_3 \oplus \mathbf{Z}_5$, where the group operation in each of $\mathbf{Z}_{15}$, $\mathbf{Z}_3$, and $\mathbf{Z}_5$ is addition.

10. Suppose that G and G' are abelian groups such that $G = H_1 \oplus H_2$ and $G' = H'_1 \oplus H'_2$. If H_1 is isomorphic to H'_1 and H_2 is isomorphic to H'_2, prove that G is isomorphic to G'.

11. Suppose a is an element of order rs in an abelian group G. Prove that if r and s are relatively prime, then a can be written in the form $a = b_1 + b_2$, where b_1 has order r and b_2 has order s.

12. (See Exercise 11.) Assume that a is an element of order $r_1 r_2 \cdots r_n$ in an abelian group, where r_i and r_j are relatively prime if $i \neq j$. Prove that a can be written in the form $a = b_1 + b_2 + \cdots + b_n$, where each b_i has order r_i.

13. Prove that if r and s are relatively prime positive integers, then any cyclic group of order rs is the direct sum of a cyclic group of order r and a cyclic group of order s.

14. Prove Theorem 4.33: If H_1, H_2, ..., H_n are subgroups of the abelian group G, then the sum $H_1 + H_2 + \cdots + H_n$ is direct if and only if the intersection of each H_j with the subgroup generated by $\bigcup_{i=1, i \neq j}^{n} H_i$ is the identity subgroup $\{0\}$.

4.7   SOME RESULTS ON FINITE ABELIAN GROUPS (OPTIONAL)

The aim of this section is to sample the flavor of more advanced work in groups while maintaining an acceptable level of rigor in the presentation. We attempt to achieve this balance by restricting our attention to proofs of results for abelian groups. There are instances where more general results hold, but their proofs are beyond the level of this text. In most instances of this sort, the more general results are stated informally and without proof.

The following definition of a p-group is fundamental to this entire section.

Definition 4.35 p-GROUP

> If p is a prime, then a group G is called a **p-group** if and only if each of its elements has an order that is a power of p.

A p-group can be finite or infinite. Although we do not prove it here, a finite group is a p-group if and only if its order is a power of p. Whether or not a group is abelian has nothing at all to do with being a p-group. This is brought out in the following example.

Example 1   With $p = 2$, we can easily exhibit three p-groups of order 8.

a. Consider first the cyclic group $C_8 = \langle a \rangle$ of order 8 generated by the permutation $a = (1, 2, 3, 4, 5, 6, 7, 8)$:

Each of a, a^3, a^5, and a^7 has order 8.
a^2 and a^6 have order 4.
a^4 has order 2.
The identity e has order 1.

Thus, C_8 is a 2-group.

b. Consider now the quaternion group $G = \{\pm 1, \pm i, \pm j, \pm k\}$ of Exercise 28 in Section 3.1:

Each of the elements $\pm i, \pm j, \pm k$ has order 4.
-1 has order 2.
1 has order 1.

Hence, G is another 2-group of order 8.

c. Last, consider the octic group $G' = \{e, \alpha, \alpha^2, \alpha^3, \beta, \gamma, \Delta, \theta\}$ of Example 8 in Section 4.4:

Each of α and α^3 has order 4.
Each of $\alpha^2, \beta, \gamma, \Delta, \theta$ has order 2.
The identity e has order 1.

Thus, G' is also a 2-group of order 8.

Of these three p-groups, C_8 is abelian while both G and G' are nonabelian. ■

It may happen that G is not a p-group, yet some of its subgroups are p-groups. In connection with that possibility, we make the following definition.

Definition 4.36 **THE SET G_p**

> If G is a finite abelian group which has order that is divisible by the prime p, then G_p is the set of all elements of G which have orders that are powers of p.

As might be expected, the set G_p turns out to be a subgroup. For the remainder of this section we write all abelian groups in additive notation.

Theorem 4.37 **p-SUBGROUPS**

The set G_p defined in Definition 4.36 is a subgroup of G.

$u \Rightarrow v$ **Proof** The identity 0 has order $1 = p^0$, so $0 \in G_p$. If $a \in G_p$, then a has order p^r for some nonnegative integer r. Since a and its inverse $-a$ have the same order, $-a$ is also in the set G_p. Let b be another element of the set G_p. Then b has order p^s for a nonnegative integer s. If t is the larger of r and s, then

$$p^t(a + b) = p^t a + p^t b$$
$$= 0 + 0$$
$$= 0.$$

This implies that the order of $a + b$ divides p^t and is therefore a power of p since p is a prime. Thus, $a + b \in G_p$, and set G_p is a subgroup of G.

Example 2   Consider the additive group $G = \mathbf{Z}_6$. The order of $\mathbf{Z}_6$ is 6, which is divisible by the primes 2 and 3. In this group:

Each of [1] and [5] has order 6.
Each of [2] and [4] has order 3.
[3] has order 2.
[0] has order 1.

For $p = 2$ or $p = 3$, the subgroups G_p are given by

$$G_2 = \{[3], [0]\}$$
$$G_3 = \{[2], [4], [0]\}.$$

The group G is not a p-group, but G_2 is a 2-subgroup of G and G_3 is a 3-subgroup of G.   ■

If a group G has p-subgroups, certain of them are given special names, as described in the following definition.

Definition 4.38   **SYLOW p-SUBGROUP**

> If p is a prime and m is a positive integer such that $p^m | o(G)$ and $p^{m+1} \nmid o(G)$, then a subgroup of G that has order p^m is called a **Sylow p-subgroup** of G.

Example 3   In Example 2, G_2 is a Sylow 2-subgroup of G and G_3 is a Sylow 3-subgroup of G. As a less trivial example, consider the octic group from Example 8 of Section 4.4:

$$H = \{e, \alpha, \alpha^2, \alpha^3, \beta, \gamma, \Delta, \theta\}$$

where

$$
\begin{array}{llll}
e = (1) & \alpha = (1, 2, 3, 4) & \alpha^2 = (1, 3)(2, 4) & \alpha^3 = (1, 4, 3, 2) \\
\beta = (1, 4)(2, 3) & \gamma = (2, 4) & \Delta = (1, 2)(3, 4) & \theta = (1, 3).
\end{array}
$$

The group H is a subgroup of order 2^3 in the symmetric group $G = S_4$, which has order $4! = 24$. Since $2^3 | o(S_4)$ and $2^4 \nmid o(S_4)$, the octic group is a Sylow 2-subgroup of S_4.   ■

Theorem 4.39   **CAUCHY'S THEOREM FOR ABELIAN GROUPS**

If G is an abelian group of order n and p is a prime such that $p | n$, then G has at least one element of order p.

Induction   **Proof**   The proof is by induction on the order n of G, using the Second Principle of Finite Induction. For $n = 1$, the theorem holds by default.

Now let k be a positive integer, assume that the theorem is true for all positive integers $n < k$, and let G be an abelian group of order k. Also, suppose that the prime p is a divisor of k.

Consider first the case where G has only the trivial subgroups $\{0\}$ and G. Then any $a \neq 0$ in G must be a generator of G, $G = \langle a \rangle$. It follows from Exercise 32 of Section 3.3 that the order k of G must be a prime. Since p divides this order, then p must equal k, and G actually has $p - 1$ elements of order p, by Theorem 3.22.

Now consider the case where G has a nontrivial subgroup H; that is, $H \neq \{0\}$ and $H \neq G$, so that $1 < o(H) < k$. If $p \mid o(H)$, then H contains an element of order p by the induction hypothesis, and the theorem is true for G. Suppose then that $p \nmid o(H)$. Since G is abelian, H is normal in G, and the quotient group G/H has order

$$o(G/H) = \frac{o(G)}{o(H)}.$$

We have

$$o(G) = o(H)o(G/H),$$

so p divides the product $o(H)o(G/H)$. Since p is a prime and $p \nmid o(H)$, p must divide $o(G/H) < o(G) = k$. Applying the induction hypothesis, the abelian group G/H has an element $b + H$ of order p. Then

$$H = p(b + H) = pb + H,$$

and therefore $pb \in H$, where $b \notin H$. Let $r = o(H)$. The order of pb must be a divisor of r so that $r(pb) = 0$ and $p(rb) = 0$. Since p is a prime and $p \nmid r$, p and r are relatively prime. Hence, there exists integers u and v such that $pu + rv = 1$.

The contention now is that the element $c = rb$ has order p. We have $pc = 0$, and we need to show that $c = rb \neq 0$. Assume the contrary, that $rb = 0$. Then

$$
\begin{aligned}
b &= 1b \\
&= (pu + rv)b \\
&= u(pb) + v(rb) \\
&= u(pb) + 0 \\
&= u(pb).
\end{aligned}
$$

Now $pb \in H$, and therefore $u(pb) \in H$. But $b \notin H$, so we have a contradiction. Thus, $c = rb \neq 0$ is an element of order p in G, and the proof is complete.

Cauchy's Theorem also holds for nonabelian groups, but we do not prove it here. The next theorem applies only to abelian groups.

Theorem 4.40 **SYLOW p-SUBGROUP**

If G is a finite abelian group and p is a prime such that $p \mid o(G)$, then G_p is a Sylow p-subgroup.

$(u \wedge v) \Rightarrow w$ **Proof**   Assume that G is a finite abelian group such that p^m divides $o(G)$ but p^{m+1} does not divide $o(G)$. Then $o(G) = p^m k$, where p and k are relatively prime. We need to prove that G_p has order p^m.

We first argue that $o(G_p)$ is a power of p. If $o(G_p)$ had a prime factor q different from p, then G_p would have to contain an element of order q, according to Cauchy's Theorem. This would contradict the very definition of G_p, so we conclude that $o(G_p)$ is a power of p. Let $o(G_p) = p^t$.

Suppose now that $o(G_p) < p^m$; that is, that $t < m$. Then the quotient group G/G_p has order $p^m k / p^t = p^{m-t} k$, which is divisible by p. Hence, G/G_p contains an element $a + G_p$ of order p, by Theorem 4.39. Then

$$G_p = p(a + G_p) = pa + G_p,$$

and this implies that $pa \in G_p$. Thus, pa has order that is a power of p. This implies that a has order a power of p, and therefore $a \in G_p$; that is, $a + G_p = G_p$. This is a contradiction to the fact that $a + G_p$ has order p. Therefore, $o(G_p) = p^m$, and G_p is a Sylow p-subgroup of G.

The next theorem shows the true significance of the Sylow p-subgroups in the structure of abelian groups.

Theorem 4.41 **DIRECT SUM OF SYLOW p-SUBGROUPS**

Let G be an abelian group of order $n = p_1^{m_1} p_2^{m_2} \cdots p_r^{m_r}$ where the p_i are distinct primes and each m_i is a positive integer. Then

$$G = G_{p_1} \oplus G_{p_2} \oplus \cdots \oplus G_{p_r}$$

where G_{p_i} is the Sylow p_i-subgroup of G that corresponds to the prime p_i.

$u \Rightarrow v$ **Proof** Assume the hypothesis of the theorem. For each prime p_i, G_{p_i} is a Sylow p-subgroup of G by Theorem 4.40. Suppose an element $a_1 \in G_{p_1}$ is also in the subgroup generated by $G_{p_2}, G_{p_3}, \ldots, G_{p_r}$. Then

$$a_1 = a_2 + a_3 + \cdots + a_r$$

where $a_i \in G_{p_i}$. Since G_{p_i} has order $p_i^{m_i}$, $p_i^{m_i} a_i = 0$ for $i = 2, \ldots, r$. Hence,

$$p_2^{m_2} p_3^{m_3} \cdots p_r^{m_r} a_1 = 0.$$

Since the order of any $a_1 \in G_{p_1}$ is a power of p_1 and p_1 is relatively prime to $p_2^{m_2} p_3^{m_3} \cdots p_r^{m_r}$, this requires that $a_1 = 0$. A similar argument shows that the intersection of any G_{p_i} with the subgroup generated by the remaining subgroups

$$G_{p_1}, G_{p_2}, \ldots, G_{p_{i-1}}, G_{p_{i+1}}, \ldots, G_{p_r}$$

is the identity subgroup $\{0\}$. Hence, the sum

$$G_{p_1} \oplus G_{p_2} \oplus \cdots \oplus G_{p_r}$$

is direct and has order equal to the product of the orders $p_i^{m_i}$:

$$o(G_{p_1} \oplus G_{p_2} \oplus \cdots \oplus G_{p_r}) = p_1^{m_1} p_2^{m_2} \cdots p_r^{m_r} = o(G).$$

Therefore,

$$G = G_{p_1} \oplus G_{p_2} \oplus \cdots \oplus G_{p_r}.$$

Example 4 In Example 2, $G = G_2 \oplus G_3$. ∎

Our next theorem is concerned with a class that is more general than finite abelian groups, the *finitely generated abelian groups*. An abelian group G is said to be **finitely generated** if there exists a set of elements $\{a_1, a_2, \ldots, a_n\}$ in G such that every $x \in G$ can be written in the form

$$x = z_1 a_1 + z_2 a_2 + \cdots + z_n a_n$$

where each z_i is an integer. The elements a_i are called **generators** of G and the set $\{a_1, a_2, \ldots, a_n\}$ is called a **generating set** for G. A finite abelian group G is surely a finitely generated group, since G itself is a generating set.

In a finitely generated group, the Well-Ordering Principle assures us that there are generating sets that have the smallest possible number of elements. Such sets are called **minimal generating sets**. The number of elements in a minimal generating set for G is called the **rank** of G.

Theorem 4.42   **DIRECT SUM OF CYCLIC GROUPS**

Any finitely generated abelian group G (and therefore any finite abelian group) is a direct sum of cyclic groups.

Induction   **Proof**   The proof is by induction on the rank of G. If G has rank 1, then G is cyclic and the theorem is true.

Assume that the theorem is true for any group of rank $k - 1$, and let G be a group of rank k. We consider two cases.

Case 1   Suppose there exists a minimal generating set $\{a_1, a_2, \ldots, a_k\}$ for G such that any relation of the form

$$z_1 a_1 + z_2 a_2 + \cdots + z_k a_k = 0$$

with $z_i \in \mathbf{Z}$ implies that $z_1 a_1 = z_2 a_2 = \cdots = z_k a_k = 0$. Then

$$G = \langle a_1 \rangle + \langle a_2 \rangle + \cdots + \langle a_k \rangle,$$

and the theorem is true for this case.

Case 2   Suppose that Case 1 does not hold. That is, for any minimal generating set $\{a_1, a_2, \ldots, a_k\}$ of G, there exists a relation of the form

$$z_1 a_1 + z_2 a_2 + \cdots + z_k a_k = 0$$

with $z_i \in \mathbf{Z}$ such that some of the $z_i a_i \neq 0$. Among all the minimal generating sets and all the relations of this form, there exists a smallest positive integer $\bar{z}_i$ that occurs as a coefficient in one of these relations. Suppose this $\bar{z}_i$ occurs in a relation with the generating set $\{b_1, b_2, \ldots, b_k\}$. If necessary, the elements in $\{b_1, b_2, \ldots, b_k\}$ can be rearranged so that this smallest positive coefficient occurs as $\bar{z}_1$ with b_1 in

$$\bar{z}_1 b_1 + \bar{z}_2 b_2 + \cdots + \bar{z}_k b_k = 0. \tag{1}$$

Now let $s_1, s_2, \ldots, s_k$ be any set of integers that occur as coefficients in a relation of the form

$$s_1 b_1 + s_2 b_2 + \cdots + s_k b_k = 0 \tag{2}$$

with these generators b_i. We shall show that $\bar{z}_1$ divides s_1. By the Division Algorithm, $s_1 = \bar{z}_1 q_1 + r_1$, where $0 \leq r_1 < \bar{z}_1$. Multiplying equation (1) by q_1 and subtracting the result from equation (2), we have

$$r_1 b_1 + (s_2 - \bar{z}_2 q_1) b_2 + \cdots + (s_k - \bar{z}_k q_1) b_k = 0.$$

The condition $0 \leq r_1 < \bar{z}_1$, forces $r_1 = 0$ by choice of $\bar{z}_1$ as the smallest positive integer in a relation of this form. Thus, $\bar{z}_1$ is a factor of s_1.

We now show that $\bar{z}_1 \mid \bar{z}_i$ for $i = 2, \ldots, k$. Consider $\bar{z}_2$, for example. By the Division Algorithm, $\bar{z}_2 = \bar{z}_1 q_2 + r_2$ where $0 \le r_2 < \bar{z}_1$. If we let $b_1' = b_1 + q_2 b_2$, then $\{b_1', b_2, \ldots, b_k\}$ is a minimal generating set for G, and

$$\bar{z}_1 b_1 + \bar{z}_2 b_2 + \cdots + \bar{z}_k b_k = 0$$
$$\Rightarrow \quad \bar{z}_1 (b_1' - q_2 b_2) + \bar{z}_2 b_2 + \cdots + \bar{z}_k b_k = 0$$
$$\Rightarrow \quad \bar{z}_1 b_1' + (\bar{z}_2 - \bar{z}_1 q_2) b_2 + \cdots + \bar{z}_k b_k = 0$$
$$\Rightarrow \quad \bar{z}_1 b_1' + r_2 b_2 + \cdots + \bar{z}_k b_k = 0.$$

Now $r_2 \ne 0$ and $0 \le r_2 < \bar{z}_1$ would contradict the choice of $\bar{z}_1$, so it must be that $r_2 = 0$ and $\bar{z}_1 \mid \bar{z}_2$. The same sort of argument can be applied to each of $\bar{z}_3, \ldots, \bar{z}_k$, so we have $\bar{z}_i = \bar{z}_1 q_i$ for $i = 2, \ldots, k$. Substituting in equation (1), we obtain

$$\bar{z}_1 b_1 + \bar{z}_1 q_2 b_2 + \cdots + \bar{z}_1 q_k b_k = 0.$$

Let $c_1 = b_1 + q_2 b_2 + \cdots + q_k b_k$, and consider the set $\{c_1, b_2, \ldots, b_k\}$. This set generates G, and we have

$$\bar{z}_1 c_1 = \bar{z}_1 b_1 + \bar{z}_1 q_2 b_2 + \cdots + \bar{z}_1 q_k b_k$$
$$= \bar{z}_1 b_1 + \bar{z}_2 b_2 + \cdots + \bar{z}_k b_k$$
$$= 0.$$

If H denotes the subgroup of G that is generated by the set $\{b_2, \ldots, b_k\}$, then $G = \langle c_1 \rangle + H$ since the set $\{c_1, b_2, \ldots, b_k\}$ is a generating set for G. We shall show that the sum is direct.

If $s_1, s_2, \ldots, s_k$ are any integers such that

$$s_1 c_1 + s_2 b_2 + \cdots + s_k b_k = 0,$$

then substitution for c_1 yields

$$s_1 b_1 + (s_1 q_2 + s_2) b_2 + \cdots + (s_1 q_k + s_k) b_k = 0.$$

This implies that $\bar{z}_1$ divides s_1, and therefore $s_1 c_1 = 0$ since $\bar{z}_1 c_1 = 0$. Hence, the sum is direct, and

$$G = \langle c_1 \rangle \oplus H.$$

Since H has rank $k - 1$, the induction hypothesis applies to H, and H is a direct sum of cyclic groups. Therefore, G is a direct sum of cyclic groups, and the theorem follows by induction.

We can now give a complete description of the structure of any finite abelian group G. As in Theorem 4.41,

$$G = G_{p_1} \oplus G_{p_2} \oplus \cdots \oplus G_{p_r},$$

where G_{p_i} is the Sylow p_i-subgroup of order $p_i^{m_i}$ corresponding to the prime p_i. Each G_{p_i} can in turn be decomposed into a direct sum of cyclic subgroups $\langle a_{i,j} \rangle$, each of which has order a power of p_i:

$$G_{p_i} = \langle a_{i,1} \rangle \oplus \langle a_{i,2} \rangle \oplus \cdots \oplus \langle a_{i,t_i} \rangle$$

where the product of the orders of the subgroups $\langle a_{i,j} \rangle$ is $p_i^{m_i}$. This description is frequently referred to as the **Fundamental Theorem on Finite Abelian Groups**. It can be used to systematically describe all the abelian groups of a given finite order, up to isomorphism.

Example 5   For n a positive integer, let C_n denote a cyclic group of order n. If G is an abelian group of order $72 = 2^3 \cdot 3^2$, then G is the direct sum of its Sylow p-subgroups G_2 of order 2^3 and G_3 of order 3^2:

$$G = G_2 \oplus G_3.$$

Each of G_2 and G_3 is a sum of cyclic groups as described in the preceding paragraph. By considering all possibilities for the decompositions of G_2 and G_3, we deduce that any abelian group of order 72 is isomorphic to one of the following direct sums of cyclic groups:

$$
\begin{array}{ll}
C_{2^3} \oplus C_{3^2} & C_{2^3} \oplus C_3 \oplus C_3 \\
C_2 \oplus C_{2^2} \oplus C_{3^2} & C_2 \oplus C_{2^2} \oplus C_3 \oplus C_3 \\
C_2 \oplus C_2 \oplus C_2 \oplus C_{3^2} & C_2 \oplus C_2 \oplus C_2 \oplus C_3 \oplus C_3.
\end{array}
$$
■

The main emphasis of this section has been on finite abelian groups, but the results presented here hardly scratch the surface. As an example of the interesting and important work that has been done on finite groups in general, we state the following theorem without proof.

Theorem 4.43   **SYLOW'S THEOREM**

Let G be a finite group, and let p be a prime integer.

a. If m is a positive integer such that $p^m | o(G)$ and $p^{m+1} \nmid o(G)$, then G has a subgroup of order p^m.
b. For the same prime p, any two Sylow p-subgroups of G are conjugate subgroups.
c. If $p | o(G)$, the number n_p of distinct Sylow p-subgroups of G satisfies $n_p \equiv 1 \pmod{p}$.

The result in part a of Theorem 4.43 can be generalized to state that if $p^m | o(G)$ and $p^{m+1} \nmid o(G)$, then G has a subgroup of order p^k for any $k \in \mathbf{Z}$ such that $0 \leq k \leq m$.

▬ EXERCISES 4.7

1. Give an example of a p-group of order 9.
2. Find two p-groups of order 4 that are not isomorphic.
3. **a.** Find all Sylow 3-subgroups of the alternating group A_4.
   b. Find all Sylow 2-subgroups of A_4.
4. Find all Sylow 3-subgroups of the symmetric group S_4.
5. For each of the following $\mathbf{Z}_n$, let G be the additive group $G = \mathbf{Z}_n$ and write G as a direct sum of cyclic groups.
   a. $\mathbf{Z}_{10}$ **b.** $\mathbf{Z}_{15}$
   c. $\mathbf{Z}_{12}$ **d.** $\mathbf{Z}_{18}$

6. For each of the following values of n, describe all the abelian groups of order n, up to isomorphism.

    a. $n = 6$ **b.** $n = 10$ **c.** $n = 12$

    d. $n = 18$ **e.** $n = 36$ **f.** $n = 100$

7. Show that $\{a_1, a_2, \ldots, a_n\}$ is a generating set for the additive abelian group G if and only if $G = \langle a_1 \rangle + \langle a_2 \rangle + \cdots + \langle a_n \rangle$.

8. Give an example where G is a finite *nonabelian* group with order that is divisble by a prime p, and where the set of all elements which have orders that are powers of p is *not* a subgroup of G.

9. If $p_1, p_2, \ldots, p_r$ are distinct primes, prove that any two abelian groups that have order $n = p_1 p_2 \cdots p_r$ are isomorphic.

10. Suppose that the abelian group G can be written as the direct sum $G = C_{2^2} \oplus C_3 \oplus C_3$, where C_n is a cyclic group of order n.

    a. Prove that G has elements of order 12, but no element of order greater than 12.

    b. Find the number of distinct elements of G that have order 12.

11. Assume that G can be written as the direct sum $G = C_2 \oplus C_2 \oplus C_3 \oplus C_3$, where C_n is a cyclic group of order n.

    a. Prove that G has elements of order 6, but no element of order greater than 6.

    b. Find the number of distinct elements of G that have order 6.

12. Suppose that G is a *cyclic* group of order p^m where p is a prime. If k is any integer such that $0 \le k \le m$, prove that G has a subgroup of order p^k.

13. Prove the result in Exercise 12 for an aribtrary *abelian* group G of order p^m where G is not necessarily cyclic.

14. Prove that if G is an abelian group of order n and s is an integer that divides n, then G has a subgroup of order s.

KEY WORDS AND PHRASES

A PIONEER IN MATHEMATICS ||||||||||||||||||||||||||||||||||||||||||||||||||||||||||||||||||||||

Augustin Louis Cauchy (1789–1857)

Augustin Louis Cauchy, a 19th-century French mathematician, has the distinction of being a major contributor to the development of modern calculus. The calculus that we know today is based substantially on his clear and precise definition of limits, which changed the whole complexion of the field. Cauchy did not limit his attention to calculus, though. In 1814, he began to develop the theory of functions of complex variables. He made significant contributions in the areas of differential equations, infinite series, probability, determinants, and mathematical physics, as well as abstract algebra. The current notation and terminology used for permutations are credited to Cauchy. A major theorem in the study of abelian groups (Theorem 4.39) was proved by Cauchy and thus named for him.

Cauchy was born in Paris on August 21, 1789. By the time he was 11 years old, French mathematicians recognized his rare talent. He went on to study civil engineering and spent the first few years of his career as an engineer in Napoleon's army, pursuing mathematical research on the side. For health reasons, he gave up engineering and began a teaching career that was mathematically fruitful in spite of political unrest in France. In 1830, Cauchy, an ardent supporter of King Charles X, refused to swear allegiance to the new government after the exile of the king. He lost his professorship and was forced to leave France for eight years. He subsequently taught in church schools and produced so many papers that the Academy of Sciences, alarmed at their printing bill (largely due to Cauchy), passed a rule limiting each paper to four pages. After the February Revolution of 1848, Cauchy was appointed professor of celestial mechanics at the École Polytechnique, a position he retained for the rest of his career.

5

RINGS, INTEGRAL DOMAINS, AND FIELDS

INTRODUCTION

Rings, integral domains, and fields are introduced in this chapter. The field of quotients of an integral domain is constructed, and ordered integral domains are considered. The development of $\mathbf{Z}_n$ continues in Section 5.1, where it appears for the first time in its proper context as a ring.

▰▰▰▰▰ 5.1 DEFINITION OF A RING

A group is one of the simpler algebraic systems because it has only one binary operation. A step upward in the order of complexity is the *ring*. A ring has two binary operations, called *addition* and *multiplication*. Conditions are made on both binary operations, but fewer are made on multiplication. A full list of the conditions is in our formal definition.

Definition 5.1a **DEFINITION OF A RING**

> Suppose R is a set in which a relation of equality, denoted by $+$, and operations of addition and multiplication, denoted by $+$ and $\cdot$, respectively, are defined. Then R is a **ring** (with respect to these operations) if the following conditions are satisfied.
>
> **1.** R is **closed** under addition: $x \in R$ and $y \in R$ imply $x + y \in R$.
> **2.** Addition in R is **associative**: $x + (y + z) = (x + y) + z$ for all x, y, z in R.
> **3.** R contains an **additive identity** 0: $x + 0 = 0 + x = x$ for all $x \in R$.
> **4.** R contains **additive inverses**: for x in R, there exists $-x$ in R such that $x + (-x) = (-x) + x = 0$.
> **5.** Addition in R is **commutative**: $x + y = y + x$ for all x, y in R.
> **6.** R is **closed** under multiplication: $x \in R$ and $y \in R$ imply $x \cdot y \in R$.
> **7.** Multiplication in R is **associative**: $x \cdot (y \cdot z) = (x \cdot y) \cdot z$ for all x, y, z in R.
> **8.** Two **distributive laws** hold in R: $x \cdot (y + z) = x \cdot y + x \cdot z$ and $(x + y) \cdot z = x \cdot z + y \cdot z$ for all x, y, z in R.
>
> The notation xy will be used interchangeably with $x \cdot y$ to indicate multiplication.

The additive identity of a ring is denoted by 0 and referred to as the **zero** of the ring. The additive inverse $-a$ is called the **negative** of a or the **opposite** of a, and **subtraction** in a ring is defined by

$$x - y = x + (-y).$$

As in elementary algebra, we adhere to the convention that *multiplication takes precedence over addition*. That is, it is understood that in any expression involving multiplication and addition, multiplications are performed first. Thus, $xy + xz$ represents $(x \cdot y) + (x \cdot z)$, not $x(y + x)z$.

The statement of the definition can be shortened to a form that is easier to remember if we note that the first five conditions amount to the requirement that R be an abelian group under addition.

Definition 5.1b **ALTERNATIVE DEFINITION OF A RING**

> Suppose R is a set in which a relation of equality, denoted by $=$, and operations of addition and multiplication, denoted by $+$ and $\cdot$, respectively, are defined. Then R is a **ring** (with respect to these operations) if these conditions hold:
>
> **1.** R forms an **abelian group** with respect to **addition**.
> **2.** R is **closed** with respect to an **associative multiplication**.
> **3.** Two **distributive laws** hold in R: $x \cdot (y + z) = x \cdot y + x \cdot z$ and $(x + y) \cdot z = x \cdot z + y \cdot z$ for all x, y, z in R.

Example 1   Some simple examples of rings are provided by the familiar number systems with their usual operations of addition and multiplication:

a. the set **Z** of all integers
b. the set **Q** of all rational numbers
c. the set **R** of all real numbers
d. the set **C** of all complex numbers.   ∎

Example 2   We shall verify that the set **E** of all even integers is a ring with respect to the usual addition and multiplication in **Z**. The following conditions of Definition 5.1a are satisfied automatically since they hold throughout the ring **Z**, which contains **E**:

2. Addition in **E** is associative.
5. Addition in **E** is commutative.
7. Multiplication in **E** is associative.
8. The two distributive laws in Definition 5.1a hold in **E**.

The remaining conditions in Definition 5.1a may be checked as follows:

1. If $x \in \mathbf{E}$ and $y \in \mathbf{E}$, then $x = 2m$ and $y = 2n$ with m and n in **Z**. For the sum, we have $x + y = 2m + 2n = 2(m + n)$, which is in **E**. Thus, **E** is closed under addition.
3. **E** contains the additive identity, since $0 = (2)(0)$.
4. For any $x = 2k$ in **E**, the additive inverse of x is in **E**, since $-x = 2(-k)$.
6. For $x = 2m$ and $y = 2n$ in **E**, the product $xy = 2(2mn)$ is in **E**, so **E** is closed under multiplication.   ∎

Definition 5.2   SUBRING

> Whenever a ring R_1 is a subset of a ring R_2 and has addition and multiplication as defined in R_2, we say that R_1 is a **subring** of R_2.

Thus, the ring **E** of even integers is a subring of the ring **Z** of all integers. From Example 1, we see that the ring **Z** is a subring of the rational numbers, the rational numbers form a subring of the real numbers, and the real numbers form a subring of the complex numbers.

Generalizing from Example 2, we may observe that conditions 2, 5, 7, and 8 of Definition 5.1a are automatically satisfied in any subset of a ring, leaving only conditions 1, 3, 4, and 6 to be verified for the subset to form a subring. A slightly more efficient characterization of subrings is given in the following theorem, the proof of which is left as an exercise.

Theorem 5.3   EQUIVALENT SET OF CONDITIONS FOR A SUBRING

A subset S of the ring R is a subring of R if and only these conditions are satisfied:

a. S is nonempty.
b. $x \in S$ and $y \in S$ imply that $x + y$ and xy are in S.
c. $x \in S$ implies $-x \in S$.

An even more efficient characterization of subrings is provided by the next theorem. The proof of this theorem is left as an exercise.

Theorem 5.4   CHARACTERIZATION OF A SUBRING

A subset S of the ring R is a subring of R if and only if these conditions are satisfied:

a. S is nonempty.
b. $x \in S$ and $y \in S$ imply that $x - y$ and xy are in S.

Example 3   Using Theorem 5.3 or Theorem 5.4, it is not difficult to verify the following examples of subrings.

a. The set of all real numbers of the form $m + n\sqrt{2}$, with $m \in \mathbf{Z}$ and $n \in \mathbf{Z}$, is a subring of the ring of all real numbers.
b. The set of all real numbers of the form $a + b\sqrt{2}$, with a and b rational numbers, is a subring of the real numbers.
c. The set of all real numbers of the form $a + b\sqrt[3]{2} + c\sqrt[3]{4}$, with $a, b,$ and c rational numbers, is a subring of the real numbers.   ■

The preceding examples of rings are all drawn from the number systems. The next example exhibits a class of rings with a different flavor: They are **finite rings** (that is, rings with a finite number of elements). The next example is also important because it presents the set $\mathbf{Z}_n$ of congruence classes modulo n for the first time in its proper context as a *ring*.

Example 4   For $n > 1$, let $\mathbf{Z}_n$ denote the congruence classes of the integers modulo n:

$$\mathbf{Z}_n = \{[0], [1], [2], \ldots, [n-1]\}.$$

We have previously seen that the rules

$$[a] + [b] = [a + b] \quad \text{and} \quad [a] \cdot [b] = [ab]$$

define binary operations of addition and multiplication in $\mathbf{Z}_n$. We have seen that $\mathbf{Z}_n$ forms an abelian group under addition, with $[0]$ as the additive identity and $[-a]$ as the additive inverse of $[a]$. It has also been noted that this multiplication is associative. For arbitrary $[a], [b], [c]$ in $\mathbf{Z}_n$, we have

$$
\begin{aligned}
[a] \cdot ([b] + [c]) &= [a] \cdot [b + c] \\
&= [a(b + c)] \\
&= [ab + ac] \\
&= [ab] + [ac] \\
&= [a] \cdot [b] + [a] \cdot [c],
\end{aligned}
$$

so the left distributive law holds in $\mathbf{Z}_n$. The right distributive law can be verified in a similar way, and $\mathbf{Z}_n$ is a ring with respect to these operations.   ■

Making use of some results from Chapter 1, we can obtain an example of a ring quite different from any of those previously discussed.

Example 5   Let U be a nonempty universal set, and let $\mathcal{P}(U)$ denote the collection of all subsets of U.

For arbitrary subsets A and B of U, let $A + B$ be defined as in Exercise 34 of Section 1.1:

$$A + B = (A \cup B) - (A \cap B)$$
$$= \{x \in U \,|\, x \in A \text{ or } B, \text{ and } x \text{ is not in both } A \text{ and } B\}.$$

This rule defines an operation of addition on the subsets of U, and this operation is associative, by Exercise 34b of Section 1.1. This addition is commutative, since $A \cup B = B \cup A$ and $A \cap B = B \cap A$. The empty set $\varnothing$ is an additive identity since

$$\varnothing + A = A + \varnothing$$
$$= (A \cup \varnothing) - (A \cap \varnothing)$$
$$= A - \varnothing$$
$$= A.$$

An unusual feature here is that each subset A of U is its own additive inverse:

$$A + A = (A \cup A) - (A \cap A)$$
$$= A - A$$
$$= \varnothing.$$

We define multiplication in $\mathcal{P}(U)$ by

$$A \cdot B = A \cap B.$$

This multiplication is associative since

$$A \cdot (B \cdot C) = A \cap (B \cap C)$$
$$= (A \cap B) \cap C$$
$$= (A \cdot B) \cdot C.$$

The left distributive law $A \cap (B + C) = (A \cap B) + (A \cap C)$ is part c of Exercise 34, Section 1.1, and the right distributive law follows from this one since forming intersections of sets is a commutative operation. Thus, $\mathcal{P}(U)$ is a ring with respect to the operations $+$ and $\cdot$ as we have defined them. ∎

Definition 5.5   RING WITH UNITY, COMMUTATIVE RING

> Let R be a ring. If there exists an element e in R such that $x \cdot e = e \cdot x = x$ for all x in R, then e is called a **unity** and R is a **ring with unity**. If multiplication in R is commutative, then R is called a **commutative ring**.

A ring may have one of the properties in Definition 5.5 without the other, it may have neither, or it may have both of the properties. These possibilities are illustrated in the following examples.

Example 6   The ring $\mathbf{Z}$ of all integers has both properties, so $\mathbf{Z}$ is a commutative ring with a unity. As other examples of this type, $\mathbf{Z}_n$ is a commutative ring with unity $[1]$, and $\mathcal{P}(U)$ is a commutative ring with the subset U as unity. ∎

Example 7   The ring $\mathbf{E}$ of all even integers is a commutative ring, but $\mathbf{E}$ does not have a unity.   ∎

Example 8   It follows from our work in Sections 1.5 and 3.2 that if $n \geq 2$, then each of the sets in the list

$$M_n(\mathbf{Z}) \subseteq M_n(\mathbf{Q}) \subseteq M_n(\mathbf{R}) \subseteq M_n(\mathbf{C})$$

is a noncummutative ring with unity I_n. Each of these four rings is a subring of every listed ring in which it is contained.   ∎

Example 9   The set

$$M_2(\mathbf{E}) = \left\{ \begin{bmatrix} a & b \\ c & d \end{bmatrix} \middle| a, b, c, \text{ and } d \text{ are in } \mathbf{E} \right\}$$

of all 2×2 matrices over the ring $\mathbf{E}$ of even integers is a noncommutative ring that does not have a unity.   ∎

The definition of a unity allows the possibility of more than one unity in a ring. However, this possibility cannot happen.

Theorem 5.6   **UNIQUENESS OF THE UNITY**

If R is a ring that has a unity, the unity is unique.

Uniqueness   **Proof**   Suppose that both e and e' are unity elements in a ring R. Consider the product $e \cdot e'$ in R. On the one hand, we have $e \cdot e' = e$, since e' is a unity. On the other hand, $e \cdot e' = e'$, since e is a unity. Thus,

$$e = e \cdot e' = e',$$

and the unity is unique.

In general discussions, we shall denote a unity by e. When a ring R has a unity, it is in order to consider the existence of multiplicative inverses.

Definition 5.7   **MULTIPLICATIVE INVERSE**

Let R be a ring with unity e, and let $a \in R$. If there is an element x in R such that $ax = xa = e$, then x is a **multiplicative inverse of a**.

As with the unity, a multiplicative inverse of an element is unique whenever it exists. The proof of this is left as an exercise.

Theorem 5.8   **UNIQUENESS OF THE MULTIPLICATIVE INVERSE**

Suppose R is a ring with unity e. If an element $a \in R$ has a multiplicative inverse, the multiplicative inverse of a is unique.

We shall use the standard notation a^{-1} to denote the multiplicative inverse of a, if the inverse exists.

Example 10   Some elements in a ring R may have multiplicative inverses while others do not. In the ring $\mathbf{Z}_{10}$, [1] and [9] are their own multiplicative inverses, while [3] and [7] are inverses of each other. All other elements of $\mathbf{Z}_{10}$ do not have multiplicative inverses.   ∎

Since every ring R forms an abelian group with respect to addition, many of our results for groups have immediate applications concerning addition in a ring. For example, Theorem 3.4 gives these results:

1. The zero element in R is unique.
2. For each x in R, $-x$ is unique.
3. For each x in R, $-(-x) = x$.
4. For any x and y in R, $-(x + y) = -y - x$.
5. If a, x, and y are in R and $a + x = a + y$, then $x = y$.

Whenever both addition and multiplication are involved, the results are not so direct, but they turn out much as we might expect. One basic result of this type is that a product is 0 if one of the factors is 0.

Theorem 5.9   ZERO PRODUCT

If R is a ring, then

$$a \cdot 0 = 0 \cdot a = 0$$

for all $a \in R$.

Proof   Let a be arbitrary in R. We reduce $a \cdot 0$ to 0 by using various conditions in Definition 5.1a, as indicated:

$$
\begin{aligned}
a \cdot 0 &= a \cdot 0 + 0 & &\text{by condition 3} \\
&= a \cdot 0 + \{a \cdot 0 + [-(a \cdot 0)]\} & &\text{by condition 4} \\
&= (a \cdot 0 + a \cdot 0) + [-(a \cdot 0)] & &\text{by condition 2} \\
&= [a \cdot (0 + 0)] + [-(a \cdot 0)] & &\text{by condition 8} \\
&= a \cdot 0 + [-(a \cdot 0)] & &\text{by condition 3} \\
&= 0 & &\text{by condition 4.}
\end{aligned}
$$

Similar steps can be used to reduce $0 \cdot a$ to 0.

Theorem 5.9 says that a product is 0 if one of the factors is 0. Note that the converse is not true: a product may be 0 when neither factor is 0. An illustration is provided by $[2] \cdot [5] = [0]$ in $\mathbf{Z}_{10}$.

Definition 5.10   ZERO DIVISOR

Let R be a ring and let $a \in R$. If $a \neq 0$, and if there exists an element $b \neq 0$ in R such that either $ab = 0$ or $ba = 0$, then a is called a **proper divisor of zero**, or a **zero divisor**.

If we compare the steps used in the proof of Theorem 5.9 to the last part of the proof of Theorem 2.2, we see that they are much the same. In the same fashion, the proof of the first part of the next theorem is parallel to another part of the proof of Theorem 2.2. The same sort of similarity exists between Exercises 1–10 of Section 2.1 and the remaining parts of Theorem 5.11. Because of this similarity, their proofs are left as exercises.

Theorem 5.11   ADDITIVE INVERSES AND PRODUCTS

For arbitrary x, y, and z in a ring R, the following equalities hold:

a. $(-x)y = -(xy)$
b. $x(-y) = -(xy)$
c. $(-x)(-y) = xy$
d. $x(y - z) = xy - xz$
e. $(x - y)z = xz - yz$.

Proof of a   Since the additive inverse $-(xy)$ of the element xy is unique, we only need to show that $xy + (-x)y = 0$. We have

$$
\begin{aligned}
xy + (-x)y &= [x + (-x)]y &&\text{by the right distributive law} \\
&= 0 \cdot y &&\text{by the definition of } -x \\
&= 0 &&\text{by Theorem 5.9.}
\end{aligned}
$$

Even though a ring does not form a group with respect to multiplication, both associative laws in a ring R can be generalized by the procedure followed in Definition 3.6 and Theorem 3.7. For any integer $n \geq 2$, the expressions $a_1 + a_2 + \cdots + a_n$ and $a_1 a_2 \cdots a_n$ are defined recursively by

$$a_1 + a_2 + \cdots + a_k + a_{k+1} = (a_1 + a_2 + \cdots + a_k) + a_{k+1}$$

and

$$a_1 a_2 \cdots a_k a_{k+1} = (a_1 a_2 \cdots a_k)a_{k+1}.$$

The details are too repetitive to present here, so we accept the following theorem without proof.

Theorem 5.12   GENERALIZED ASSOCIATIVE LAWS

Let $n \geq 2$ be a positive integer, and let $a_1, a_2, \ldots, a_n$ denote elements of a ring R. For any positive integer m such that $1 \leq m < n$,

$$(a_1 + a_2 + \cdots + a_m) + (a_{m+1} + \cdots + a_n) = a_1 + a_2 + \cdots + a_n$$

and

$$(a_1 a_2 \cdots a_m)(a_{m+1} \cdots a_n) = a_1 a_2 \cdots a_n.$$

Generalized distributive laws also hold in an arbitrary ring. This fact is stated in the following theorem, with the proofs left as exercises.

Theorem 5.13 **GENERALIZED DISTRIBUTIVE LAWS**

Let $n \geq 2$ be a positive integer, and let $b, a_1, a_2, \ldots, a_n$ denote elements of a ring R. Then we have

a. $b(a_1 + a_2 + \cdots + a_n) = ba_1 + ba_2 + \cdots + ba_n$, and
b. $(a_1 + a_2 + \cdots + a_n)b = a_1b + a_2b + \cdots + a_nb$.

EXERCISES 5.1

1. Confirm the statements made in Example 3 by proving that the following sets are subrings of the ring of all real numbers.
   a. the set of all real numbers of the form $m + n\sqrt{2}$, with $m \in \mathbf{Z}$ and $n \in \mathbf{Z}$
   b. the set of all real numbers of the form $a + b\sqrt{2}$, with a and b rational numbers
   c. the set of all real numbers of the form $a + b\sqrt[3]{2} + c\sqrt[3]{4}$, with a, b, and c rational numbers

2. Decide whether each of the following sets is a ring with respect to the usual operations of addition and multiplication. If it is not a ring, state at least one condition in Definition 5.1a that fails to hold.
   a. the set of all integers that are multiples of 5
   b. the set of all real numbers of the form $m + n\sqrt{3}$ with $m \in \mathbf{Z}$ and $n \in \mathbf{Z}$
   c. the set of all real numbers of the form $a + b\sqrt[3]{5}$ where a and b are rational numbers
   d. the set of all real numbers of the form $a + b\sqrt[3]{5} + c\sqrt[3]{25}$ where a, b, and c are rational numbers
   e. the set of all positive real numbers
   f. the set of all complex numbers of the form $m + ni$ where $m \in \mathbf{Z}$ and $n \in \mathbf{Z}$ (This set is known as the **Gaussian integers**.)
   g. the set of all real numbers of the form $m + n\sqrt{2}$ where $m \in \mathbf{E}$ and $n \in \mathbf{Z}$
   h. the set of all real numbers of the form $m + n\sqrt{2}$ where $m \in \mathbf{Z}$ and $n \in \mathbf{E}$

3. Let $U = \{a, b\}$. Using addition and multiplication as they are defined in Example 5, construct addition and multiplication tables for the ring $\mathcal{P}(U)$ that consists of the elements $\varnothing$, $A = \{a\}, B = \{b\}, U$.

4. Follow the instructions in Exercise 3 and use the universal set $U = \{a, b, c\}$.

5. Let $U = \{a, b\}$. Define addition and multiplication in $\mathcal{P}(U)$ by $C + D = C \cup D$ and $CD = C \cap D$. Decide if $\mathcal{P}(U)$ is a ring with respect to these operations. If it is not, state a condition in Definition 5.1a that fails to hold.

6. Work Exercise 5 using $U = \{a\}$.

7. Find all zero divisors in $\mathbf{Z}_n$ for the following values of n.
   a. $n = 6$　　　　　　　　　　b. $n = 8$
   c. $n = 10$　　　　　　　　　d. $n = 12$
   e. $n = 14$　　　　　　　　　f. n a prime integer

8. For the given value of n, find the elements of $\mathbf{Z}_n$ that have multiplicative inverses.
   a. $n = 6$　　　　　　　　　　b. $n = 8$
   c. $n = 16$　　　　　　　　　d. $n = 12$
   e. $n = 14$　　　　　　　　　f. n a prime integer

9. Prove Theorem 5.3: A subset S of the ring R is a subring of R if and only if these conditions are satisfied:
   a. S is nonempty.
   b. $x \in S$ and $y \in S$ imply that $x + y$ and xy are in S.
   c. $x \in S$ implies $-x \in S$.

10. Prove Theorem 5.4: A subset S of the ring R is a subring of R if and only if these conditions are satisfied:
    a. S is nonempty.
    b. $x \in S$ and $y \in S$ imply that $x - y$ and xy are in S.

11. Assume R is a ring with unity e. Prove Theorem 5.8: If $a \in R$ has a multiplicative inverse, the multiplicative inverse of a is unique.

12. (See Example 4.) Prove the right distributive law in $\mathbf{Z}_n$:

$$([a] + [b]) \cdot [c] = [a] \cdot [c] + [b] \cdot [c].$$

13. Complete the proof of Theorem 5.9 by showing that $0 \cdot a = 0$ for any a in a ring R.

14. Let R be a ring, and let x, y, and z be arbitrary elements of R. Complete the proof of Theorem 5.11 by proving the following statements.
    a. $x(-y) = -(xy)$
    b. $(-x)(-y) = xy$
    c. $x(y - z) = xy - xz$
    d. $(x - y)z = xz - yz$

15. Suppose that G is a group with respect to addition, with identity element 0. Define a multiplication in G by $ab = 0$ for all $a, b \in G$. Show that G forms a ring with respect to these operations.

16. If R_1 and R_2 are subrings of the ring R, prove that $R_1 \cap R_2$ is a subring of R.

17. Find subrings R_1 and R_2 of $\mathbf{Z}$ such that $R_1 \cup R_2$ is not a subring of $\mathbf{Z}$.

18. Find a specific example of two elements a and b in a ring R such that $ab = 0$ and $ba \neq 0$.

19. Define a new operation of addition in $\mathbf{Z}$ by $x \oplus y = x + y - 1$ and a new multiplication in $\mathbf{Z}$ by $x \odot y = x + y - xy$. Verify that $\mathbf{Z}$ forms a ring with respect to these operations.

20. If R is a ring with unity, the elements of R that have multiplicative inverses are called **units** (or **invertible elements**) of R. Prove that the set of all units of R forms a group with respect to multiplication.

21. Prove that if the element a of a ring R with unity has a multiplicative inverse in R, then a is not a zero divisor in R.

22. Suppose that a, b, and c are elements of a ring R such that $ab = ac$. Prove that if a has a multiplicative inverse, then $b = c$.

23. For a fixed element a of a ring R, prove that the set $\{x \in R | ax = 0\}$ is a subring of R.

24. Consider the set $R = \{[0], [2], [4], [6], [8]\} \subseteq \mathbf{Z}_{10}$.
    a. Construct addition and multiplication tables for R, using the operations as defined in $\mathbf{Z}_{10}$.
    b. Observe that R is a commutative ring with unity [6].
    c. Is R a subring of $\mathbf{Z}_{10}$? If not, give a reason.
    d. Does R have zero divisors?
    e. Which elements of R have multiplicative inverses?

25. Consider the set $S = \{[0], [2], [4], [6], [8], [10], [12], [14], [16]\} \subseteq \mathbf{Z}_{18}$. Using addition and multiplication as defined in $\mathbf{Z}_{18}$, consider the following questions.
    a. Is S a ring? If not, give a reason.
    b. Is S a commutative ring with unity? If not, give a reason.
    c. Is S a subring of $\mathbf{Z}_{18}$? If not, give a reason.
    d. Does S have zero divisors?
    e. Which elements of S have multiplicative inverses?

Sec. 6.2, #12 ≪

26. The addition table and part of the multiplicative table for the ring $R = \{a, b, c\}$ are given in Figure 5-1. Use the distributive laws to complete the multiplication table.

Figure 5-1

+	a	b	c
a	a	b	c
b	b	c	a
c	c	a	b

·	a	b	c
a	a	a	a
b	a	c	
c	a		

Sec. 6.2, #13 ≪

27. The addition table and part of the multiplication table for the ring $R = \{a, b, c, d\}$ are given in Figure 5-2. Use the distributive laws to complete the multiplication table.

Figure 5-2

+	a	b	c	d
a	a	b	c	d
b	b	c	d	a
c	c	d	a	b
d	d	a	b	c

·	a	b	c	d
a	a	a	a	a
b	a	c		
c	a		a	
d	a		a	c

28. Give an example of a zero divisor in the ring $M_2(\mathbf{Z})$.

Sec. 3.1, #39 ≫
Sec. 5.2, #13 ≪

29. (See Exercise 39 of Section 3.1.) An element x in a ring is called **idempotent** if $x^2 = x$. Find two different idempotent elements in $M_2(\mathbf{Z})$.

30. (See Exercise 29.) Show that the set of all idempotent elements of a commutative ring is closed under multiplication.

31. Decide whether each of the following sets S is a subring of the ring $M_2(\mathbf{Z})$. If a set is not a subring, give a reason why it is not.

a. $S = \left\{ \begin{bmatrix} x & 0 \\ x & 0 \end{bmatrix} \,\middle|\, x \in \mathbf{Z} \right\}$

b. $S = \left\{ \begin{bmatrix} x & x \\ 0 & 0 \end{bmatrix} \,\middle|\, x \in \mathbf{Z} \right\}$

c. $S = \left\{ \begin{bmatrix} x & y \\ x & y \end{bmatrix} \,\middle|\, x, y \in \mathbf{Z} \right\}$

d. $S = \left\{ \begin{bmatrix} x & 0 \\ y & z \end{bmatrix} \,\middle|\, x, y, z \in \mathbf{Z} \right\}$

32. Consider the set S that consists of all 2×2 matrices of the form $\begin{bmatrix} a & b \\ 0 & c \end{bmatrix}$ where $a, b, c \in \mathbf{Z}$, with the same rules for addition and multiplication as in $M_2(\mathbf{Z})$.

a. Show that S is a noncommutative ring with unity.

Sec. 6.1, #15 ≪

b. Which elements of S have multiplicative inverses?

33. Consider the set T of all 2×2 matrices of the form $\begin{bmatrix} a & a \\ b & b \end{bmatrix}$ where a and b are real numbers, with the same rules for addition and multiplication as in $M_2(\mathbf{R})$.

a. Show that T is a ring that does not have a unity.

b. Show that T is not a commutative ring.

34. Prove the following equalities in an arbitrary ring R.

a. $(x + y)(z + w) = (xz + xw) + (yz + yw)$

b. $(x + y)(z - w) = (xz + yz) - (xw + yw)$

c. $(x - y)(z - w) = (xz + yw) - (xw + yz)$

d. $(x + y)(x - y) = (x^2 - y^2) + (yx - xy)$

35. Prove Theorem 5.13a.

36. Prove Theorem 5.13b.

Sec. 6.1, #21 ≪
37. An element a of a ring R is called **nilpotent** if $a^n = 0$ for some positive integer n. Prove that the set of all nilpotent elements in a commutative ring R forms a subring of R.

Sec. 5.2, #16 ≪
Sec. 6.2, #14 ≪
Sec. 6.3, #1 ≪
Sec. 6.4, #21 ≪
Sec. 6.4, #22 ≪
38. Let R and S be arbitrary rings. In the Cartesian product $R \times S$ of R and S, define

$$(r, s) = (r', s') \quad \text{if and only if} \quad r = r' \text{ and } s = s',$$
$$(r_1, s_1) + (r_2, s_2) = (r_1 + r_2, s_1 + s_2),$$
$$(r_1, s_1)(r_2, s_2) = (r_1 r_2, s_1 s_2).$$

Prove that the Cartesian product is a ring with respect to these operations. It is called the **direct sum** of R and S and is denoted by $R \oplus S$.

39. (See Exercise 38.) Write out the elements of $\mathbf{Z}_2 \oplus \mathbf{Z}_2$ and construct addition and multiplication tables for this ring. (*Suggestion:* Write 0 for [0], 1 for [1] in $\mathbf{Z}_2$.) ∘

40. Suppose R is a ring in which all elements x satisfy $x^2 = x$. (Such a ring is called a **Boolean ring**.)

Sec. 5.2, #17 ≪
  a. Prove that $x = -x$ for each $x \in R$. [*Hint:* Consider $(x + x)^2$.]
  b. Prove that R is commutative. [*Hint:* Consider $(x + y)^2$.]

5.2 INTEGRAL DOMAINS AND FIELDS

In the preceding section we defined the terms *ring with unity, commutative ring,* and *zero divisors.* All three of these terms are used in defining an *integral domain.*

Definition 5.14 **INTEGRAL DOMAIN**

> Let D be a ring. Then D is an **integral domain** provided these conditions hold:
>
> **1.** D is a commutative ring.
> **2.** D has a unity e, and $e \neq 0$.
> **3.** D has no zero divisors.

Note that the requirement $e \neq 0$ means that an integral domain must have at least two elements.

Example 1 The ring $\mathbf{Z}$ of all integers is an integral domain, but the ring $\mathbf{E}$ of all even integers is not an integral domain, because it does not contain a unity. As familiar examples of integral domains, we can list the set of all rational numbers, the set of all real numbers, and the set of all complex numbers—all of these with their usual operations. ∎

Example 2 The ring $\mathbf{Z}_{10}$ is a commutative ring with a unity, but the presence of zero divisors such as [2] and [5] prevents $\mathbf{Z}_{10}$ from being an integral domain. Considered as a possible integral domain, the ring M of all 2×2 matrices with real numbers as elements fails on two counts: Multiplication is not commutative, and it has zero divisors. ∎

In Example 4 of Section 5.1, we saw that $\mathbf{Z}_n$ is a ring for every value of $n > 1$. Moreover, $\mathbf{Z}_n$ is a commutative ring since

$$[a] \cdot [b] = [ab] = [ba] = [b] \cdot [a]$$

for all $[a], [b]$ in $\mathbf{Z}_n$. Since $\mathbf{Z}_n$ has $[1]$ as the unity, $\mathbf{Z}_n$ is an integral domain if and only if it has no zero divisors. The following theorem characterizes these $\mathbf{Z}_n$, and it provides us with a large class of *finite integral domains* (that is, integral domains that have a finite number of elements).

Theorem 5.15   **THE INTEGRAL DOMAIN $\mathbf{Z}_n$ WHEN n IS PRIME**

For $n > 1$, $\mathbf{Z}_n$ is an integral domain if and only if n is a prime.

$p \Leftarrow q$ **Proof**   From the previous discussion, it is clear that we need only prove that $\mathbf{Z}_n$ has no zero divisors if and only if n is a prime.

Suppose first that n is a prime. Let $[a] \neq [0]$ in $\mathbf{Z}_n$, and suppose $[a][b] = [0]$ for some $[b]$ in $\mathbf{Z}_n$. Now $[a][b] = [0]$ implies that $[ab] = [0]$, and therefore $n|ab$. However, $[a] \neq [0]$ means that $n \nmid a$. Thus, $n|ab$ and $n \nmid a$. Since n is a prime, this implies that $n|b$, by Theorem 2.16; that is, $[b] = [0]$. We have shown that if $[a] \neq [0]$, the only way that $[a][b]$ can be $[0]$ is for $[b]$ to be $[0]$. Therefore, $\mathbf{Z}_n$ has no zero divisors and is an integral domain.

$\sim p \Leftarrow \sim q$ Suppose now that n is not a prime. Then n has divisors other than ± 1 and $\pm n$, so there are integers a and b such that

$$n = ab \quad \text{where } 1 < a < n \text{ and } 1 < b < n.$$

This means that $[a] \neq [0], [b] \neq [0]$, but

$$[a][b] = [ab] = [n] = [0].$$

Therefore, $[a]$ is a zero divisor in $\mathbf{Z}_n$, and $\mathbf{Z}_n$ is not an integral domain.

Combining the two cases, we see that n is a prime if and only if $\mathbf{Z}_n$ is an integral domain.

One direct consequence of the absence of zero divisors in an integral domain is that the cancellation law for multiplication must hold.

Theorem 5.16   **CANCELLATION LAW FOR MULTIPLICATION**

If a, b, and c are elements of an integral domain D such that $a \neq 0$ and $ab = ac$, then $b = c$.

$(p \wedge q) \Rightarrow r$ **Proof**   Suppose a, b, and c are elements of an integral domain D such that $a \neq 0$ and $ab = ac$. Now

$$ab = ac \Rightarrow ab - ac = 0$$
$$\Rightarrow a(b - c) = 0.$$

Since $a \neq 0$ and D has no zero divisors, it must be true that $b - c = 0$, and hence $b = c$.

It can be shown that if the cancellation law holds in a commutative ring, then the ring cannot have zero divisors. The proof of this is left as an exercise.

To require that a ring have no zero divisors is equivalent to requiring that a product of nonzero elements must always be different from 0. Or, stated another way, a product that is 0 must have at least one factor equal to 0.

A *field* is another special type of ring, and we shall examine the relationship between a field and an integral domain. We begin with a definition.

Definition 5.17 FIELD

Let F be a ring. Then F is a **field** provided these conditions hold:

1. F is a commutative ring.
2. F has a unity e, and $e \neq 0$.
3. Every nonzero element of F has a multiplicative inverse.

The rational numbers, the real numbers, and the complex numbers are familiar examples of fields. We shall see in Corollary 5.20 that if p is a prime, then $\mathbf{Z}_p$ is a field. Other and less familiar examples of fields are found in the exercises for this section.

Part of the relation between fields and integral domains is stated in the following theorem.

Theorem 5.18 FIELDS AND INTEGRAL DOMAINS

Every field is an integral domain.

$p \Rightarrow q$ **Proof** Let F be a field. To prove that F is an integral domain, we only need to show that F has no zero divisors. Suppose a and b are elements of F such that $ab = 0$. If $a \neq 0$, then $a^{-1} \in F$ and

$$
\begin{aligned}
ab = 0 &\Rightarrow a^{-1}(ab) = a^{-1} \cdot 0 \\
&\Rightarrow (a^{-1}a)b = 0 \\
&\Rightarrow eb = 0 \\
&\Rightarrow b = 0.
\end{aligned}
$$

Similarly, if $b \neq 0$, then $a = 0$. Therefore, F has no zero divisors and is an integral domain.

It is certainly not true that every integral domain is a field. For example, the set $\mathbf{Z}$ of all integers forms an integral domain, and the integers 1 and -1 are the only elements of $\mathbf{Z}$ that have multiplicative inverses. It is perhaps surprising, but an integral domain with a finite number of elements is always a field. This is the other part of the relationship between a field and an integral domain.

Theorem 5.19 FINITE INTEGRAL DOMAINS AND FIELDS

Every finite integral domain is a field.

$p \Rightarrow q$ **Proof** Assume D is a finite integral domain. Let n be the number of distinct elements in D; say,

$$
D = \{d_1, d_2, \ldots, d_n\}
$$

where the d_i are the distinct elements of D. Now let a be any nonzero element of D, and consider the set of products

$$
\{ad_1, ad_2, \ldots, ad_n\}.
$$

These products are all distinct, for $a \neq 0$ and $ad_r = ad_s$ would imply $d_r = d_s$, by Theorem 5.16, and the d_i are all distinct. These n products are all contained in D, and no two of them are equal. Hence, they are the same as the elements of D, except possibly for order. This means that every element of D appears somewhere in the list

$$ad_1, ad_2, \ldots, ad_n.$$

In particular, the unity e is one of these products. That is, $ad_k = e$ for some d_k. Since multiplication is commutative in D, we have $d_k a = ad_k = e$, and d_k is a multiplicative inverse of a. Thus, D is a field.

Corollary 5.20   THE FIELD Z_n WHEN n IS PRIME

Z_n is a field if and only if n is a prime.

Proof   This follows at once from Theorems 5.15, 5.18, and 5.19.

We have seen that the elements of a ring form an abelian group with respect to addition. A similar comparison can be made for the nonzero elements of a field. It is readily seen that the nonzero elements form an abelian group with respect to multiplication. The definition of a field can thus be reformulated as follows: A **field** is a set of elements in which equality, addition, and multiplication are defined such that the following conditions hold:

1. F forms an abelian group with respect to addition.
2. The nonzero elements of F form an abelian group with respect to multiplication.
3. The distributive law $x(y + z) = xy + xz$ holds for all x, y, z in F.

The last example in this section points out that some of our most familiar rings do not form integral domains.

Example 3   For $n \geq 2$, each of the rings

$$M_n(\mathbf{Z}), \quad M_n(\mathbf{Q}), \quad M_n(\mathbf{R}), \quad M_n(\mathbf{C})$$

is not an integral domain, since multiplication in each of them is not commutative. It is also true that each of them contains zero divisors if $n \geq 2$. For $n = 2$, the product

$$\begin{bmatrix} 1 & 0 \\ 1 & 0 \end{bmatrix} \begin{bmatrix} 0 & 0 \\ 1 & 1 \end{bmatrix} = \begin{bmatrix} 0 & 0 \\ 0 & 0 \end{bmatrix}$$

illustrates this statement. Similar examples can easily be constructed for $n > 2$.   ∎

EXERCISES 5.2

1. Decide which of the following are integral domains and which are fields with respect to the usual operations of addition and multiplication. State a reason for each one that fails to be an integral domain or a field.
   a. the set of all real numbers of the form $m + n\sqrt{2}$ where m and n are integers
   b. the set of all real numbers of the form $a + b\sqrt{2}$ where a and b are rational numbers

   c. the set of all real numbers of the form $a + b\sqrt[3]{2}$ where a and b are rational numbers

   d. the set of all real numbers of the form $a + b\sqrt[3]{2} + c\sqrt[3]{4}$ where a, b, and c are rational numbers

   e. the **Gaussian integers**, that is, the set of all complex numbers of the form $m + ni$ where $m \in \mathbf{Z}$ and $n \in \mathbf{Z}$

   f. the set of all complex numbers of the form $m + ni$ where $m \in \mathbf{E}$ and $n \in \mathbf{E}$ ($\mathbf{E}$ is the ring of all even integers.)

   g. the set of all complex numbers of the form $a + bi$ where a and b are rational numbers

   h. the set of all real numbers of the form $m + n\sqrt{2}$ where $m \in \mathbf{Z}$ and $n \in \mathbf{E}$

2. Consider the set $R = \{[0], [2], [4], [6], [8]\} \subseteq \mathbf{Z}_{10}$, with addition and multiplication as defined in $\mathbf{Z}_{10}$.

   a. Is R an integral domain? If not, give a reason.

   b. Is R a field? If not, give a reason.

3. Consider the set $S = \{[0], [2], [4], [6], [8], [10], [12], [14], [16]\} \subseteq \mathbf{Z}_{18}$, with addition and multiplication as defined in $\mathbf{Z}_{18}$.

   a. Is S an integral domain? If not, give a reason.

   b. Is S a field? If not, give a reason.

4. Let $S = \{(0, 0), (1, 1), (0, 1), (1, 0)\}$ where $0 = [0]$ and $1 = [1]$ are the elements of $\mathbf{Z}_2$. Equality, addition, and multiplication are defined in S as follows:

$$(a, b) = (c, d) \quad \text{if and only if} \quad a = c \text{ and } b = d \text{ in } \mathbf{Z}_2,$$
$$(a, b) + (c, d) = (a + c, b + d),$$
$$(a, b) \cdot (c, d) = (ad + bc + bd, ad + bc + ac).$$

   a. Prove that multiplication in S is associative.

   Assume that S is a ring and consider these questions, giving a reason for any negative answers.

   b. Is S a commutative ring?

   c. Does S have a unity?

   d. Is S an integral domain?

   e. Is S a field?

5. Let W be the set of all ordered pairs (x, y) of integers x and y. Equality, addition, and multiplication are defined as follows:

$$(x, y) = (z, w) \quad \text{if and only if} \quad x = z \text{ and } y = w \text{ in } \mathbf{Z},$$
$$(x, y) + (z, w) = (x + z, y + w),$$
$$(x, y) \cdot (z, w) = (xz - yw, xw + yz).$$

Given that W is a ring, determine whether W is commutative and whether W has a unity. Justify your decisions.

Sec. 5.3, #9 ≪

6. Let S be the set of all 2×2 matrices of the form $\begin{bmatrix} x & 0 \\ x & 0 \end{bmatrix}$ where x is a real number. Assume that S is a ring with respect to matrix addition and multiplication. Answer the following questions, and give a reason for any negative answers.

   a. Is S a commutative ring?

   b. Does S have a unity? If so, identify the unity.

   c. Is S an integral domain?

   d. Is S a field?

7. Work Exercise 6 using S as the set of all 2×2 matrices of the form $\begin{bmatrix} x & x \\ 0 & 0 \end{bmatrix}$ where x is a real number.

8. Let R be the set of all matrices of the form $\begin{bmatrix} a & -b \\ b & a \end{bmatrix}$ where a and b are integers. Assume that R is a ring with respect to matrix addition and multiplication. Determine whether R is commutative and identify the unity if R has one.

Sec. 5.3, #9 ≪

9. Let R be the set of all matrices of the form $\begin{bmatrix} a & -b \\ b & a \end{bmatrix}$ where a and b are real numbers. Assume that R is a ring with respect to matrix addition and multiplication. Answer the following questions and give a reason for any negative answers.

   a. Is R a commutative ring?

   b. Does R have a unity? If so, identify the unity.

   c. Is R an integral domain?

Sec. 5.3, #10 ≪   **d.** Is R a field?

10. Let R be a commutative ring in which the cancellation law for multiplication holds. That is, if a, b, and c are elements of R, then $a \neq 0$ and $ab = ac$ always imply $b = c$. Prove that R has no zero divisors.

11. Prove that if a subring R of an integral domain D contains the unity element of D, then R is an integral domain.

12. If e is the unity in an integral domain D, prove that $(-e)a = -a$ for all $a \in D$.

Sec. 5.1, #29 ≫
Sec. 7.2, #33 ≪   13. (See Exercise 29 in Section 5.1.) Prove that the only idempotent elements in an integral domain are 0 and e.

14. **a.** Give an example where a and b are not zero divisors in a ring R, but the sum $a + b$ is a zero divisor.

   b. Prove that the set of all elements in a ring R that are not zero divisors is closed under multiplication.

15. Find the multiplicative inverse of the given element. (See Example 4 of Section 2.6.)

   a. $[11]$ in $\mathbf{Z}_{317}$ **b.** $[11]$ in $\mathbf{Z}_{138}$ **c.** $[9]$ in $\mathbf{Z}_{242}$ **d.** $[6]$ in $\mathbf{Z}_{319}$

Sec. 5.1, #38 ≫   16. (See Exercise 38 in Section 5.1.) Prove that if R and S are integral domains, then the direct sum $R \oplus S$ is *not* an integral domain.

Sec. 5.1, #40 ≫   17. (See Exercise 40 of Section 5.1.) Let R be a Boolean ring with unity e. Prove that every element of R except 0 and e is a zero divisor.

18. If $a \neq 0$ in a field F, prove that for every $b \in F$ the equation $ax = b$ has a unique solution x in F.

19. Suppose S is a subset of a field F that contains at least two elements and satisfies the following conditions:

   a. $x \in S$ and $y \in S$ imply $x - y \in S$;

   b. $x \in S$ and $y \neq 0 \in S$ imply $xy^{-1} \in S$.

   Prove that S is a field.

5.3   THE FIELD OF QUOTIENTS OF AN INTEGRAL DOMAIN

The example of an integral domain that is most familiar to us is the set $\mathbf{Z}$ of all integers, and the most familiar example of a field is the set of all rational numbers. There is a very natural and intimate relationship between these two systems. In fact, a rational number is by definition a quotient a/b of integers a and b, with $b \neq 0$; that is, the set of rational numbers is the set of all quotients of integers with nonzero denominators. For this reason, the set of rational numbers is frequently referred to as "the quotient field of the integers." In this section, we shall see that an analogous field of quotients can be constructed for an arbitrary integral domain.

Before we present this construction, let us review the basic definitions of equality, addition, and multiplication in the rational numbers. We recall that, for rational numbers $\dfrac{a}{b}$ and $\dfrac{c}{d}$,

$$\frac{a}{b} = \frac{c}{d} \quad \text{if and only if} \quad ad = bc$$

$$\frac{a}{b} + \frac{c}{d} = \frac{ad + bc}{bd}$$

$$\frac{a}{b} \cdot \frac{c}{d} = \frac{ac}{bd}.$$

Note that the definitions of equality, addition, and multiplication for rational numbers are based on the corresponding definitions for the integers. These definitions guide our construction of the quotient field for an arbitrary integral domain D.

Our first step in this construction is the following definition.

Definition 5.21   A RELATION ON ORDERED PAIRS

Let D be an integral domain and let S be the set of all ordered pairs (a, b) of elements of D with $b \neq 0$:

$$S = \{(a, b) \mid a, b \in D \text{ and } b \neq 0\}.$$

The relation $\sim$ is defined on S by

$$(a, b) \sim (c, d) \quad \text{if and only if} \quad ad = bc.$$

The relation $\sim$ is an obvious imitation of the equality of rational numbers, and we can show that it is indeed an equivalence relation on S.

Lemma 5.22   THE EQUIVALENCE RELATION $\sim$

The relation $\sim$ in Definition 5.21 is an equivalence relation on S.

Proof   We shall show that $\sim$ is reflexive, symmetric, and transitive. Let (a, b), (c, d), and (f, g) be arbitrary elements of S.

Reflexive   **1.** $(a, b) \sim (a, b)$, since the commutative multiplication in D implies that $ab = ba$.

Symmetric   **2.** $(a, b) \sim (c, d) \Rightarrow ad = bc$ by definition of $\sim$

$\qquad\qquad\qquad \Rightarrow da = cb \quad \text{or} \quad cb = da$ since multiplication is commutative in D

$\qquad\qquad\qquad \Rightarrow (c, d) \sim (a, b)$ by definition of $\sim$

Transitive   **3.** Assume that $(a, b) \sim (c, d)$ and $(c, d) \sim (f, g)$.

$$\left. \begin{array}{l} (a, b) \sim (c, d) \Rightarrow ad = bc \Rightarrow adg = bcg \\ (c, d) \sim (f, g) \Rightarrow cg = df \Rightarrow bcg = bdf \end{array} \right\} \Rightarrow adg = bdf$$

Using the commutative property of multiplication in D once again, we have[†]

$$dag = dbf$$

[†]It is tempting here to use $ad = bc$ and $cg = df$ to obtain $(ad)(cg) = (bc)(df)$, but this would not imply that $ag = bf$ because c might be zero.

where $d \neq 0$, and therefore

$$ag = bf$$

by Theorem 5.16. According to Definition 5.21, this implies that $(a, b) \sim (f, g)$. Thus, $\sim$ is an equivalence relation on S.

The next definition reveals the basic plan for our construction of the quotient field of D.

Definition 5.23   THE SET OF QUOTIENTS

> Let D, S, and $\sim$ be the same as in Definition 5.21 and Lemma 5.22. For each (a, b) in S, let $[a, b]$ denote the equivalence class in S that contains (a, b), and let Q denote the set of all equivalence classes $[a, b]$:
>
> $$Q = \{[a, b] \mid (a, b) \in S\}.$$
>
> The set Q is called the **set of quotients** for D.

We shall at times need the fact that, for any $x \neq 0$ in D and any $[a, b]$ in Q,

$$[a, b] = [ax, bx].$$

This follows at once from the equality $a(bx) = b(ax)$ in the integral domain D.

Lemma 5.24   ADDITION AND MULTIPLICATION IN Q

The following rules define binary operations on Q. **Addition** in Q is defined by

$$[a, b] + [c, d] = [ad + bc, bd],$$

and **multiplication** in Q is defined by

$$[a, b] \cdot [c, d] = [ac, bd].$$

Proof   We shall verify that the rule stated for addition defines a binary operation on Q. For arbitrary $[a, b]$ and $[c, d]$ in Q, we have $b \neq 0$ and $d \neq 0$ in D. Since D is an integral domain, $b \neq 0$ and $d \neq 0$ imply $bd \neq 0$, so $[a, b] + [c, d] = [ad + bc, bd]$ is an element of Q.

To show that the sum of two elements is unique (or well-defined), suppose that $[a, b] = [x, y]$ and $[c, d] = [z, w]$ in Q. We need to show $[a, b] + [c, d] = [x, y] + [z, w]$. Now

$$[a, b] + [c, d] = [ad + bc, bd]$$

and

$$[x, y] + [z, w] = [xw + yz, yw].$$

To prove these elements equal, we need

$$(ad + bc)yw = bd(xw + yz)$$

or

$$adyw + bcyw = bdxw + bdyz.$$

We have

$$[a, b] = [x, y] \Rightarrow ay = bx$$
$$\Rightarrow (ay)(dw) = (bx)(dw)$$
$$\Rightarrow adyw = bdxw$$

and

$$[c, d] = [z, w] \Rightarrow cw = dz$$
$$\Rightarrow (cw)(by) = (dz)(by)$$
$$\Rightarrow bcyw = bdyz.$$

By adding corresponding sides of equations, we obtain

$$adyw + bcyw = bdxw + bdyz.$$

Thus, $[a, b] + [c, d] = [x, y] + [z, w]$.

It can be similarly shown that multiplication as defined by the given rule is a binary operation on Q.

It is important to note that the set of all ordered pairs of the form $(0, x)$, where $x \neq 0$, forms a complete equivalence class that can be written as $[0, b]$ for any nonzero element b of D. With these preliminaries out of the way, we can now state our theorem.

Theorem 5.25   **THE QUOTIENT FIELD**

Let D be an integral domain. The set Q as given in Definition 5.23 is a field, called the **quotient field** of D with respect to the operations defined in Lemma 5.24.

Proof   We first consider the postulates for addition. It is left as an exercise to prove that addition is associative. The zero element of Q is the class $[0, b]$, since

$$[x, y] + [0, b] = [x \cdot b + y \cdot 0, y \cdot b] = [xb, yb] = [x, y],$$

and similar steps show that

$$[0, b] + [x, y] = [x, y].$$

The equality $[xb, yb] = [x, y]$ follows from the fact that $b \neq 0$, as was pointed out just after Definition 5.23. Routine calculations show that $[-a, b]$ is the additive inverse of $[a, b]$ in Q, and that addition in Q is commutative. The verification of the associative property for multiplication is left as an exercise.

We shall verify the left distributive property and leave the other as an exercise. Let $[x, y], [z, w]$, and $[u, v]$ denote arbitrary elements of Q. We have

$$[x, y] \cdot ([z, w] + [u, v]) = [x, y][zv + wu, wv]$$
$$= [xzv + xwu, ywv]$$

and

$$[x, y] \cdot [z, w] + [x, y] \cdot [u, v] = [xz, yw] + [xu, yv]$$
$$= [xyzv + xywu, y^2wv]$$
$$= [y(xzv + xwu), y(ywv)].$$

Comparing the results of these two calculations, we see that the last one differs from the first only in that both elements in the pair have been multiplied by y. Since $[x, y]$ in Q requires $y \neq 0$, these results are equal.

Since multiplication in D is commutative, we have

$$[a, b] \cdot [c, d] = [ac, bd]$$
$$= [ca, db]$$
$$= [c, d] \cdot [a, b].$$

Thus, Q is a commutative ring.

Let $b \neq 0$ in D, and consider the element $[b, b]$ in Q. For any $[x, y]$ in Q we have

$$[x, y] \cdot [b, b] = [xb, yb]$$
$$= [x, y],$$

so $[b, b]$ is a right identity for multiplication. Since multiplication is commutative, $[b, b]$ is a nonzero unity for Q.

We have seen that the zero element of Q is the class $[0, b]$. Thus, any nonzero element has the form $[c, d]$, with both c and d nonzero. But then $[d, c]$ is also in Q, and

$$[c, d] \cdot [d, c] = [cd, dc]$$
$$= [d, d],$$

so $[d, c]$ is the multiplicative inverse of $[c, d]$ in Q. This completes the proof that Q is a field.

Note that in the proof of Theorem 5.25, the unity e in D did not appear explicitly anywhere. In fact, the construction yields a field if we start with a commutative ring that has no zero divisors instead of with an integral domain. However, we make use of the unity of D in Theorem 5.27.

The concept of an isomorphism can be applied to rings as well as to groups. The definition is a very natural extension of the concept of a group isomorphism. Since there are two binary operations involved in the definition of a ring, we simply require that both operations be preserved.

Definition 5.26   **RING ISOMORPHISM**

> Let R and R' denote two rings. A mapping ϕ: $R \to R'$ is a **ring isomorphism** from R to R' provided the following conditions hold:
>
> **1.** ϕ is a one-to-one correspondence from R to R'.
> **2.** $\phi(x + y) = \phi(x) + \phi(y)$ for all x and y in R.
> **3.** $\phi(x \cdot y) = \phi(x) \cdot \phi(y)$ for all x and y in R.
>
> If an isomorphism from R to R' exists, we say that R is **isomorphic** to R'.

Of course, the term *ring isomorphism* may be applied to systems that are more than a ring; that is, there may be a ring isomorphism that involves integral domains or fields. The relation of being isomorphic is reflexive, symmetric, and transitive on rings, just as it was with groups.

The field of quotients Q of an integral domain D has a significant feature that has not yet been brought to light. In the sense of isomorphism, it contains the integral domain D. More precisely, Q contains a subring D' that is isomorphic to D.

Theorem 5.27 **SUBRING OF Q ISOMORPHIC TO D**

Let D and Q be as given in Definition 5.23, and let e denote the unity of D. The set D' that consists of all elements of Q that have the form $[x, e]$ is a subring of Q, and D is isomorphic to D'.

Proof   Referring to Definition 5.1a, we see that condition 2, 5, 7, and 8 are automatically satisfied in D', and we only need to check conditions 1, 3, 4, and 6.

For arbitrary $[x, e]$ and $[y, e]$ in D', we have

$$[x, e] + [y, e] = [x \cdot e + y \cdot e, e \cdot e]$$
$$= [x + y, e],$$

and D' is closed under addition. The element $[0, e]$ is in D', so D' contains the zero element of Q. For $[x, e]$ in D', the additive inverse is $[-x, e]$, an element of D'. Finally, the calculation

$$[x, e] \cdot [y, e] = [xy, e]$$

shows that D' is closed under multiplication. Thus, D' is a subring of Q.

To prove that D is isomorphic to D', we use the natural mapping $\phi: D \to D'$ defined by

$$\phi(x) = [x, e].$$

The mapping ϕ is obviously a one-to-one correspondence. Since

$$\phi(x + y) = [x + y, e]$$
$$= [x, e] + [y, e]$$
$$= \phi(x) + \phi(y)$$

and

$$\phi(x \cdot y) = [xy, e]$$
$$= [x, e] \cdot [y, e]$$
$$= \phi(x) \cdot \phi(y),$$

ϕ is a ring isomorphism from D to D'.

Thus, the quotient field Q contains D in the sense of isomorphism. We say that D is **embedded** in Q or that Q is an **extension** of D. More generally, if S is a ring that contains a subring R' that is isomorphic to a given ring R, we say that R is **embedded** in S or that S is an **extension** of R.

There is one more observation about Q that should be made. For any nonzero $[b, e]$ in D', the multiplicative inverse of $[b, e]$ in Q is $[b, e]^{-1} = [e, b]$, and every element of Q can be written in the form

$$[a, b] = [a, e] \cdot [e, b] = [a, e] \cdot [b, e]^{-1}.$$

If the isomorphism ϕ in the proof of Theorem 5.27 is used to identify x in D with $[x, e]$ in D', then every element of Q can be identified as a quotient ab^{-1} of elements a and b of D, with $b \neq 0$.

From this, it follows that any field F that contains the integral domain D must also contain Q because F must contain b^{-1} for each $b \neq 0$ in D and must also contain the product ab^{-1} for all $a \in D$. Thus, Q is the smallest field that contains D.

If the construction presented in this section is carried out beginning with $D = \mathbf{Z}$, the field $\mathbf{Q}$ of rational numbers is obtained, with the elements written as $[a, b]$ instead of a/b. The isomorphism ϕ in the proof of Theorem 5.27 maps an integer x onto $[x, 1]$, which is playing the role of $x/1$ in the notation, and we end up with the integers embedded in the rational numbers. The construction of the rational numbers from the integers is in this way a special case of the procedure described here.

EXERCISES 5.3

1. Prove that the multiplication defined in Lemma 5.24 is a binary operation on Q.
2. Prove that addition is associative in Q.
3. Show that $[-a, b]$ is the additive inverse of $[a, b]$ in Q.
4. Prove that addition is commutative in Q.
5. Prove that multiplication is associative in Q.
6. Prove the right distributive property in Q:

$$([x, y] + [z, w]) \cdot [u, v] = [x, y] \cdot [u, v] + [z, w] \cdot [u, v].$$

7. Prove that, on a given set of rings, the relation of being isomorphic has the reflexive, symmetric, and transitive properties.
8. Assume that the ring R is isomorphic to the ring R'. Prove that if R is commutative, then R' is commutative.

Sec. 5.2, #5 ≫
Sec. 5.2, #8 ≫

9. Let W be the ring in Exercise 5 of Section 5.2, and let R be the ring in Exercise 8 of the same section. Given that W and R are isomorphic rings, define an isomorphism from W to R and prove that your mapping is an isomorphism.

Sec. 5.2, #9 ≫

10. Assume that the set R in Exercise 9 of Section 5.2 is a field, and let $\mathbf{C}$ be the field of all complex numbers $a + bi$, where a and b are real numbers and $i^2 = -1$. Given that R and $\mathbf{C}$ are isomorphic fields, define an isomorphism from $\mathbf{C}$ to R and prove that your mapping is an isomorphism.

11. Since this section presents a method for constructing a field of quotients for an arbitrary integral domain D, we might ask what happens if D is already a field. As an example, consider the situation when $D = \mathbf{Z}_3$.
    a. With $D = \mathbf{Z}_3$, write out all the elements of S, sort these elements according to the relation $\sim$, and then list all the distinct elements of Q.
    b. Exhibit an isomorphism from D to Q.

12. Work Exercise 11 with $D = \mathbf{Z}_5$.

13. Prove that if D is a field to begin with, the field of quotients Q is isomorphic to D.

14. Just after the end of the proof of Theorem 5.25, we noted that the construction in the proof yields a field if we start with a commutative ring that has no zero divisors. Assume this is true, and let F denote the field of quotients of the ring $\mathbf{E}$ of all even integers. Prove that F is isomorphic to the field of rational numbers.

15. Let D be the set of all complex numbers of the form $m + ni$, where $m \in \mathbf{Z}$ and $n \in \mathbf{Z}$. Carry out the construction of the quotient field Q for this integral domain, and show that this quotient field is isomorphic to the set of all complex numbers of the form $a + bi$, where a and b are rational numbers.

16. Prove that any field that contains an integral domain D must contain a subfield isomorphic to the quotient field Q of D.

17. Assume R is a ring, and let S be the set of all ordered pairs (m, x) where $m \in \mathbf{Z}$ and $x \in R$. Equality in S is defined by

$$(m, x) = (n, y) \quad \text{if and only if} \quad m = n \text{ and } x = y.$$

Addition and multiplication in S are defined by

$$(m, x) + (n, y) = (m + n, x + y)$$

and

$$(m, x) \cdot (n, y) = (mn, my + nx + xy),$$

where my and nx are *multiples* of y and x in the ring R.
   a. Prove that S is a ring with unity.
   b. Prove that $\phi: R \to S$ defined by $\phi(x) = (0, x)$ is an isomorphism from R to a subring R' of S. This result shows that any ring can be embedded in a ring that has a unity.

18. Let T be the smallest subring of the field $\mathbf{Q}$ of rational numbers that contains $\frac{1}{2}$. Find a description for a typical element of T.

5.4   ORDERED INTEGRAL DOMAINS

In Section 2.1 we assumed that the set $\mathbf{Z}$ of all integers satisfied a list of five postulates. The last two of these postulates led to the introduction of the order relation "greater than" in $\mathbf{Z}$, and to the proof of the Well-Ordering Theorem (Theorem 2.7). In this section we follow a development along similar lines in a more general setting.

Definition 5.28   ORDERED INTEGRAL DOMAIN

An integral domain D is an **ordered integral domain** if D contains a subset D^+ that has the following properties.

1. D^+ is closed under addition.
2. D^+ is closed under multiplication.
3. For each $x \in D$, one and only one of the following statements is true:

$$x \in D^+, \quad x = 0, \quad -x \in D^+.$$

Such a subset D^+ is called a **set of positive elements** for D.

Analogous to the situation in **Z**, condition 3 in Definition 5.28 is referred to as the **law of trichotomy**, and an element $x \in D$ such that $-x \in D^+$ is called a **negative element** of D.

Example 1   The integral domain **Z** is, of course, an example of an ordered integral domain. With their usual sets of positive elements, the set of all rational numbers and the set of all real numbers furnish two other examples of ordered integral domains.   ∎

Later, we shall see that not all integral domains are ordered integral domains.

Following the same sort of procedure that we followed with the integers, we can use the set of positive elements in an ordered integral domain D to define the order relation "greater than" in D.

Definition 5.29   **GREATER THAN**

> Let D be an ordered integral domain with D^+ as the set of positive elements. The relation **greater than**, denoted by $>$, is defined on elements x and y of D by
>
> $$x > y \quad \text{if and only if} \quad x - y \in D^+.$$

The symbol $>$ is read "greater than." Similarly, $<$ is read "is less than." We define $x < y$ if and only if $y > x$. As direct consequences of the definition, we have

$$x > 0 \quad \text{if and only if} \quad x \in D^+$$

and

$$x < 0 \quad \text{if and only if} \quad -x \in D^+.$$

The three properties of D^+ in Definition 5.28 translate at once into the following properties of $>$ in D.

1. If $x > 0$ and $y > 0$, then $x + y > 0$.
2. If $x > 0$ and $y > 0$, then $xy > 0$.
3. For each $x \in D$, one and only one of the following statements is true:

$$x > 0, \quad x = 0, \quad x < 0.$$

The other basic properties of $>$ are stated in the next theorem. We prove the first two and leave the proofs of the others as exercises.

Theorem 5.30   **PROPERTIES OF $>$**

Suppose that D is an ordered integral domain. The relation $>$ has the following properties, where x, y, and z are arbitrary elements of D.

a. If $x > y$, then $x + z > y + z$.
b. If $x > y$ and $z > 0$, then $xz > yz$.
c. If $x > y$ and $y > z$, then $x > z$.

d. One and only one of the following statements is true:

$$x > y, \quad x = y, \quad x < y.$$

$p \Rightarrow q$ **Proof of a**   If $x > y$, then $x - y \in D^+$, by Definition 5.29. Since

$$(x + z) - (y + z) = x + z - y - z$$
$$= x - y,$$

this means that $(x + z) - (y + z) \in D^+$, and therefore $x + z > y + z$.

$(p \land q) \Rightarrow r$ **Proof of b**   Suppose $x > y$ and $z > 0$. Then $x - y \in D^+$ and $z \in D^+$. Condition 2 of Definition 5.28 requires that D^+ be closed under multiplication, so the product $(x - y)z$ must be in D^+. Since $(x - y)z = xz - yz$, we have $xz - yz \in D^+$, and therefore $xz > yz$.

Our main goal in this section is to characterize the integers as an ordered integral domain that has a certain type of set of positive elements. As a first step in this direction, we prove the following simple theorem, which may be compared to Theorem 2.5.

Theorem 5.31   **SQUARE OF A NONZERO ELEMENT**

For any $x \neq 0$ in an ordered integral domain D, $x^2 \in D^+$.

$p \Rightarrow q$ **Proof**   Suppose $x \neq 0$ in D. By condition 3 of Definition 5.28, either $x \in D^+$ or $-x \in D^+$. If $x \in D^+$, then $x^2 = x \cdot x$ is in D^+ since D^+ is closed under multiplication. If $-x \in D^+$, then $x^2 = x \cdot x = (-x)(-x)$ is in D^+, again by closure of D^+ under multiplication. In either case, we have $x^2 \in D^+$.

Corollary 5.32   **THE UNITY ELEMENT**

In any ordered integral domain, $e \in D^+$.

Proof   This follows from the fact that $e = e^2$.

The preceding theorem and its corollary can be used to show that the set **C** of all complex numbers does not form an ordered integral domain. Suppose, to the contrary, that **C** does contain a set $\mathbf{C}^+$ of positive elements. By Corollary 5.32, $1 \in \mathbf{C}^+$, and therefore $-1 \notin \mathbf{C}^+$ by the law of trichotomy. Theorem 5.31 requires, however, that $i^2 = -1$ be in $\mathbf{C}^+$, and we have a contradiction. Therefore, **C** does not contain a set of positive elements. In other words, *it is impossible to impose an order relation on the set of complex numbers.*

In the next definition, we use the symbol $\leq$ with is usual meaning. Similarly, we later use the symbol $\geq$ with its usual meaning and without formal definition.

Definition 5.33   **WELL-ORDERED SUBSET**

A nonempty subset S of an ordered integral domain D is **well-ordered** if for every nonempty subset T of S, there is an element $m \in T$ such that $m \leq x$ for all $x \in T$. Such an element m is called a **least element** of T.

Thus, $S \neq \varnothing$ in D is well-ordered if every nonempty subset of S contains a least element. We proved in Theorem 2.7 that the set of all positive integers is well-ordered.

The next step toward our characterization of the integers is the following theorem.

Theorem 5.34   WELL-ORDERED D^+

If D is an ordered integral domain in which the set D^+ of positive elements is well-ordered, then

   a. e is the least element of D^+, and
   b. $D^+ = \{ne \,|\, n \in \mathbf{Z}^+\}$.

$p \Rightarrow q$ **Proof** We have $e \in D^+$ by Corollary 5.32. To prove that e is the least element of D^+, let T be the set of all $x \in D^+$ such that $e > x > 0$, and assume that T is nonempty. Since D^+ is well-ordered, T has a least element m, and

$$e > m > 0.$$

Using Theorem 5.30b and multiplying by m, we have

$$m \cdot e > m^2 > m \cdot 0.$$

That is,

$$m > m^2 > 0,$$

and this contradicts the choice of m as the least element of T. Therefore, T is empty and e is the least element of D^+.

$p \Rightarrow r$ Now let S be the set of all $m \in \mathbf{Z}^+$ such that $me \in D^+$. We have $1 \in S$ since $1e = e \in D^+$. Assume that $k \in S$. Then $ke \in D^+$, and this implies that

$$(k + 1)e = ke + e$$

is in S, since D^+ is closed under addition. Thus, $k \in S$ implies $k + 1 \in S$, and $S = \mathbf{Z}^+$ by the induction postulate for the positive integers. This proves that

$$D^+ \supseteq \{ne \,|\, n \in \mathbf{Z}^+\}.$$

In order to prove that $D^+ \subseteq \{ne \,|\, n \in \mathbf{Z}^+\}$, let L be the set of all elements of D^+ that are not of the form ne with $ne \in \mathbf{Z}^+$, and suppose that L is nonempty. Since D^+ is well-ordered, L has a least element ℓ. It must be true that

$$\ell > e,$$

since e is the least element of D^+, and therefore $\ell - e > 0$. Now

$$
\begin{aligned}
e > 0 &\Rightarrow e + (-e) > 0 + (-e) \quad \text{by Theorem 5.30a} \\
&\Rightarrow 0 > -e \\
&\Rightarrow \ell > \ell - e \qquad\qquad\quad \text{by Theorem 5.30a.}
\end{aligned}
$$

We thus have $\ell > \ell - e > 0$. By choice of ℓ as least element of L, $\ell - e \notin L$, so

$$\ell - e = pe \quad \text{for some } p \in \mathbf{Z}^+.$$

This implies that

$$\ell = pe + e$$
$$= (p + 1)e, \quad \text{where } p + 1 \in \mathbf{Z}^+,$$

and we have a contradiction to the fact that ℓ is an element that cannot be written in the form ne with $n \in \mathbf{Z}^+$. Therefore, $L = \varnothing$, and

$$D^+ = \{ne \mid n \in \mathbf{Z}^+\}.$$

We can now give the characterization of the integers toward which we have been working.

Theorem 5.35   ISOMORPHIC IMAGES OF Z

If D is an ordered integral domain in which the set D^+ of positive elements is well-ordered, then D is isomorphic to the ring $\mathbf{Z}$ of all integers.

$(p \wedge q) \Rightarrow r$ **Proof**   We first show that

$$D = \{ne \mid n \in \mathbf{Z}\}.$$

For an arbitrary $x \in D$, the law of trichotomy requires that exactly one of the following holds:

$$x \in D^+, \quad x = 0, \quad -x \in D^+.$$

If $x \in D^+$, then $x = ne$ for some $n \in \mathbf{Z}^+$, by Theorem 5.34b. If $x = 0$, then $x = 0e$. Finally, if $-x \in D^+$, then $-x = me$ for $m \in \mathbf{Z}^+$, and therefore[†] $x = -(me) = (-m)e$, where $-m \in \mathbf{Z}$. Hence, $D = \{ne \mid n \in \mathbf{Z}\}$.

Consider now the rule defined by

$$\phi(ne) = n,$$

for any ne in D. To demonstrate that this rule is well-defined, it is sufficient to show that each element of D can be written as ne in only one way. To do this, suppose $me = ne$. Without loss of generality, we may assume that $m \geq n$. Now

$$me = ne \Rightarrow me - ne = 0$$
$$\Rightarrow (m - n)e = 0.$$

If $m - n > 0$, then $(m - n)e \in D^+$, by Theorem 5.34b. Therefore, it must be that $m - n = 0$ and $m = n$. This shows that the rule $\phi(ne) = n$ defines a mapping ϕ from D to $\mathbf{Z}$.

If $\phi(me) = \phi(ne)$, then $m = n$, so $me = ne$. Hence, ϕ is one-to-one. An arbitrary $n \in \mathbf{Z}$ is the image of $ne \in D$ under ϕ, so ϕ is an onto mapping.

To show that ϕ is a ring isomorphism, we need to verify that

$$\phi(me + ne) = \phi(me) + \phi(ne)$$

[†]The equality $-(me) = (-m)e$ is the additive form of the familiar property of exponents $(a^m)^{-1} = a^{-m}$ in a group.

and also that

$$\phi(me \cdot ne) = \phi(me) \cdot \phi(ne).$$

From the laws of multiples in Section 3.2, we know that $me + ne = (m + n)e$, and it follows that

$$\phi(me + ne) = \phi((m + n)e)$$
$$= m + n$$
$$= \phi(me) + \phi(ne).$$

To show that ϕ preserves multiplication, we need the fact that $me \cdot ne = (mn)e$. This fact is a consequence of the generalized distributive laws stated in Theorem 5.13 and other results from Section 5.1. We leave the details of this proof as Exercise 9 at the end of this section. Using $me \cdot ne = (mn)e$, we have

$$\phi(me \cdot ne) = \phi[(mn)e]$$
$$= mn$$
$$= \phi(me) \cdot \phi(ne).$$

EXERCISES 5.4

1. Complete the proof of Theorem 5.30 by proving the following statements, where x, y, and z are arbitrary elements of an ordered integral domain D.
   a. If $x > y$ and $y > z$, then $x > z$.
   b. One and only one of the following statements is true:

   $$x > y, \quad x = y, \quad x < y.$$

2. Prove the following statements for arbitrary elements x, y, z of an ordered integral domain D.
   a. If $x > y$ and $z < 0$, then $xz < yz$.
   b. If $x > y$ and $z > w$, then $x + z > y + w$.
   c. If $x > y > 0$, then $x^2 > y^2$.
   d. If $x \neq 0$ in D, then $x^{2n} > 0$ for every positive integer n.
   e. If $x > 0$ and $xy > xz$, then $y > z$.

3. Prove the following statements for arbitrary elements in an ordered integral domain.
   a. $a > b$ implies $-b > -a$.
   b. $a > e$ implies $a^2 > a$.
   c. If $a > b$ and $c > d$, where a, b, c, and d are all positive elements, then $ac > bd$.

4. If a and b have multiplicative inverses in an ordered integral domain and $a > b > 0$, prove that $b^{-1} > a^{-1} > 0$.

5. Prove that the equation $x^2 + e = 0$ has no solution in an ordered integral domain.

6. Prove that if a is any element of an ordered integral domain D, then there exists an element $b \in D$ such that $b > a$. (Thus, D has no greatest element, and *no finite integral domain can be an ordered integral domain.*)

7. For an element x of an ordered integral domain D, the **absolute value** $|x|$ is defined by

$$|x| = \begin{cases} x & \text{if } x \geq 0 \\ -x & \text{if } 0 > x. \end{cases}$$

a. Prove that $-|x| \le x \le |x|$ for all $x \in D$.

b. Prove that $|xy| = |x| \cdot |y|$ for all $x, y \in D$.

Sec. 7.3, #26 ≪ **c.** Prove that $|x + y| \le |x| + |y|$ for all $x, y \in D$.

8. If x and y are elements of an ordered integral domain D, prove the following inequalities.

a. $x^2 - 2xy + y^2 \ge 0$

b. $x^2 + y^2 \ge xy$

c. $x^2 + y^2 \ge -xy$

9. If e denotes the unity element in an integral domain D, prove that $me \cdot ne = (mn)e$ for all $m, n \in \mathbf{Z}$.

10. An **ordered field** is an ordered integral domain that is also a field. In the quotient field Q of an ordered integral domain D, define Q^+ by

$$Q^+ = \{[a, b] \mid ab \in D^+\}.$$

Prove that Q^+ is a set of positive elements for Q, and hence that Q is an ordered field.

11. (See Exercise 10.) According to Definition 5.29, $>$ is defined in Q by $[a, b] > [c, d]$ if and only if $[a, b] - [c, d] \in Q^+$. Show that $[a, b] > [c, d]$ if and only if $abd^2 - cdb^2 \in D^+$.

12. (See Exercises 10 and 11.) If each $x \in D$ is identified in $[x, e]$ in Q, prove that $D^+ \subseteq Q^+$. (This means that the order relation defined in Exercise 10 coincides in D with the original order relation in D. We say that the ordering in Q is an **extension** of the ordering in D.)

13. Prove that if x and y are rational numbers such that $x > y$, then there exists a rational number z such that $x > z > y$. (This means that between any two distinct rational numbers there is another rational number.)

14. a. If D is an ordered integral domain, prove that each element in the quotient field Q of D can be written in the form $[a, b]$ with $b > 0$ in D.

b. If $[a, b] \in Q$ with $b > 0$ in D, prove that $[a, b] \in Q^+$ if and only if $a > 0$ in D.

15. (See Exercise 14.) If $[a, b]$ and $[c, d] \in Q$ with $b > 0$ and $d > 0$ in D, prove that $[a, b] > [c, d]$ if and only if $ad > bc$ in D.

16. If x and y are positive rational numbers, prove that there exists a positive integer n such that $nx > y$. This property is called the **Archimedean Property** of the rational numbers. (*Hint:* Write $x = a/b$ and $y = c/d$ with each of $a, b, c, d \in \mathbf{Z}^+$.)

KEY WORDS AND PHRASES

A Pioneer in Mathematics |||||||||||||||||||||||||||||||||||||||||||||||||||||||||||||||||||||||||||||
Richard Dedekind (1831–1916)

Julius Wilhelm Richard Dedekind, born on October 6, 1831, in Brunswick, Germany, has been called "the effective founder of abstract algebra" by the mathematics historian Morris Kline. Dedekind introduced the concepts of a ring and an ideal; in fact, he coined the terms *ring, ideal,* and *field.* His *Dedekind cuts* provided a technique for construction of the real numbers. Far ahead of his time, he built a foundation for further developments in ring and ideal theory by the famous algebraist, Emmy Noether (1882–1935).

At the age of 21, Dedekind earned his doctorate degree in mathematics working under Carl Friedrich Gauss (1777–1855) at the University of Göttingen. He taught at the university for a few years and presented the first formal lectures on Galois theory to an audience of two students. For four years, beginning in 1858, he was a professor in Zurich, Switzerland. Dedekind spent the next 50 years of his life in Brunswick, teaching in a technical high school that he had once attended. He died on February 12, 1916.

MORE ON RINGS

INTRODUCTION

The basic theorems on quotient rings and ring homomorphisms are presented in this chapter, along with a section on the characteristic of a ring and a section on maximal ideals. The development of $\mathbf{Z}_n$ culminates in Section 6.1 with the final description of $\mathbf{Z}_n$ as a quotient ring of the integers by the principal ideal (n).

6.1 IDEALS AND QUOTIENT RINGS

In this chapter we develop some theory of rings that parallels the theory of groups presented in Chapters 3 and 4. We shall see that the concept of an *ideal* in a ring is analogous to that of a *normal subgroup* in a group.

Definition 6.1a **DEFINITION OF AN IDEAL**

> The subset I of a ring R is an **ideal** of R if the following conditions hold:
>
> **1.** I is a subring of R.
> **2.** $x \in I$ and $r \in R$ imply that xr and rx are in I.

Note that the second condition in this definition requires more than closure of I under multiplication. It requires that I "absorb" multiplication by arbitrary elements of R, both on the right and on the left.

In more advanced study of rings, the type of subring described in Definition 6.1a is referred to as a "two-sided" ideal, and terms that are more specialized are introduced: a **right ideal** of R is a subring S of R such that $xr \in S$ for all $x \in S, r \in R$, and a **left ideal** of R is a subring S of R such that $rx \in S$ for all $x \in S, r \in R$. We only mention these terms in passing here, and observe that these distinctions cannot be made in a commutative ring.

The subrings $I = \{0\}$ and $I = R$ are always ideals of a ring R. These ideals are labeled **trivial**.

If R is a ring with unity e and I is an ideal of R that contains e, then

$$e \in I \quad \text{and} \quad r \in R \implies er = r = re \quad \text{is in } I,$$

so it must be true that $I = R$. That is, the only ideal of R that contains e is the ring R itself.

Example 1   In Section 5.1, we saw that the set **E** of all even integers is a subring of the ring **Z** of all integers. To show that condition 2 of Definition 6.1a holds, let $x \in \mathbf{E}$ and $m \in \mathbf{Z}$. Since $x \in \mathbf{E}$, $x = 2k$ for some integer k. We have

$$xm = mx = m(2k) = 2(mk),$$

so $xm = mx$ is in **E**. Thus, **E** is an ideal of **Z**.

It is worth noting that **E** is also a subring of the ring **Q** of all rational numbers, but **E** is not an ideal of **Q**. Condition 2 fails with $x = 4 \in \mathbf{E}$ and $r = \frac{1}{3} \in \mathbf{Q}$, but $xr = \frac{4}{3}$ is not in **E**.   ∎

In combination with Theorem 5.3, Definition 6.1a provides the following checklist of conditions that must be satisfied in order that a subset I of a ring R be an ideal:

1. I is nonempty.
2. $x \in I$ and $y \in I$ imply that $x + y$ and xy are in I.
3. $x \in I$ implies $-x \in I$.
4. $x \in I$ and $r \in R$ imply that xr and rx are in I.

The multiplicative closure in the second condition is implied by the fourth condition, so it may be deleted to obtain an alternate form of the definition of an ideal.

Definition 6.1b   **ALTERNATIVE DEFINITION OF AN IDEAL**

A subset I of a ring R is an **ideal** of R provided the following conditions are satisfied:

1. I is nonempty.
2. $x \in I$ and $y \in I$ imply $x + y \in I$.
3. $x \in I$ implies $-x \in I$.
4. $x \in I$ and $r \in R$ imply that xr and rx are in I.

A more efficient checklist is given in Exercise 1 at the end of this section.

Example 2   In Exercise 32 of Section 5.1, we saw that the set

$$S = \left\{ \begin{bmatrix} a & b \\ 0 & c \end{bmatrix} \middle| a, b, c \in \mathbf{Z} \right\}$$

forms a noncommutative ring with respect to the operations of matrix addition and multiplication. In this ring S, consider the subset

$$I = \left\{ \begin{bmatrix} 0 & b \\ 0 & 0 \end{bmatrix} \middle| b \in \mathbf{Z} \right\},$$

which is clearly nonempty. Since

$$\begin{bmatrix} 0 & x \\ 0 & 0 \end{bmatrix} + \begin{bmatrix} 0 & y \\ 0 & 0 \end{bmatrix} = \begin{bmatrix} 0 & x + y \\ 0 & 0 \end{bmatrix},$$

I is closed under addition. And since

$$-\begin{bmatrix} 0 & b \\ 0 & 0 \end{bmatrix} = \begin{bmatrix} 0 & -b \\ 0 & 0 \end{bmatrix},$$

I contains the additive inverse of each of its elements. For arbitrary $\begin{bmatrix} x & y \\ 0 & z \end{bmatrix}$ in S, we have

$$\begin{bmatrix} 0 & b \\ 0 & 0 \end{bmatrix}\begin{bmatrix} x & y \\ 0 & z \end{bmatrix} = \begin{bmatrix} 0 & bz \\ 0 & 0 \end{bmatrix} \quad \text{and} \quad \begin{bmatrix} x & y \\ 0 & z \end{bmatrix}\begin{bmatrix} 0 & b \\ 0 & 0 \end{bmatrix} = \begin{bmatrix} 0 & xb \\ 0 & 0 \end{bmatrix},$$

and both of these products are in I. Thus, I is an ideal of S.   ∎

Example 3   Example 8 of Section 5.1 introduced the ring $M = M_2(\mathbf{R})$ of all 2×2 matrices over the real numbers $\mathbf{R}$, and Exercise 33 of Section 5.1 introduced the subring T of M, given by

$$T = \left\{ \begin{bmatrix} a & a \\ b & b \end{bmatrix} \middle| a, b \in \mathbf{R} \right\}.$$

For arbitrary $\begin{bmatrix} a & a \\ b & b \end{bmatrix} \in T$, $\begin{bmatrix} x & y \\ z & w \end{bmatrix} \in M$, the product

$$\begin{bmatrix} x & y \\ z & w \end{bmatrix}\begin{bmatrix} a & a \\ b & b \end{bmatrix} = \begin{bmatrix} xa + yb & xa + yb \\ za + wb & za + wb \end{bmatrix}$$

is in T, so T absorbs multiplication on the left by elements of M. However, the product

$$\begin{bmatrix} a & a \\ b & b \end{bmatrix}\begin{bmatrix} x & y \\ z & w \end{bmatrix} = \begin{bmatrix} ax + az & ay + aw \\ bx + bz & by + bw \end{bmatrix}$$

is *not* always in T, and T does not absorb multiplication on the right by elements of M. This failure keeps T from being an ideal[†] of M.   ■

Example 1 may be generalized to the set of all multiplies of any fixed integer n. That is, the set $\{nk \,|\, k \in \mathbf{Z}\}$ of all multiples of n is an ideal of $\mathbf{Z}$. Instead of proving this fact, we establish the following more general result.

Example 4   Let R be a commutative ring with unity e. For any fixed $a \in R$, we shall show that the set

$$(a) = \{ar \,|\, r \in R\}$$

is an ideal of R.

This set is nonempty, since $a = ae$ is in (a). Let $x = ar$ and $y = as$ be arbitrary elements of (a), where $r \in R$, $s \in R$. Then

$$x + y = ar + as = a(r + s)$$

where $r + s \in R$, so (a) is closed under addition. We also have

$$-x = -(ar) = a(-r)$$

where $-r \in R$, so (a) contains additive inverses. For arbitrary $t \in R$,

$$tx = xt = (ar)t = a(rt)$$

where $rt \in R$. Thus, $tx = xt$ is in (a) for arbitrary $x \in (a)$, $t \in R$, and (a) is an ideal of R.   ■

This example leads to the following definition.

Definition 6.2   PRINCIPAL IDEAL

> If a is a fixed element of the commutative ring R with unity, the ideal
>
> $$(a) = \{ar \,|\, r \in R\},$$
>
> which consists of all multiples of a by elements r of R, is called the **principal ideal** generated by a in R.

The next theorem gives an indication of the importance of principal ideals.

[†]T could be said to be a *left ideal* of M.

Theorem 6.3   IDEALS IN **Z**

In the ring **Z** of integers, every ideal is a principal ideal.

$p \Rightarrow q$ **Proof**   The trivial ideal $\{0\}$ is certainly a principal ideal, $\{0\} = (0)$. Consider then an ideal I of **Z** such that $I \neq \{0\}$. Since $I \neq \{0\}$, I contains an integer $m \neq 0$. And since I contains both m and $-m$, it must contain some positive integers. Let n be the least positive integer in I. (Such an n exists, by the Well-Ordering Theorem.) For an arbitrary $k \in I$, the Division Algorithm asserts that there are integers q and r such that

$$k = nq + r \quad \text{with} \quad 0 \leq r < n.$$

Solving for r, we have

$$r = k - nq,$$

and this equation shows that $r \in I$, since k and n are in I and I is an ideal. That is, r is an element of I such that $0 \leq r < n$, where n is the *least positive element* of I. This forces the equality $r = 0$, and therefore $k = nq$. It follows that every element of I is a multiple of n, and therefore $I = (n)$.

Part of the analogy between ideals of a ring and normal subgroups of a group lies in the fact that ideals form the basis for a quotient structure much like the quotient group formed from the cosets of a normal subgroup.

To begin with, a ring R is an abelian group under addition, and any ideal I of R is a normal subgroup of this additive group. Thus, we may consider the additive quotient group R/I that consists of all the cosets

$$r + I = I + r = \{r + x \,|\, x \in I\}$$

of I in R. From our work in Chapter 4, we know that

$$a + I = b + I \quad \text{if and only if} \quad a - b \in I,$$
$$(a + I) + (b + I) = (a + b) + I,$$

and that R/I is an abelian group with respect to this operation of addition.

STRATEGY ▶

> If the defining rule for a possible binary operation is stated in terms of a certain type of representation for the elements, then the rule does not define a binary operation unless the result is independent of the representation for the elements; that is, unless the rule is well-defined.

In order to make a ring from the cosets in R/I, we consider a multiplication defined by

$$(a + I)(b + I) = ab + I.$$

We must show that this multiplication is well-defined. That is, we need to show that if

$$a + I = a' + I \quad \text{and} \quad b + I = b' + I,$$

then

$$ab + I = a'b' + I.$$

Now

$$a + I = a' + I \implies a = a' + x \quad \text{where } x \in I$$
$$b + I = b' + I \implies b = b' + y \quad \text{where } y \in I.$$

Thus,

$$ab = (a' + x)(b' + y) = a'b' + a'y + xb' + xy.$$

Since $x \in I$, $y \in I$, and I is an ideal, each of $a'y$, xb', and xy are in I. Therefore, their sum

$$z = a'y + xb' + xy$$

is in I, and $z + I = I$. This gives

$$ab + I = a'b' + z + I = a'b' + I$$

and our product is well-defined.

Theorem 6.4   THE RING OF COSETS

Let I be an ideal of the ring R. Then the set R/I of additive cosets $r + I$ of I in R forms a ring with respect to coset addition

$$(a + I) + (b + I) = (a + b) + I$$

and coset multiplication

$$(a + I)(b + I) = ab + I.$$

Proof   Assume I is an ideal of R. We noted earlier that the additive quotient group R/I is an abelian group with respect to addition.

We have already proved that the product

$$(a + I)(b + I) = ab + I$$

is well-defined in R/I, and closure under multiplication is automatic from the definition of this product. That the product is associative follows from

$$
\begin{aligned}
(a + I)[(b + I)(c + I)] &= (a + I)(bc + I) \\
&= a(bc) + I \\
&= (ab)c + I \quad \text{since multiplication is associative in } R \\
&= (ab + I)(c + I) \\
&= [(a + I)(b + I)](c + I).
\end{aligned}
$$

Verifying the left distributive law, we have

$$
\begin{aligned}
(a + I)[(b + I) + (c + I)] &= (a + I)[(b + c) + I] \\
&= a(b + c) + I \\
&= (ab + ac) + I \quad \text{from the left distributive law in } R \\
&= (ab + I) + (ac + I) \\
&= (a + I)(b + I) + (a + I)(c + I).
\end{aligned}
$$

The proof of the right distributive law is similar. Leaving that as an exercise, we conclude that R/I is a ring.

Definition 6.5   **QUOTIENT RING**

> If I is an ideal of the ring R, the ring R/I described in Theorem 6.4 is called the **quotient ring** of R by I.[†]

Example 5   In the ring $\mathbf{Z}$ of integers, consider the principal ideal

$$(4) = \{4k \mid k \in \mathbf{Z}\}.$$

The distinct elements of the ring $\mathbf{Z}/(4)$ are

$$(4) = \{\ldots, -8, -4, 0, 4, 8, \ldots\}$$
$$1 + (4) = \{\ldots, -7, -3, 1, 5, 9, \ldots\}$$
$$2 + (4) = \{\ldots, -6, -2, 2, 6, 10, \ldots\}$$
$$3 + (4) = \{\ldots, -5, -1, 3, 7, 11, \ldots\}.$$

We see, then, that these cosets are the same as the elements of $\mathbf{Z}_4$:

$$(4) = [0], \quad 1 + (4) = [1], \quad 2 + (4) = [2], \quad 3 + (4) = [3].$$

Moreover, the addition

$$\{a + (4)\} + \{b + (4)\} = \{a + b\} + (4)$$

agrees exactly with

$$[a] + [b] = [a + b]$$

in $\mathbf{Z}_4$, and the multiplication

$$\{a + (4)\}\{b + (4)\} = ab + (4)$$

agrees exactly with

$$[a][b] = [ab]$$

in $\mathbf{Z}_4$. Thus, $\mathbf{Z}/(4)$ is our old friend $\mathbf{Z}_4$. Put another way, $\mathbf{Z}_4$ is the quotient ring of the integers $\mathbf{Z}$ by the ideal (4).   ∎

The specific case in Example 5 generalizes at once to an arbitrary integer $n > 1$, and we see that $\mathbf{Z}_n$ is the quotient ring of $\mathbf{Z}$ by the ideal (n). This is our final and best description of $\mathbf{Z}_n$.

As a final remark to this section, we note that

$$(a + I)(b + I) = ab + I$$
$$\neq \{xy \mid x \in a + I \text{ and } y \in b + I\}.$$

[†] R/I is also known as "the ring of residue classes modulo the ideal I."

As a particular instance, consider $I = (4)$ as in Example 5. We have

$$(0 + I)(0 + I) = 0 + I = I,$$

but

$$\{xy \mid x \in 0 + I \text{ and } y \in 0 + I\} = \{16r \mid r \in \mathbf{Z}\},$$

since $x = 4p$ and $y = 4q$ for $p, q \in \mathbf{Z}$ imply $xy = 16pq$.

EXERCISES 6.1

1. Let I be a subset of the ring R. Prove that I is an ideal of R if and only if I is nonempty and $x - y$, xr, and rx are in I for all x and $y \in I, r \in R$.

2. Complete the proof of Theorem 6.4 by proving the right distributive law in R/I.

3. If I_1 and I_2 are two ideals of the ring R, prove that $I_1 \cap I_2$ is an ideal of R.

4. If $\{I_\lambda\}$, $\lambda \in \mathscr{L}$, is an arbitrary collection of ideals I_λ of the ring R, prove that $\bigcap_{\lambda \in \mathscr{L}} I_\lambda$ is an ideal of R.

5. Find two ideals I_1 and I_2 of the ring $\mathbf{Z}$ such that
   a. $I_1 \cup I_2$ is *not* an ideal of $\mathbf{Z}$.
   b. $I_1 \cup I_2$ is an ideal of $\mathbf{Z}$.

6. If I_1 and I_2 are two ideals of the ring R, prove that the set
   $$I_1 + I_2 = \{x + y \mid x \in I_1, y \in I_2\}$$
   is an ideal of R that contains each of I_1 and I_2.

7. Prove that if R is a field, then R has only the trivial ideals $\{0\}$ and R.

8. Let I be an ideal in a commutative ring R with unity. Prove that if I contains an element a that has a multiplicative inverse, then $I = R$.

9. In the ring $\mathbf{Z}$ of integers, prove that every subring is an ideal.

10. Let m and n be nonzero integers. Prove that $(m) \subseteq (n)$ if and only if n divides m.

11. If a and b are nonzero integers and m is the least common multiple of a and b, prove that $(a) \cap (b) = (m)$.

Sec. 6.2, #16 ≪   12. Prove that every ideal of $\mathbf{Z}_n$ is a principal ideal. (*Hint:* See Corollary 3.23.)

13. Find all distinct principal ideals of $\mathbf{Z}_n$ for the given value of n.
    a. $n = 7$
    b. $n = 11$
    c. $n = 12$
    d. $n = 18$
    e. $n = 20$
    f. $n = 24$

14. If R is a commutative ring and a is a fixed element of R, prove that the set $I_a = \{x \in R \mid ax = 0\}$ is an ideal of R.

Sec. 5.1, #32 ≫   15. (See Exercise 32 of Section 5.1.) Given that the set
   $$S = \left\{ \begin{bmatrix} x & y \\ 0 & z \end{bmatrix} \,\middle|\, x, y, z \in \mathbf{Z} \right\}$$
   is a ring with respect to matrix addition and multiplication, show that
   $$I = \left\{ \begin{bmatrix} a & b \\ 0 & 0 \end{bmatrix} \,\middle|\, a, b \in \mathbf{Z} \right\}$$

Sec. 6.2, #4 ≪   is an ideal of S.

16. Show that the set

$$I = M_2(\mathbf{E}) = \left\{ \begin{bmatrix} a & b \\ c & d \end{bmatrix} \middle| a, b, c, \text{ and } d \text{ are in } \mathbf{E} \right\}$$

of all 2×2 matrices over the ring $\mathbf{E}$ of even integers is an ideal of the ring $M_2(\mathbf{Z})$.

17. With S as in Exercise 15, decide whether or not the set

$$U = \left\{ \begin{bmatrix} a & b \\ 0 & a \end{bmatrix} \middle| a, b \in \mathbf{Z} \right\}$$

is an ideal of S, and justify your answer.

18. **a.** Show that the set

$$R = \left\{ \begin{bmatrix} x & 0 \\ y & 0 \end{bmatrix} \middle| x, y \in \mathbf{Z} \right\}$$

is a ring with respect to matrix addition and multiplication.

   b. Is R commutative?
   c. Does R have a unity?
   d. Decide whether or not the set

$$U = \left\{ \begin{bmatrix} 0 & 0 \\ a & 0 \end{bmatrix} \middle| a \in \mathbf{Z} \right\}$$

Sec. 6.2, #5 ≪   is an ideal of R, and justify your answer.

19. For a fixed element a of a commutative ring R, prove that the set $I = \{ar | r \in R\}$ is an ideal of R. (*Hint:* Compare this with Example 4, and note that the element a itself may not be in this set I.)

Sec. 6.4, #2 ≪
Sec. 6.4, #3 ≪
20. Let R be a commutative ring that does not have a unity. For a fixed $a \in R$, prove that the set
Sec. 6.4, #4 ≪
Sec. 6.4, #14 ≪

$$(a) = \{na + ra | n \in \mathbf{Z}, r \in R\}$$

Sec. 6.4, #15 ≪
Sec. 6.4, #17 ≪
Sec. 6.4, #18 ≪   is an ideal of R that contains the element a. (This ideal is called the **principal ideal** of R that is **generated** by a.)

Sec. 5.1, #37 ≫
21. An element a of a ring R is called **nilpotent** if $a^n = 0$ for some positive integer n. Show that the set of all nilpotent elements in a commutative ring R forms an ideal of R.

22. If I is an ideal of R, prove that the set

$$K_I = \{x \in R | xa = 0 \text{ for all } a \in I\}$$

is an ideal of R.

23. Let R be a commutative ring with unity whose only ideals are $\{0\}$ and R itself. Prove that R is a field. (*Hint:* See Exercise 19.)

24. Suppose that R is a commutative ring with unity and that I is an ideal of R. Prove that the set of all $x \in R$ such that $x^n \in I$ for some positive integer n is an ideal of R.

6.2   RING HOMOMORPHISMS

We turn our attention now to *ring homomorphisms* and their relations to ideals and quotient rings.

Definition 6.6   **RING HOMOMORPHISM**

> If R and R' are rings, a **ring homomorphism** from R to R' is a mapping $\theta: R \to R'$ such that
>
> $$\theta(x + y) = \theta(x) + \theta(y) \quad \text{and} \quad \theta(xy) = \theta(x)\theta(y)$$
>
> for all x and y in R.

That is, a ring homomorphism is a mapping from one ring to another that preserves both ring operations. This situation is analogous to the one where a homomorphism from one group to another preserves the group operation, and it explains the use of the term "homomorphism" in both situations. It is sometimes desirable to use one of the terms *group homomorphism* or *ring homomorphism* for clarity, but in many cases, the context makes the meaning clear for the single word *homomorphism*. If only groups are under consideration, then "homomorphism" means group homomorphism, and if rings are under consideration, "homomorphism" means ring homomorphism.

Some terminology for a special type of homomorphism is given in the following definition.

Definition 6.7   **RING EPIMORPHISM, ISOMORPHISM**

> Let θ be a homomorphism from the ring R to the ring R'.
> 1. If θ is onto, then θ is called an **epimorphism** and R' is called a **homomorphic image** of R.
> 2. If θ is a one-to-one correspondence (both onto and one-to-one), then θ is an **isomorphism**.

Example 1   Consider the mapping $\theta: \mathbf{Z} \to \mathbf{Z}_n$ defined by

$$\theta(a) = [a].$$

Since

$$\theta(a + b) = [a + b] = [a] + [b] = \theta(a) + \theta(b)$$

and

$$\theta(ab) = [ab] = [a][b] = \theta(a)\theta(b)$$

for all a and b in $\mathbf{Z}$, θ is a homomorphism from $\mathbf{Z}$ to $\mathbf{Z}_n$. In fact, θ is an *epimorphism* and $\mathbf{Z}_n$ is a *homomorphic image* of $\mathbf{Z}$. ∎

Example 2   Consider $\theta: \mathbf{Z}_6 \to \mathbf{Z}_6$ defined by

$$\theta([a]) = 4[a].$$

It follows from

$$\begin{aligned}
\theta([a] + [b]) + 4([a] + [b]) \\
= 4[a] + 4[b] \\
= \theta([a]) + \theta([b])
\end{aligned}$$

that θ preserves addition. For multiplication, we have

$$\theta([a][b]) = \theta([ab]) = 4[ab] = [4ab]$$

and

$$\theta([a])\theta([b]) = (4[a])(4[b]) = 16[ab] = [16ab] = [4ab],$$

since $[16] = [4]$ in $\mathbf{Z}_6$. Thus, θ is a homomorphism. It can be verified that $\theta(\mathbf{Z}_6) = \{[0], [2], [4]\}$, and we see that θ is neither onto nor one-to-one.   ■

Theorem 6.8   IMAGES OF ZERO AND ADDITIVE INVERSES

If θ is a homomorphism from the ring R to the ring R', then
a. $\theta(0) = 0$, and
b. $\theta(-r) = -\theta(r)$ for all $r \in R$.

$p \Rightarrow q$ **Proof**   The statement in part a follows from

$$\begin{aligned}
\theta(0) &= \theta(0) + 0 \\
&= \theta(0) + \theta(0) - \theta(0) \\
&= \theta(0 + 0) - \theta(0) \\
&= \theta(0) - \theta(0) \\
&= 0.
\end{aligned}$$

$(p \wedge q) \Rightarrow r$ To prove part **b**, we observe that

$$\begin{aligned}
\theta(r) + \theta(-r) &= \theta[r + (-r)] \\
&= \theta(0) \\
&= 0.
\end{aligned}$$

Since the additive inverse is unique in the additive group of R',

$$-\theta(r) = \theta(-r).$$

Under a ring homomorphism, images of subrings are subrings, and inverse images of subrings are also subrings. This is the content of the next theorem.

Theorem 6.9   IMAGES AND INVERSE IMAGES OF SUBRINGS

Suppose θ is a homomorphism from the ring R to the ring R'.
a. If S is a subring of R, then $\theta(S)$ is a subring of R'.
b. If S' is a subring of R', then $\theta^{-1}(S')$ is a subring of R.

$(p \wedge q) \Rightarrow r$ **Proof**   To prove part **a**, suppose S is a subring of R. We shall verify that the conditions of Theorem 5.3 are satisfied by $\theta(S)$. The element $\theta(0) = 0$ is in $\theta(S)$, so $\theta(S)$ is nonempty. Let x' and y' be arbitrary elements of $\theta(S)$. Then there exist elements x, $y \in S$ such that $\theta(x) = x'$ and $\theta(y) = y'$. Since S is a subring, $x + y$ and xy are in S. Therefore,

$$\begin{aligned}
\theta(x + y) &= \theta(x) + \theta(y) \\
&= x' + y'
\end{aligned}$$

and

$$\theta(xy) = \theta(x)\theta(y) = x'y'$$

are in $\theta(S)$, and $\theta(S)$ is closed under addition and multiplication. Since $-x$ is in S and

$$\theta(-x) = -\theta(x) = -x',$$

we have $-x' \in \theta(S)$, and it follows that $\theta(S)$ is a subring of R'.

$(p \wedge q) \Rightarrow r$ To prove part b, assume that S' is a subring of R'. We have 0 in $\theta^{-1}(S')$ since $\theta(0) = 0$, so $\theta^{-1}(S')$ is nonempty. Let $x \in \theta^{-1}(S')$ and $y \in \theta^{-1}(S')$. This implies that $\theta(x) \in S'$ and $\theta(y) \in S'$. Hence, $\theta(x) + \theta(y) = \theta(x + y)$ and $\theta(x)\theta(y) = \theta(xy)$ are in S', since S' is a subring. Now

$$\theta(x + y) \in S' \Rightarrow x + y \in \theta^{-1}(S')$$

and

$$\theta(xy) \in S' \Rightarrow xy \in \theta^{-1}(S').$$

We also have

$$\theta(x) \in S' \Rightarrow -\theta(x) = \theta(-x) \in S'$$
$$\Rightarrow -x \in \theta^{-1}(S'),$$

and $\theta^{-1}(S')$ is a subring of R by Theorem 5.3.

Definition 6.10   KERNEL

> If θ is a homomorphism from the ring R to the ring R', the **kernel** of θ is the set
>
> $$\ker \theta = \{x \in R \mid \theta(x) = 0\}.$$

Example 3   In Example 1, the epimorphism $\theta: \mathbf{Z} \to \mathbf{Z}_n$ is defined by $\theta(a) = [a]$. Now $\theta(a) = [0]$ if and only if a is a multiple of n, so

$$\ker \theta = \{\dots, -2n, -n, 0, n, 2n, \dots\}$$

for this θ.

In Example 2, the homomorphism $\theta: \mathbf{Z}_6 \to \mathbf{Z}_6$ defined by $\theta([a]) = 4[a]$ has kernel given by

$$\ker \theta = \{[0], [3]\}. \qquad \blacksquare$$

In these two examples, $\ker \theta$ is an ideal of the domain of θ. This is true in general for homomorphisms, according to the following theorem.

Theorem 6.11   KERNEL OF A RING HOMOMORPHISM

If θ is any homomorphism from the ring R to the ring R', then $\ker \theta$ is an ideal of R, and $\ker \theta = \{0\}$ if and only if θ is one-to-one.

$p \Rightarrow q$ **Proof**   Under the hypothesis, we know that $\ker \theta$ is a subring of R from Theorem 6.9. For any $x \in \ker \theta$ and $r \in R$, we have

$$\theta(xr) = \theta(x)\theta(r)$$
$$= 0 \cdot \theta(r) = 0,$$

and similarly $\theta(rx) = 0$. Thus, xr and rx are in ker θ, and ker θ is an ideal of R.

$u \Leftarrow v$ Suppose θ is one-to-one. Then $x \in$ ker θ implies $\theta(x) = 0 = \theta(0)$, and therefore $x = 0$. Hence, ker $\theta = \{0\}$ if θ is one-to-one.

$u \Rightarrow v$ Conversely, if ker $\theta = \{0\}$, then

$$\theta(x) = \theta(y) \Rightarrow \theta(x) - \theta(y) = 0$$
$$\Rightarrow \theta(x - y) = 0$$
$$\Rightarrow x - y = 0$$
$$\Rightarrow x = y.$$

This means that θ is one-to-one if ker $\theta = \{0\}$, and the proof is complete.

Example 4   This example illustrates the last part of Theorem 6.11 and provides a nice example of a ring isomorphism.

For the set $U = \{a, b\}$, the power set of U is $\mathscr{P}(U) = \{\varnothing, A, B, U\}$ where $A = \{a\}$ and $B = \{b\}$. With addition defined by

$$X + Y = (X \cup Y) - (X \cap Y)$$

and multiplication by

$$X \cdot Y = X \cap Y,$$

$\mathscr{P}(U)$ forms a ring, as we saw in Example 5 of Section 5.1. Addition and multiplication tables for $\mathscr{P}(U)$ are given in Figure 6-1.

Figure 6-1

+	$\varnothing$	A	B	U
$\varnothing$	$\varnothing$	A	B	U
A	A	$\varnothing$	U	B
B	B	U	$\varnothing$	A
U	U	B	A	$\varnothing$

$\cdot$	$\varnothing$	A	B	U
$\varnothing$	$\varnothing$	$\varnothing$	$\varnothing$	$\varnothing$
A	$\varnothing$	A	$\varnothing$	A
B	$\varnothing$	$\varnothing$	B	B
U	$\varnothing$	A	B	U

The ring $R = \mathbf{Z}_2 \oplus \mathbf{Z}_2$ was introduced in Exercises 38 and 39 of Section 5.1. If we write 0 for $[0]$ and 1 for $[1]$ in $\mathbf{Z}_2$, the elements of R are given by $R = \{(0, 0), (1, 0), (0, 1), (1, 1)\}$. Tables for R are displayed in Figure 6-2.

Figure 6-2

+	$(0,0)$	$(1,0)$	$(0,1)$	$(1,1)$
$(0,0)$	$(0,0)$	$(1,0)$	$(0,1)$	$(1,1)$
$(1,0)$	$(1,0)$	$(0,0)$	$(1,1)$	$(0,1)$
$(0,1)$	$(0,1)$	$(1,1)$	$(0,0)$	$(1,0)$
$(1,1)$	$(1,1)$	$(0,1)$	$(1,0)$	$(0,0)$

$\cdot$	$(0,0)$	$(1,0)$	$(0,1)$	$(1,1)$
$(0,0)$	$(0,0)$	$(0,0)$	$(0,0)$	$(0,0)$
$(1,0)$	$(0,0)$	$(1,0)$	$(0,0)$	$(1,0)$
$(0,1)$	$(0,0)$	$(0,0)$	$(0,1)$	$(0,1)$
$(1,1)$	$(0,0)$	$(1,0)$	$(0,1)$	$(1,1)$

Consider the mapping $\theta: \mathscr{P}(U) \to R$ defined by

$$\theta(\varnothing) = (0,0), \qquad \theta(A) = (1,0), \qquad \theta(B) = (0,1), \qquad \theta(U) = (1,1).$$

If each element x in the tables for $\mathcal{P}(U)$ is replaced by $\theta(x)$, the resulting tables agree completely with those in Figure 6-2. Thus, θ is an isomorphism. We note that the kernel of θ consists of the zero element in $\mathcal{P}(U)$.   ∎

We know now that every kernel of a homomorphism from a ring R is an ideal of R. The next theorem shows that every ideal of R is a kernel of a homomorphism from R. This means that the ideals of R and the kernels of the homomorphisms from R to another ring are the same subrings of R.

Theorem 6.12   QUOTIENT RING $\Rightarrow$ HOMOMORPHIC IMAGE

If I is an ideal of the ring R, the mapping $\theta: R \rightarrow R/I$ defined by

$$\theta(r) = r + I$$

is an epimorphism from R to R/I with kernel I.

$p \Rightarrow q$ **Proof** It is clear that the rule $\theta(r) = r + I$ defines an onto mapping θ from R to R/I, and that ker $\theta = I$. Since

$$\begin{aligned} \theta(x + y) &= (x + y) + I \\ &= (x + I) + (y + I) \\ &= \theta(x) + \theta(y) \end{aligned}$$

and

$$\begin{aligned} \theta(xy) &= xy + I \\ &= (x + I)(y + I) \\ &= \theta(x)\theta(y), \end{aligned}$$

θ is indeed an epimorphism from R to R/I.

The last theorem shows that every quotient ring of a ring R is a homomorphic image of R. A result in the opposite direction is given in the next theorem.

STRATEGY ▶   In the proof of Theorem 6.13, it is shown that a certain rule defines a mapping ϕ. When the defining rule for a possible mapping is stated in terms of a certain type of representation for the elements, the rule does not define a mapping unless the result is independent of the representation of the elements; that is, unless the rule is well-defined.

Theorem 6.13   HOMOMORPHIC IMAGE $\Rightarrow$ QUOTIENT RING

If a ring R' is a homomorphic image of the ring R, then R' is isomorphic to a quotient ring of R.

$p \Rightarrow q$ **Proof** Suppose θ is an epimorphism from R to R', and let $K = \ker \theta$. For each $a + K$ in R/K, define $\phi(a + K)$ by

$$\phi(a + K) = \theta(a).$$

To prove that this rule defines a mapping, let $a + K$ and $b + K$ be arbitrary elements of R/K. Then

$$a + K = b + K \Leftrightarrow a - b \in K$$
$$\Leftrightarrow \theta(a - b) = 0$$
$$\Leftrightarrow \theta(a) = \theta(b)$$
$$\Leftrightarrow \phi(a + K) = \phi(b + K).$$

This shows that ϕ is well-defined and one-to-one as well. From the definition of ϕ, it follows that $\phi(R/K) = \theta(R)$. But $\theta(R) = R'$, since θ is an epimorphism. Thus, ϕ is onto and, consequently, is a one-to-one correspondence from R/K to R'.

For arbitrary $a + K$ and $b + K$ in R/K,

$$\phi[(a + K) + (b + K)] = \phi[(a + b) + K]$$
$$= \theta(a + b)$$
$$= \theta(a) + \theta(b) \quad \text{since } \theta \text{ is an epimorphism}$$
$$= \phi(a + K) + \phi(b + K)$$

and

$$\phi[(a + K)(b + K)] = \phi(ab + K)$$
$$= \theta(ab)$$
$$= \theta(a)\theta(b) \quad \text{since } \theta \text{ is an epimorphism}$$
$$= \phi(a + K)\phi(b + K).$$

Thus, ϕ is an isomorphism from R/K to R'.

As an immediate consequence of the proof of this theorem, we have the following **Fundamental Theorem of Ring Homomorphisms**.

Theorem 6.14   **FUNDAMENTAL THEOREM OF RING HOMOMORPHISMS**

If θ is an epimorphism from the ring R to the ring R', then R' is isomorphic to $R/\ker \theta$.

We now see that, in the sense of isomorphism, the homomorphic images of a ring R are the same as the quotient rings of R. This gives a systematic way to search for all the homomorphic images of a given ring. To illustrate the usefulness of this method, we shall find all the homomorphic images of the ring $\mathbf{Z}$ of integers.

Example 5   In order to find all homomorphic images of $\mathbf{Z}$, we shall find all possible ideals of $\mathbf{Z}$ and form all possible quotient rings. According to Theorem 6.3, every ideal of $\mathbf{Z}$ is a principal ideal.

For the trivial ideal $(0) = \{0\}$, we obtain the quotient ring $\mathbf{Z}/(0)$, which is isomorphic to $\mathbf{Z}$, since $a + (0) = b + (0)$ if and only if $a = b$. For the other trivial ideal $(1) = \mathbf{Z}$, we obtain the quotient ring $\mathbf{Z}/\mathbf{Z}$, which has only one element and is isomorphic to $\{0\}$. As shown in the proof of Theorem 6.3, any nontrivial ideal I of $\mathbf{Z}$ has the form $I = (n)$ for some positive integer $n > 1$. For these ideals, we obtain the quotient rings[†] $\mathbf{Z}/(n) = \mathbf{Z}_n$. Thus, the homomorphic images of $\mathbf{Z}$ are $\mathbf{Z}$ itself, $\{0\}$, and the rings $\mathbf{Z}_n$.   ■

Exercises 6.2

Unless otherwise stated, R and R' denote arbitrary rings throughout this set of exercises.

1. Suppose θ is an epimorphism from R to R'. Prove that R' is commutative if R is commutative.
2. Prove that if θ is an epimorphism from R to R' and if R has a unity e, then $\theta(e)$ is a unity in R'.
3. (See Exercise 2.) Suppose that θ is an epimorphism from R to R' and that R has a unity. Prove that if a^{-1} exists for $a \in R$, then $[\theta(a)]^{-1}$ exists, and $[\theta(a)]^{-1} = \theta(a^{-1})$.

Sec. 6.1, #15 ≫ 4. (See Exercise 15 of Section 6.1.) Assume that the set

$$S = \left\{ \begin{bmatrix} x & y \\ 0 & z \end{bmatrix} \middle| x, y, z \in \mathbf{Z} \right\}$$

is a ring with respect to matrix addition and multiplication.

   a. Verify that the mapping $\theta: S \to \mathbf{Z}$ defined by $\theta\left(\begin{bmatrix} x & y \\ 0 & z \end{bmatrix} \right) = z$ is an epimorphism from S to $\mathbf{Z}$.

   b. Describe $\ker \theta$ and exhibit an isomorphism from $S/\ker \theta$ to $\mathbf{Z}$.

Sec. 6.1, #18 ≫ 5. (See Exercise 18 of Section 6.1.) Assume that the set

$$R = \left\{ \begin{bmatrix} x & 0 \\ y & 0 \end{bmatrix} \middle| x, y \in \mathbf{Z} \right\}$$

is a ring with respect to matrix addition and multiplication.

   a. Verify that the mapping $\theta: R \to \mathbf{Z}$ defined by $\theta\left(\begin{bmatrix} x & 0 \\ y & 0 \end{bmatrix} \right) = x$ is an epimorphism from R to $\mathbf{Z}$.

   b. Describe $\ker \theta$ and exhibit an isomorphism from $R/\ker \theta$ to $\mathbf{Z}$.

6. For any $a \in \mathbf{Z}$, let $[a]_6$ denote $[a]$ in $\mathbf{Z}_6$ and $[a]_2$ denote $[a]$ in $\mathbf{Z}_2$.
   a. Prove that the mapping $\theta: \mathbf{Z}_6 \to \mathbf{Z}_2$ defined by $\theta([a]_6) = [a]_2$ is a homomorphism.
   b. Find $\ker \theta$.

7. In the field $\mathbf{C}$ of complex numbers, show that the mapping θ that maps each complex number onto its conjugate, $\theta(a + bi) = a - bi$, is an isomorphism from $\mathbf{C}$ to $\mathbf{C}$.

8. (See Example 3 of Section 5.1.) Let S denote the subring of the real numbers that consists of all real numbers of the form $m + n\sqrt{2}$, with $m \in \mathbf{Z}$ and $n \in \mathbf{Z}$. Prove that $\theta(m + n\sqrt{2}) = m - n\sqrt{2}$ defines an isomorphism from S to S.

9. Define $\theta: M_2(\mathbf{Z}) \to M_2(\mathbf{Z}_2)$ by

$$\theta\left(\begin{bmatrix} a & b \\ c & d \end{bmatrix} \right) = \begin{bmatrix} [a] & [b] \\ [c] & [d] \end{bmatrix}.$$

Prove that θ is a homomorphism, and describe $\ker \theta$.

10. Assume that

$$R = \left\{ \begin{bmatrix} m & 2n \\ n & m \end{bmatrix} \middle| m, n \in \mathbf{Z} \right\}$$

[†]See the paragraph immediately following Example 5 in Section 6.1.

and

$$R' = \{m + n\sqrt{2} \mid m, n \in \mathbf{Z}\}$$

are rings with respect to their usual operations, and prove that R and R' are isomorphic rings.

11. Consider the mapping $\theta: \mathbf{Z}_{12} \to \mathbf{Z}_{12}$ defined by $\theta([a]) = 4[a]$. Decide whether θ is a homomorphism, and justify your answer.

Sec. 5.1, #26 ≫   12. Figure 6-3 gives addition and multiplication tables for the ring $R = \{a, b, c\}$ in Exercise 26 of Section 5.1. Use these tables together with addition and multiplication tables for $\mathbf{Z}_3$ to find an isomorphism from R to $\mathbf{Z}_3$.

Figure 6-3

+	a	b	c		·	a	b	c
a	a	b	c		a	a	a	a
b	b	c	a		b	a	c	b
c	c	a	b		c	a	b	c

Sec. 5.1, #27 ≫   13. Figure 6-4 gives addition and multiplication tables for the ring $R = \{a, b, c, d\}$ in Exercise 27 of Section 5.1. Construct addition and multiplication tables for the subring $R' = \{[0], [2], [4], [6]\}$ of $\mathbf{Z}_8$, and find an isomorphism from R to R'.

Figure 6-4

+	a	b	c	d		·	a	b	c	d
a	a	b	c	d		a	a	a	a	a
b	b	c	d	a		b	a	c	a	c
c	c	d	a	b		c	a	a	a	a
d	d	a	b	c		d	a	c	a	c

Sec. 5.1, #38 ≫   14. (See Exercise 38 of Section 5.1.) Let R_1 be the subring of $R \oplus R'$ that consists of all elements of the form $(r, 0)$ where $r \in R$. Prove that R_1 is isomorphic to R.

15. Each of the following rules determines a mapping $\theta: \mathbf{R} \to \mathbf{R}$, where $\mathbf{R}$ is the field of real numbers. Decide in each case whether θ preserves addition, whether θ preserves multiplication, and whether θ is a homomorphism.
   a. $\theta(x) = |x|$ b. $\theta(x) = 2x$
   c. $\theta(x) = -x$ d. $\theta(x) = x^2$
   e. $\theta(x) = \begin{cases} 0 & \text{if } x = 0 \\ \dfrac{1}{x} & \text{if } x \neq 0 \end{cases}$ f. $\theta(x) = x + 1$

Sec. 6.1, #12 ≫   16. For each given value of n, find all homomorphic images of $\mathbf{Z}_n$. (*Hint:* See Exercise 12 of Section 6.1.)
   a. $n = 6$ b. $n = 10$ c. $n = 12$
   d. $n = 18$ e. $n = 8$ f. $n = 20$

17. Suppose F is a field and θ is an epimorphism from F to a ring S such that $\ker \theta \neq F$. Prove that θ is an isomorphism and that S is a field.

18. Assume that θ is an epimorphism from R to R'. Prove the following statements.
   a. If I is an ideal of R, then $\theta(I)$ is an ideal of R'.
   b. If I' is an ideal of R', then $\theta^{-1}(I')$ is an ideal of R.

c. The mapping $I \rightarrow \theta(I)$ is a bijection from the set of ideals I of R that contain ker θ to the set of all ideals of R'.

19. In the ring **Z** of integers, let new operations of addition and multiplication be defined by

$$x \oplus y = x + y + 1 \quad \text{and} \quad x \odot y = xy + x + y,$$

where x and y are arbitrary integers and $x + y$ and xy denote the usual addition and multiplication in **Z**.
   a. Prove that the integers form a ring R' with respect to $\oplus$ and $\odot$.
   b. Identify the zero element and unity of R'.
   c. Prove that **Z** is isomorphic to R'.

Sec. 4.5, #26 ≫ **20.** Let K and I be ideals of the ring R. Prove that $K/K \cap I$ is isomorphic to $(K + I)/I$. (*Hint:* See Exercise 26 of Section 4.5.)

6.3 THE CHARACTERISTIC OF A RING

In this section we focus on the fact that the elements of a ring R form an abelian group under addition.

When the binary operation in a group G is multiplication, each element a of G generates a cyclic group $\langle a \rangle$ that consists of all integral powers of a. If there are positive integers n such that $a^n = e$ and m is the smallest such positive integer, then m is the (multiplicative) *order* of a.

When the binary operation in a group is addition, the cyclic subgroup $\langle a \rangle$ consists of all integral multiples ka of a. If there are positive integers n such that $na = 0$, and m is the smallest such positive integer, then m is the (additive) *order* of a. In a sense, the characteristic of a ring is a generalization from this idea.

Definition 6.15   **CHARACTERISTIC**

> If there are positive integers n such that $nx = 0$ for *all* x in the ring R, then the smallest positive integer m such that $mx = 0$ for all $x \in R$ is called the **characteristic** of R. If no such positive integer exists, then R is said to be of **characteristic zero**.

It is logical in the last case to call zero the characteristic of R since $n = 0$ is the only integer such that $nx = 0$ for all $x \in R$.

Example 1   The ring **Z** of integers has characteristic zero since $nx = 0$ for all $x \in \mathbf{Z}$ requires that $n = 0$. For the same reason, the field **R** of real numbers and the field **C** of complex numbers each have characteristic zero. ∎

Example 2   Consider the ring $\mathbf{Z}_6$. For the various elements of $\mathbf{Z}_6$, we have

$$1[0] = [0] \quad 6[1] = [0] \quad 3[2] = [0]$$
$$2[3] = [0] \quad 3[4] = [0] \quad 6[5] = [0].$$

Although smaller positive integers work for some individual elements of $\mathbf{Z}_6$, the smallest positive integer m such that $m[a] = [0]$ for all $[a] \in \mathbf{Z}_6$ is $m = 6$. Thus, $\mathbf{Z}_6$ has characteristic 6. This example generalizes readily, and we see that $\mathbf{Z}_n$ has characteristic n. ∎

Theorem 6.16   CHARACTERISTIC OF A RING

Let R be a ring with unity e. If e has finite additive order m, then m is the characteristic of R.

$p \Rightarrow q$ **Proof**   Suppose R is a ring with unity e and that e has finite additive order m. Then m is the least positive integer such that $me = 0$. For arbitrary $x \in R$,

$$mx = m(ex) = (me)x = 0 \cdot x = 0.$$

Thus, $mx = 0$ for all $x \in R$, and m is the smallest positive integer for which this is true. By Definition 6.15, R has characteristic m.

In connection with the last theorem, we note that if R has a unity e and e does not have finite additive order, then R has characteristic zero. In either case, the characteristic can be determined simply by investigating the additive order of e.

Theorem 6.17   CHARACTERISTIC OF AN INTEGRAL DOMAIN

The characteristic of an integral domain is either zero or a prime integer.

$\sim p \Leftarrow (\sim q \wedge \sim r)$ **Proof**   Let D be an integral domain. As mentioned before, D has characteristic zero if the additive order of the unity e is not finite. Suppose, then, that e has finite additive order m. By Theorem 6.16, D has characteristic m, and we only need to show that m is a prime integer. Assume, to the contrary, that m is not a prime and $m = rs$ for positive integers r and s such that $1 < r < m$ and $1 < s < m$. Then we have $re \neq 0$ and $se \neq 0$, but

$$(re)(se) = (rs)e^2 = (rs)e = me = 0.$$

This is a contradiction to the fact that D is an integral domain. Therefore, m is a prime integer, and the proof is complete.

If the characteristic of a ring R is zero, it follows that R has an infinite number of elements. However, the converse is not true. R may have an infinite number of elements and not have characteristic zero. This is illustrated in the next example.

Example 3   Consider the ring $\mathscr{P}(\mathbf{Z})$ of all subsets of the integers $\mathbf{Z}$, with operations

$$X + Y = (X \cup Y) - (X \cap Y)$$
$$X \cdot Y = X \cap Y$$

for all $X,\ Y$ in $\mathscr{P}(\mathbf{Z})$. The ring $\mathscr{P}(\mathbf{Z})$ has an infinite number of elements, yet

$$X + X = (X \cup X) - (X \cap X)$$
$$= X - X$$
$$= \varnothing,$$

where $\varnothing$ is the zero element for $\mathscr{P}(\mathbf{Z})$. Thus, $\mathscr{P}(\mathbf{Z})$ has characteristic 2. ∎

Theorem 6.18   **INTEGRAL DOMAINS, Z, AND $\mathbf{Z}_p$**

An integral domain with characteristic zero contains a subring that is isomorphic to **Z**, and an integral domain with positive characteristic p contains a subring that is isomorphic to $\mathbf{Z}_p$.

Proof   Let D be an integral domain with unity e. Define the mapping $\theta \colon \mathbf{Z} \to D$ by

$$\theta(n) = ne$$

for each $n \in \mathbf{Z}$. Since

$$\theta(m + n) = (m + n)e = me + ne = \theta(m) + \theta(n)$$

and

$$\theta(mn) = (mn)e = mne^2 = (me)(ne) = \theta(m)\theta(n),$$

θ is a homomorphism from **Z** to D. By Theorem 6.9a, $\theta(\mathbf{Z})$ is a subring of D.

$r \Rightarrow s$ Suppose D has characteristic zero. Then $ne = 0$ if and only if $n = 0$, and it follows that ker $\theta = \{0\}$. According to Theorem 6.11, this means that θ is one-to-one and therefore an isomorphism from **Z** to the subring $\theta(\mathbf{Z})$ of D.

$u \Rightarrow v$ Suppose now that D has characteristic p. Then p is the additive order of e, and $ne = 0$ if and only if $p|n$, by Theorem 3.17b. In this case, we have ker $\theta = (p)$, the set of all multiples of p in **Z**. By Theorem 6.14, the subring $\theta(\mathbf{Z})$ of D is isomorphic to $\mathbf{Z}/(p) = \mathbf{Z}_p$.

The terms *embedded* and *extension* were introduced in connection with quotient fields in Section 5.3. Stated in these terms, Theorem 6.18 says that any integral domain with characteristic zero has **Z** embedded in it, and any integral domain with characteristic p has $\mathbf{Z}_p$ embedded in it.

In Exercise 17 of Section 5.3, a construction was given by which an arbitrary ring can be embedded in a ring with unity. The next theorem is an improvement on that statement.

Theorem 6.19   **EMBEDDING A RING IN A RING WITH UNITY**

Any ring R can be embedded in a ring S with unity that has the same characteristic as R.

$u \Rightarrow (v \wedge w)$ **Proof**   If R has characteristic zero, Exercise 17 of Section 5.3 gives a construction whereby R can be embedded in a ring S with unity. To see that the ring S has characteristic zero, we observe that

$$n(1, 0) = (n, 0) = (0, 0)$$

if and only if $n = 0$.

Suppose now that R has characteristic n. We follow the same type of construction as before, with **Z** replaced by $\mathbf{Z}_n$. Let S be the set of all ordered pairs $([m], x)$ where $[m] \in \mathbf{Z}_n$ and $x \in R$. Equality in S is defined by

$$([m], x) = ([k], y) \quad \text{if and only if} \quad [m] = [k] \quad \text{and} \quad x = y.$$

Addition and multiplication are defined by

$$([m], x) + ([k], y) = ([m + k], x + y)$$

and

$$([m], x) \cdot ([k], y) = ([mk], my + kx + xy).$$

It is straightforward to show that S forms an abelian group with respect to addition, the zero element being $([0], 0)$. This is left as an exercise (see Exercise 16 at the end of this section).

The rule for multiplication yields an element of S, but we need to show that this element is unique. To do this, let $([m_1], x_1) = ([m_2], x_2)$ and $([k_1], y_1) = ([k_2], y_2)$. Then $[m_1] = [m_2]$, $x_1 = x_2$, $[k_1] = [k_2]$, and $y_1 = y_2$ from the definition of equality. Using the definition of multiplication and these equalities, we get

$$([m_1], x_1) \cdot ([k_1], y_1) = ([m_1 k_1], m_1 y_1 + k_1 x_1 + x_1 y_1)$$

and

$$([m_2], x_2) \cdot ([k_2], y_2) = ([m_2 k_2], m_2 y_2 + k_2 x_2 + x_2 y_2)$$
$$= ([m_1 k_1], m_2 y_1 + k_2 x_1 + x_1 y_1).$$

Comparing the results of these two computations, we see that we need

$$m_2 y_1 + k_2 x_1 = m_1 y_1 + k_1 x_1$$

to conclude that the results are equal. Now

$$[m_1] = [m_2] \Rightarrow m_2 - m_1 = pn \quad \text{for some } p \in \mathbf{Z}$$
$$\Rightarrow m_2 = m_1 + pn.$$

Therefore,

$$m_2 y_1 = (m_1 + pn)y_1 = m_1 y_1 + npy_1$$
$$= m_1 y_1,$$

since py_1 is in R and R has characteristic n. Similarly, $k_2 x_1 = k_1 x_1$, and we conclude that the product is well-defined.

Verifying that multiplication is associative, we have

$$([m], x)\{([k], y)([r], z)\} = ([m], x)([kr], kz + ry + yz)$$
$$= ([mkr], mkz + mry + myz + krx + kxz$$
$$\quad + rxy + xyz)$$
$$= ([mk], my + kx + xy) \cdot ([r], z)$$
$$= \{([m], x)([k], y)\}([r], z).$$

The left distributive law follows from

$$([m], x)\{([k], y) + ([r], z)\} = ([m], x)([k + r], y + z)$$
$$= ([mk + mr], my + mz + kx + rx + xy + xz)$$
$$= ([mk], my + kx + xy) + ([mr], mz + rx + xz)$$
$$= ([m], x)([k], y) + ([m], x)([r], z).$$

The verification of the right distributive law is similar to this and is left as an exercise.

The argument up to this point shows that S is a ring. Since each of $\mathbf{Z}_n$ and R has characteristic n,

$$n([m], x) = (n[m], nx) = ([0], 0)$$

for all $([m], x)$ in S, and n is the least positive integer for which this is true. Thus, S has characteristic n.

Consider now the mapping $\theta\colon R \to S$ defined by $\theta(x) = ([0], x)$ for all $x \in R$. Since

$$\theta(x) = \theta(y) \iff ([0], x) = ([0], y) \iff x = y,$$

θ is a one-to-one correspondence from R to $\theta(R)$. Now

$$\theta(x + y) = ([0], x + y) = ([0], x) + ([0], y) = \theta(x) + \theta(y)$$

and

$$\theta(xy) = ([0], xy) = ([0], x)([0], y) = \theta(x)\theta(y),$$

so θ is an isomorphism from R to $\theta(R)$, and $\theta(R)$ is a subring of S by Theorem 6.9a. This shows that R is embedded in S.

■ EXERCISES 6.3

Sec. 5.1, #38 ≫

1. Find the characteristic of the following rings. ($R \oplus S$ is defined in Exercise 38 of Section 5.1.)
   a. $\mathbf{Z}_2 \oplus \mathbf{Z}_2$ **b.** $\mathbf{Z}_3 \oplus \mathbf{Z}_3$ **c.** $\mathbf{Z}_2 \oplus \mathbf{Z}_3$
   d. $\mathbf{Z}_2 \oplus \mathbf{Z}_4$ **e.** $\mathbf{Z}_4 \oplus \mathbf{Z}_6$

2. Let D be an integral domain with positive characteristic. Prove that all nonzero elements of D have the same additive order.

3. Show by example that the statement in Exercise 2 is no longer true if "an integral domain" is replaced by "a ring".

4. Suppose that R and S are rings with positive characteristics m and n, respectively. If k is the least common multiple of m and n, prove that $R \oplus S$ has characteristic k.

5. Prove that if both R and S in Exercise 4 are integral domains, then $R \oplus S$ has characteristic mn if $m \neq n$.

6. Prove that the characteristic of a field is either 0 or a prime.

7. Let D be an integral domain with four elements, $D = \{0, e, a, b\}$ where e is the unity.
   a. Prove that D has characteristic 2.
   b. Construct an addition table for D.

8. Prove that $\mathbf{Z}_n$ has a nonzero element whose additive order is less than n if and only if n is not a prime integer.

9. Let R be a ring with more than one element that has no zero divisors. Prove that the characteristic of R is either zero or a prime integer.

10. A **Boolean ring** is a ring in which all elements x satisfy $x^2 = x$. Prove that every Boolean ring has characteristic 2.

11. Suppose R is a ring with positive characteristic n. Prove that if I is any ideal of R, then n is a multiple of the characteristic of I.

12. If F is a field with positive characteristic p, prove that the set

$$\{0e = 0, e, 2e, 3e, \dots, (p-1)e\}$$

of multiples of the unity e forms a subfield of F.

13. If p is a positive prime integer, prove that any field with p elements is isomorphic to $\mathbf{Z}_p$.

14. Let I be the set of all elements of a ring R that have finite additive order. Prove that I is an ideal of R.

15. Prove that if a ring R has a finite number of elements, then the characteristic of R is a positive integer.

16. As in the proof of Theorem 6.19, let $S = \{([m], x) | [m] \in \mathbf{Z}_n \text{ and } x \in R\}$. Prove that S forms an abelian group with respect to addition.

17. With S as in Exercise 16, prove that the right distributive law holds in S.

18. With S as in Exercise 16, prove that the set $R' = \{([0], x) | x \in R\}$ is an ideal of S.

19. Prove that every ordered integral domain has characteristic zero.

6.4 MAXIMAL IDEALS (OPTIONAL)

We conclude this chapter with a brief study of certain ideals that yield very special quotient rings. We are interested primarily in commutative rings R with unity, and we consider the question of when a quotient ring R/I is a field. (The question of when R/I is an integral domain is treated very briefly in the exercises for this section.)

Definition 6.20   **MAXIMAL IDEAL**

> Let M be an ideal of the commutative ring R. Then M is a **maximal ideal** of R if M is not a proper subset[†] of any ideal except R itself.

Thus, an ideal M is a maximal ideal of R if and only if $M \subset I \subseteq R$ and I an ideal imply $I = R$.

Example 1   Consider the commutative ring $R = \mathbf{Z}$. According to Theorem 6.3, every ideal of $\mathbf{Z}$ is a principal ideal (n). We shall show that if $n \neq 1$, then (n) is a maximal ideal of $\mathbf{Z}$ if and only if n is a prime.

Suppose first that $n = p$ where p is a prime integer, and let I be an ideal of $\mathbf{Z}$ such that $(p) \subset I \subseteq \mathbf{Z}$. Then there exists an integer k in I such that $k \notin (p)$. That is, k is not a multiple of p. Since p is a prime, this implies that k and p are relatively prime and there exist integers u and v such that

$$1 = uk + vp.$$

Now $uk \in I$, since $k \in I$. We also have $vp \in I$, since $p \in I$. Therefore, $uk + vp = 1$ is in I, since I is an ideal. But $1 \in I$ implies immediately that $I = \mathbf{Z}$, and this proves that (p) is a maximal ideal if p is a prime.

Suppose now that n is not a prime integer. Since $n \neq 1$, there are integers a and b such that

$$n = ab \quad \text{where} \quad 1 < a < n \quad \text{and} \quad 1 < b < n.$$

Consider the ideal $I = (a)$. We have $(n) \subset I$, since $a < n$. Also, we have $I \subset \mathbf{Z}$, since $1 < a$. Thus, $(n) \subset I \subset \mathbf{Z}$, and (n) is not a maximal ideal if n is not a prime.   ■

[†]The term *proper subset* is defined in Definition 1.3.

Example 2   Example 1 shows that the ideal (4) is not maximal in **Z**. However, (4) is a maximal ideal of the ring **E** of all even integers. To see that this is true, let I be an ideal of **E** such that $(4) \subset I \subseteq \mathbf{E}$. Let x be any element of I that is not in (4). Then x has the form

$$x = 4k + 2 = 2(2k + 1)$$

where $k \in \mathbf{Z}$. Since I is an ideal,

$$x \in I \quad \text{and} \quad 4k \in I \Rightarrow x - 4k = 2 \in I.$$

But $2 \in I$ implies $I = \mathbf{E}$. Thus, (4) is a maximal ideal of E.   ∎

The importance of maximal ideals is evident from the result of the following theorem.

Theorem 6.21   **QUOTIENT RINGS THAT ARE FIELDS**

Let R be a commutative ring with unity, and let M be an ideal of R. Then R/M is a field if and only if M is a maximal ideal of R.

Proof   Let R be a commutative ring with unity e, and let M be an ideal of R. It follows immediately from Theorem 6.4 that R/M is a commutative ring with unity $e + M$. Thus, R/M is a field if and only if every nonzero element of R/M has a multiplicative inverse in R/M.

$p \Leftarrow q$ Assume first that M is a maximal ideal, and let $a + M$ be a nonzero element of R/M. That is, $a + M \neq M$ and $a \notin M$. Let

$$I = \{ar + m \mid r \in R, m \in M\}.$$

It is clear that each element $a \cdot 0 + m = m$ of M is in I, and that $a = ae + 0$ is in I but not in M. Thus, $M \subset I$. We shall show that I is an ideal of R.

Let $x = ar_1 + m_1$ and $y = ar_2 + m_2$ be arbitrary elements of I with $r_i \in R$ and $m_i \in M$. Then

$$x + y = a(r_1 + r_2) + (m_1 + m_2)$$

where $r_1 + r_2 \in R$ and $m_1 + m_2 \in M$, since M is an ideal. Thus, $x + y \in I$. Also,

$$-x = a(-r_1) + (-m_1)$$

is in I, since $-r_1 \in R$ and $-m_1 \in M$. For any element r of R,

$$rx = xr = a(r_1 r) + (m_1 r)$$

is in I, since $r_1 r \in R$ and $m_1 r \in M$. Thus, I is an ideal of R.

Since M is a maximal ideal and $M \subset I$, it must be true that $I = R$. Therefore, there exist $r \in R$ and $m \in M$ such that

$$ar + m = e.$$

Hence,

$$\begin{aligned} e + M &= (ar + m) + M \\ &= ar + M \qquad \text{since } m \in M \end{aligned}$$

$$= (a + M)(r + M),$$

and this means that $r + M$ is the multiplicative inverse of $a + M$ in R/M. We have thus shown that R/M is a field if M is a maximal ideal.

$p \Rightarrow q$ Assume now that R/M is a field, and let I be an ideal of R such that $M \subset I \subseteq R$. Since $M \subset I$, there exists an element $a \in I$ such that $a \notin M$.

We shall show that $I = R$. To this end, let b be an arbitrary element of R. Since R/M is a field and $a + M$ is not zero in R/M, there exists[†] an element $x + M$ in R/M such that

$$(a + M)(x + M) = b + M$$

or

$$ax + M = b + M.$$

Therefore, $ax - b = m$ for some $m \in M$, and

$$b = ax - m.$$

Now $ax \in I$, since $a \in I$, $x \in R$, and I is an ideal of R. Also, $m \in I$ since $M \subset I$. Hence, $b = ax - m \in I$. Since b was an arbitrary element of R, we have proved that $R \subseteq I$, and therefore $I = R$. It follows that M is a maximal ideal of R.

Example 3   We showed in Example 1 of this section that (n) is a maximal ideal of $\mathbf{Z}$ if and only if n is a prime. It follows from Theorem 6.21 that $\mathbf{Z}/(n)$ is a field if and only if n is a prime. However, this fact is not new to us. In connection with Example 5 of Section 6.1, we saw that $\mathbf{Z}_n$ was the same as $\mathbf{Z}/(n)$, and we know from Corollary 5.20 that $\mathbf{Z}_n$ is a field if and only if n is a prime.   ■

Example 4   We saw in Example 2 of this section that (4) is a maximal ideal of the ring $\mathbf{E}$ of all even integers. The distinct elements of the quotient ring $\mathbf{E}/(4)$ are given by

$$(4) = \{\dots, -8, -4, 0, 4, 8, \dots\}$$
$$2 + (4) = \{\dots, -6, -2, 2, 6, 10, \dots\}.$$

Now $\mathbf{E}/(4)$ is not a field, since $2 + (4)$ is not zero in $\mathbf{E}/(4)$, but

$$[2 + (4)][2 + (4)] = 4 + (4) = (4),$$

and (4) is the zero in $\mathbf{E}/(4)$. At first glance, this seems to contradict Theorem 6.21. However, $\mathbf{E}$ does not have the unity that is required in the hypothesis of Theorem 6.21.   ■

EXERCISES 6.4

1. According to part **a** of Example 3 in Section 5.1, the set

$$R = \{m + n\sqrt{2} \,|\, m \in \mathbf{Z}, n \in \mathbf{Z}\}$$

[†]See Exercise 18 of Section 5.2.

is a ring. Assume that the set

$$I = \{a + b\sqrt{2} \mid a \in \mathbf{E}, b \in \mathbf{E}\}$$

is an ideal of R, and show that I is not a maximal ideal of R.

Sec. 6.1, #20 ≫   **2.** (See Exercise 20 of Section 6.1.) Let R be as in Exercise 1 and show that the principal ideal

$$I = (\sqrt{2}) = \{2n + m\sqrt{2} \mid n \in \mathbf{Z}, m \in \mathbf{Z}\}$$

is a maximal ideal of R.

Sec. 6.1, #20 ≫   **3.** (See Exercise 20 of Section 6.1.) Show that the ideal $I = (6)$ is a maximal ideal of $\mathbf{E}$.

Sec. 6.1, #20 ≫   **4.** (See Exercise 20 of Section 6.1.) Show that the ideal $I = (10)$ is a maximal ideal of $\mathbf{E}$.

5. Let R and I be as in Exercise 1 and write out the distinct elements of R/I.

6. Let R and I be as in Exercise 2 and write out the distinct elements of R/I.

7. With I as in Exercise 3, write out the distinct elements of $\mathbf{E}/I$.

8. With I as in Exercise 4, write out the distinct elements of $\mathbf{E}/I$.

9. Find all maximal ideals of $\mathbf{Z}_{12}$.

10. Find all maximal ideals of $\mathbf{Z}_{18}$.

11. An ideal I of a commutative ring R is a **prime ideal** if $I \neq R$ and if $ab \in I$ implies either $a \in I$ or $b \in I$. Let R be a commutative ring with unity, and suppose that I is an ideal of R such that $I \neq R$ and $I \neq \{0\}$. Prove that R/I is an integral domain if and only if I is a prime ideal.

12. Prove that, for $n \neq 1$ and $(n) \neq \{0\}$, an ideal (n) of $\mathbf{Z}$ is a prime ideal if and only if n is a prime integer.

13. Show that the ideal I in Exercise 1 is not a prime ideal of R.

Sec. 6.1, #20 ≫   **14.** (See Exercise 20 of Section 6.1.) Show that the ideal (4) of $\mathbf{E}$ is not a prime ideal of $\mathbf{E}$.

Sec. 6.1, #20 ≫   **15.** (See Exercise 20 of Section 6.1.) Show that the ideal (6) in Exercise 3 is a prime ideal of $\mathbf{E}$.

16. Show that the ideal I in Exercise 2 is a prime ideal of R.

Sec. 6.1, #20 ≫   **17.** (See Exercise 20 of Section 6.1.) Show that (10) is a prime ideal of $\mathbf{E}$.

Sec. 6.1, #20 ≫   **18.** (See Exercise 20 of Section 6.1.) Show that (14) is a prime ideal of $\mathbf{E}$.

19. Find all prime ideals of $\mathbf{Z}_{12}$.

20. Find all prime ideals of $\mathbf{Z}_{18}$.

Sec. 5.1, #38 ≫   **21.** (See Exercise 38 of Section 5.1.) Show that $\mathbf{Z} \oplus \mathbf{E}$ is a maximal ideal of $\mathbf{Z} \oplus \mathbf{Z}$.

Sec. 5.1, #38 ≫   **22.** (See Exercise 38 of Section 5.1.) Show that $\mathbf{Z} \oplus \{0\}$ is a prime ideal of $\mathbf{Z} \oplus \mathbf{Z}$ but is not a maximal ideal of $\mathbf{Z} \oplus \mathbf{Z}$.

23. If R is a commutative ring with unity, prove that any maximal ideal of R is also a prime ideal.

KEY WORDS AND PHRASES

characteristic of a ring, 255
epimorphism, 247
Fundamental Theorem of Ring
   Homomorphisms, 252
homomorphic image, 247

ideal, 239, 240
isomorphism, 247
kernel, 249
maximal ideal, 260
prime ideal, 263

principal ideal, 241, 246
quotient ring, 244
ring homomorphism, 247
trivial ideals, 239

A Pioneer in Mathematics ||||||||||||||||||||||||||||||||||||||||||||||||||||||||||||||||||||||||||||

Amalie Emmy Noether (1882–1935)

Amalie Emmy Noether, born on March 23, 1882, in Erlangen, Germany, is considered the foremost female mathematician up to her time. She overcame numerous obstacles to receive her education and to be permitted to work as a mathematician in a university environment. Yet her contributions revolutionized abstract algebra and subsequently influenced mathematics as a whole.

Even though university policy stated that admission of women would "overthrow all academic order,"[†] in 1900, Noether and one other woman were given special permission to audit classes at the University of Erlangen along with 1000 regularly enrolled male students. It wasn't until 1904 that Noether was allowed to enroll formally and enjoy the same privileges as her male counterparts. Three years later, she completed her doctoral dissertation.

Between 1908 and 1915, Noether was allowed only to substitute teach at Erlangen whenever her father was ill. In 1915, she was brought to the University of Göttingen by David Hilbert (1862–1943) to help in his study of the mathematics involved in the general theory of relativity. Hilbert tried to secure a teaching position for Noether but met strong opposition from the faculty in requesting to hire a woman. According to David M. Burton,[†] Hilbert, in a faculty senate meeting held to discuss her appointment, exploded in frustration, "I do not see that the sex of the candidate is an argument against her admission as a Privatdozent. After all we are a university, not a bathhouse." Her appointment was voted down, but Hilbert allowed her to lecture in courses that were listed under his own name.

At Göttingen, Noether eventually became a lecturer in algebra and earned a modest salary. Göttingen was an international center of mathematics during this time. From her students, the "Noether boys," came some of the brightest mathematical talents of the era.

Noether, a Jew, was forced to leave Germany in 1933 when Hitler came into power. She fled to the United States, where she accepted a position as visiting professor at Bryn Mawr College in Pennsylvania. She also worked at the Institute for Advanced Study in Princeton, New Jersey. Eighteen months later, at the height of her creative career, she died unexpectedly after an operation.

[†]David M. Burton, *Abstract Algebra* (Cincinnati: William C. Brown, 1988), p. 242.

7

REAL AND COMPLEX NUMBERS

INTRODUCTION

The material in this chapter is included for the benefit of those who would not see it in some other course. However, it may be skipped by some instructors. It is possible to cover Chapter 8 before this one, and some instructors use this option.

▌ 7.1   THE FIELD OF REAL NUMBERS

At this point it is possible to fit some of the familiar number systems into the structures developed in the preceding chapters.

In Theorem 5.35, the ring $\mathbf{Z}$ of all integers was characterized as an ordered integral domain in which the set of positive elements is well-ordered. By "characterized," we mean that any ordered integral domain in which the set of positive elements is well-ordered must be isomorphic to the ring $\mathbf{Z}$ of all integers.

At the end of Section 5.3 we noted that the construction of the rational numbers from $\mathbf{Z}$ is a special case of the procedure described in that section. That is, the set $\mathbf{Q}$ of all rational numbers is the quotient field of $\mathbf{Z}$ and therefore is the smallest field that contains $\mathbf{Z}$. From a more abstract point of view, the field of rational numbers can be characterized as the smallest ordered field. That is, any ordered field must contain a subfield that is isomorphic to $\mathbf{Q}$. (See Exercises 22–24 at the end of this section.)

The main goal of this section is to present a similar characterization for the field of real numbers. The following definition is essential.

Definition 7.1   **UPPER BOUND, LEAST UPPER BOUND**

> Let S be a nonempty subset of an ordered field F. An element u of F is an **upper bound** of S if $u \geq x$ for all $x \in S$. An element u of F is a **least upper bound** of S if these conditions are satisfied:
>
> **1.** u is an upper bound of S.
> **2.** If $b \in F$ is an upper bound of S, then $b \geq u$.
>
> The phrase *least upper bound* is abbreviated l.u.b.

Example 1   Let $F = \mathbf{Q}$ be the field of rational numbers, and let S be the set of all negative rational numbers.

If a is any negative rational number, then there exists $b \in \mathbf{Q}$ such that $0 > b > a$, by Exercise 13 of Section 5.4. Thus, no negative number is an upper bound of S. However, any positive rational number u is an upper bound of S, since

$$u > 0 > x \qquad \text{for all } x \in S.$$

The rational number 0 is also an upper bound of S, since $0 > x$ for all $x \in S$. In fact, 0 is a least upper bound of S in $\mathbf{Q}$.   ■

If $u \in F$ and $v \in F$ are both least upper bounds of the nonempty subset S of an ordered field F, then the second condition in Definition 7.1 requires both $v \geq u$ and $u \geq v$. Therefore, $u = v$ and the least upper bound of S in F is unique whenever it exists.

Later we shall exhibit a nonempty subset of $\mathbf{Q}$ that has an upper bound in $\mathbf{Q}$ but does not have a least upper bound in $\mathbf{Q}$. The following theorem will be needed.

Theorem 7.2 $\sqrt{2}$ **IS NOT RATIONAL**

There is no rational number x such that $x^2 = 2$.

Contradiction   **Proof**   Assume that the theorem is false. That is, assume a rational number x exists such that $x^2 = 2$. We may assume, without loss of generality, that $x = p/q$ is expressed in *lowest terms* as a quotient of integers p and q. That is,

$$\left(\frac{p}{q}\right)^2 = 2$$

with 1 as the greatest common divisor of p and q. This implies that

$$p^2 = 2q^2.$$

Hence, 2 divides p^2, and since 2 is a prime, this implies that 2 divides p, by Theorem 2.16. Let $p = 2r$ where $r \in \mathbf{Z}$. Then we have

$$(2r)^2 = 2q^2$$
$$4r^2 = 2q^2$$

and therefore

$$2r^2 = q^2.$$

This implies, however, that 2 divides q, by another application of Theorem 2.16. Thus, 2 is a common divisor of p and q, and we have a contradiction to the fact that 1 is the greatest common divisor of p and q. This contradiction establishes the theorem.

Example 2   Let

$$S = \{x \in \mathbf{Q} \mid x > 0 \text{ and } x^2 \leq 2\}.$$

We shall show that S is a nonempty subset of $\mathbf{Q}$ that has an upper bound in $\mathbf{Q}$ but does not have a l.u.b. in $\mathbf{Q}$.

The set S is nonempty since 1 is in S. The rational number 3 is an upper bound of S in $\mathbf{Q}$ since $x \geq 3$ requires $x^2 \geq 9$ by Exercise 2c of Section 5.4.

It is not so easy to show that S does not have a l.u.b. in $\mathbf{Q}$. As a start, we shall prove the following two statements for *positive* $u \in \mathbf{Q}$:

1. If u is not an upper bound of S, then $u^2 < 2$.
2. If $u^2 < 2$, then u is not an upper bound of S.

Consider statement 1. If $u \in \mathbf{Q}$ is not an upper bound of S, then there exists $x \in S$ such that $0 < u < x$. By Exercise 2c of Section 5.4, this implies that $u^2 < x^2$. Since $x^2 \leq 2$ for all $x \in S$, we have $u^2 < 2$.

To prove statement 2, suppose that $u \in \mathbf{Q}$ is positive and $u^2 < 2$. Then $\dfrac{2 - u^2}{2u + 1}$ is a positive rational number. By Exercise 13 of Section 5.4, there exists a rational number d such that

$$0 < d < \min\left\{1, \frac{2 - u^2}{2u + 1}\right\}$$

where $\min\left\{1, \dfrac{2 - u^2}{2u + 1}\right\}$ denotes the smaller of the two numbers in braces. If we now put $v = u + d$, then v is a positive rational number, $v > u$, and

$$v^2 = u^2 + 2ud + d^2$$
$$< u^2 + 2ud + d \qquad \text{since } 0 < d < 1 \text{ implies } 0 < d^2 < d$$
$$= u^2 + (2u + 1)d$$
$$< u^2 + (2u + 1) \cdot \frac{2 - u^2}{2u + 1} \qquad \text{since } d < \frac{2 - u^2}{2u + 1}$$
$$= 2.$$

Thus, v is an element of S such that $v > u$, and u is not an upper bound of S.

Having established statements 1 and 2, we may combine them with Theorem 7.2 and obtain the following statement:

3. A positive $u \in \mathbf{Q}$ is an upper bound of S if and only if $u^2 > 2$.

With this fact at hand, we can now show that S does not have a l.u.b. in $\mathbf{Q}$.

Suppose $u \in \mathbf{Q}$ is an upper bound of S. Then u is positive, since all elements of S are positive, and $u^2 > 2$ by statement 3. Let

$$w = u - \frac{u^2 - 2}{2u}$$
$$= \frac{u^2 + 2}{2u} = \frac{u}{2} + \frac{1}{u}.$$

Then w is a positive rational number. We also have $w < u$, since $\frac{u^2 - 2}{2u}$ is positive.

Now

$$w^2 = \left(u - \frac{u^2 - 2}{2u} \right)^2$$
$$= u^2 - (u^2 - 2) + \left(\frac{u^2 - 2}{2u} \right)^2$$
$$= 2 + \left(\frac{u^2 - 2}{2u} \right)^2$$
$$> 2,$$

so w is an upper bound of S by statement 3. Since $w < u$, we have that u is not a least upper bound of S. Since u was an arbitrary upper bound of S in $\mathbf{Q}$, this proves that S does not have a l.u.b. in $\mathbf{Q}$. ∎

Example 2 establishes a very significant deficiency in the field $\mathbf{Q}$ of rational numbers: Some nonempty sets of rational numbers have an upper bound in $\mathbf{Q}$ but fail to have a least upper bound in $\mathbf{Q}$. The next definition gives a designation for those ordered fields that do not have this deficiency.

Definition 7.3   **COMPLETE ORDERED FIELD**

> Let F be an ordered field. Then F is **complete** if every nonempty subset of F that has an upper bound in F has a least upper bound in F.

The basic difference between the field of rational numbers and the field of real numbers is that the real number field is complete. It is possible to construct the field of real numbers from the field of rational numbers, but this construction is too lengthy and difficult to be included here. It is more properly a part of that area of mathematics known as *analysis*. The method of construction most commonly used is one that is credited to Richard Dedekind (1831–1916) and utilizes what are called *Dedekind cuts*. In our treatment, we shall assume the validity of the following theorem.

Theorem 7.4   **THE FIELD OF REAL NUMBERS**

There exists a field **R**, called the **field of real numbers**, that is a complete ordered field. Any complete ordered field F has the following properties.

a. F is isomorphic to **R**.
b. F contains a subfield that is isomorphic to the field **Q** of rational numbers, and the ordering in F is an extension of the ordering in this subfield.

The set of all real numbers may be represented geometrically by setting up a one-to-one correspondence between real numbers and the points on a straight line. To begin, we select a point on a horizontal line, designate it as the *origin*, and let this point correspond to the number 0. A second point is now chosen to the right of the origin, and we let this point correspond to the number 1. The distance between the two points corresponding to 0 and 1 is now taken as one unit of measure. Points on the line located successively one unit farther to the right are made to correspond to the positive integers 2, 3, 4, . . . in succession. With the same unit of measure and beginning at the origin, points on the line located successively one unit farther to the left are made to correspond to the negative integers $-1, -2, -3, \ldots$ (see Figure 7-1). This sets up a one-to-one correspondence between the set **Z** of all integers and some of the points on the line.

Figure 7-1

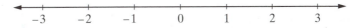

Points on the line that correspond to nonintegral rational numbers are now located by using distances proportional to their expressions as quotients a/b of integers a and b, and by using directions to the right for positive numbers and to the left for negative numbers. For example, the point corresponding to $\frac{3}{2}$ is located midway between the points that correspond to 1 and 2, whereas the point corresponding to $-\frac{3}{2}$ is located midway between those that correspond to -1 and -2. In this manner, a one-to-one correspondence is established between the set **Q** of rational numbers and a subset of the points on the line.

It is not very difficult to demonstrate that there are points on the line that do not correspond to any rational number. This can be done by considering a right triangle with each leg one unit in length (see Figure 7-2). By the Pythagorean Theorem, the length h of the hypotenuse of the triangle in Figure 7-2 satisfies the equation $h^2 = 2$. There is a point on the line located at a distance h units to the right of the origin, but this point cannot correspond to a rational number, by Theorem 7.2.

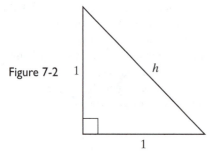

Figure 7-2

The foregoing demonstration shows that there are gaps in the rational numbers, even though any two distinct rational numbers have another rational number located between them (see Exercise 13 of Section 5.4). We *assume* now that the one-to-one correspondence that we have set up between the rational numbers and points on the line can be extended to the set of all real numbers and the set of all points on the line. The points that do not correspond to rational numbers are assumed to correspond to real numbers that are not rational; that is, to **irrational** numbers. For example, the discussion in the preceding paragraph located the point that corresponds to the irrational number $h = \sqrt{2}$.

One more aspect of the real numbers is worthy of mention: the **decimal representation** of real numbers. Here we assume that each real number can be represented by a decimal expression that either terminates, such as

$$\frac{9}{8} = 1.125,$$

or continues without end, as does the repeating decimal[†]

$$\frac{14}{11} = 1.272727\ldots = 1.\overline{27}$$

and the nonrepeating decimal

$$\sqrt{2} = 1.41421356\ldots$$

The decimal expression for a rational number a/b may be found by long division. For example, for the rational number $\frac{14}{11}$, long division yields the following.

$$
\begin{array}{r}
1.27 \\
11\overline{)14.00} \\
\underline{11} \\
30 \\
\underline{22} \\
80 \\
\underline{77} \\
3
\end{array}
$$

[†]The bar above 27 indicates that the digits 27 repeat endlessly.

The repetition of the remainder 3 at this point makes it clear that we have the repeating decimal expression

$$\frac{14}{11} = 1.272727\ldots = 1.\overline{27}.$$

A terminating decimal expression may be regarded as a repeating pattern where zeros repeat endlessly. For example,

$$\frac{9}{8} = 1.125000\ldots = 1.125\overline{0}.$$

With this point of view, the decimal expression for any rational number a/b will always have a repeating pattern. This can be seen from the long division algorithm: Each remainder satisfies $0 \leq r < b$, so there are only b distinct possibilities for the remainders, and the expression starts repeating whenever a remainder occurs for the second time.

Rational numbers that have a terminating decimal expression can be represented in another way by changing the range on the remainders in the long division from $0 \leq r < b$ to $0 < r \leq b$. If we perform the long division for $\frac{9}{8}$ in this way, it appears as follows.

$$
\begin{array}{r}
1.1249 \\
8\ \overline{)9.0000} \\
\underline{8} \\
10 \\
\underline{8} \\
20 \\
\underline{16} \\
40 \\
\underline{32} \\
80 \\
\underline{72} \\
8
\end{array}
$$

At this point, the remainder 8 has occurred twice, and the repeating pattern is seen to be

$$\frac{9}{8} = 1.124999\ldots = 1.124\overline{9}.$$

It can be proved that every repeating decimal expression represents a rational number, but we shall not explore that result here. The next example provides some insight into why this assertion is true.

Example 3   We shall express $2.1\overline{34}$ as a quotient of integers. If

$$x = 2.1343434\ldots,$$

then

$$10x = 21.343434\ldots$$

and

$$1000x = 2134.3434\ldots$$

Thus,

$$1000x - 10x = 2134.\overline{34} - 21.\overline{34}$$
$$990x = 2113$$
$$x = \frac{2113}{990}.$$ ■

This discussion of decimal representations is not intended to be a rigorous presentation. Its purpose is to make the following remarks appear plausible:

1. Each real number can be represented by a decimal expression.
2. Decimal expressions that repeat or terminate represent rational numbers.
3. Decimal expressions that do not repeat and do not terminate represent irrational numbers.

EXERCISES 7.1

Find the decimal representation for each of the numbers in Exercises 1–6.

1. $\frac{5}{9}$ **2.** $\frac{7}{33}$ **3.** $\frac{80}{81}$

4. $\frac{16}{7}$ **5.** $\frac{22}{7}$ **6.** $\frac{19}{11}$

Express each of the numbers in Exercises 7–12 as a quotient of integers, reduced to lowest terms.

7. $3.\overline{4}$ **8.** $1.\overline{6}$ **9.** $0.\overline{12}$

10. $0.\overline{63}$ **11.** $2.5\overline{1}$ **12.** $3.21\overline{321}$

13. Prove that $\sqrt{3}$ is irrational. (That is, prove there is no rational number x such that $x^2 = 3$.)

14. Prove that $\sqrt[3]{2}$ is irrational.

15. Prove that if p is a prime integer, then $\sqrt{p}$ is irrational.

16. Prove that if a is rational and b is irrational, then $a + b$ is irrational.

17. Prove that if a is a nonzero rational number and b is irrational, then ab is irrational.

18. Prove that if a is an irrational number, then a^{-1} is an irrational number.

19. Prove that if a is a nonzero rational number and ab is irrational, then b is irrational.

20. Give counterexamples for the following statements.
  a. If a and b are irrational, then $a + b$ is irrational.
  b. If a and b are irrational, then ab is irrational.

21. Let S be a nonempty subset of an ordered field F.
  a. Write definitions for **lower bound** of S and **greatest lower bound** of S.
  b. Prove that if F is a complete ordered field and the nonempty subset S has a lower bound in F, then S has a greatest lower bound in F.

22. Prove that if F is an ordered field with F^+ as its set of positive elements, then $F^+ \supseteq \{ne|n \in \mathbf{Z}^+\}$, where e denotes the multiplicative identity in F. (*Hint:* See Theorem 5.34 and its proof.)

23. If F is an ordered field, prove that F contains a subring that is isomorphic to $\mathbf{Z}$. (*Hint:* See Theorem 5.35 and its proof.)

24. Prove that any ordered field must contain a subfield that is isomorphic to the field $\mathbf{Q}$ of rational numbers.

25. If a and b are positive real numbers, prove that there exists a positive integer n such that $na > b$. This property is called the **Archimedean Property** of the real numbers. (*Hint*: If $ma \leq b$ for all $m \in \mathbf{Z}^+$, then b is an upper bound for the set $S = \{ma \mid m \in \mathbf{Z}^+\}$. Use the completeness property of $\mathbf{R}$ to arrive at a contradiction.)

26. Prove that if a and b are real numbers such that $a > b$, then there exists a rational number m/n, such that $a > m/n > b$. (*Hint*: Use Exercise 25 to obtain $n \in \mathbf{Z}^+$ such that $a - b > 1/n$. Then choose m to be the least integer such that $m > nb$. With these choices of m and n, show that $(m - 1)/n \leq b$ and then that $a > m/n > b$.)

7.2 COMPLEX NUMBERS AND QUATERNIONS

The fact that negative real numbers do not have square roots in $\mathbf{R}$ is a serious deficiency of the field of real numbers, but it is one that can be overcome by the introduction of complex numbers.

Although we do not present a characterization of the field of complex numbers until Section 8.4, it is possible to construct the complex numbers from the real numbers. Such a construction is the main purpose of this section.

In our construction, complex numbers appear first as ordered pairs (a, b) and later in the more familiar form $a + bi$. The operations given in the following definition will seem more natural if they are compared with the usual operations on complex numbers in the form $a + bi$.

Definition 7.5 COMPLEX NUMBERS

> Let $\mathbf{C}$ be the set of all ordered pairs (a, b) of real numbers a and b. Equality, addition, and multiplication are defined in $\mathbf{C}$ by
>
> $$(a, b) = (c, d) \quad \text{if and only if} \quad a = c \quad \text{and} \quad b = d$$
> $$(a, b) + (c, d) = (a + c, b + d)$$
> $$(a, b)(c, d) = (ac - bd, ad + bc).$$
>
> The elements of $\mathbf{C}$ are called **complex numbers**.

It is easy to see that the stated rules for addition and multiplication do in fact define binary operations on $\mathbf{C}$.

Theorem 7.6 THE FIELD OF COMPLEX NUMBERS

With addition and multiplication as given in Definition 7.5, $\mathbf{C}$ is a field. The set of all elements of the form $(a, 0)$ in $\mathbf{C}$ forms a subfield of $\mathbf{C}$ that is isomorphic to the field $\mathbf{R}$ of real numbers.

Proof Closure of $\mathbf{C}$ under addition follows at once from the fact that $\mathbf{R}$ is closed under addition. It is left for the exercises to prove that addition is associative and commutative, that $(0, 0)$ is the additive identity in $\mathbf{C}$, and that the additive inverse of $(a, b) \in \mathbf{C}$ is $(-a, -b) \in \mathbf{C}$.

Since $\mathbf{R}$ is closed under multiplication and addition, each of $ac - bd$ and $ad + bc$ is in $\mathbf{R}$ whenever (a, b) and (c, d) are in $\mathbf{C}$. Thus, $\mathbf{C}$ is closed under multiplication.

For the remainder of the proof, let $(a, b), (c, d)$, and (e, f) represent arbitrary elements of $\mathbf{C}$. The associative property of multiplication is verified by the following computations:

$$
\begin{aligned}
(a, b)[(c, d)(e, f)] &= (a, b)(ce - df, cf + de) \\
&= [a(ce - df) - b(cf + de), a(cf + de) + b(ce - df)] \\
&= (ace - adf - bcf - bde, acf + ade + bce - bdf) \\
&= [(ac - bd)e - (ad + bc)f, (ac - bd)f + (ad + bc)e] \\
&= (ac - bd, ad + bc)(e, f) \\
&= [(a, b)(c, d)](e, f).
\end{aligned}
$$

Before considering the distributive laws, we shall show that multiplication is commutative in $\mathbf{C}$. This follows from

$$
\begin{aligned}
(c, d)(a, b) &= (ca - db, cb + da) \\
&= (ca - db, da + cb) \\
&= (ac - bd, ad + bc) \\
&= (a, b)(c, d).
\end{aligned}
$$

We shall verify the left distributive property and leave the proof of the right distributive property as an exercise:

$$
\begin{aligned}
(a, b)[(c, d) + (e, f)] &= (a, b)(c + e, d + f) \\
&= [a(c + e) - b(d + f), a(d + f) + b(c + e)] \\
&= (ac + ae - bd - bf, ad + af + bc + be) \\
&= (ac - bd, ad + bc) + (ae - bf, af + be) \\
&= (a, b)(c, d) + (a, b)(e, f).
\end{aligned}
$$

To this point, we have established that $\mathbf{C}$ is a commutative ring.

The computation

$$
(1, 0)(a, b) = (1 \cdot a - 0 \cdot b, 1 \cdot b + 0 \cdot a) = (a, b)
$$

shows that $(1, 0)$ is a left identity for multiplication in $\mathbf{C}$. Since multiplication in $\mathbf{C}$ is commutative, it follows that $(1, 0)$ is a nonzero unity in $\mathbf{C}$.

If $(a, b) \neq (0, 0)$ in $\mathbf{C}$, then at least one of the real numbers a or b is nonzero, and it follows that $a^2 + b^2$ is a positive real number. Hence,

$$
\left(\frac{a}{a^2 + b^2}, \frac{-b}{a^2 + b^2} \right)
$$

is an element of $\mathbf{C}$. The multiplication

$$
(a, b)\left(\frac{a}{a^2 + b^2}, \frac{-b}{a^2 + b^2} \right) = \left(\frac{a^2 + b^2}{a^2 + b^2}, \frac{-ab + ba}{a^2 + b^2} \right) = (1, 0)
$$

shows that

$$
(a, b)^{-1} = \left(\frac{a}{a^2 + b^2}, \frac{-b}{a^2 + b^2} \right),
$$

since multiplication is commutative in $\mathbf{C}$. This completes the proof that $\mathbf{C}$ is a field.

Consider now that the set R' that consists of all elements of $\mathbf{C}$ that have the form $(a, 0)$:

$$
R' = \{(a, 0) \,|\, a \in \mathbf{R}\}.
$$

The proof that R' is a subfield of $\mathbf{C}$ is left as an exercise. The mapping $\theta \colon \mathbf{R} \to R'$ defined by

$$\theta(a) = (a, 0)$$

is a one-to-one correspondence, since $(a, 0) = (b, 0)$ if and only if $a = b$. For arbitrary a and b in $\mathbf{R}$,

$$
\begin{aligned}
\theta(a + b) &= (a + b, 0) \\
&= (a, 0) + (b, 0) \\
&= \theta(a) + \theta(b)
\end{aligned}
$$

and

$$
\begin{aligned}
\theta(ab) &= (ab, 0) \\
&= (a, 0)(b, 0) \\
&= \theta(a)\theta(b).
\end{aligned}
$$

Thus, θ preserves both operations and is an isomorphism from $\mathbf{R}$ to R'.

We shall use the isomorphism θ in the preceding proof to identify $a \in \mathbf{R}$ with $(a, 0)$ in R'. We write a instead of $(a, 0)$, and consider $\mathbf{R}$ to be a subset of $\mathbf{C}$. The calculation

$$
\begin{aligned}
(0, 1)(0, 1) &= (0 \cdot 0 - 1 \cdot 1, 0 \cdot 1 + 1 \cdot 0) \\
&= (-1, 0) \\
&= -1
\end{aligned}
$$

shows that the equation $x^2 = -1$ has a solution $x = (0, 1)$ in $\mathbf{C}$.

To obtain the customary notation for complex numbers, we define the number i by

$$i = (0, 1).$$

This makes i a number such that $i^2 = -1$. We now note that any $(a, b) \in \mathbf{C}$ can be written in the form

$$
\begin{aligned}
(a, b) &= (a, 0) + (0, b) \\
&= (a, 0) + b(0, 1) \\
&= a + bi,
\end{aligned}
$$

and this gives us the familiar form for complex numbers.

Using the field properties freely, we may rewrite the rules for addition and multiplication in $\mathbf{C}$ as follows:

$$
\begin{aligned}
(a + bi) + (c + di) &= a + c + bi + di \\
&= (a + c) + (b + d)i
\end{aligned}
$$

and

$$
\begin{aligned}
(a + bi)(c + di) &= (a + bi)c + (a + bi)di \\
&= ac + bci + adi + bdi^2 \\
&= (ac - bd) + (ad + bc)i,
\end{aligned}
$$

where the last step was obtained by replacing i^2 with -1.

The fact that $i^2 = -1$ was used in Section 5.4 to prove that it is impossible to impose an order relation on **C**. Hence, **C** *is not an ordered field.*

It is easy to show that all negative real numbers have square roots in **C**. For any positive real number a, the negative real number $-a$ has both $\sqrt{a}i$ and $-\sqrt{a}i$ as square roots, since

$$(\sqrt{a}i)^2 = (\sqrt{a})^2 i^2 = a(-1) = -a$$

and

$$(-\sqrt{a}i)^2 = (-\sqrt{a})^2 i^2 = a(-1) = -a.$$

We shall see later in this chapter that every nonzero complex number has two distinct square roots in **C**.

Example 1   The following results illustrate some calculations with complex numbers.

a. $(1 + 2i)(3 - 5i) = 3 + 6i - 5i - 10i^2 = 13 + i$
b. $(2 + 3i)(2 - 3i) = 4 - 9i^2 = 13$
c. $(-3 + 4i)(3 + 4i) = -9 + 16i^2 = -25$
d. $(1 - i)^2 = 1 - 2i + i^2 = -2i$
e. $i^4 = (i^2)^2 = (-1)^2 = 1$ ■

In connection with part **b** of Example 1, we note that

$$(a + bi)(a - bi) = a^2 - b^2 i^2$$
$$= a^2 + b^2$$

for any complex number $a + bi$. The number $a^2 + b^2$ is always real, and it is positive if $a + bi$ is nonzero.

Definition 7.7   CONJUGATE

> For any a, b in **R**, the **conjugate** of the complex number $a + bi$ is the number $a - bi$. The notation $\bar{z}$ indicates the conjugate of z: If $z = a + bi$ with a and b real, then $\bar{z} = a - bi$.

Using the bar notation of Definition 7.7, we can write

$$\bar{z}z = z\bar{z} = a^2 + b^2,$$

and the multiplicative inverse of a nonzero z is given by

$$z^{-1} = \left(\frac{1}{\bar{z}z}\right)\bar{z}.$$

Division of complex numbers may be accomplished by multiplying the numerator and denominator of a quotient by the conjugate of the denominator.

Example 2   We have the following illustrations of division.

a. $\dfrac{3 + 7i}{2 - 3i} = \dfrac{3 + 7i}{2 - 3i} \cdot \dfrac{2 + 3i}{2 + 3i} = \dfrac{6 + 23i - 21}{4 + 9} = -\dfrac{15}{13} + \dfrac{23}{13}i$

b. $\dfrac{1}{2+i} = \dfrac{1}{2+i} \cdot \dfrac{2-i}{2-i} = \dfrac{2-i}{5} = \dfrac{2}{5} - \dfrac{1}{5}i$ ∎

By using the techniques illustrated in Examples 1 and 2, we can write the result of any calculation involving the field operations with complex numbers in the form $a + bi$, with a and b real numbers. This form is called the **standard form** of the complex number. If $b \neq 0$, the number is called **imaginary**. If $a = 0$, the number is called **pure imaginary**.

The construction of the complex numbers by use of ordered pairs was first accomplished by Hamilton (see the biographical section at the end of this chapter). Eventually, he was able to use ordered quadruples (x, y, z, w) of real numbers in order to extend the complex numbers to a larger set that he called the **quaternions**. His quaternions satisfy all the postulates for a field except the requirement that multiplication be commutative. A system with these properties is called a **division ring**, or a **skew field**, and Hamilton's quaternions was the first known example of a division ring.

Example 3 In this example we outline the development of the quaternions as the set

$$H = \{(x, y, z, w) \mid x, y, z, w \in \mathbf{R}\},$$

with most of the details left as exercises.

Equality and addition are defined in H by

$(x, y, z, w) = (r, s, t, u)$ if and only if $x = r, y = s, z = t,$ and $w = u$;
$(x, y, z, w) + (r, s, t, u) = (x + r, y + s, z + t, w + u).$

It is easy to see that this addition is a binary operation on H, and $(0, 0, 0, 0)$ in H is the additive identity. Also, each (x, y, z, w) in H has an additive inverse $(-x, -y, -z, -w)$ in H. The proofs that addition is associative and commutative are left as exercises. Thus, H forms an abelian group with respect to addition.

When the definition of multiplication in H is presented in the same manner as multiplication of complex numbers in Definition 7.5, it has the following complicated appearance:

$$(x, y, z, w)(r, s, t, u) = (xr - ys - zt - wu, xs + yr + zu - wt,$$
$$xt - yu + zr + ws, xu + yt - zs + wr).$$

This multiplication is a binary operation on H, and it is easy to verify that $(1, 0, 0, 0)$ is a unity in H. Laborious computations will show that multiplication is associative in H and that both distributive laws hold. These verifications are left as exercises and they lead to the conclusion that H is a ring.

At this point, it can be shown that the set

$$R' = \{(a, 0, 0, 0) \mid a \in \mathbf{R}\}$$

is a field contained in H and that the mapping $\theta: \mathbf{R} \to R'$ defined by

$$\theta(a) = (a, 0, 0, 0)$$

is an isomorphism. Similar to the identification of a with $(a, 0)$ in $\mathbf{C}$, we can identify a in $\mathbf{R}$ with $(a, 0, 0, 0)$ in R' and consider $\mathbf{R}$ to be a subring of H.

Some other notational changes can be used to give the elements of H a more natural appearance. We let

$$i = (0, 1, 0, 0), \qquad j = (0, 0, 1, 0), \qquad \text{and} \qquad k = (0, 0, 0, 1).$$

Then an arbitrary element (x, y, z, w) in H can be written as

$$(x, y, z, w) = (x, 0, 0, 0) + (y, 0, 0, 0)i + (z, 0, 0, 0)j + (w, 0, 0, 0)k$$
$$= x + yi + zj + wk.$$

Routine calculations confirm the equations

$$(-1)^2 = 1$$
$$i^2 = j^2 = k^2 = -1$$
$$(-1)a = a(-1) = -a \quad \text{for all } a \in \{\pm 1, \pm i, \pm j, \pm k\}$$

$$ij = -ji = k$$
$$jk = -kj = i$$
$$ki = -ik = j.$$

In fact, this multiplication agrees with the table constructed for the quaternion group in Exercise 28 of Section 3.1. The circular order of multiplication observed previously is also valid in H (see Figure 7-3). With a positive (counterclockwise) rotation, the product of two consecutive elements is the third one on the circle, and the sign changes with a negative (clockwise) rotation.

Figure 7-3

Computations such as $ij = k$ and $ji = -k$ show that multiplication in H is not commutative and H is not a field.

With the i, j, k notation, H can be written in the form

$$H = \{x + yi + zj + wk \mid x, y, z, w \in \mathbf{R}\},$$

with addition and multiplication appearing as

$$(x + yi + zj + wk) + (r + si + tj + uk) = (x + r) + (y + s)i + (z + t)j$$
$$+ (w + u)k;$$
$$(x + yi + zj + wk)(r + si + tj + uk) = (xr - ys - zt - wu)$$
$$+ (xs + yr + zu - wt)i$$
$$+ (xt - yu + zr + ws)j$$
$$+ (xu + yt - zs + wr)k.$$

Multiplication can thus be performed by using the distributive laws and other natural ring properties, with two exceptions:

1. Multiplication is not commutative.
2. Products of $i, j,$ or k are simplified using the equations in the preceding paragraph.

The most outstanding feature of H is that each nonzero element has a multiplicative inverse. For each $q = x + yi + zj + wk$ in H, we imitate conjugates in $\mathbf{C}$ and write

$$\bar{q} = x - yi - zj - wk.$$

It is left as an exercise to verify that

$$\bar{q}q = q\bar{q} = x^2 + y^2 + z^2 + w^2.$$

If $q \neq 0$, then $\bar{q}q \neq 0$, and

$$q^{-1} = \left(\frac{1}{\bar{q}q}\right)\bar{q}.$$

Thus, H has all the field properties except commutative multiplication.   ∎

EXERCISES 7.2

Perform the computations in Exercises 1–12 and express the results in standard form $a + bi$.

1. $(2 - 3i)(-1 + 4i)$

2. $(5 - 3i)(2 - 4i)$

3. i^{15}

4. i^{87}

5. $(2 - i)^3$

6. $i(2 + i)^2$

7. $\dfrac{1}{2 - i}$

8. $\dfrac{1}{3 + i}$

9. $\dfrac{2 - i}{8 - 6i}$

10. $\dfrac{1 - i}{1 + 3i}$

11. $\dfrac{5 + 2i}{5 - 2i}$

12. $\dfrac{4 - 3i}{4 + 3i}$

13. Find two square roots of each given number.

    a. -9 **b.** -16 **c.** -25

    d. -36 **e.** -13 **f.** -8

14. With addition as given in Definition 7.5, prove the following statements.

    a. Addition is associative in $\mathbf{C}$.

    b. Addition is commutative in $\mathbf{C}$.

    c. $(0, 0)$ is the additive identity in $\mathbf{C}$.

    d. The additive inverse of $(a, b) \in \mathbf{C}$ is $(-a, -b) \in \mathbf{C}$.

15. With addition and multiplication as in Definition 7.5, prove that the right distributive property holds in $\mathbf{C}$.

16. With $\mathbf{C}$ as given in Definition 7.5, prove that $R' = \{(a, 0) | a \in \mathbf{R}\}$ is a subfield of $\mathbf{C}$.

17. Let θ be the mapping $\theta: \mathbf{C} \to \mathbf{C}$ defined for $z = a + bi$ in standard form by

$$\theta(z) = a - bi.$$

Prove that θ is a ring isomorphism.

18. It follows from Exercise 17 that $\overline{z_1 + z_2} = \bar{z}_1 + \bar{z}_2$ and $\overline{z_1 z_2} = \bar{z}_1 \bar{z}_2$ for all z_1, z_2 in $\mathbf{C}$. Prove the following statements concerning conjugates of complex numbers.

    a. $\overline{(\bar{z})} = z$

    b. If $z \neq 0$, $(\bar{z})^{-1} = \overline{(z^{-1})}$.

    c. $z + \bar{z} \in \mathbf{R}$

    d. $z = \bar{z}$ if and only if $z \in \mathbf{R}$.

19. Assume that $\theta: \mathbf{C} \to \mathbf{C}$ is an isomorphism and $\theta(a) = a$ for all $a \in \mathbf{R}$. Prove that if θ is not the identity mapping, then $\theta(z) = \bar{z}$ for all $z \in \mathbf{C}$.

20. (See Example 8 of Section 5.1.) Show that the mapping θ defined by

$$\theta(a + bi) = \begin{bmatrix} a & -b \\ b & a \end{bmatrix} \quad \text{for} \quad a, b \in \mathbf{R}$$

is an isomorphism from $\mathbf{C}$ to a subring of the ring of all 2×2 matrices over $\mathbf{R}$.

21. With addition as given in Example 3 of this section, prove the following statements.
   a. Addition is associative in H.
   b. Addition is commutative in H.

22. Prove that multiplication in the set H of Example 3 has the associative property.

23. With addition and multiplication as defined in Example 3, prove that both distributive laws hold in H.

Exercises 24–28 are stated using the notation in the last paragraph of Example 3.

24. Prove that $\overline{(\bar{q})} = q$ for all $q \in H$.

25. Prove that $\overline{q_1 + q_2} = \bar{q}_1 + \bar{q}_2$ for all $q_1, q_2 \in H$.

26. Prove that $\overline{q_1 q_2} = \bar{q}_2\, \bar{q}_1$ for all $q_1, q_2 \in H$.

27. Prove or disprove: $\overline{q_1 q_2} = \bar{q}_1\, \bar{q}_2$ for all $q_1, q_2 \in H$.

28. Verify that $\bar{q}q = q\bar{q} = x^2 + y^2 + z^2 + w^2$ for arbitrary $q = x + yi + zj + wk$ in H.

29. (See Exercise 28.) For arbitrary $q = x + yi + zj + wk$ in H, we define the **absolute value** of q by $|q| = \sqrt{x^2 + y^2 + z^2 + w^2}$. Verify that $|q_1 q_2| = |q_1| \cdot |q_2|$.

30. a. With H as defined in Example 3, prove that the set

$$R' = \{(a, 0, 0, 0) | a \in \mathbf{R}\}$$

is a field that is contained in H.
   b. Prove that the mapping $\theta: \mathbf{R} \to R'$ defined by $\theta(a) = (a, 0, 0, 0)$ is an isomorphism.

31. Assume that

$$C' = \{(a, b, 0, 0) | a, b \in \mathbf{R}\}$$

is a subring of the quaternions in Example 3 when H is regarded as a set of quadruples. Prove that the mapping $\theta: \mathbf{C} \to C'$ defined by $\theta(a + bi) = (a, b, 0, 0)$ is an isomorphism from the field $\mathbf{C}$ of complex numbers to C'. (Thus, we can consider $\mathbf{C}$ to be a subring of H.)

32. Let S be the subset of $M_2(\mathbf{C})$ given by

$$S = \left\{ \begin{bmatrix} x & y \\ -\bar{y} & \bar{x} \end{bmatrix} \Big| x, y \in \mathbf{C} \right\}.$$

   a. Prove that S is a subring of $M_2(\mathbf{C})$.
   b. Prove that the mapping $\theta: H \to S$ defined by

$$\theta(a + bi + cj + dk) = \begin{bmatrix} a + bi & c + di \\ -(c - di) & a - bi \end{bmatrix}$$

is an isomorphism from the ring of quaternions H to S. [Note that $a + bi + cj + dk = (a + bi) + (c + di)j$.]

Sec. 5.2, #13 ≫   **33.** An element a in a ring R is **idempotent** if $a^2 = a$. Prove that a division ring must contain exactly two idempotent elements.

34. Prove that a finite ring R with unity and no zero divisors is a division ring.

▆▆▆ 7.3   DE MOIVRE'S THEOREM AND ROOTS OF COMPLEX NUMBERS

We have seen that real numbers may be represented geometrically by the points on a straight line. In much the same way, it is possible to represent complex numbers by the points in a plane. We begin with a conventional rectangular coordinate system in the plane (see Figure 7-4). With each complex number $x + yi$ in standard form, we associate the point that has coordinates (x, y). This association establishes a one-to-one correspondence from the set **C** of complex numbers to the set of all points in the plane.

Figure 7-4

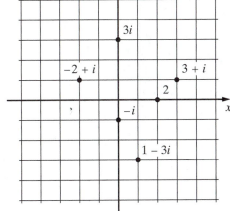

The point in the plane that corresponds to a complex number is called the **graph** of the number, and the complex number that corresponds to a point in the plane is called the **coordinate** of the point. Points on the horizontal axis have coordinates $a + 0i$ that are real numbers. Consequently the horizontal axis is referred to as the **real axis**. Points on the vertical axis have coordinates $0 + bi$ that are pure imaginary numbers, so the vertical axis is called the **imaginary axis**. Several points are labeled with their coordinates in Figure 7-4.

Figure 7-5

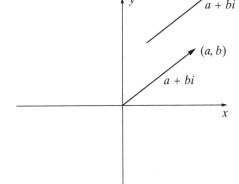

Complex numbers are sometimes represented geometrically by directed line segments called **vectors**. In this approach, the complex number $a + bi$ is represented by the directed line segment from the origin of the coordinate system to the point with rectangular coordinates (a, b), or by any directed line segment with the same length and direction as this one. This is shown on the preceding page in Figure 7-5.

In this book we have little use for the vector representation of complex numbers. We simply note that, in this interpretation, addition of complex numbers corresponds to the usual "parallelogram rule" for adding vectors. This is illustrated in Figure 7-6.

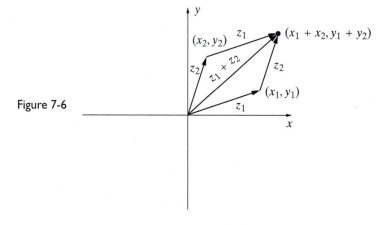

Figure 7-6

Returning now to the representation of complex numbers by points in the plane, we observe that any point P in the plane can be located by designating its *distance r* from the origin O and an *angle θ* in standard position that has OP as its terminal side. Figure 7-7 shows r and θ for a complex number $x + yi$ in standard form.

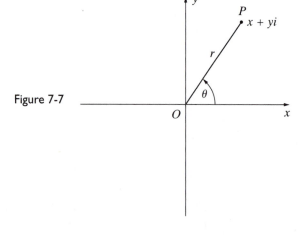

Figure 7-7

From Figure 7-7, we see that r and θ are related to x and y by the equations

$$x = r \cos \theta, \qquad y = r \sin \theta, \qquad r = \sqrt{x^2 + y^2}.$$

The complex number $x + yi$ can thus be written in the form

$$x + yi = r(\cos \theta + i \sin \theta).$$

Definition 7.8   TRIGONOMETRIC FORM

> When a complex number in standard form $x + yi$ is written as
>
> $$x + yi = r(\cos \theta + i \sin \theta),$$
>
> the expression[†] $r(\cos \theta + i \sin \theta)$ is called the **trigonometric form** (or **polar form**) of $x + yi$. The number $r = \sqrt{x^2 + y^2}$ is called the **absolute value** (or **modulus**) of $x + yi$, and the angle θ is called the **argument** (or **amplitude**) of $x + yi$.

The usual notation is used for the absolute value of a complex number:

$$|x + yi| = r = \sqrt{x^2 + y^2}.$$

The absolute value, r, is unique, but the angle θ is not unique since there are many angles in standard position with P on their terminal side. This is illustrated in the next example.

Example 1   Expressing the complex number $-1-i$ in trigonometric form[‡], we have

$$-1 - i = \sqrt{2}\left(-\frac{1}{\sqrt{2}} - \frac{1}{\sqrt{2}}i\right)$$

$$= \sqrt{2}\left(\cos \frac{5\pi}{4} + i \sin \frac{5\pi}{4}\right)$$

$$= \sqrt{2}\left[\cos \left(-\frac{3\pi}{4}\right) + i \sin \left(-\frac{3\pi}{4}\right)\right]$$

$$= \sqrt{2}\left(\cos \frac{13\pi}{4} + i \sin \frac{13\pi}{4}\right).$$

Many other such expressions are possible.   ■

Although the argument θ is not unique, an equation of the form

$$r_1(\cos \theta_1 + i \sin \theta_1) = r_2(\cos \theta_2 + i \sin \theta_2)$$

does require that $r_1 = r_2$ and that θ_1 and θ_2 be coterminal. Hence,

$$\theta_2 = \theta_1 + k(2\pi)$$

for some integer k.

[†]The expression $\cos \theta + i \sin \theta$ is sometimes abbreviated as cis θ.
[‡]We choose to use radian measure for angles. Degree measure could also be used.

The next theorem gives a hint as to the usefulness of the trigonometric form of complex numbers. In proving the theorem, we shall use the following identities from trigonometry:

$$\cos(A + B) = \cos A \cos B - \sin A \sin B$$
$$\sin(A + B) = \sin A \cos B + \cos A \sin B.$$

Theorem 7.9   **PRODUCT OF COMPLEX NUMBERS**

If

$$z_1 = r_1(\cos \theta_1 + i \sin \theta_1)$$

and

$$z_2 = r_2(\cos \theta_2 + i \sin \theta_2)$$

are arbitrary complex numbers in trigonometric form, then

$$z_1 z_2 = r_1 r_2 [\cos(\theta_1 + \theta_2) + i \sin(\theta_1 + \theta_2)].$$

In words, the absolute value of the product of two complex numbers is the product of their absolute values, and an argument of the product is the sum of their arguments.

Proof   The statement of the theorm follows from

$$
\begin{aligned}
z_1 z_2 &= [r_1(\cos \theta_1 + i \sin \theta_1)][r_2(\cos \theta_2 + i \sin \theta_2)] \\
&= r_1 r_2 [(\cos \theta_1 \cos \theta_2 - \sin \theta_1 \sin \theta_2) \\
&\quad + i(\cos \theta_1 \sin \theta_2 + \sin \theta_1 \cos \theta_2)] \\
&= r_1 r_2 [\cos (\theta_1 + \theta_2) + i \sin (\theta_1 + \theta_2)].
\end{aligned}
$$

The preceding result leads to the next theorem, which begins to reveal the true usefulness of the trigonometric form.

Theorem 7.10   **DE MOIVRE'S THEOREM**

If n is a positive integer and

$$z = r(\cos \theta + i \sin \theta)$$

is a complex number in trigonometric form, then

$$z^n = r^n(\cos n\theta + i \sin n\theta).$$

Induction   **Proof**   For $n = 1$, the statement is trivial. Assume that it is true for $n = k$; that is, that

$$z^k = r^k(\cos k\theta + i \sin k\theta).$$

Using Theorem 7.9, we have

$$
\begin{aligned}
z^{k+1} &= z^k \cdot z \\
&= [r^k(\cos k\theta + i \sin k\theta)][r(\cos \theta + i \sin \theta)] \\
&= r^{k+1}[\cos(k\theta + \theta) + i \sin(k\theta + \theta)] \\
&= r^{k+1}[\cos(k + 1)\theta + i \sin(k + 1)\theta].
\end{aligned}
$$

Thus, the theorem is true for $n = k + 1$, and it follows by induction that the theorem is true for all positive integers.

Example 2   Some applications of De Moivre's Theorem are shown in the following computations.

a. $(-2 + 2i)^4 = \left[2\sqrt{2}\left(-\frac{1}{\sqrt{2}} + \frac{1}{\sqrt{2}}i \right) \right]^4$

$$= \left[2\sqrt{2}\left(\cos\frac{3\pi}{4} + i\sin\frac{3\pi}{4} \right) \right]^4$$

$$= 64(\cos 3\pi + i\sin 3\pi)$$

$$= 64(-1 + 0i) = -64$$

b. $\left(\frac{\sqrt{3}}{2} + \frac{1}{2}i \right)^{40} = \left[1\left(\cos\frac{\pi}{6} + i\sin\frac{\pi}{6} \right) \right]^{40}$

$$= 1^{40}\left(\cos\frac{20\pi}{3} + i\sin\frac{20\pi}{3} \right)$$

$$= \cos\left(\frac{2\pi}{3} + 6\pi \right) + i\sin\left(\frac{2\pi}{3} + 6\pi \right)$$

$$= \cos\frac{2\pi}{3} + i\sin\frac{2\pi}{3}$$

$$= -\frac{1}{2} + \frac{\sqrt{3}}{2}i \qquad\qquad ■$$

If n is a positive integer greater than 1 and $u^n = z$ for complex numbers u and z, then u is called an **nth root** of z. We shall prove that every nonzero complex number has exactly n nth roots in **C**.

Theorem 7.11 nth ROOTS OF A COMPLEX NUMBER

For each integer $n \geq 1$, any nonzero complex number

$$z = r(\cos\theta + i\sin\theta)$$

has exactly n distinct nth roots in **C**, and these are given by

$$r^{1/n}\left(\cos\frac{\theta + 2k\pi}{n} + i\sin\frac{\theta + 2k\pi}{n} \right), \quad k = 0, 1, 2, \ldots, n - 1$$

where $r^{1/n} = \sqrt[n]{r}$ denotes the positive real nth root of r.

Proof   For an arbitrary integer k, let

$$v = r^{1/n}\left(\cos\frac{\theta + 2k\pi}{n} + i\sin\frac{\theta + 2k\pi}{n} \right).$$

Then

$$v^n = (r^{1/n})^n\left(\cos\frac{n(\theta + 2k\pi)}{n} + i\sin\frac{n(\theta + 2k\pi)}{n} \right)$$

$$= r[\cos(\theta + 2k\pi) + i\sin(\theta + 2k\pi)]$$

$$= r(\cos\theta + i\sin\theta) = z,$$

and v is an nth root of z. The n angles

$$\frac{\theta}{n}, \quad \frac{\theta + 2\pi}{n}, \quad \frac{\theta + 2(2\pi)}{n}, \quad \dots, \quad \frac{\theta + (n-1)(2\pi)}{n}$$

are equally spaced $\frac{2\pi}{n}$ radians apart, so no two of them have the same terminal side. Thus, the n values of v obtained by letting $k = 0, 1, 2, \dots, n-1$ are distinct, and we have shown that z has at least n distinct nth roots in $\mathbf{C}$.

To show there are no other nth roots of z in $\mathbf{C}$, suppose $v = t(\cos \phi + i \sin \phi)$ is the trigonometric form of a complex number v such that $v^n = z$. Then

$$t^n(\cos n\phi + i \sin n\phi) = r(\cos \theta + i \sin \theta),$$

by De Moivre's Theorem. It follows from this that

$$t^n = r, \quad \cos n\phi = \cos \theta, \quad \text{and} \quad \sin n\phi = \sin \theta.$$

Since r and t are positive, it must be true that $t = r^{1/n}$. The other equations require that $n\phi$ and θ be coterminal, and hence they differ by a multiple of 2π:

$$n\phi = \theta + m(2\pi)$$

for some integer m. By the Division Algorithm,

$$m = qn + k$$

where $k \in \{0, 1, 2, \dots, n-1\}$. Thus,

$$n\phi = \theta + (qn + k)(2\pi)$$

and

$$\phi = \frac{\theta + 2k\pi}{n} + q(2\pi).$$

This equation shows that ϕ is coterminal with the angle $\frac{(\theta + 2k\pi)}{n}$, and hence v is one of the nth roots listed in the statement of the theorem.

Example 3   We shall find the three cube roots of $8i$ and express each in standard form $a + bi$. Expressing $8i$ in trigonometric form, we have

$$8i = 8\left(\cos \frac{\pi}{2} + i \sin \frac{\pi}{2}\right).$$

By the formula in Theorem 7.11, the cube roots of $z = 8i$ are given by

$$8^{1/3}\left(\cos \frac{\frac{\pi}{2} + 2k\pi}{3} + i \sin \frac{\frac{\pi}{2} + 2k\pi}{3}\right), \quad k = 0, 1, 2.$$

Each of these roots has absolute value $8^{1/3} = 2$, and they are equally spaced $\frac{2\pi}{3}$ radians apart, with the first one at $\frac{\pi}{6}$. Thus, the three cube roots of $8i$ are

$$2\left(\cos \frac{\pi}{6} + i \sin \frac{\pi}{6}\right) = 2\left(\frac{\sqrt{3}}{2} + \frac{1}{2}i\right) = \sqrt{3} + i$$

$$2\left(\cos\frac{5\pi}{6} + i\sin\frac{5\pi}{6}\right) = 2\left(-\frac{\sqrt{3}}{2} + \frac{1}{2}i\right) = -\sqrt{3} + i$$

$$2\left(\cos\frac{3\pi}{2} + i\sin\frac{3\pi}{2}\right) = 2(0 - i) = -2i.$$

These results may be checked by direct multiplication. ■

EXERCISES 7.3

1. Graph each of the following complex numbers and express each in trigonometric form.
   a. $-2 + 2\sqrt{3}i$ b. $2 + 2i$ c. $3 - 3i$
   d. $\sqrt{3} + i$ e. $1 + \sqrt{3}i$ f. $-1 - i$
   g. -4 h. $-5i$

2. Find each of the following products. Write each result in both trigonometric and standard form.
   a. $[4(\cos\frac{\pi}{8} + i\sin\frac{\pi}{8})][\cos\frac{5\pi}{8} + i\sin\frac{5\pi}{8}]$
   b. $[3(\cos\frac{7\pi}{6} + i\sin\frac{7\pi}{6})][\cos\frac{2\pi}{3} + i\sin\frac{2\pi}{3}]$
   c. $[2(\cos\frac{5\pi}{6} + i\sin\frac{5\pi}{6})]\{3[\cos(-\frac{\pi}{6}) + i\sin(-\frac{\pi}{6})]\}$
   d. $[6(\cos\frac{5\pi}{3} + i\sin\frac{5\pi}{3})]\{5[\cos(-\frac{\pi}{3}) + i\sin(-\frac{\pi}{3})]\}$

3. Use De Moivre's Theorem to find the value of each of the following. Leave your answers in standard form $a + bi$.
   a. $(\sqrt{3} + i)^7$
   b. $\left(\frac{\sqrt{3}}{2} + \frac{1}{2}i\right)^{21}$
   c. $(-\sqrt{3} + i)^{10}$
   d. $\left(\frac{\sqrt{3}}{2} - \frac{1}{2}i\right)^{18}$
   e. $\left(-\frac{1}{2} + \frac{\sqrt{3}}{2}i\right)^{18}$
   f. $(\sqrt{2} + \sqrt{2}i)^9$
   g. $(1 - \sqrt{3}i)^8$
   h. $\left(-\frac{1}{2} + \frac{\sqrt{3}}{2}i\right)^{12}$

4. Show that the n distinct nth roots of 1 are equally spaced around a circle with center at the origin and radius 1.

5. If $\omega = \cos\frac{2\pi}{n} + i\sin\frac{2\pi}{n}$, show that the distinct nth roots of 1 are given by $\omega, \omega^2, \ldots, \omega^{n-1}$, $\omega^n = 1$.

6. Find the indicated roots of 1 in standard form $a + bi$ and graph them on a unit circle with center at the origin.
   a. cube roots of 1   b. fourth roots of 1
   c. eighth roots of 1   d. sixth roots of 1

7. Find all the indicated roots of the given number. Leave your results in trigonometric form.
   a. cube roots of $\frac{\sqrt{3}}{2} + \frac{1}{2}i$ b. cube roots of $-1 + i$
   c. fourth roots of $-\frac{\sqrt{3}}{2} + \frac{1}{2}i$ d. fourth roots of $\frac{1}{2} - \frac{\sqrt{3}}{2}i$
   e. fifth roots of $-16\sqrt{2} - 16\sqrt{2}i$ f. sixth roots of $32\sqrt{3} - 32i$

8. Find all complex numbers that are solutions of the given equation. Leave your answers in standard form $a + bi$.
   a. $z^3 + 27 = 0$ b. $z^8 - 16 = 0$

c. $z^3 - i = 0$ **d.** $z^3 + 8i = 0$

e. $z^4 + \dfrac{1}{2} - \dfrac{\sqrt{3}}{2}i = 0$ **f.** $z^4 + 1 - \sqrt{3}\,i = 0$

g. $z^4 + \dfrac{1}{2} + \dfrac{\sqrt{3}}{2}i = 0$ **h.** $z^4 + 8 + 8\sqrt{3}\,i = 0$

9. If $\omega = \cos\frac{2\pi}{n} + i\sin\frac{2\pi}{n}$, and u is any nth root of $z \in \mathbf{C}$, show that the nth roots of z are given by $\omega u, \omega^2 u, \dots, \omega^{n-1}u, \omega^n u = u$.

10. Prove that for a fixed value of n, the set U_n of all nth roots of 1 forms a group with respect to multiplication.

In Exercises 11–14, take U_n to be the group in Exercise 10 and find all elements of the subgroup $\langle a \rangle$ generated by the given a. Leave your results in trigonometric form.

11. $a = \cos\frac{2\pi}{3} + i\sin\frac{2\pi}{3}$ in U_6

12. $a = \cos\frac{3\pi}{2} + i\sin\frac{3\pi}{2}$ in U_8

13. $a = \cos\frac{5\pi}{3} + i\sin\frac{5\pi}{3}$ in U_6

14. $a = \cos\frac{5\pi}{4} + i\sin\frac{5\pi}{4}$ in U_8

15. Prove that the group in Exercise 10 is cyclic, with $\omega = \cos\frac{2\pi}{n} + i\sin\frac{2\pi}{n}$ as a generator.

16. Any generator of the group in Exercise 10 is called a **primitive nth root** of 1. Prove that

$$\cos\frac{2k\pi}{n} + i\sin\frac{2k\pi}{n}$$

is a primitive nth root of 1 if and only if k and n are relatively prime.

17. a. Find all primitive sixth roots of 1.
    b. Find all primitive eighth roots of 1.

18. Prove that the set of *all* roots of 1 forms a group with respect to multiplication.

19. Prove that the sum of all the distinct nth roots of 1 is 0.

20. Prove that the product of all the distinct nth roots of 1 is $(-1)^{n+1}$.

21. Prove the following statements concerning absolute values of complex numbers. (As in Definition 7.7, $\bar{z}$ denotes the conjugate of z.)

    a. $|\bar{z}| = |z|$ **b.** $z\bar{z} = |z|^2$

    c. If $z \neq 0$, then $z^{-1} = \dfrac{\bar{z}}{|z|^2}$.   **d.** If $z_2 \neq 0$, then $\left|\dfrac{z_1}{z_2}\right| = \dfrac{|z_1|}{|z_2|}$.

    e. $|z_1 + z_2| \leq |z_1| + |z_2|$

22. Prove that the set of all complex numbers that have absolute value 1 forms a group with respect to multiplication.

23. Prove that if $z = r(\cos\theta + i\sin\theta)$ is a nonzero complex number in trigonometric form, then $z^{-1} = r^{-1}[\cos(-\theta) + i\sin(-\theta)]$.

24. Prove that if n is a positive integer and $z = r(\cos\theta + i\sin\theta)$ is a nonzero complex number in trigonometric form, then $z^{-n} = r^{-n}[\cos(-n\theta) + i\sin(-n\theta)]$.

25. Prove that if $z_1 = r_1(\cos\theta_1 + i\sin\theta_1)$ and $z_2 = r_2(\cos\theta_2 + i\sin\theta_2)$ are complex numbers in trigonometric form and z_2 is nonzero, then

$$\frac{z_1}{z_2} = \frac{r_1}{r_2}[\cos(\theta_1 - \theta_2) + i\sin(\theta_1 - \theta_2)].$$

Sec. 5.4, #7 ≫ **26.** In the ordered field $\mathbf{R}$, absolute value is defined according to Exercise 7 of Section 5.4 by

$$|a| = \begin{cases} a & \text{if } a \geq 0 \\ -a & \text{if } a < 0. \end{cases}$$

For $a \in \mathbf{R}$, show that the absolute value of $a + 0i$ according to Definition 7.8 agrees with the definition from Chapter 5. (Keep in mind, however, that $\mathbf{C}$ is not an ordered field, as was shown in Section 5.4.)

KEY WORDS AND PHRASES

absolute value, 280, 283
amplitude, 283
argument, 283
complete ordered field, 268
complex numbers, 273
conjugate of a complex number, 276
decimal representation, 270
De Moivre's Theorem, 284

division ring, 277
imaginary number, 277
irrational numbers, 270
least upper bound, 266
modulus, 283
nth roots, 285
polar form, 283
pure imaginary number, 277

quaternions, 277
rational numbers, 269
real numbers, 269
skew field, 277
standard form of a complex number, 277
trigonometric form, 283
upper bound, 266

A PIONEER IN MATHEMATICS ||||||||||||||||||||||||||||||||||||||||||||||||||||||||||||||||

William Rowan Hamilton (1805–1865)

William Rowan Hamilton, born in Dublin, Ireland, on August 3, 1805, became Ireland's greatest mathematician. He was the fourth of nine children and did not attend school. Instead, he was tutored by an uncle. By the age of 3 he showed amazing ability in reading and arithmetic; he had mastered 13 languages by age 13. His interest turned to mathematics in 1813, when he placed only second in a public contest of arithmetic skills. This humbling incident led him to a study of the classical mathematics texts in their original languages of Greek, Latin, and French. In 1823, he was the top student entering Trinity College at Dublin. He was knighted in 1835 for obtaining significant results in the field of optics.

In 1833, Hamilton initiated a new line of thought about complex numbers by treating them as ordered pairs. He spent the next ten years of his life trying to generalize this treatment of ordered pairs to ordered triples. One day, while walking and chatting with his wife along the Royal Canal on the way to a meeting, he became preoccupied with his own thoughts about the ordered triples and suddenly made a dramatic discovery. He realized that if he considered quadruples (the "quaternions") instead of triples and compromised the commutative law for multiplication, he would have the generalization that he had been seeking for several years. Hamilton became so excited about his discovery that he recorded it in a pocket book and impulsively carved it in a stone on the Brougham Bridge. A tablet there marks the spot of Hamilton's discovery of the quaternions.

Hamilton's approach to complex numbers and their four-dimensional generalization, the quaternions, revolutionized algebraic thought. He spent the last 22 years of his life studying the theory of quaternions and reporting his results.

8

POLYNOMIALS

INTRODUCTION

The elementary theory of polynomials over a field is
presented in this chapter. Topics included are the
division algorithm, the greatest common divisor,
factorization theorems, simple algebraic extensions,
and splitting fields for polynomials. This chapter
may be studied independently of Chapter 7.

8.1 POLYNOMIALS OVER A RING

Starting with beginning algebra courses, a great deal of time is devoted to developing skills in various manipulations with polynomials. Procedures are learned for the basic operations of addition, subtraction, multiplication, and division of polynomials. By the time a student begins an abstract algebra course, polynomials are a very familiar topic.

Much of this prior experience involved polynomials in a single letter, such as $5 + 4t + t^2$, where the letter usually represented a variable with domain a subset of the real numbers. In this section our point of view is very different. We wish to start with a commutative ring R with unity[†] 1 and *construct* a ring that contains both R and a given element x. More precisely, we want to construct a *smallest* ring that contains R and x in this sense: Any ring that contains both R and x would necessarily contain the constructed ring. We assume that x is *not* an element of R, but nothing more than this. For the time being, the letter x will be a formal symbol subject only to the definitions that are made as we proceed. The letter x is referred to as an **indeterminate** in order to emphasize its role here. Later, we shall consider other possible roles for x.

Definition 8.1   **POLYNOMIAL IN x OVER R**

> Let R be a commutative ring with unity 1, and let x be an indeterminate. A **polynomial in x with coefficients in R**, or a **polynomial in x over R**, is an expression of the form
>
> $$a_0 x^0 + a_1 x^1 + a_2 x^2 + \cdots + a_n x^n$$
>
> where n is a nonnegative integer and each a_i is an element of R. The set of all polynomials in x over R is denoted by $R[x]$.

The construction that we shall carry out will be guided by our previous experience with polynomials. Consistent with this, we adopt the familiar language of elementary algebra and refer to the parts $a_i x^i$ of the expression in Definition 8.1 as **terms** of the polynomial and to a_i as the **coefficient** of x^i in the term $a_i x^i$. As a notational convenience, we shall use functional notations such as $f(x)$ for shorthand names of polynomials. That is, we shall write things such as

$$f(x) = a_0 x^0 + a_1 x^1 + \cdots + a_n x^n,$$

but this indicates only that $f(x)$ is a symbolic name for the polynomial. It does *not* indicate a function or a function value.

Example 1   Some examples of polynomials in x over the ring **Z** of integers are listed here.

a. $f(x) = 2x^0 + (-4)x^1 + 0x^2 + 5x^3$
b. $g(x) = 1x^0 + 2x^1 + (-1)x^2$
c. $h(x) = (-5)x^0 + 0x^1 + 0x^2$ ∎

[†]Throughout this chapter, the unity is denoted by 1 rather than e. A similar construction can be made with fewer restrictions on R, but such generality results in complications that are avoided here.

We have not yet defined equality of polynomials. (The preceding use of $=$ only indicated that certain polynomials had been given shorthand names.) To be consistent with prior experience, it is desirable to define equality of polynomials so that terms with zero coefficients can be deleted with equality retained. With this goal in mind, we make the following (somewhat cumbersome) definition.

Definition 8.2a   **EQUALITY OF POLYNOMIALS**

> Suppose that R is a commutative ring with unity, that x is an indeterminate, and that
>
> $$f(x) = a_0 x^0 + a_1 x^1 + \cdots + a_n x^n$$
>
> and
>
> $$g(x) = b_0 x^0 + b_1 x^1 + \cdots + b_m x^m$$
>
> are polynomials in x over R. Then $f(x)$ and $g(x)$ are **equal polynomials**, $f(x) = g(x)$, if and only if the following conditions hold for all i that occur as a subscript on a coefficient in either $f(x)$ or $g(x)$:
>
> **1.** If one of a_i, b_i is zero, then the other is either omitted or is also zero.
> **2.** If one of a_i, b_i is not zero, then the other is not omitted, and $a_i = b_i$.

Example 2   According to Definition 8.2a, the following equalities are valid in the set $\mathbf{Z}[x]$ of all polynomials in x over $\mathbf{Z}$.

a. $2x^0 + (-4)x^1 + 0x^2 + 5x^3 = 2x^0 + (-4)x^1 + 5x^3$
b. $(-5)x^0 + 0x^1 + 0x^2 = (-5)x^0$ ∎

The compact sigma notation is useful when we work with polynomials. The polynomial

$$f(x) = a_0 x^0 + a_1 x^1 + \cdots + a_n x^n$$

may be written compactly using the sigma notation as

$$f(x) = \sum_{i=0}^{n} a_i x^i.$$

After the convention concerning zero coefficients has been clarified and agreed upon as stated in conditions 1 and 2 of Definition 8.2a, the definition of *equality of polynomials* may be shortened as follows.

Definition 8.2b   **ALTERNATIVE DEFINITION, EQUALITY OF POLYNOMIALS**

> If R is a commutative ring with unity, and $f(x) = \sum_{i=0}^{n} a_i x^i$ and $g(x) = \sum_{i=0}^{m} b_i x^i$ are polynomials in x over R, then $f(x) = g(x)$ if and only if $a_i = b_i$ for all i.

It is understood, of course, that any polynomial over R has only a finite number of nonzero terms. The notational agreements that have been made allow us to make concise definitions of addition and multiplication in $R[x]$.

Definition 8.3   **ADDITION AND MULTIPLICATION OF POLYNOMIALS**

Let R be a commutative ring with unity. For any $f(x) = \sum_{i=0}^{n} a_i x^i$ and $g(x) = \sum_{i=0}^{m} b_i x^i$ in $R[x]$, we define **addition** in $R[x]$ by

$$f(x) + g(x) = \sum_{i=0}^{k} (a_i + b_i)x^i,$$

where k is the larger of the two integers n, m. We define **multiplication** in $R[x]$ by

$$f(x)g(x) = \sum_{i=0}^{n+m} c_i x^i,$$

where $c_i = \sum_{j=0}^{i} a_j b_{i-j}$.

The expanded expression for c_i appears as

$$c_i = a_0 b_i + a_1 b_{i-1} + a_2 b_{i-2} + \cdots + a_{i-2} b_2 + a_{i-1} b_1 + a_i b_0.$$

We shall see presently that this formula agrees with previous experience in the multiplication of polynomials.

To introduce some novelty in our next example, we consider the sum and product of two polynomials over the ring $\mathbf{Z}_6$.

Example 3   We shall follow a convention that has been used on some earlier occasions, and write a for $[a]$ in $\mathbf{Z}_6$. Let

$$f(x) = \sum_{i=0}^{3} a_i x^i = 1x^0 + 5x^1 + 3x^3$$

and

$$g(x) = \sum_{i=0}^{1} b_i x^i = 4x^0 + 2x^1$$

in $\mathbf{Z}_6[x]$. According to our agreement regarding zero coefficients, these polynomials may be written as

$$f(x) = 1x^0 + 5x^1 + 0x^2 + 3x^3$$
$$g(x) = 4x^0 + 2x^1 + 0x^2 + 0x^3,$$

and the definition of addition yields

$$\begin{aligned}
f(x) + g(x) &= \sum_{i=0}^{3} (a_i + b_i)x^i \\
&= (1 + 4)x^0 + (5 + 2)x^1 + (0 + 0)x^2 + (3 + 0)x^3 \\
&= 5x^0 + 1x^1 + 0x^2 + 3x^3 \\
&= 5x^0 + 1x^1 + 3x^3,
\end{aligned}$$

since $5 + 2 = 1$ in $\mathbf{Z}_6$. The definition of multiplication gives

$$f(x)g(x) = \sum_{i=1}^{4} c_i x^i,$$

where

$$c_0 = a_0 b_0 = 1 \cdot 4 = 4$$
$$c_1 = a_0 b_1 + a_1 b_0 = 1 \cdot 2 + 5 \cdot 4 = 2 + 2 = 4$$
$$c_2 = a_0 b_2 + a_1 b_1 + a_2 b_0 = 1 \cdot 0 + 5 \cdot 2 + 0 \cdot 4 = 4$$
$$c_3 = a_0 b_3 + a_1 b_2 + a_2 b_1 + a_3 b_0 = 1 \cdot 0 + 5 \cdot 0 + 0 \cdot 2 + 3 \cdot 4 = 0$$
$$c_4 = a_0 b_4 + a_1 b_3 + a_2 b_2 + a_3 b_1 + a_4 b_0$$
$$= 1 \cdot 0 + 5 \cdot 0 + 0 \cdot 0 + 3 \cdot 2 + 0 \cdot 4 = 0.$$

Thus,

$$f(x)g(x) = (1x^0 + 5x^1 + 3x^3)(4x^0 + 2x^1)$$
$$= 4x^0 + 4x^1 + 4x^2 + 0x^3 + 0x^4$$
$$= 4x^0 + 4x^1 + 4x^2$$

in $\mathbf{Z}_6[x]$. This product, obtained by using Definition 8.3, agrees with the result obtained by the usual multiplication procedure based on the distributive laws:

$$f(x)g(x) = (1x^0 + 5x^1 + 3x^3)(4x^0) + (1x^0 + 5x^1 + 3x^3)(2x^1)$$
$$= (4x^0 + 2x^1 + 0x^3) + (2x^1 + 4x^2 + 0x^4)$$
$$= 4x^0 + 4x^1 + 4x^2. \qquad \blacksquare$$

The expanded forms of the c_i in Example 3 illustrate how the coefficient of x^i in the product is the sum of all products of the form $a_p b_q$ with $p + q = i$. In general, it is true that

$$c_i = \sum_{j=0}^{i} a_j b_{i-j}$$
$$= a_0 b_i + a_1 b_{i-1} + a_2 b_{i-2} + \cdots + a_{i-1} b_1 + a_i b_0$$
$$= \sum_{p+q=i} a_p b_q.$$

This observation is useful in the proof of our next theorem.

Theorem 8.4   THE RING OF POLYNOMIALS OVER R

Let R be a commutative ring with unity. With addition and multiplication as given in Definition 8.3, $R[x]$ forms a commutative ring with unity.

Proof   Let

$$f(x) = \sum_{i=0}^{n} a_i x^i, \quad g(x) = \sum_{i=0}^{m} b_i x^i, \quad h(x) = \sum_{i=0}^{k} c_i x^i$$

represent arbitrary elements of $R[x]$, and let s be the greatest of the integers n, m, and k.

It follows immediately from Definition 8.3 that the sum $f(x) + g(x)$ is a well-defined element of $R[x]$, and $R[x]$ is closed under addition. Addition in $R[x]$ is associative since

$$f(x) + [g(x) + h(x)] = \sum_{i=0}^{n} a_i x^i + \sum_{i=0}^{s} (b_i + c_i)x^i$$
$$= \sum_{i=0}^{s} [a_i + (b_i + c_i)]x^i$$

$$= \sum_{i=0}^{s} [(a_i + b_i) + c_i]x^i \text{ since addition is associative in } R$$

$$= \sum_{i=0}^{s} (a_i + b_i)x^i + \sum_{i=0}^{k} c_i x^i$$

$$= [f(x) + g(x)] + h(x).$$

The polynomial $0x^0$ is an additive identity in $R[x]$ since

$$f(x) + 0x^0 = 0x^0 + f(x) = f(x)$$

for all $f(x)$ in $R[x]$. The additive inverse of $f(x)$ is $\sum_{i=0}^{n}(-a_i)x^i$ since

$$f(x) + \sum_{i=0}^{n} (-a_i)x^i = \sum_{i=0}^{n} [a_i + (-a_i)]x^i = 0x^0,$$

and $\sum_{i=0}^{n} (-a_i)x^i + f(x) = 0x^0$ in similar fashion. Addition in $R[x]$ is commutative since

$$f(x) + g(x) = \sum_{i=0}^{s} (a_i + b_i)x^i = \sum_{i=0}^{s} (b_i + a_i)x^i = g(x) + f(x).$$

Thus, $R[x]$ is an abelian group with respect to addition.

It is clear from Definition 8.3 that $R[x]$ is closed under the binary operation of multiplication. To see that multiplication is associative in $R[x]$, we first note that the coefficient of x^i in $f(x)[g(x)h(x)]$ is given by

$$\sum_{p+q+r=i} (a_p b_q) c_r.$$

the sum of all products $a_p(b_q c_r)$ of coefficients a_p, b_q, c_r such that the subscripts sum to i. Similarly, in $[f(x)g(x)]h(x)$, the coefficient of x^i is

$$\sum_{p+q+r=i} (a_p b_q) c_r.$$

Now $a_p(b_q c_r) = (a_p b_q)c_r$ since multiplication is associative in R, and therefore $f(x)[g(x)h(x)] = [f(x)g(x)]h(x)$.

Before considering the distributive laws, we shall establish that multiplication in $R[x]$ is commutative. This follows from the equalities

$$f(x)g(x) = \sum_{i=0}^{n+m} \left(\sum_{p+q=i} a_p b_q \right) x^i$$

$$= \sum_{i=0}^{m+n} \left(\sum_{q+p=i} b_q a_p \right) x^i \text{ since multiplication is commutative in } R$$

$$= g(x)f(x).$$

Let t be the greater of the integers m and k, and consider the left distributive law. We have

$$f(x)[g(x) + h(x)] = \sum_{i=0}^{n} a_i x^i \left[\sum_{i=0}^{t} (b_i + c_i) x^i \right]$$

$$= \sum_{i=0}^{n+t} \left[\sum_{p+q=i} a_p (b_q + c_q) \right] x^i$$

$$= \sum_{i=0}^{n+t} \left[\sum_{p+q=i} (a_p b_q + a_p c_q) \right] x^i$$

$$= \sum_{i=0}^{n+t} \left(\sum_{p+q=i} a_p b_q \right) x^i + \sum_{i=0}^{n+t} \left(\sum_{p+q=i} a_p c_q \right) x^i$$

$$= \sum_{i=0}^{n+m} \left(\sum_{p+q=i} a_p b_q \right) x^i + \sum_{i=0}^{n+k} \left(\sum_{p+q=i} a_p c_q \right) x^i$$

$$= f(x)\, g(x) + f(x)\, h(x),$$

and the left distributive property is established. The right distributive property is now easy to prove:

$$[f(x) + g(x)]h(x) = h(x)[f(x) + g(x)] \quad \text{since multiplication is commutative in } R[x]$$

$$= h(x)f(x) + h(x)g(x) \quad \text{by the left distributive law}$$

$$= f(x)h(x) + g(x)h(x) \quad \text{since multiplication is commutative in } R[x].$$

The element $1x^0$ is a unity in $R[x]$ since

$$1x^0 \cdot f(x) = f(x) \cdot 1x^0 = \sum_{i=0}^{n} (a_i \cdot 1)x^i = \sum_{i=0}^{n} a_i x^i = f(x).$$

This completes the proof that $R[x]$ is a commutative ring with unity.

Theorem 8.4 justifies referring to $R[x]$ as the **ring of polynomials over R**, or as the **ring of polynomials with coefficients in R**.

Theorem 8.5   **SUBRING OF $R[x]$ ISOMORPHIC TO R**

For any commutative ring R with unity, the ring $R[x]$ of polynomials over R contains a subring R' that is isomorphic to R.

Proof   Let R' be the subset of $R[x]$ that consists of all elements of the form ax^0. We shall show that R' is a subring by utilizing Theorem 5.4.

The subset R' contains elements such as the additive identity $0x^0$ and the unity $1x^0$ of $R[x]$. For arbitrary ax^0 and bx^0 in R',

$$ax^0 - bx^0 = (a - b)x^0$$

and

$$(ax^0)(bx^0) = (ab)x^0$$

are in R', and therefore R' is a subring of $R[x]$ by Theorem 5.4.

Guided by our previous experience with polynomials, we define $\theta \colon R \to R'$ by

$$\theta(a) = ax^0$$

for all $a \in R$. This rule defines a one-to-one correspondence since θ is onto and

$$\theta(a) = \theta(b) \iff ax^0 = bx^0 \iff a = b.$$

Moreover, θ is an isomorphism, since

$$\theta(a + b) = (a + b)x^0 = ax^0 + bx^0 = \theta(a) + \theta(b)$$

and

$$\theta(ab) = (ab)x^0 = (ax^0)(bx^0) = \theta(a)\theta(b).$$

Thus, R is embedded in $R[x]$. We can use the isomorphism θ to identify $a \in R$ with ax^0 in $R[x]$, and from now on we shall write a in place of ax^0. In particular, 0 may denote the zero polynomial $0x^0$, and 1 may denote the unity $1x^0$ in $R[x]$. We write an arbitrary polynomial

$$f(x) = a_0 x^0 + a_1 x^1 + a_2 x^2 + \cdots + a_n x^n$$

as

$$f(x) = a_0 + a_1 x^1 + a_2 x^2 + \cdots + a_n x^n.$$

Actually, we want to carry this notational simplification a bit farther, writing x for x^1, x^i for $1x^i$, and $-a_i x^i$ for $(-a_i)x^i$. This allows us to use all the conventional polynomial notations for the elements of $R[x]$. Also, we can now regard each term $a_i x^i$ with $i \geq 1$ as a product:

$$a_i x^i = a_i \cdot x \cdot x \cdots \cdots x$$

with i factors of x in the product.

Having made the agreements described in the last paragraph, we may observe that our major goal for this section has been achieved. We have constructed a "smallest" ring $R[x]$ that contains R and x. It is "smallest" because any ring that contained both R and x would have to contain all polynomials

$$f(x) = a_0 + a_1 x + a_2 x^2 + \cdots + a_n x^n$$

as a consequence of the closure properties.

It is now appropriate to pick up some more of the language that is customarily used in work with polynomials.

Definition 8.6   **DEGREE, LEADING COEFFICIENT, CONSTANT TERM**

> Let R be a commutative ring with unity, and let
>
> $$f(x) = a_0 + a_1 x + \cdots + a_n x^n$$
>
> be a *nonzero element* of $R[x]$. Then the **degree** of $f(x)$ is the largest integer k such that the coefficient of x^k is not zero, and this coefficient a_k is called the **leading coefficient** of $f(x)$. The term a_0 of $f(x)$ is called the **constant term** of $f(x)$, and elements of R are referred to as **constant polynomials**.

The degree of $f(x)$ will be abbreviated deg $f(x)$. Note that degree is *not defined* for the zero polynomial. (The reason for this will be clear later.) Note also that the *polynomials of degree zero* are the same as the *nonzero elements of R*.

Example 4   The polynomials $f(x)$ and $g(x)$ in Example 3 can now be written as

$$f(x) = 1 + 5x + 3x^3 = 3x^3 + 5x + 1$$
$$g(x) = 4 + 2x = 2x + 4.$$

a. The constant term of $f(x)$ is 1, and the leading coefficient of $f(x)$ is 3.
b. The polynomial $g(x)$ has constant term 4 and leading coefficient 2.
c. deg $f(x) = 3$ and deg $g(x) = 1$.
d. In Example 3, we found that

$$f(x)\,g(x) = 4 + 4x + 4x^2,$$

so that deg $(f(x)\,g(x)) = 2$. In connection with the next theorem, we note that

$$\deg(f(x)g(x)) \neq \deg f(x) + \deg g(x)$$

in this instance.   ∎

Theorem 8.7   **DEGREE OF A PRODUCT**

If R is an integral domain and $f(x)$ and $g(x)$ are nonzero elements of $R[x]$, then

$$\deg(f(x)g(x)) = \deg f(x) + \deg g(x).$$

$(p \wedge q) \Rightarrow r$ **Proof**   Let R be an integral domain, and suppose that

$$f(x) = \sum_{i=0}^{n} a_i x^i \text{ has degree } n$$

and

$$g(x) = \sum_{i=0}^{m} b_i x^i \text{ has degree } m$$

in $R[x]$. Then $a_n \neq 0$ and $b_m \neq 0$, and this implies that $a_n b_m \neq 0$ since R is an integral domain. But $a_n b_m$ is the leading coefficient in $f(x)g(x)$ since

$$f(x)\,g(x) = \sum_{i=0}^{n+m} \left(\sum_{j=0}^{i} a_j b_{i-j} \right) x^i$$

by Definition 8.3. Therefore,

$$\deg(f(x)\,g(x)) = n + m = \deg f(x) + \deg g(x).$$

Corollary 8.8   **POLYNOMIALS OVER AN INTEGRAL DOMAIN**

$R[x]$ is an integral domain if and only if R is an integral domain.

$p \Leftarrow q$ **Proof**   Assume that R is an integral domain. If $f(x)$ and $g(x)$ are arbitrary nonzero elements of $R[x]$, then both $f(x)$ and $g(x)$ have degrees. According to Theorem 8.7, $f(x)g(x)$ has a degree that is the sum of deg $f(x)$ and deg $g(x)$. Therefore, $f(x)g(x)$ is not the zero polynomial and this shows that $R[x]$ is an integral domain.
$p \Rightarrow q$ If $R[x]$ is an integral domain, however, then R must also be an integral domain since R is a commutative ring with unity and $R \subseteq R[x]$.

We make some final observations concerning Theorem 8.7. Since the product of the zero polynomial and any polynomial always yields the zero polynomial, the equation in Theorem 8.7 cannot hold when one of the factors is a zero polynomial. This is justification for not defining degree for the zero polynomial. We also note that the reason the conclusion of Theorem 8.7 fails to hold in Example 4 is that $\mathbf{Z}_6$ is *not* an integral domain.

EXERCISES 8.1

1. Write the following polynomials in expanded form.

   a. $\displaystyle\sum_{i=0}^{3} c_i x^i$
   b. $\displaystyle\sum_{j=0}^{4} d_j x^j$

   c. $\displaystyle\sum_{k=1}^{3} a_k x^k$
   d. $\displaystyle\sum_{k=2}^{4} x^k$

2. Express the following polynomials by using sigma notation.
   a. $c_0 x^0 + c_1 x^1 + c_2 x^2$
   b. $d_2 x^2 + d_3 x^3 + d_4 x^4$
   c. $x + x^2 + x^3 + x^4$
   d. $x^3 + x^4 + x^5$

3. Consider the following polynomials over $\mathbf{Z}_8$, where a is written for $[a]$ in $\mathbf{Z}_8$:

   $$f(x) = 2x^3 + 7x + 4, \quad g(x) = 4x^2 + 4x + 6, \quad h(x) = 6x^2 + 3.$$

   Find each of the following polynomials with all coefficients in $\mathbf{Z}_8$.
   a. $f(x) + g(x)$
   b. $g(x) + h(x)$
   c. $f(x) g(x)$
   d. $g(x) h(x)$
   e. $f(x) g(x) + h(x)$
   f. $f(x) + g(x) h(x)$
   g. $f(x) g(x) + f(x) h(x)$
   h. $f(x) h(x) + g(x) h(x)$

4. Consider the following polynomials over $\mathbf{Z}_9$, where a is written for $[a]$ in $\mathbf{Z}_9$:

   $$f(x) = 2x^3 + 7x + 4, \quad g(x) = 4x^2 + 4x + 6, \quad h(x) = 6x^2 + 3.$$

   Find each of the following polynomials with all coefficients in $\mathbf{Z}_9$.
   a. $f(x) + g(x)$
   b. $g(x) + h(x)$
   c. $f(x) g(x)$
   d. $g(x) h(x)$
   e. $f(x) g(x) + h(x)$
   f. $f(x) + g(x) h(x)$
   g. $f(x) g(x) + f(x) h(x)$
   h. $f(x) h(x) + g(x) h(x)$

5. Decide whether each of the following subsets is a subring of $R[x]$, and justify your decision in each case.
   a. the set of all polynomials with zero constant term
   b. the set of all polynomials that have zero coefficients for all *even* powers of x
   c. the set of all polynomials that have zero coefficients for all *odd* powers of x
   d. the set consisting of the zero polynomial together with all polynomials that have degree 2 or less

6. Determine which of the subsets in Exercise 5 are ideals of $R[x]$ and justify your choices.

7. Let R be a commutative ring with unity. Prove that

   $$\deg (f(x) g(x)) \le \deg f(x) + \deg g(x)$$

   for all nonzero $f(x), g(x)$ in $R[x]$, even if R is not an integral domain.

8. List all the polynomials in $\mathbf{Z}_3[x]$ that have degree 2.

9. **a.** How many polynomials of degree 2 are there in $\mathbf{Z}_n[x]$?
   b. If m is a positive integer, how many polynomials of degree m are there in $\mathbf{Z}_n[x]$?

10. a. Suppose that R is a commutative ring with unity, and define $\theta: R[x] \to R$ by

$$\theta(a_0 + a_1 x + \cdots + a_n x^n) = a_0$$

for all $a_0 + a_1 x + \cdots + a_n x^n$ in $R[x]$. Prove that θ is an epimorphism from $R[x]$ to R.

b. Describe the kernel of the epimorphism in part a.

11. Let R be a commutative ring with unity and let I be the principal ideal $I = (x)$ in $R[x]$. Prove that $R[x]/I$ is isomorphic to R.

12. In the integral domain $\mathbf{Z}[x]$, let $(\mathbf{Z}[x])^+$ denote the set of all $f(x)$ in $\mathbf{Z}[x]$ that have a positive integer as a leading coefficient. Prove that $\mathbf{Z}[x]$ is an ordered integral domain by proving that $(\mathbf{Z}[x])^+$ is a set of positive elements for $\mathbf{Z}[x]$.

13. Consider the mapping $\phi: \mathbf{Z}[x] \to \mathbf{Z}_k[x]$ defined by

$$\phi(a_0 + a_1 x + \cdots + a_n x^n) = [a_0] + [a_1]x + \cdots + [a_n]x^n,$$

where $[a_i]$ denotes the congruence class of $\mathbf{Z}_k$ that contains a_i. Prove that θ is an epimorphism from $\mathbf{Z}[x]$ to $\mathbf{Z}_k[x]$.

14. Describe the kernel of the epimorphism ϕ in Exercise 13.

15. Assume that each of R and S is a commutative ring with unity and that $\theta: R \to S$ is an epimorphism from R to S. Let $\phi: R[x] \to S[x]$ be defined by

$$\phi(a_0 + a_1 x + \cdots + a_n x^n) = \theta(a_0) + \theta(a_1)x + \cdots + \theta(a_n)x^n.$$

Prove that ϕ is an epimorphism from $R[x]$ to $S[x]$.

16. Describe the kernel of the epimorphism ϕ in Exercise 15.

17. For each $f(x) = \sum_{i=0}^{n} a_i x^i$ in $R[x]$, the **formal derivative** of $f(x)$ is the polynomial

$$f'(x) = \sum_{i=1}^{n} i a_i x^{i-1}.$$

(For $n = 0, f'(x) = 0$ by definition.)

a. Prove that $[f(x) + g(x)]' = f'(x) + g'(x)$.

b. Prove that $[f(x)g(x)]' = f(x)g'(x) + f'(x)g(x)$.

8.2   DIVISIBILITY AND GREATEST COMMON DIVISOR

If a ring R is not an integral domain, the division of polynomials over R is not a very satisfactory subject for study because of the possible presence of zero divisors. In order to obtain the results that we need on division of polynomials, the ring of coefficients actually needs to be a field. For this reason, we confine our attention for the rest of this chapter to rings of polynomials $F[x]$ where F is a *field*. This assures us that $F[x]$ is an integral domain (Corollary 8.8), and that every nonzero element of F has a multiplicative inverse.

The definitions, theorems, and even the proofs in this section are very similar to corresponding statements in Chapter 2 concerning division in the integral domain $\mathbf{Z}$.

Definition 8.9   Divisor, Multiple

> If $f(x)$ and $g(x)$ are in $F[x]$, then $f(x)$ **divides** $g(x)$ if there exists $h(x)$ in $F[x]$ such that $g(x) = f(x)h(x)$.
>
> If $f(x)$ divides $g(x)$, we write $f(x)|g(x)$, and we say that $g(x)$ is a **multiple** of $f(x)$, that $f(x)$ is a **factor** of $g(x)$, or that $f(x)$ is a **divisor** of $g(x)$. We write $f(x) \nmid g(x)$ to indicate that $f(x)$ does not divide $g(x)$.

Polynomials of degree zero (the nonzero elements of F) have two special properties that are worth noting. First, any nonzero element a of F is a factor of every $f(x) \in F[x]$, since $a^{-1}f(x)$ is in $F[x]$ and

$$f(x) = a[a^{-1}f(x)].$$

Second, if $f(x)|g(x)$, then $af(x)|g(x)$ for all nonzero $a \in F$, since the equation

$$g(x) = f(x)h(x)$$

implies that

$$g(x) = [af(x)][a^{-1}h(x)].$$

The Division Algorithm for integers has the following analogue in $F[x]$.

Theorem 8.10   THE DIVISION ALGORITHM

Let $f(x)$ and $g(x)$ be elements of $F[x]$, with $f(x)$ a nonzero polynomial. There exist unique elements $q(x)$ and $r(x)$ in $F[x]$ such that

$$g(x) = f(x)q(x) + r(x)$$

with either $r(x) = 0$ or deg $r(x) <$ deg $f(x)$.

Existence   **Proof**   We postpone the proof of uniqueness until existence of the required $q(x)$ and $r(x)$ in $F[x]$ has been proven. There are two trivial cases that we shall dispose of first.

1. If $g(x) = 0$ or if deg $g(x) <$ deg $f(x)$, then we see from the equality

$$g(x) = f(x) \cdot 0 + g(x)$$

that $q(x) = 0$ and $r(x) = g(x)$ satisfy the required conditions.

2. If deg $f(x) = 0$, then $f(x) = c$ for some nonzero constant c. The equality

$$g(x) = c[c^{-1}g(x)] + 0$$

shows that $q(x) = c^{-1}g(x)$ and $r(x) = 0$ satisfy the required conditions.

Induction   Suppose now that $g(x) \neq 0$ and $1 \leq$ deg $f(x) \leq$ deg $g(x)$. The proof is by induction on $n =$ deg $g(x)$, using the second principle of finite induction. For each positive integer n, let S_n be the statement that if $g(x) \in F[x]$ has degree n and $1 \leq$ deg $f(x) \leq$ deg $g(x)$, then there exist $q(x)$ and $r(x) \in F[x]$ such that $g(x) = f(x)q(x) + r(x)$, with either $r(x) = 0$ or deg $r(x) <$ deg $f(x)$.
       If $n = 1$, then the condition $1 \leq$ deg $f(x) \leq$ deg $g(x) = n$ requires that both $f(x)$ and $g(x)$ have degree 1; say,

$$f(x) = ax + b, \qquad g(x) = cx + d,$$

where $a \neq 0$ and $c \neq 0$. The equality

$$cx + d = (ax + b)(ca^{-1}) + (d - bca^{-1})$$

shows that $q(x) = ca^{-1}$ and $r(x) = d - bca^{-1}$ satisfy the required conditions, and S_1 is true.

Now assume that k is a positive integer such that S_m is true for all positive integers $m < k$. To prove that S_k is true, let $g(x) \in F[x]$ with deg $g(x) = k$ and $f(x) \in F[x]$ with $1 \le$ deg $f(x) \le$ deg $g(x)$. Then

$$f(x) = ax^j + \cdots, \qquad g(x) = cx^k + \cdots$$

with $a \ne 0, c \ne 0$, and $j \le k$. The first step in the usual long division of $g(x)$ by $f(x)$ is shown in Figure 8-1.

Figure 8-1

$$
\begin{array}{r}
ca^{-1}x^{k-j} \\
ax^j + \cdots \overline{\smash{\big)}\, cx^k + \cdots} \\
ca^{-1}x^{k-j}f(x) \\
\hline
g(x) - ca^{-1}x^{k-j}f(x)
\end{array}
$$

This first step in long division yields

$$g(x) = ca^{-1}x^{k-j}f(x) + [g(x) - ca^{-1}x^{k-j}f(x)].$$

Let $h(x) = g(x) - ca^{-1}x^{k-j}f(x)$. Then the coefficient of x^k in $h(x)$ is zero, and deg $h(x) < k$. By the induction hypothesis, there exist polynomials $q_0(x)$ and $r(x)$ such that

$$h(x) = f(x)q_0(x) + r(x)$$

with either $r(x) = 0$ or deg $r(x) <$ deg $f(x)$. This gives the equality

$$
\begin{aligned}
g(x) &= ca^{-1}x^{k-j}f(x) + h(x) \\
&= ca^{-1}x^{k-j}f(x) + f(x)q_0(x) + r(x) \\
&= f(x)[ca^{-1}x^{k-j} + q_0(x)] + r(x),
\end{aligned}
$$

which shows that $q(x) = ca^{-1}x^{k-j} + q_0(x)$ and $r(x)$ are polynomials that satisfy the required conditions. Therefore, S_k is true and the existence part of the theorem follows from the second principle of finite induction.

Uniqueness   To prove uniqueness, suppose that $g(x) = f(x)q_1(x) + r_1(x)$ and $g(x) = f(x)q_2(x) + r_2(x)$ where either $r_i(x) = 0$ or deg $r_i(x) <$ deg $f(x)$ for $i = 1, 2$. Then

$$
\begin{aligned}
r_1(x) - r_2(x) &= [g(x) - f(x)q_1(x)] - [g(x) - f(x)q_2(x)] \\
&= f(x)[q_2(x) - q_1(x)].
\end{aligned}
$$

The right member of this equation, $f(x)[q_2(x) - q_1(x)]$, is either zero or has degree greater than or equal to deg $f(x)$, by Theorem 8.7. However, the left member, $r_1(x) - r_2(x)$, is either zero or has degree less than deg $f(x)$, since deg $r_1(x) <$ deg $f(x)$ and deg $r_2(x) <$ deg $f(x)$. Therefore, both members must be zero, and this requires that $r_1(x) = r_2(x)$ and $q_1(x) = q_2(x)$ since $f(x)$ is nonzero. Therefore, $q(x)$ and $r(x)$ are unique and the proof is complete.

In the Division Algorithm, the polynomial $q(x)$ is called the **quotient** and $r(x)$ is called the **remainder** in the division of $g(x)$ by $f(x)$. For any field F, the quotient and remainder in $F[x]$ can be found by the familiar long division procedures. An illustration is given in the next example.

Example I   Let $f(x) = 3x^2 + 2$ and $g(x) = 4x^4 + 2x^3 + 6x^2 + 4x + 5$ in $\mathbf{Z}_7[x]$. We shall find $q(x)$ and $r(x)$ by the long division procedure. Referring to Figure 8-1, we have $a = 3$ in $f(x)$, $c = 4$ in $g(x)$, and $ca^{-1} = 3(4^{-1}) = 3(2) = 6$ in the first step.

$$
\require{enclose}
\begin{array}{r}
6x^2 + 3x \ \ + 5 \\
3x^2 + 2 \enclose{longdiv}{4x^4 + 2x^3 + 6x^2 + 4x + 5} \\
4x^4 + \quad 5x^2 \\
2x^3 + x^2 \\
2x^3 + \quad 6x \\
x^2 + 5x \\
x^2 \quad + 3 \\
5x + 2
\end{array}
$$

Thus, the quotient is $q(x) = 6x^2 + 3x + 5$ and the remainder is $r(x) = 5x + 2$ in the division of $g(x)$ by $f(x)$. ∎

Our next objective in this section is to prove that any two nonzero polynomials over F have a greatest common divisor in $F[x]$. We saw earlier that if $f(x)$ is a divisor of $g(x)$, then $af(x)$ is also a divisor of $g(x)$ for every nonzero $a \in F$. By choosing a to be the multiplicative inverse of the leading coefficient of $f(x)$, the leading coefficient in $af(x)$ can be made equal to 1. This means that in the consideration of common divisors of two polynomials, there is no loss of generality if attention is restricted to polynomials having 1 as their leading coefficient.

Definition 8.11   MONIC POLYNOMIAL

> A polynomial with 1 as its leading coefficient is called a **monic** polynomial.

One of the conditions that we place on a greatest common divisor of two polynomials is that it be monic. Without this condition, the greatest common divisor of two polynomials would not be unique.

Definition 8.12   GREATEST COMMON DIVISOR

> Let $f(x)$ and $g(x)$ be nonzero polynomials in $F[x]$. A polynomial $d(x)$ in $F[x]$ is a **greatest common divisor** of $f(x)$ and $g(x)$ if these conditions are satisfied:
>
> 1. $d(x)$ is a monic polynomial.
> 2. $d(x)|f(x)$ and $d(x)|g(x)$.
> 3. $h(x)|f(x)$ and $h(x)|g(x)$ imply that $h(x)|d(x)$.

The next theorem shows that any two nonzero elements $f(x), g(x)$ of $F[x]$ have a unique greatest common divisor $d(x)$.

STRATEGY ▶

The proof of Theorem 8.13 is obtained by making minor adjustments in the proof of Theorem 2.12, and it shows that $d(x)$ is a **linear combination** of $f(x)$ and $g(x)$; that is, $d(x)$ can be written in the form

$$d(x) = f(x)s(x) + g(x)t(x)$$

for some $s(x), t(x) \in F[x]$.

Theorem 8.13   GREATEST COMMON DIVISOR

Let $f(x)$ and $g(x)$ be nonzero polynomials over F. Then there exists a unique greatest common divisor $d(x)$ of $f(x)$ and $g(x)$ in $F[x]$. Moreover, $d(x)$ can be expressed as

$$d(x) = f(x)s(x) + g(x)t(x)$$

for $s(x)$ and $t(x)$ in $F[x]$, and $d(x)$ is the monic polynomial of least degree that can be written in this form.

Existence   **Proof**   Consider the set S of all polynomials in $F[x]$ that can be written in the form

$$f(x)u(x) + g(x)v(x)$$

with $u(x)$ and $v(x)$ in $F[x]$. Since $f(x) = f(x) \cdot 1 + g(x) \cdot 0 \neq 0$, the set of nonzero polynomials in S is nonempty. Let

$$d_1(x) = f(x)u_1(x) + g(x)v_1(x)$$

be a polynomial of least degree among the nonzero elements of S. If c is the leading coefficient of $d_1(x)$, then

$$d(x) = c^{-1}d_1(x) = f(x)[c^{-1}u_1(x)] + g(x)[c^{-1}v_1(x)]$$

is a monic polynomial of least degree in S. Letting $s(x) = c^{-1}u_1(x)$ and $t(x) = c^{-1}v_1(x)$, we have a polynomial

$$d(x) = f(x)s(x) + g(x)t(x),$$

which is expressed in the required form and satisfies the first condition in Definition 8.12.

We shall show that $d(x)|f(x)$. By the Division Algorithm, there are elements $q(x)$ and $r(x)$ of $F[x]$ such that

$$f(x) = d(x)q(x) + r(x)$$

with either $r(x) = 0$ or deg $r(x) <$ deg $d(x)$. Since

$$\begin{aligned}
r(x) &= f(x) - d(x)q(x) \\
&= f(x) - [f(x)s(x) + g(x)t(x)]q(x) \\
&= f(x)[1 - s(x)q(x)] + g(x)[-t(x)q(x)],
\end{aligned}$$

$r(x)$ is an element of S. By choice of $d(x)$ as having smallest possible degree among the nonzero elements of S, it cannot be true that deg $r(x) <$ deg $d(x)$. Therefore, $r(x) = 0$ and $d(x)|f(x)$. A similar argument shows that $d(x)|g(x)$, and hence $d(x)$ satisfies condition 2 in Definition 8.12.

If $h(x)|f(x)$ and $h(x)|g(x)$, then $f(x) = h(x)p_1(x)$ and $g(x) = h(x)p_2(x)$ for $p_i(x) \in F[x]$. Therefore,

$$d(x) = f(x)s(x) + g(x)t(x)$$
$$= h(x)p_1(x)s(x) + h(x)p_2(x)t(x)$$
$$= h(x)[p_1(x)s(x) + p_2(x)t(x)],$$

and this shows that $h(x)|d(x)$. By Definition 8.12, $d(x)$ is a greatest common divisor of $f(x)$ and $g(x)$.

Uniqueness   To show uniqueness, suppose that $d_1(x)$ and $d_2(x)$ are both greatest common divisors of $f(x)$ and $g(x)$. Then $d_1(x)|d_2(x)$ and also $d_2(x)|d_1(x)$. Since both $d_1(x)$ and $d_2(x)$ are monic polynomials, this means that $d_1(x) = d_2(x)$. (See Exercise 16 at the end of this section.)

If $f(x)$ and $g(x)$ are nonzero polynomials such that $f(x)|g(x)$, the greatest common divisor of $f(x)$ and $g(x)$ is simply the product of $f(x)$ and the multiplicative inverse of its leading coefficient. If $f(x) \nmid g(x)$, the Euclidean Algorithm extends readily to polynomials, furnishing a systematic method for finding the greatest common divisor of $f(x)$ and $g(x)$ and for finding $s(x)$ and $t(x)$ in the equation

$$d(x) = f(x)s(x) + g(x)t(x).$$

The Euclidean Algorithm consists of repeated application of the Division Algorithm to yield the following sequence, where $r_n(x)$ is the last nonzero remainder.

Euclidean Algorithm

$$g(x) = f(x)q_0(x) + r_1(x), \qquad \deg r_1(x) < \deg f(x)$$
$$f(x) = r_1(x)q_1(x) + r_2(x), \qquad \deg r_2(x) < \deg r_1(x)$$
$$r_1(x) = r_2(x)q_2(x) + r_3(x), \qquad \deg r_3(x) < \deg r_2(x)$$
$$\vdots \qquad\qquad\qquad \vdots$$
$$r_{n-2}(x) = r_{n-1}(x)q_{n-1}(x) + r_n(x), \qquad \deg r_n(x) < \deg r_{n-1}(x)$$
$$r_{n-1}(x) = r_n(x)q_n(x)$$

Suppose that a is the leading coefficient of the last nonzero remainder, $r_n(x)$. It is left as an exercise to prove that $a^{-1}r_n(x)$ is the greatest common divisor of $f(x)$ and $g(x)$.

Example 2   We shall find the greatest common divisor of $f(x) = 3x^3 + 5x^2 + 6x$ and $g(x) = 4x^4 + 2x^3 + 6x^2 + 4x + 5$ in $\mathbf{Z}_7[x]$. Long division of $g(x)$ by $f(x)$ yields a quotient of $q_0(x) = 6x$ and a remainder of $r_1(x) = 5x^2 + 4x + 5$, so we have

$$g(x) = f(x) \cdot (6x) + (5x^2 + 4x + 5).$$

Dividing $f(x)$ by $r_1(x)$, we obtain

$$f(x) = r_1(x) \cdot (2x + 5) + (4x + 3),$$

so $q_1(x) + 2x + 5$ and $r_2(x) = 4x + 3$ in the Euclidean Algorithm. Division of $r_1(x)$ by $r_2(x)$ then yields

$$r_1(x) = r_2(x) \cdot (3x + 4).$$

Thus, $r_2(x) = 4x + 3$ is the last nonzero remainder, and the greatest common divisor of $f(x)$ and $g(x)$ in $\mathbf{Z}_7[x]$ is

$$\begin{aligned} d(x) &= 4^{-1}(4x + 3) \\ &= 2(4x + 3) \\ &= x + 6. \end{aligned}$$
■

As mentioned earlier, the Euclidean Algorithm can also be used to find polynomials $s(x)$ and $t(x)$ such that

$$d(x) = f(x)s(x) + g(x)t(x).$$

This is illustrated in Example 3.

Example 3   As in Example 2, let $f(x) = 3x^3 + 5x^2 + 6x$ and $g(x) = 4x^4 + 2x^3 + 6x^2 + 4x + 5$. From Example 2, the greatest common divisor of $f(x)$ and $g(x)$ is $d(x) = x + 6$. To find polynomials $s(x)$ and $t(x)$ such that

$$d(x) = f(x)s(x) + g(x)t(x),$$

we first solve for the remainders in the Euclidean Algorithm (see Example 2) as follows:

$$\begin{aligned} r_2(x) &= f(x) - r_1(x)(2x + 5) \\ r_1(x) &= g(x) - f(x)(6x). \end{aligned}$$

Substituting for $r_1(x)$ in the first equation, we have

$$\begin{aligned} r_2(x) &= f(x) - [g(x) - f(x)(6x)](2x + 5) \\ &= f(x) + f(x)(6x)(2x + 5) - g(x)(2x + 5) \\ &= f(x)[1 + (6x)(2x + 5)] + g(x)(-2x - 5) \\ &= f(x)(5x^2 + 2x + 1) + g(x)(5x + 2). \end{aligned}$$

To express $d(x) = 4^{-1}r_2(x) = 2r_2(x)$ as a linear combination of $f(x)$ and $g(x)$, we multiply both members of the last equation by $4^{-1} = 2$:

$$\begin{aligned} d(x) = 2r_2(x) &= f(x)(2)(5x^2 + 2x + 1) + g(x)(2)(5x + 2) \\ d(x) &= f(x)(3x^2 + 4x + 2) + g(x)(3x + 4). \end{aligned}$$

The desired polynomials are given by $s(x) = 3x^2 + 4x + 2$ and $t(x) = 3x + 4$.   ■

EXERCISES 8.2

For $f(x)$, $g(x)$, and $\mathbf{Z}_n[x]$ given in Exercises 1–6, find $q(x)$ and $r(x)$ in $\mathbf{Z}_n[x]$ that satisfy the conditions in the Division Algorithm.

1. $f(x) = 3x + 1, g(x) = 2x^3 + 3x^2 + 4x + 1$, in $\mathbf{Z}_5[x]$
2. $f(x) = 2x + 2, g(x) = x^3 + 2x^2 + 2$, in $\mathbf{Z}_3[x]$
3. $f(x) = x^3 + x^2 + 2x + 2, g(x) = x^4 + 2x^2 + x + 1$, in $\mathbf{Z}_3[x]$
4. $f(x) = x^3 + 2x^2 + 2, g(x) = 2x^5 + 2x^4 + x^2 + 2$, in $\mathbf{Z}_3[x]$
5. $f(x) = 3x^2 + 2, g(x) = x^4 + 5x^2 + 2x + 2$, in $\mathbf{Z}_7[x]$

6. $f(x) = 3x^2 + 2, g(x) = 4x^4 + 2x^3 + 6x^2 + 4x + 5$, in $\mathbf{Z}_7[x]$

For $f(x), g(x)$, and $\mathbf{Z}_n[x]$ given in Exercises 7–10, find the greatest common divisor $d(x)$ of $f(x)$ and $g(x)$ in $\mathbf{Z}_n[x]$.

7. $f(x) = x^3 + x^2 + 2x + 2, g(x) = x^4 + 2x^2 + x + 1$, in $\mathbf{Z}_3[x]$

8. $f(x) = x^3 + 2x^2 + 2, g(x) = 2x^5 + 2x^4 + x^2 + 2$, in $\mathbf{Z}_3[x]$

9. $f(x) = 3x^2 + 2, g(x) = x^4 + 5x^2 + 2x + 2$, in $\mathbf{Z}_7[x]$

10. $f(x) = 3x^2 + 2, g(x) = 4x^4 + 2x^3 + 6x^2 + 4x + 5$, in $\mathbf{Z}_7[x]$

For $f(x), g(x)$ and $\mathbf{Z}_n[x]$ given in Exercises 11–14, find $s(x)$ and $t(x)$ in $\mathbf{Z}_n[x]$ such that $d(x) = f(x)s(x) + g(x)t(x)$ is the greatest common divisor of $f(x)$ and $g(x)$.

11. $f(x) = 2x^3 + 2x^2 + x + 1, g(x) = x^4 + 2x^2 + x + 1$, in $\mathbf{Z}_3[x]$

12. $f(x) = 2x^3 + x^2 + 1, g(x) = x^5 + x^4 + 2x^2 + 1$, in $\mathbf{Z}_3[x]$

13. $f(x) = 3x^2 + 2, g(x) = x^4 + 5x^2 + 2x + 2$, in $\mathbf{Z}_7[x]$

14. $f(x) = 3x^2 + 2, g(x) = 4x^4 + 2x^3 + 6x^2 + 4x + 5$, in $\mathbf{Z}_7[x]$

15. Prove that if $f(x)$ and $g(x)$ are nonzero elements of $F[x]$ such that $f(x)|g(x)$ and $g(x)|f(x)$, then $f(x) = ag(x)$ for some nonzero $a \in F$.

16. Prove that if $d_1(x)$ and $d_2(x)$ are monic polynomials over the field F such that $d_1(x)|d_2(x)$ and $d_2(x)|d_1(x)$, then $d_1(x) = d_2(x)$.

17. Prove that if $h(x)|f(x)$ and $h(x)|g(x)$ in $F[x]$, then $h(x)|[f(x)u(x) + g(x)v(x)]$ for all $u(x)$ and $v(x)$ in $F[x]$.

18. In the statement of the Division Algorithm (Theorem 8.10), prove that the greatest common divisor of $g(x)$ and $f(x)$ is equal to the greatest common divisor of $f(x)$ and $r(x)$.

19. With the notation used in the description of the Euclidean Algorithm, prove that $a^{-1}r_n(x)$ is the greatest common divisor of $f(x)$ and $g(x)$.

20. Prove that every nonzero remainder $r_j(x)$ in the Euclidean Algorithm is a linear combination of $f(x)$ and $g(x)$: $r_j(x) = f(x)s_j(x) + g(x)t_j(x)$ for some $s_j(x)$ and $t_j(x)$ in $F[x]$.

21. Prove that the only elements of $F[x]$ that have multiplicative inverses are the nonzero elements of the field F. (Hence, $F[x]$ is *not* a field.)

8.3 FACTORIZATION IN $F[x]$

Let $f(x) = a_0 + a_1x + a_2x^2 + \cdots + a_nx^n$ denote an arbitrary polynomial over the field F. For any $c \in F, f(c)$ is defined by the equation

$$f(c) = a_0 + a_1c + a_2c^2 + \cdots + a_nc^n.$$

That is, $f(c)$ is obtained by replacing the indeterminate x in $f(x)$ by the element c. For each $c \in F$, this replacement rule yields a unique value $f(c) \in F$, and hence the pairing $(c, f(c))$ defines a mapping from F to F. A mapping obtained in this manner is called a **polynomial mapping**, or a **polynomial function**, from F to F.

Definition 8.14   **ZERO, ROOT, SOLUTION**

> Let $f(x)$ be a polynomial over the field F. If c is an element of F such that $f(c) = 0$, then c is called a **zero** of $f(x)$, and we say that c is a **root**, or a **solution**, of the equation $f(x) = 0$.

Example 1   Consider $f(x) = x^2 + 1$ in $\mathbf{Z}_5[x]$. Since

$$f(2) = 2^2 + 1 = 0$$

in $\mathbf{Z}_5$, 2 is a *zero* of $x^2 + 1$. Also, 2 is a *root*, or a *solution*, of $x^2 + 1 = 0$ over $\mathbf{Z}_5$.   ■

For arbitrary polynomials $f(x)$ and $g(x)$ over a field F, let $h(x) = f(x) + g(x)$ and $p(x) = f(x)g(x)$. Two consequences of the definitions of addition and multiplication in $F[x]$ are that

$$h(c) = f(c) + g(c) \quad \text{and} \quad p(c) = f(c)g(c)$$

for all c in F. We shall use these results quite freely, with their justifications left as exercises.

The following example is of some interest in connection with the discussion in the preceding paragraph.

Example 2   Consider the polynomials $f(x) = 3x^5 - 4x^2$ and $g(x) = x^2 + 3x$ in $\mathbf{Z}_5[x]$. By direct computation, we find that

$$f(0) = 0 = g(0) \qquad f(1) = 4 = g(1) \qquad f(2) = 0 = g(2)$$
$$f(3) = 3 = g(3) \qquad f(4) = 3 = g(4).$$

Thus, $f(c) = g(c)$ for all c in $\mathbf{Z}_5$, but $f(x) \neq g(x)$ in $\mathbf{Z}_5[x]$.   ■

The next two theorems are two of the simplest and most useful results on factorization in $F[x]$.

Theorem 8.15   **THE REMAINDER THEOREM**

If $f(x)$ is a polynomial over the field F and $c \in F$, the remainder in the division of $f(x)$ by $x - c$ is $f(c)$.

$(u \wedge v) \Rightarrow w$ **Proof**   Since $x - c$ has degree 1, the remainder r in

$$f(x) = (x - c)q(x) + r$$

is a constant. Replacing x with c, we obtain

$$f(c) = (c - c)q(c) + r$$
$$= 0 \cdot q(c) + r$$
$$= r.$$

Thus, $r = f(c)$.

Theorem 8.16   **THE FACTOR THEOREM**

A polynomial $f(x)$ over the field F has a factor $x - c$ in $F[x]$ if and only if $c \in F$ is a zero of $f(x)$.

$p \Leftrightarrow q$ **Proof**   From the Remainder Theorem, we have

$$f(x) = (x - c)q(x) + f(c).$$

Thus, $x - c$ is a factor of $f(x)$ if and only if $f(c) = 0$.

The Factor Theorem can be extended as follows.

Theorem 8.17   FACTORIZATION OF $f(x)$ WITH DISTINCT ZEROS

Let $f(x)$ be a polynomial over the field F that has positive degree n and leading coefficient a. If $c_1, c_2, \ldots, c_n$ are n distinct zeros of $f(x)$ in F, then

$$f(x) = a(x - c_1)(x - c_2) \cdots (x - c_n).$$

Induction   **Proof**   The proof is by induction on $n = \deg f(x)$. For each positive integer n, let S_n be the statement of the theorem.

For $n = 1$, suppose that $f(x)$ has degree 1 and leading coefficient a, and let c_1 be a zero of $f(x)$ in F. Then $f(x) = ax + b$, where $a \neq 0$ and $f(c_1) = 0$. This implies that $ac_1 + b = 0$ and $b = -ac_1$. Therefore, $f(x) = ax - ac_1 = a(x - c_1)$, and S_1 is true.

Assume now that S_k is true, and let $f(x)$ be a polynomial with leading coefficient a and degree $k + 1$ that has $k + 1$ distinct zeros $c_1, c_2, \ldots, c_k, c_{k+1}$ in F. Since c_{k+1} is a zero of $f(x)$,

$$f(x) = (x - c_{k+1})q(x)$$

by the Factor Theorem. By Theorem 8.7, $q(x)$ must have degree k. Since the factor $x - c_{k+1}$ is monic, $q(x)$ and $f(x)$ have the same leading coefficient. For $i = 1, 2, \ldots, k$, we have

$$(c_i - c_{k+1})q(c_i) = f(c_i) = 0,$$

where $c_i - c_{k+1} \neq 0$, since the zeros $c_1, c_2, \ldots, c_k, c_{k+1}$ are distinct. Therefore, $q(c_i) = 0$ for $i = 1, 2, \ldots, k$. That is, $c_1, c_2, \ldots, c_k$ are k distinct zeros of $q(x)$ in F. By the induction hypothesis,

$$q(x) = a(x - c_1)(x - c_2) \cdots (x - c_k).$$

Substitution of this factored expression for $q(x)$ in $f(x) = (x - c_{k+1})q(x)$ yields

$$f(x) = a(x - c_1)(x - c_2) \cdots (x - c_k)(x - c_{k+1}).$$

Therefore, S_{k+1} is true whenever S_k is true, and it follows by induction that S_n is true for all positive integers n.

The proof of the following corollary is left as an exercise.

Corollary 8.18   NUMBER OF DISTINCT ZEROS

A polynomial of positive degree n over the field F has at most n distinct zeros in F.

In the factorization of polynomials over a field F, the concept of an irreducible polynomial is analogous to the concept of a prime integer in the factorization of integers.

Definition 8.19   IRREDUCIBLE, PRIME, REDUCIBLE

A polynomial $f(x)$ in $F[x]$ is **irreducible** (or **prime**) over F if $f(x)$ has positive degree and $f(x)$ *cannot* be expressed as a product $f(x) = g(x)h(x)$ with both $g(x)$ and $h(x)$ of positive degree in $F[x]$. If $f(x)$ is not irreducible, then $f(x)$ is said to be **reducible**.

Example 3   Note that whether or not a given polynomial is irreducible over F depends on the field F. For instance, $x^2 + 1$ is irreducible over the field of real numbers, but it is reducible over the field $\mathbf{C}$ of complex numbers, since $x^2 + 1$ can be factored as

$$x^2 + 1 = (x - i)(x + i)$$

in $\mathbf{C}[x]$.   ∎

If $g(x)$ and $h(x)$ are polynomials of positive degree, their product $g(x)h(x)$ has degree at least 2. Therefore, all polynomials of degree 1 are irreducible. Constant polynomials, however, are never irreducible because they do not have positive degree.

It is usually not easy to decide whether or not a given polynomial is irreducible over a certain field. However, the following theorem is sometimes quite helpful for polynomials with degree less than 4.

Theorem 8.20   **POLYNOMIALS OF DEGREE 2 OR 3**

$p \Leftrightarrow q$ If $f(x)$ is a polynomial of degree 2 or 3 over the field F, then $f(x)$ is irreducible over F if and only if $f(x)$ has no zeros in F.

Proof   Let $f(x)$ be a polynomial of degree 2 or 3 over the field F.

$\sim p \Leftrightarrow \sim q$ We shall prove the theorem in this form: $f(x)$ is reducible over F if and only if $f(x)$ has at least one zero in F.

$\sim p \Leftarrow \sim q$ Suppose first that $f(x)$ has a zero c in F. By the Factor Theorem,

$$f(x) = (x - c)q(x),$$

where $q(x)$ has degree 1 less than that of $f(x)$ by Theorem 8.7. This factorization shows that $f(x)$ is reducible over F.

$\sim p \Rightarrow \sim q$ Assume, conversely, that $f(x)$ is reducible over F. That is, there are polynomials $g(x)$ and $h(x)$ in $F[x]$ such that $f(x) = g(x)h(x)$, with both $g(x)$ and $h(x)$ of positive degree. By Theorem 8.7,

$$\deg f(x) = \deg g(x) + \deg h(x).$$

Since $\deg f(x)$ is either 2 or 3, one of the factors $g(x)$ and $h(x)$ must have degree 1. Without loss of generality, we may assume that this factor is $g(x)$, and we have

$$f(x) = (ax + b)h(x),$$

where $a \neq 0$. It follows at once from this equation that $-a^{-1}b$ is a zero of $f(x)$ in F, and the proof is complete.

Example 4   Let us determine whether each of the following polynomials is irreducible over $\mathbf{Z}_5$.

a. $f(x) = x^3 + 2x^2 - 3x + 4$
b. $g(x) = x^2 + 3x + 4$

Routine computations show that

$$f(0) = 4, \qquad f(1) = 4, \qquad f(2) = 4, \qquad f(3) = 0, \qquad f(4) = 3.$$

Thus, 3 is a zero of $f(x)$ in $\mathbf{Z}_5$, and $f(x)$ is reducible over $\mathbf{Z}_5$. However, $g(x)$ is irreducible over $\mathbf{Z}_5$ since $g(x)$ has no zeros in $\mathbf{Z}_5$:

$$g(0) = 4, \qquad g(1) = 3, \qquad g(2) = 4, \qquad g(3) = 2, \qquad g(4) = 2. \qquad \blacksquare$$

Irreducible polynomials play a role in the factorization of polynomials corresponding to the role prime integers play in the factorization of integers. This is illustrated by the next theorem.

Theorem 8.21   IRREDUCIBLE FACTORS
_____

If $p(x)$ is an irreducible polynomial over the field F and $p(x)$ divides $f(x)g(x)$ in $F[x]$, then either $p(x)|f(x)$ or $p(x)|g(x)$ in $F[x]$.

$(u \wedge v) \Rightarrow (w \vee z)$ **Proof**   Assume that $p(x)$ is irreducible over F and that $p(x)$ divides $f(x)g(x)$; say,

$$f(x)g(x) = p(x)q(x)$$

for some $q(x)$ in $F[x]$. If $p(x)|f(x)$, the conclusion is satisfied. Suppose, then, that $p(x)$ does not divide $f(x)$. This means that 1 is the greatest common divisor of $f(x)$ and $p(x)$, since the only divisors of $p(x)$ with positive degree are constant multiples of $p(x)$. By Theorem 8.13, there exist $s(x)$ and $t(x)$ in $F[x]$ such that

$$1 = f(x)s(x) + p(x)t(x),$$

and this implies that

$$\begin{aligned}
g(x) &= g(x)[f(x)s(x) + p(x)t(x)] \\
&= f(x)g(x)s(x) + p(x)g(x)t(x) \\
&= p(x)q(x)s(x) + p(x)g(x)t(x),
\end{aligned}$$

since $f(x)g(x) = p(x)q(x)$. Factoring $p(x)$ from the two terms in the right member, we see that $p(x)|g(x)$:

$$g(x) = p(x)[q(x)s(x) + g(x)t(x)].$$

Thus, $p(x)$ divides $g(x)$ if it does not divide $f(x)$.
_____

A comparison of Theorem 8.21 with Theorem 2.16 provides an indication of how closely the theory of divisibility in $F[x]$ resembles the theory of divisibility in the integers. This analogy carries over to the proofs as well. For this reason, the proofs of the remaining results in this section are left as exercises.

Theorem 8.22
_____

If $p(x)$ is an irreducible polynomial over the field F and $p(x)$ divides a product $f_1(x)f_2(x) \cdots f_n(x)$ in $F[x]$, then $p(x)$ divides some $f_j(x)$.
_____

Just as with integers, two nonzero polynomials $f(x)$ and $g(x)$ over the field F are called **relatively prime** over F if their greatest common divisor in $F[x]$ is 1.

Theorem 8.23

If $f(x)$ and $g(x)$ are relatively prime polynomials over the field F and if $f(x)|g(x)h(x)$ in $F[x]$, then $f(x)|h(x)$ in $F[x]$.

Theorem 8.24   **UNIQUE FACTORIZATION THEOREM**

Every polynomial of positive degree over the field F can be expressed as a product of its leading coefficient and a finite number of monic irreducible polynomials over F. This factorization is unique except for the order of the factors.

Of course, the monic irreducible polynomials involved in the factorization of $f(x)$ over F may not all be distinct. If $p_1(x), p_2(x), \ldots, p_r(x)$ are the *distinct* monic irreducible factors of $f(x)$, then all repeated factors may be collected together and expressed by use of exponents to yield

$$f(x) = a[p_1(x)]^{m_1}[p_2(x)]^{m_2} \cdots [p_r(x)]^{m_r},$$

where each m_i is a positive integer.

In the last expression for $f(x)$, m_i is called the **multiplicity** of the factor $p_i(x)$. More generally, if $g(x)$ is an arbitrary polynomial of positive degree such that $[g(x)]^m$ divides $f(x)$ and no higher power of $g(x)$ divides $f(x)$ in $F[x]$, then $g(x)$ is said to be a factor of $f(x)$ over $F[x]$ with **multiplicity m**. Also, if c is an element of the field F such that $(x - c)^m$ divides $f(x)$ for some positive integer m but no higher power of $x - c$ divides $f(x)$, then c is called a **zero of multiplicity m**.

Example 5   We shall find the factorization that is described in the Unique Factorization Theorem for the polynomial

$$f(x) = 2x^4 + x^3 + 3x^2 + 2x + 4$$

over the field $\mathbf{Z}_5$.

We first determine the zeros of $f(x)$ in $\mathbf{Z}_5$:

$$f(0) = 4, \quad f(1) = 2, \quad f(2) = 0, \quad f(3) = 1, \quad f(4) = 1.$$

Thus, 2 is the only zero of $f(x)$ in $\mathbf{Z}_5$, and the Factor Theorem assures us that $x - 2$ is a factor of $f(x)$. Dividing by $x - 2$, we get

$$f(x) = (x - 2)(2x^3 + 3x + 3).$$

By Exercise 10 at the end of this section, the zeros of $f(x)$ are 2 and the zeros of $g(x) = 2x^3 + 3x + 3$. We therefore need to determine the zeros of $g(x)$, and the only possibility is 2, since this is the only zero of $f(x)$ in $\mathbf{Z}_5$. We find that $g(2) = 0$, and this indicates that $x - 2$ is a factor of $g(x)$. Performing the required division, we obtain

$$2x^3 + 3x + 3 = (x - 2)(2x^2 + 4x + 1)$$

and

$$\begin{aligned} f(x) &= (x - 2)(x - 2)(2x^2 + 4x + 1) \\ &= (x - 2)^2(2x^2 + 4x + 1). \end{aligned}$$

We now find that $2x^2 + 4x + 1$ is irreducible over $\mathbf{Z}_5$, since it has no zeros in $\mathbf{Z}_5$. To arrive at the desired factorization, we only need to factor the leading coefficient of $f(x)$ from the factor $2x^2 + 4x + 1$:

$$
\begin{aligned}
f(x) &= (x - 2)^2(2x^2 + 4x + 1) \\
&= (x - 2)^2[2x^2 + 4x + (2)(3)] \\
&= 2(x - 2)^2(x^2 + 2x + 3).
\end{aligned}
$$
∎

EXERCISES 8.3

1. Let $\mathbf{Q}$ denote the field of rational numbers, $\mathbf{R}$ the field of real numbers, and $\mathbf{C}$ the field of complex numbers. Determine whether each of the following polynomials is irreducible over each of the indicated fields.
   a. $x^2 - 2$ over $\mathbf{Q}, \mathbf{R}$, and $\mathbf{C}$
   b. $x^2 + 1$ over $\mathbf{Q}, \mathbf{R}$, and $\mathbf{C}$
   c. $x^2 + x - 2$ over $\mathbf{Q}, \mathbf{R}$, and $\mathbf{C}$
   d. $x^2 + 2x + 2$ over $\mathbf{Q}, \mathbf{R}$, and $\mathbf{C}$
   e. $x^2 + x + 2$ over $\mathbf{Z}_3, \mathbf{Z}_5$, and $\mathbf{Z}_7$
   f. $x^2 + 2x + 2$ over $\mathbf{Z}_3, \mathbf{Z}_5$, and $\mathbf{Z}_7$
   g. $x^3 - x^2 + 2x + 2$ over $\mathbf{Z}_3, \mathbf{Z}_5$, and $\mathbf{Z}_7$
   h. $x^4 + 2x^2 + 1$ over $\mathbf{Z}_3, \mathbf{Z}_5$, and $\mathbf{Z}_7$

2. Find all monic irreducible polynomials of degree 2 over $\mathbf{Z}_3$.

3. Write each of the following polynomials as a product of its leading coefficient and a finite number of monic irreducible polynomials over $\mathbf{Z}_5$.
   a. $2x^3 + 1$ b. $3x^3 + 2x^2 + x + 2$
   c. $3x^3 + x^2 + 2x + 4$ d. $2x^3 + 4x^2 + 3x + 1$
   e. $2x^4 + x^3 + 3x + 2$ f. $3x^4 + 3x^3 + x + 3$
   g. $x^4 + x^3 + x^2 + 2x + 3$ h. $x^4 + x^3 + 2x^2 + 3x + 2$

4. Prove Corollary 8.18: A polynomial of positive degree n over the field F has at most n distinct zeros in F.

5. Let $f(x)$ and $g(x)$ be two polynomials over the field F, both of degree n or less. Prove that if $m > n$ and if there exist m distinct elements $c_1, c_2, \ldots, c_m$ of F such that $f(c_i) = g(c_i)$ for $i = 1, 2, \ldots, m$, then $f(x) = g(x)$.

6. Let p be a prime integer and consider the polynomials $f(x) = x^p$ and $g(x) = x$ over the field $\mathbf{Z}_p$. Prove that $f(c) = g(c)$ for all c in $\mathbf{Z}_p$. (This result is known as **Fermat's Little Theorem**: $n^p \equiv n \pmod{p}$. To prove it, consider the multiplicative group of nonzero elements of $\mathbf{Z}_p$.)

7. Give an example of a polynomial of degree 4 over the field $\mathbf{R}$ of real numbers that is reducible over $\mathbf{R}$ and yet has no zeros in the real numbers.

8. If $f(x)$ and $g(x)$ are polynomials over the field F, and $h(x) = f(x) + g(x)$, prove that $h(c) = f(c) + g(c)$ for all c in F.

9. If $f(x)$ and $g(x)$ are polynomials over the field F, and $p(x) = f(x)g(x)$, prove that $p(c) = f(c)g(c)$ for all c in F.

10. Let $f(x)$ be a polynomial of positive degree n over the field F, and assume that $f(x) = (x - c)q(x)$ for some $c \in F$ and $q(x)$ in $F[x]$. Prove that
    a. c and the zeros of $q(x)$ in F are zeros of $f(x)$
    b. $f(x)$ has no other zeros in F.

11. Suppose that $f(x), g(x)$, and $h(x)$ are polynomials over the field F, each of which has positive degree, and that $f(x) = g(x)h(x)$. Prove that the zeros of $f(x)$ in F consist of the zeros of $g(x)$ in F together with the zeros of $h(x)$ in F.

12. Prove that a polynomial $f(x)$ of positive degree n over the field F has at most n (not necessarily distinct) zeros in F.

13. Prove Theorem 8.22: If $p(x)$ is an irreducible polynomial over the field F and $p(x)$ divides a product $f_1(x)f_2(x) \cdots f_n(x)$ in $F[x]$, then $p(x)$ divides some $f_j(x)$.

14. Prove Theorem 8.23: If $f(x)$ and $g(x)$ are relatively prime polynomials over the field F and if $f(x)|g(x)h(x)$ in $F[x]$, then $f(x)|h(x)$ in $F[x]$.

15. Prove the Unique Factorization Theorem in $F[x]$ (Theorem 8.24).

8.4 ZEROS OF A POLYNOMIAL

We now focus our interest on polynomials that have their coefficients in either the field $\mathbf{C}$ of complex numbers, the field $\mathbf{R}$ of real numbers, or the field $\mathbf{Q}$ of rational numbers. Our results are concerned with the zeros of these polynomials and the related property of irreducibility over these fields.

The statement in Theorem 8.25 is so important that it is known as the Fundamental Theorem of Algebra. It was first proved in 1799 by the great German mathematician Carl Friedrich Gauss (1777–1855). Unfortunately, all known proofs of this theorem require theories that we do not have at our disposal, so we are forced to accept the theorem without proof.

Theorem 8.25 THE FUNDAMENTAL THEOREM OF ALGEBRA

If $f(x)$ is a polynomial of positive degree over the field of complex numbers, then $f(x)$ has a zero in the complex numbers.

The Fundamental Theorem opens the door to a complete decomposition of any polynomial over $\mathbf{C}$, as described in the following theorem.

Theorem 8.26 FACTORIZATION OVER $\mathbf{C}$

If $f(x)$ is a polynomial of positive degree n over the field $\mathbf{C}$ of complex numbers, then $f(x)$ can be factored as

$$f(x) = a(x - c_1)(x - c_2) \cdots (x - c_n),$$

where a is the leading coefficient of $f(x)$ and $c_1, c_2, \ldots, c_n$ are n (not necessarily distinct) complex numbers that are zeros of $f(x)$.

Induction **Proof** For each positive integer n, let S_n be the statement of the theorem.

If $n = 1$, then $f(x) = ax + b$, where $a \neq 0$. The complex number $c_1 = -a^{-1}b$ is a zero of $f(x)$, and

$$f(x) = ax + b = ax - ac_1 = a(x - c_1).$$

Thus, S_1 is true.

Assume that S_k is true, and let $f(x)$ be a polynomial of degree $k + 1$ over $\mathbf{C}$. By the Fundamental Theorem of Algebra, $f(x)$ has a zero c_1 in the complex numbers, and the Factor Theorem asserts that

$$f(x) = (x - c_1)q(x)$$

for some polynomial $q(x)$ over $\mathbf{C}$. Since $x - c_1$ is monic, $q(x)$ has the same leading coefficient as $f(x)$, and Theorem 8.7 implies that $q(x)$ has degree k. By the induction hypothesis, $q(x)$ can be factored as the product of its leading coefficient and k factors of the form $x - c_i$:

$$q(x) = a(x - c_2)(x - c_3) \cdots (x - c_{k+1}).$$

Therefore,

$$\begin{aligned} f(x) &= (x - c_1)q(x) \\ &= a(x - c_1)(x - c_2) \cdots (x - c_{k+1}), \end{aligned}$$

and S_{k+1} is true. It follows that the theorem is true for all positive integers n.

As noted in the statement of Theorem 8.26, the zeros c_i are not necessarily distinct in the factorization of $f(x)$ that is described there. If the repeated factors are collected together, we have

$$f(x) = a(x - c_1)^{m_1}(x - c_2)^{m_2} \cdots (x - c_r)^{m_r}$$

as a standard form for the unique factorization of a polynomial over the complex numbers. In particular, we observe that *the only irreducible polynomials over* $\mathbf{C}$ *are the first-degree polynomials.*

With such a simple description of the irreducible polynomials over $\mathbf{C}$, it is natural to ask which polynomials are irreducible over the real numbers. For polynomials of degree 2 (quadratic polynomials), an answer to this question is readily available from the *quadratic formula.* According to the **quadratic formula**, the zeros of a polynomial

$$f(x) = ax^2 + bx + c$$

with real coefficients[†] and $a \neq 0$ are given by

$$r_1 = \frac{-b + \sqrt{b^2 - 4ac}}{2a} \quad \text{and} \quad r_2 = \frac{-b - \sqrt{b^2 - 4ac}}{2a}.$$

These zeros are not real numbers if and only if the *discriminant,* $b^2 - 4ac$, is negative. Thus, a quadratic polynomial is irreducible over the real numbers if and only if it has a negative discriminant.

If we introduce some appropriate terminology, a meaningful characterization of the field of complex numbers can now be formulated. If F and E are fields such that $F \subseteq E$, then E is called an **extension** of F. An element $a \in E$ is called **algebraic** over F

[†] The quadratic formula is also valid if the coefficients are complex numbers, but at the moment we are interested only the the real case.

if a is the zero of a polynomial $f(x)$ with coefficients in F, and E is an **algebraic extension** of F if every element of E is algebraic over F. E is **algebraically closed** if every polynomial over E has a zero in E.

The field **C** of complex numbers can be characterized as a field with the following properties:

1. **C** is an algebraic extension of the field **R** of real numbers.
2. **C** is algebraically closed.

If $z = a + bi$ with $a, b \in \mathbf{R}$, then z is a zero of the polynomial

$$f(x) = [x - (a + bi)][x - (a - bi)]$$
$$= x^2 - 2ax + (a^2 + b^2)$$

over **R**. Thus, z is algebraic over **R**, and property 1 is established. The Fundamental Theorem of Algebra (Theorem 8.25) asserts that **C** is algebraically closed. It can be proved that any field that is an algebraic extension of **R** and is algebraically closed must be isomorphic to **C**. The proof of this assertion is beyond the scope of this text.

If a and b are real numbers, the *conjugate* of the complex number $z = a + bi$ is the complex number $\bar{z} = a - bi$. Note that the zeros r_1 and r_2 given by the quadratic formula are conjugates of each other when the coefficients are real and $b^2 - 4ac < 0$.

In the exercises at the end of this section, proofs are requested for the following facts concerning conjugates:

$$\overline{z_1 + z_2 + \cdots + z_n} = \bar{z}_1 + \bar{z}_2 + \cdots + \bar{z}_n$$
$$\overline{z_1 \cdot z_2 \cdot \cdots \cdot z_n} = \bar{z}_1 \cdot \bar{z}_2 \cdot \cdots \cdot \bar{z}_n.$$

That is, the conjugate of a sum of terms is the sum of the conjugates of the individual terms, and the conjugate of a product of factors is the product of the conjugates of the individual factors. As a special case for products,

$$\overline{(z^n)} = (\bar{z})^n.$$

These properties of conjugates are used in the proof of the next theorem.

Theorem 8.27   CONJUGATE ZEROS

Suppose that $f(x)$ is a polynomial that has all its coefficients in the real numbers. If the complex number z is a zero of $f(x)$, then its conjugate $\bar{z}$ is also a zero of $f(x)$.

$p \Rightarrow q$ **Proof**   Let $f(x) = \sum_{i=0}^{n} a_i x^i$, where all a_i are real, and assume that z is a zero of $f(x)$. Then $f(z) = 0$, and therefore,

$$\begin{aligned}
0 &= \overline{f(z)} \\
&= \overline{a_0 + a_1 z + a_2 z^2 + \cdots + a_n z^n} \\
&= \bar{a}_0 + \overline{a_1 z} + \overline{a_2 z^2} + \cdots + \overline{a_n z^n} \\
&= \bar{a}_0 + \bar{a}_1 \bar{z} + \bar{a}_2 (\bar{z})^2 + \cdots + \bar{a}_n (\bar{z})^n \\
&= a_0 + a_1 \bar{z} + a_2 (\bar{z})^2 + \cdots + a_n (\bar{z})^n,
\end{aligned}$$

where the last equality follows from the fact that each a_i is a real number. We thus have $f(\bar{z}) = 0$, and the theorem is proved.

Example 1   The monic polynomial of least degree over the complex numbers that has $1 - i$ and $2i$ as zeros is

$$f(x) = [x - (1 - i)][x - 2i]$$
$$= x^2 - (1 + i)x + 2 + 2i.$$

However, a polynomial *with real coefficients* that has $1 - i$ and $2i$ as zeros must also have $1 + i$ and $-2i$ as zeros. Thus, the monic polynomial of least degree with real coefficients that has $1 - i$ and $2i$ as zeros is

$$g(x) = [x - (1 - i)][x - (1 + i)][x - 2i][x + 2i]$$
$$= (x^2 - 2x + 2)(x^2 + 4)$$
$$= x^4 - 2x^3 + 6x^2 - 8x + 8. \qquad \blacksquare$$

Example 2   Suppose that it is known that $1 - 2i$ is a zero of the fourth-degree polynomial $f(x) = x^4 - 3x^3 + x^2 + 7x - 30$ and that we wish to find all the zeros of $f(x)$. From Theorem 8.27 we know that $1 + 2i$ is also a zero of $f(x)$. The Factor Theorem then assures us that $x - (1 - 2i)$ and $x - (1 + 2i)$ are factors of $f(x)$:

$$f(x) = [x - (1 - 2i)][x - (1 + 2i)]q(x).$$

To find $q(x)$, we divide $f(x)$ by the polynomial

$$[x - (1 - 2i)][x - (1 + 2i)] = x^2 - 2x + 5$$

and obtain $q(x) = x^2 - x - 6$. Thus,

$$f(x) = [x - (1 - 2i)][x - (1 + 2i)](x^2 - x - 6)$$
$$= [x - (1 - 2i)][x - (1 + 2i)](x - 3)(x + 2).$$

It is now evident that the zeros of $f(x)$ are $1 - 2i, 1 + 2i, 3$, and -2. $\qquad \blacksquare$

The results obtained thus far prepare for the next theorem, which describes a standard form for the unique factorization of a polynomial over the real numbers. The proof of this theorem is left as an exercise.

Theorem 8.28   **FACTORIZATION OVER R**

Every polynomial of positive degree over the field **R** of real numbers can be factored as the product of its leading coefficient and a finite number of monic irreducible polynomials over **R**, each of which is either quadratic or of first degree.

We restrict our attention now to the rational zeros of polynomials with rational coefficients, and to the irreducibility of such polynomials. Neither the zeros of a polynomial nor its irreducibility are changed when it is multiplied by a nonzero constant, so we lose no generality be restricting our attention to polynomials with coefficients that are all integers.

Theorem 8.29 **RATIONAL ZEROS**

Let

$$f(x) = a_0 + a_1 x + \cdots + a_{n-1} x^{n-1} + a_n x^n$$

be a polynomial of positive degree n with coefficients that are all integers, and let p/q be a rational number that has been written in lowest terms. If p/q is a zero of $f(x)$, then p divides a_0 and q divides a_n.

$u \Rightarrow (v \wedge w)$

$u \Rightarrow w$

Proof Suppose that p/q is a rational number in lowest terms that is a zero of $f(x) = \sum_{i=0}^{n} a_i x^i$. Then

$$a_0 + a_1 \left(\frac{p}{q} \right) + \cdots + a_{n-1} \left(\frac{p}{q} \right)^{n-1} + a_n \left(\frac{p}{q} \right)^n = 0.$$

Multiplying both sides of this equality by q^n gives

$$a_0 q^n + a_1 p q^{n-1} + \cdots + a_{n-1} p^{n-1} q + a_n p^n = 0.$$

Subtracting $a_n p^n$ from both sides, we have

$$a_0 q^n + a_1 p q^{n-1} + \cdots + a_{n-1} p^{n-1} q = -a_n p^n,$$

and hence

$$q(a_0 q^{n-1} + a_1 p q^{n-2} + \cdots + a_{n-1} p^{n-1}) = -a_n p^n.$$

This shows that q divides $a_n p^n$, and therefore $q | a_n$, since q and p are relatively prime.

$u \Rightarrow v$

Similarly, the equation

$$a_1 p q^{n-1} + \cdots + a_{n-1} p^{n-1} q + a_n p^n = -a_0 q^n$$

can be used to show that $p | a_0$.

It is important to note that Theorem 8.29 only restricts the possibilities of the rational zeros. It does not guarantee that any of these possibilities is actually a zero of $f(x)$.

It may happen that when some of the rational zeros of a polynomial have been found, the remaining zeros may be obtained by use of the quadratic formula. This is illustrated in the next example.

Example 3 We shall obtain all zeros of the polynomial

$$f(x) = 2x^4 - 5x^3 + 3x^2 + 4x - 6$$

by first finding the rational zeros of $f(x)$. According to Theorem 8.29, any rational zero p/q of $f(x)$ that is in lowest terms must have a numerator p that divides the constant term and a denominator q that divides the leading coefficient. This means that

$$p \in \{\pm 1, \pm 2, \pm 3, \pm 6\}$$
$$q \in \{\pm 1, \pm 2\}$$
$$\frac{p}{q} \in \left\{ \pm \frac{1}{2}, \pm 1, \pm \frac{3}{2}, \pm 2, \pm 3, \pm 6 \right\}.$$

Testing the positive possibilities systematically, we get

$$f\left(\frac{1}{2}\right) = -\frac{15}{4}, \ f(1) = -2, \ f\left(\frac{3}{2}\right) = 0.$$

We could continue to test the remaining possibilities, but chances are that it is worthwhile to divide $f(x)$ by $x - (3/2)$ and then work with the quotient. Performing the division, we obtain

$$\begin{aligned} f(x) &= (x - \tfrac{3}{2})(2x^3 - 2x^2 + 4) \\ &= (2x - 3)(x^3 - x^2 + 2). \end{aligned}$$

From this factorization, we see that the other zeros of $f(x)$ are the zeros of the factor $q(x) = x^3 - x^2 + 2$. Since this factor is monic, the only possible rational zeros are the divisors of 2. We already know that 1 is not a zero, since $f(1) = -2$. Thus the remaining possibilities are $2, -1$, and -2. We find that

$$q(2) = 6, \ q(-1) = 0.$$

Therefore, $x + 1$ is a factor of $x^3 - x^2 + 2$. Division by $x + 1$ yields

$$x^3 - x^2 + 2 = (x + 1)(x^2 - 2x + 2)$$

and

$$f(x) = (2x - 3)(x + 1)(x^2 - 2x + 2).$$

The remaining zeros of $f(x)$ can be found by using the quadratic formula on the factor $x^2 - 2x + 2$:

$$x = \frac{2 \pm \sqrt{4 - 8}}{2} = 1 \pm i.$$

Thus, the zeros of $f(x)$ are $3/2, -1, 1 + i$, and $1 - i$. ∎

The results concerning irreducibility over the field $\mathbf{Q}$ of rational numbers are not nearly as neat or complete as those we have obtained for the fields $\mathbf{C}$ and $\mathbf{R}$. The best known result for $\mathbf{Q}$ is a theorem that states what is known as *Eisenstein's Irreducibility Criterion*. To establish this result is the goal of the rest of this section. We need the following definition and two intermediate theorems to reach our objective.

Definition 8.30   **PRIMITIVE POLYNOMIAL**

> Let $f(x) = \sum_{i=0}^{n} a_i x^i$ be a polynomial in which all coefficients are integers. Then $f(x)$ is a **primitive** polynomial if the greatest common divisor of $a_0, a_1, \ldots, a_n$ is 1.

That is, a polynomial is primitive if and only if there is no prime integer that divides all of its coefficients.

Our first intermediate result simply asserts that the product of two primitive polynomials is primitive.

Theorem 8.31   **PRODUCT OF PRIMITIVE POLYNOMIALS**

$(p \wedge q) \Rightarrow r$

If $g(x)$ and $h(x)$ are primitive polynomials, then $g(x)h(x)$ is a primitive polynomial.

$(p \wedge q \wedge \sim r) \Rightarrow$
$(\sim p \vee \sim q)$

Proof   We shall assume that the theorem is false and arrive at a contradiction. Suppose that $g(x)$ and $h(x)$ are primitive polynomials, but the product $f(x) = g(x)h(x)$ is not primitive. Then there is a prime integer p that divides every coefficient of $f(x) = \sum_{i=0}^{n} a_i x^i$. The mapping $\phi: \mathbf{Z}[x] \to \mathbf{Z}_p[x]$ defined by

$$\phi(a_0 + a_1 x + \cdots + a_n x^n) = [a_0] + [a_1]x + \cdots + [a_n]x^n$$

is an epimorphism from $\mathbf{Z}[x]$ to $\mathbf{Z}_p[x]$, by Exercise 13 of Section 8.1. Since every coefficient of $f(x)$ is a multiple of p, $\phi(f(x)) = [0]$ in $\mathbf{Z}_p[x]$. Therefore,

$$\begin{aligned}
\phi(g(x)) \cdot \phi(h(x)) &= \phi(g(x)h(x)) \\
&= \phi(f(x)) \\
&= [0]
\end{aligned}$$

in $\mathbf{Z}_p[x]$. Since p is a prime, $\mathbf{Z}_p[x]$ is an integral domain, and either $\phi(g(x)) = [0]$ or $\phi(h(x)) = [0]$. Consequently, either p divides every coefficient of $g(x)$, or p divides every coefficient of $h(x)$. In either case, we have a contradiction to the supposition that $g(x)$ and $h(x)$ are primitive polynomials. This contradiction establishes the theorem.

The following theorem is credited to the same mathematician who first proved the Fundamental Theorem of Algebra.

Theorem 8.32   **GAUSS'S LEMMA**

Let $f(x)$ be a primitive polynomial. If $f(x)$ can be factored as $f(x) = g(x)h(x)$, where $g(x)$ and $h(x)$ have rational coefficients and positive degree, then $f(x)$ can be factored as $f(x) = G(x)H(x)$, where $G(x)$ and $H(x)$ have integral coefficients and positive degree.

$p \Rightarrow q$

Proof   Suppose that $f(x) = g(x)h(x)$ as described in the hypothesis. Let b be the least common denominator of the coefficients of $g(x)$, so that $g(x)$ can be expressed as $g(x) = \frac{1}{b} g_1(x)$, where $g_1(x)$ has integral coefficients. Now let a be the greatest common divisor of the coefficients of $g_1(x)$, so that $g_1(x) = aG(x)$, where $G(x)$ is a primitive polynomial. Then we have $g(x) = \frac{a}{b} G(x)$, where a and b are integers and $G(x)$ is primitive and of the same degree as $g(x)$. Similarly, we may write $h(x) = \frac{c}{d} H(x)$, where c and d are integers and $H(x)$ is primitive and of the same degree as $h(x)$. Substituting these expressions for $g(x)$ and $h(x)$, we obtain

$$f(x) = \frac{a}{b} G(x) \cdot \frac{c}{d} H(x),$$

and therefore

$$bdf(x) = acG(x)H(x).$$

Since $f(x)$ is primitive, the greatest common divisor of the coefficients of the left member of this equation is bd. By Theorem 8.31, $G(x)H(x)$ is primitive, and therefore the greatest common divisor of the coefficients of the right member is ac. Hence, $bd = ac$, and this implies that $f(x) = G(x)H(x)$, where $G(x)$ and $H(x)$ have integral coefficients and positive degrees.

We are now in a position to prove Eisenstein's result.

Theorem 8.33   **EISENSTEIN'S IRREDUCIBILITY CRITERION**

Let $f(x) = a_0 + a_1x + \cdots + a_nx^n$ be a polynomial of positive degree with integral coefficients. If there exists a prime integer p such that $p|a_i$ for $i = 0, 1, \ldots, n - 1$ but $p \nmid a_n$ and $p^2 \nmid a_0$, then $f(x)$ is irreducible over the field of rational numbers.

Contradiction   **Proof**   Dividing out the greatest common divisor of the coefficients of a polynomial would have no effect on whether or not the criterion was satisfied by a prime p because of the requirement that $p \nmid a_n$. Therefore, we may restrict our attention to the case where $f(x)$ is a primitive polynomial.

Let $f(x) = \sum_{i=0}^{n} a_i x^i$ be a primitive polynomial, and assume there exists a prime integer p that satisfies the hypothesis. At the same time, assume that the conclusion is false, so that $f(x)$ factors over the rational numbers as a product of two polynomials of positive degree. Then $f(x)$ can be factored as the product of two polynomials of positive degree that have integral coefficients, by Theorem 8.32. Suppose that

$$f(x) = (b_0 + b_1x + \cdots + b_rx^r)(c_0 + c_1x + \cdots + c_sx^s),$$

where all the coefficients are integers and $r > 0$, $s > 0$. Then $a_0 = b_0c_0$, and hence $p|b_0c_0$, but $p^2 \nmid b_0c_0$ by the hypothesis. This implies that either $p|b_0$ or $p|c_0$, but p does not divide both b_0 and c_0. Without loss of generality, we may assume that $p|b_0$ and $p \nmid c_0$. If all of the b_i were divisible by p, then p would divide all the coefficients in the product, $f(x)$. Since $p \nmid a_n$, some of the b_i are not divisible by p. Let k be the smallest subscript such that $p \nmid b_k$, and consider

$$a_k = b_0c_k + b_1c_{k-1} + \cdots + b_{k-1}c_1 + b_kc_0.$$

By the choice of k, p divides each of $b_0, b_1, \ldots, b_{k-1}$, and therefore

$$p|(b_0c_k + b_1c_{k-1} + \cdots + b_{k-1}c_1).$$

Also, $p|a_k$, since $k < n$. Hence, p divides the difference:

$$p|[a_k - (b_0c_k + b_1c_{k-1} + \cdots + b_{k-1}c_1)].$$

That is, $p|b_kc_0$. This is impossible, however, since $p \nmid b_k$ and $p \nmid c_0$. We have arrived at a contradiction, and therefore $f(x)$ is irreducible over the rational numbers.   ∎

Example 4   Consider the polynomial

$$f(x) = 10 - 15x + 25x^2 - 7x^4.$$

The prime integer $p = 5$ divides all of the coefficients in $f(x)$ except the leading coefficient $a_n = -7$, and 5^2 does not divide the constant term $a_0 = 10$. Therefore, $f(x)$ is irreducible over the rational numbers, by Eisenstein's Criterion.   ∎

Exercises 8.4

1. Find a monic polynomial $f(x)$ of least degree over $\mathbf{C}$ that has the given numbers as zeros, and a monic polynomial $g(x)$ of least degree with real coefficients that has the given numbers as zeros.
   a. $2i, 3$
   b. $-3i, 4$
   c. $2, 1 - i$
   d. $3, 2 - i$
   e. $3i, 1 + 2i$
   f. $i, 2 - i$
   g. $2 + i, -i$, and 1
   h. $3 - i, i$, and 2

2. One of the zeros is given for each of the following polynomials. Find the other zeros in the field of complex numbers.
   a. $x^3 - 4x^2 + 6x - 4$; $1 - i$ is a zero.
   b. $x^3 + x^2 - 4x + 6$; $1 - i$ is a zero.
   c. $x^4 + x^3 + 2x^2 + x + 1$; $-i$ is a zero.
   d. $x^4 + 3x^3 + 6x^2 + 12x + 8$; $2i$ is a zero.

Find all rational zeros of each of the polynomials in Exercises 3–6.

3. $2x^3 - x^2 - 8x - 5$
4. $3x^3 + 19x^2 + 30x + 8$
5. $2x^4 - x^3 - x^2 - x - 3$
6. $2x^4 + x^3 - 8x^2 + x - 10$

In Exercises 7–12, find all zeros of the given polynomial.

7. $x^3 + x^2 - x + 2$
8. $3x^3 - 7x^2 + 8x - 2$
9. $3x^3 + 2x^2 - 7x + 2$
10. $3x^3 - 2x^2 - 7x - 2$
11. $6x^3 + 11x^2 + x - 4$
12. $9x^3 + 27x^2 + 8x - 20$

Factor each of the polynomials in Exercises 13–16 as a product of its leading coefficient and a finite number of monic irreducible polynomials over the field of rational numbers.

13. $x^4 - x^3 - 2x^2 + 6x - 4$
14. $2x^4 - x^3 - 13x^2 + 5x + 15$
15. $2x^4 + 5x^3 - 7x^2 - 10x + 6$
16. $6x^4 + x^3 + 3x^2 - 14x - 8$

17. Show that each of the following polynomials is irreducible over the field of rational numbers.
   a. $3 + 9x + x^3$
   b. $7 - 14x + 28x^2 + x^3$
   c. $3 - 27x^2 + 2x^5$
   d. $6 + 12x^2 - 27x^3 + 10x^5$

18. Prove that $\overline{z_1 + z_2 + \cdots + z_n} = \overline{z}_1 + \overline{z}_2 + \cdots + \overline{z}_n$ for complex numbers $z_1, z_2, \ldots, z_n$.

19. Prove that $\overline{z_1 \cdot z_2 \cdot \cdots \cdot z_n} = \overline{z}_1 \cdot \overline{z}_2 \cdot \cdots \cdot \overline{z}_n$ for complex numbers $z_1, z_2, \ldots, z_n$.

20. Prove that for every positive integer n there exist polynomials of degree n that are irreducible over the rational numbers. (*Hint:* Consider $x^n - 2$.)

21. Let $f(x) = a_0 + a_1 x + \cdots + a_{n-1} x^{n-1} + x^n$ monic polynomial of positive degree n with coefficients that are all integers. Prove that any rational zero of $f(x)$ is an integer that divides the constant term a_0.

22. Derive the quadratic formula for the zeros of $ax^2 + bx + c$, where a, b, and c are complex numbers and $a \neq 0$

23. Prove Theorem 8.28. [*Hint:* In the factorization described in Theorem 8.26, pair those factors of the form $x - (a + bi)$ and $x - (a - bi)$.]

24. Prove that any polynomial of odd degree that has real coefficients must have a zero in the field of real numbers.

8.5   ALGEBRAIC EXTENSIONS OF A FIELD

Some of the results in Chapter 6 concerning ideals and quotient rings are put to good use in this section. Starting with an irreducible polynomial $p(x)$ over a field F, these results are used in the construction of a field which is an extension of F that contains a zero of $p(x)$.

As a special case of Definition 6.2, if $p(x)$ is a fixed polynomial over the field F, the *principal ideal* generated by $p(x)$ in $F[x]$ is the set

$$P = (p(x)) = \{f(x)p(x) \mid f(x) \in F[x]\},$$

which consists of all multiples of $p(x)$ by elements $f(x)$ of $F[x]$. Most of our work in this section is related to quotient rings of the form $F[x]/(p(x))$.

Theorem 8.34   **THE QUOTIENT RINGS $F[x]/(p(x))$**

Let $p(x)$ be a polynomial of positive degree over the field F. Then the quotient ring $F[x]/(p(x))$ is a commutative ring with unity that contains a subring that is isomorphic to F.

Proof   For a fixed polynomial $p(x)$ in $F[x]$, let $P = (p(x))$. According to Theorem 6.4, the set $F[x]/P$ forms a ring with respect to addition defined by

$$[f(x) + P] + [g(x) + P] = (f(x) + g(x)) + P$$

and multiplication defined by

$$[f(x) + P][g(x) + P] = f(x)g(x) + P.$$

The ring $F[x]/P$ is commutative, since $f(x)g(x) = g(x)f(x)$ in $F[x]$, and $1 + P$ is the unity in $F[x]$.

Consider the nonempty subset F' of $F[x]/P$ that consists of all cosets of the form $a + P$ with $a \in F$:

$$F' = \{a + P \mid a \in F\}.$$

For arbitrary elements $a + P$ and $b + P$ of F', the elements

$$(a + P) - (b + P) = (a - b) + P$$

and

$$(a + P)(b + P) = ab + P$$

are in F' since $a - b$ and ab are in F. Thus, F' is a subring of $F[x]/P$, by Theorem 5.4. The unity $1 + P$ is in F', and every nonzero element $a + P$ of F' has the multiplicative inverse $a^{-1} + P$ in F'. Hence, F' is a field.

The mapping $\theta: F \to F'$ defined by

$$\theta(a) = a + P$$

is a homomorphism, since

$$\theta(a + b) = (a + b) + P$$
$$= (a + P) + (b + P)$$
$$= \theta(a) + \theta(b)$$

and

$$\theta(ab) = ab + P$$
$$= (a + P)(b + P)$$
$$= \theta(a)\theta(b).$$

It follows from the definition of F' that θ is an epimorphism. Since $p(x)$ has positive degree, 0 is the only element of F that is contained in P, and therefore

$$\theta(a) = \theta(b) \Leftrightarrow a + P = b + P$$
$$\Leftrightarrow a - b \in P$$
$$\Leftrightarrow a = b.$$

Thus, θ is an isomorphism from F to the subring F' of $F[x]/(p(x))$.

As we have done in similar situations in the past, we can now use the isomorphism θ in the preceding proof to identify $a \in F$ with $a + P$ in $F[x]/(p(x))$. This identification allows us to regard F as a subset of $F[x]/(p(x))$. This point of view is especially advantageous when the quotient ring $F[x]/(p(x))$ is a field.

Theorem 8.35 $F[x]/(p(x))$ WITH $p(x)$ IRREDUCIBLE

Let $p(x)$ be a polynomial of positive degree over the field F. Then the ring $F[x]/(p(x))$ is a field if and only if $p(x)$ is an irreducible polynomial over F.

$u \Leftarrow v$ **Proof**   As in the proof of Theorem 8.34, let $P = (p(x))$. Assume first that $p(x)$ is an irreducible polynomial over F. In view of Theorem 8.34, we only need to show that any nonzero element $f(x) + P$ in $F[x]/P$ has a multiplicative inverse in $F[x]/P$. If $f(x) + P \neq P$, then $f(x)$ is not a multiple of $p(x)$, and this means that the greatest common divisor of $f(x)$ and $p(x)$ is 1, since $p(x)$ is irreducible. By Theorem 8.13, there exist $s(x)$ and $t(x)$ in $F[x]$ such that

$$f(x)s(x) + p(x)t(x) = 1.$$

Now $p(x)t(x) \in P$, so $p(x)t(x) + P = 0 + P$, and hence

$$1 + P = [f(x)s(x) + p(x)t(x)] + P$$
$$= [f(x)s(x) + P] + [p(x)t(x) + P]$$
$$= [f(x)s(x) + P] + [0 + P]$$
$$= f(x)s(x) + P$$
$$= [f(x) + P][s(x) + P].$$

Thus, $s(x) + P = [f(x) + P]^{-1}$, and we have proved that $F[x]/P$ is a field.

$\sim u \Leftarrow \sim v$ Suppose now that $p(x)$ is reducible over F. Then there exist polynomials $g(x)$ and $h(x)$ of positive degree in $F[x]$ such that $p(x) = g(x)h(x)$. Since $\deg p(x) = \deg g(x) + \deg h(x)$ and all these degrees are positive, it must be true that $\deg g(x) < \deg p(x)$ and $\deg h(x) < \deg p(x)$. Therefore, neither $g(x)$ nor $h(x)$ is a multiple of $p(x)$. That is,

$$g(x) + P \neq P \qquad \text{and} \qquad h(x) + P \neq P,$$

but

$$[g(x) + P][h(x) + P] = g(x)h(x) + P$$
$$= p(x) + P$$
$$= P.$$

We have $g(x) + P$ and $h(x) + P$ as two nonzero elements of $F[x]/P$ whose product is zero. Hence, $F[x]/P$ is not a field in this case, and the proof is complete.

If F and E are fields such that $F \subseteq E$, then E is called an **extension field** of F. With the identification that we have made between F and F', the preceding theorem shows that $F[x]/(p(x))$ is an extension field of F if and only if $p(x)$ is an irreducible polynomial over F. The main significance of all this becomes clear in the proof of the next theorem, which is credited to the German mathematician Leopold Kronecker (1823–1891).

Theorem 8.36   **EXTENSION FIELD CONTAINING A ZERO**

If $p(x)$ is an irreducible polynomial over the field F, there exists an extension field of F that contains a zero of $p(x)$.

$u \Rightarrow v$ **Proof**   For a given irreducible polynomial

$$p(x) = p_0 + p_1x + p_2x^2 + \cdots + p_nx^n$$

over the field F, let $P = (p(x))$ in $F[x]$ and let $\alpha = x + P$ in $F[x]/P$. From the definition of multiplication in $F[x]/P$, it follows that

$$\alpha^2 = (x + P)(x + P) = x^2 + P$$

and that

$$\alpha^i = x^i + P$$

for every positive integer i. By using the identification of $a \in F$ with $a + P$ in $F[x]/P$, the polynomial

$$p(x) = p_0 + p_1x + p_2x^2 + \cdots + p_nx^n$$

may be written in the form

$$p(x) = (p_0 + P) + (p_1 + P)x + (p_2 + P)x^2 + \cdots + (p_n + P)x^n.$$

Hence,

$$\begin{aligned}
p(\alpha) &= (p_0 + P) + (p_1 + P)\alpha + (p_2 + P)\alpha^2 + \cdots + (p_n + P)\alpha^n \\
&= (p_0 + P) + (p_1 + P)(x + P) + (p_2 + P)(x^2 + P) \\
&\quad + \cdots + (p_n + P)(x^n + P) \\
&= (p_0 + P) + (p_1x + P) + (p_2x^2 + P) + \cdots + (p_nx^n + P) \\
&= (p_0 + p_1x + p_2x^2 + \cdots + p_nx^n) + P \\
&= p(x) + P \\
&= 0 + P.
\end{aligned}$$

Thus, $p(\alpha)$ is the zero element of $F[x]/P$, and α is a zero of $p(x)$ in $F[x]/P$.

For a particular polynomial $p(x)$, explicit standard forms for the elements of the ring $F[x]/(p(x))$ can be given. Before going into this, we note that the ring $F[x]/(p(x))$ is unchanged if $p(x)$ is replaced by a multiple of the form $cp(x)$, with $c \neq 0$ in F. This follows from the fact that the ideal $P = (p(x))$, which consists of the set of all multiples of $p(x)$ in $F[x]$, is the same as the set of all multiples of $cp(x)$ in $F[x]$. In particular, c can be chosen to be the multiplicative inverse of the leading coefficient of $p(x)$, thereby obtaining a monic polynomial that gives the same ring $F[x]/P$ as $p(x)$ does. Thus, there is no loss of generality in assuming from now on that $p(x)$ is a *monic* polynomial over F.

Before considering the general situation, we examine some particular cases in the following examples.

Example 1   Consider the monic irreducible polynomial

$$p(x) = x^2 + 2x + 2$$

over the field $\mathbf{Z}_3$. We shall determine all the elements of the field $\mathbf{Z}_3[x]/(p(x))$, and at the same time construct addition and multiplication tables for this field.

Let $P = (p(x))$ and $\alpha = x + P$ in $\mathbf{Z}_3[x]/P$. We start construction of the addition table for $\mathbf{Z}_3[x]/P$ with the elements $0 = 0 + P, 1 = 1 + P, 2 = 2 + P$, and α. Filling out the table until closure is obtained, we pick up the new elements $\alpha + 1, \alpha + 2, 2\alpha$, $2\alpha + 1$, and $2\alpha + 2$. The completed table in Figure 8-2 shows that the set

$$\{0, 1, 2, \alpha, \alpha + 1, \alpha + 2, 2\alpha, 2\alpha + 1, 2\alpha + 2\}$$

is closed under addition.

Figure 8-2

$+$	0	1	2	α	$\alpha + 1$	$\alpha + 2$	2α	$2\alpha + 1$	$2\alpha + 2$
0	0	1	2	α	$\alpha + 1$	$\alpha + 2$	2α	$2\alpha + 1$	$2\alpha + 2$
1	1	2	0	$\alpha + 1$	$\alpha + 2$	α	$2\alpha + 1$	$2\alpha + 2$	2α
2	2	0	1	$\alpha + 2$	α	$\alpha + 1$	$2\alpha + 2$	2α	$2\alpha + 1$
α	α	$\alpha + 1$	$\alpha + 2$	2α	$2\alpha + 1$	$2\alpha + 2$	0	1	2
$\alpha + 1$	$\alpha + 1$	$\alpha + 2$	α	$2\alpha + 1$	$2\alpha + 2$	2α	1	2	0
$\alpha + 2$	$\alpha + 2$	α	$\alpha + 1$	$2\alpha + 2$	2α	$2\alpha + 1$	2	0	1
2α	2α	$2\alpha + 1$	$2\alpha + 2$	0	1	2	α	$\alpha + 1$	$\alpha + 2$
$2\alpha + 1$	$2\alpha + 1$	$2\alpha + 2$	2α	1	2	0	$\alpha + 1$	$\alpha + 2$	α
$2\alpha + 2$	$2\alpha + 2$	2α	$2\alpha + 1$	2	0	1	$\alpha + 2$	α	$\alpha + 1$

Turning now to multiplication, we start with the same nine elements that occur in the addition table. In constructing this table, we make use of the fact that α is a zero of $p(x) = x^2 + 2x + 2$ in the following manner:

$$\alpha^2 + 2\alpha + 2 = 0 \implies \alpha^2 = -2\alpha - 2 = \alpha + 1.$$

That is, whenever α^2 occurs in a product, it is replaced by $\alpha + 1$. As an illustration, we have

$$(2\alpha + 1)(\alpha + 2) = 2\alpha^2 + 2\alpha + 2$$
$$= 2(\alpha + 1) + 2\alpha + 2$$
$$= 2\alpha + 2 + 2\alpha + 2$$
$$= \alpha + 1.$$

The completed table is shown in Figure 8-3.

Figure 8-3

·	0	1	2	α	$\alpha + 1$	$\alpha + 2$	2α	$2\alpha + 1$	$2\alpha + 2$
0	0	0	0	0	0	0	0	0	0
1	0	1	2	α	$\alpha + 1$	$\alpha + 2$	2α	$2\alpha + 1$	$2\alpha + 2$
2	0	2	1	2α	$2\alpha + 2$	$2\alpha + 1$	α	$\alpha + 2$	$\alpha + 1$
α	0	α	2α	$\alpha + 1$	$2\alpha + 1$	1	$2\alpha + 2$	2	$\alpha + 2$
$\alpha + 1$	0	$\alpha + 1$	$2\alpha + 2$	$2\alpha + 1$	2	α	$\alpha + 2$	2α	1
$\alpha + 2$	0	$\alpha + 2$	$2\alpha + 1$	1	α	$2\alpha + 2$	2	$\alpha + 1$	2α
2α	0	2α	α	$2\alpha + 2$	$\alpha + 2$	2	$\alpha + 1$	1	$2\alpha + 1$
$2\alpha + 1$	0	$2\alpha + 1$	$\alpha + 2$	2	2α	$\alpha + 1$	1	$2\alpha + 2$	α
$2\alpha + 2$	0	$2\alpha + 2$	$\alpha + 1$	$\alpha + 2$	1	2α	$2\alpha + 1$	α	2

∎

Example 2 The polynomial $p(x) = x^2 + 1$ is not irreducible over the field $\mathbf{Z}_2$ since $p(1) = 0$. We follow the same procedure as in Example 1 and construct addition and multiplication tables for the ring $\mathbf{Z}_2[x]/(p(x))$.

As before, let $P = (p(x))$ and $\alpha = x + P$ in $\mathbf{Z}_2[x]/P$. Extending an addition table until closure is obtained, we arrive at the table shown in Figure 8-4.

Figure 8-4

+	0	1	α	$\alpha + 1$
0	0	1	α	$\alpha + 1$
1	1	0	$\alpha + 1$	α
α	α	$\alpha + 1$	0	1
$\alpha + 1$	$\alpha + 1$	α	1	0

In making the multiplication table shown in Figure 8-5, we use the fact that $p(\alpha) = 0$ in this way:

$$a^2 + 1 = 0 \implies \alpha^2 = -1$$
$$\implies \alpha^2 = 1.$$

Figure 8-5

·	0	1	α	$\alpha + 1$
0	0	0	0	0
1	0	1	α	$\alpha + 1$
α	0	α	1	$\alpha + 1$
$\alpha + 1$	0	$\alpha + 1$	$\alpha + 1$	0

Theorem 8.35 assures us that $\mathbf{Z}_2[x]/P$ is not a field, and the multiplication table confirms this fact by showing that $\alpha + 1$ does not have a multiplicative inverse. ∎

The next theorem and its corollary set forth the standard forms for the elements of the ring $F[x]/(p(x))$ that were referred to earlier.

Theorem 8.37 **ELEMENTS OF $F[x]/(p(x))$**

Let $p(x)$ be a polynomial of positive degree n over the field F, and let $P = (p(x))$ in $F[x]$. Then each element of the ring $F[x]/P$ can be expressed uniquely in the form

$$(a_0 + a_1x + a_2x^2 + \cdots + a_{n-1}x^{n-1}) + P.$$

$u \Rightarrow v$ **Proof**   Assume the hypothesis and let $f(x) + P$ be an arbitrary element in $F[x]/P$. By the Division Algorithm, there exist $q(x)$ and $r(x)$ in $F[x]$ such that

$$f(x) = p(x)q(x) + r(x),$$

where either $r(x) = 0$ or $\deg r(x) < n = \deg p(x)$. In either case, we may write

$$r(x) = a_0 + a_1x + a_2x^2 + \cdots + a_{n-1}x^{n-1}.$$

Since $p(x)q(x)$ is in P, $p(x)q(x) + P = 0 + P$, and therefore

$$\begin{aligned}
f(x) + P &= [p(x)q(x) + P] + [r(x) + P] \\
&= [0 + P] + [r(x) + P] \\
&= r(x) + P \\
&= (a_0 + a_1x + \cdots + a_{n-1}x^{n-1}) + P.
\end{aligned}$$

Uniqueness   To show uniqueness, suppose that $f(x) + P = r(x) + P$ as before and also that $f(x) + P = g(x) + P$, where

$$g(x) = b_0 + b_1x + b_2x^2 + \cdots + b_{n-1}x^{n-1}.$$

Then $r(x) + P = g(x) + P$, and therefore $r(x) - g(x)$ is in P. Each of $r(x)$ and $g(x)$ is either zero or has degree less than n, and this implies that the difference $r(x) - g(x)$ is either zero or has degree less than n. Since $P = (p(x))$ contains no polynomials with degree less than n, it must be true that $r(x) - g(x) = 0$, and $r(x) = g(x)$.

Corollary 8.38   **ELEMENTS OF $F[x]/P$ AS POLYNOMIALS**

For a polynomial $p(x)$ of positive degree n over the field F, let $P = (p(x))$ in $F[x]$ and let $\alpha = x + P$ in $F[x]/P$. Then each element of the ring $F[x]/P$ can be expressed uniquely in the form

$$a_0 + a_1\alpha + a_2\alpha^2 + \cdots + a_{n-1}\alpha^{n-1}.$$

$u \Rightarrow v$ **Proof**   From the theorem, each $f(x) + P$ in $F[x]/P$ can be expressed uniquely in the form

$$\begin{aligned}
f(x) + P &= (a_0 + a_1x + \cdots + a_{n-1}x^{n-1}) + P \\
&= (a_0 + P) + (a_1 + P)(x + P) + \cdots + (a_{n-1} + P)(x^{n-1} + P) \\
&= (a_0 + P) + (a_1 + P)\alpha + \cdots + (a_{n-1} + P)\alpha^{n-1} \\
&= a_0 + a_1\alpha + \cdots + a_{n-1}\alpha^{n-1},
\end{aligned}$$

where the last equality follows from the identification of a_i in F with $a_i + P$ in $F[x]/P$.

In Example 1, the polynomials $f(x)$ in $\mathbf{Z}_3[x]$ and the cosets $f(x) + P$ in $\mathbf{Z}_3[x]/P$ receded into the background once the notation $\alpha = x + P$ was introduced, and we ended up with a field whose elements had the form $a_0 + a_1\alpha$, with $a_i \in \mathbf{Z}_3$. This field $\mathbf{Z}_3(\alpha)$ of nine elements, given by

$$\mathbf{Z}_3(\alpha) = \{0, 1, 2, \alpha, \alpha + 1, \alpha + 2, 2\alpha, 2\alpha + 1, 2\alpha + 2\},$$

is called *the field obtained by adjoining a zero α of $x^2 + 2x + 2$ to $\mathbf{Z}_3$.*

In general, if $p(x)$ is an irreducible polynomial over the field F, the smallest field that contains both F and a zero α of $p(x)$ is denoted by $F(\alpha)$ and is referred to as the **field[†] obtained by adjoining α to the field F.** A field $F(\alpha)$ of this type is called a **simple algebraic extension** of F, and F is referred to as the **ground field.** Corollary 8.38 describes the standard form for the elements of $F(\alpha)$.

Example 3   The polynomial $p(x) = x^3 + 2x^2 + 4x + 2$ is irreducible over $\mathbf{Z}_5$, since

$$p(0) = 2, \qquad p(1) = 4, \qquad p(2) = 1, \qquad p(3) = 4, \qquad p(4) = 4.$$

In the field $\mathbf{Z}_5(\alpha)$ obtained by adjoining a zero α of $p(x)$ to $\mathbf{Z}_5$, we shall obtain a formula for the product of two arbitrary elements $a_0 + a_1\alpha + a_2\alpha^2$ and $b_0 + b_1\alpha + b_2\alpha^2$.

In order to accomplish this objective, we first express α^3 and α^4 as polynomials in α with degrees less than 3. Since $p(\alpha) = 0$, we have

$$\alpha^3 + 2\alpha^2 + 4\alpha + 2 = 0 \Rightarrow \alpha^3 = -2\alpha^2 - 4\alpha - 2$$
$$= 3\alpha^2 + \alpha + 3.$$

Hence,

$$\begin{aligned}
\alpha^4 &= \alpha(3\alpha^2 + \alpha + 3) \\
&= 3\alpha^3 + \alpha^2 + 3\alpha \\
&= 3(3\alpha^2 + \alpha + 3) + \alpha^2 + 3\alpha \\
&= 4\alpha^2 + 3\alpha + 4 + \alpha^2 + 3\alpha = \alpha + 4.
\end{aligned}$$

Using these results, we get

$$\begin{aligned}
(a_0 + a_1\alpha &+ a_2\alpha^2)(b_0 + b_1\alpha + b_2\alpha^2) \\
&= a_0 b_0 + (a_0 b_1 + a_1 b_0)\alpha + (a_0 b_2 + a_1 b_1 + a_2 b_0)\alpha^2 \\
&\quad + (a_1 b_2 + a_2 b_1)\alpha^3 + a_2 b_2 \alpha^4 \\
&= a_0 b_0 + (a_0 b_1 + a_1 b_0)\alpha + (a_0 b_2 + a_1 b_1 + a_2 b_0)\alpha^2 \\
&\quad + (a_1 b_2 + a_2 b_1)(3\alpha^2 + \alpha + 3) + a_2 b_2(\alpha + 4) \\
&= (a_0 b_0 + 3a_1 b_2 + 3a_2 b_1 + 4a_2 b_2) \\
&\quad + (a_0 b_1 + a_1 b_0 + a_1 b_2 + a_2 b_1 + a_2 b_2)\alpha \\
&\quad + (a_0 b_2 + a_1 b_1 + a_2 b_0 + 3a_1 b_2 + 3a_2 b_1)\alpha^2.
\end{aligned}$$ ■

Example 4   With $\mathbf{Z}_5(\alpha)$ as in Example 3, suppose that we wish to find the multiplicative inverse of the element $\alpha^2 + 3\alpha + 1$ in the field $\mathbf{Z}_5(\alpha)$.

The polynomials $f(x) = x^2 + 3x + 1$ and $p(x) = x^3 + 2x^2 + 4x + 2$ are relatively prime over $\mathbf{Z}_5$, so there exists $s(x)$ and $t(x)$ in $\mathbf{Z}_5[x]$ such that

$$f(x)s(x) + p(x)t(x) = 1,$$

[†] The existence of such a field $F(\alpha)$ is assured by Theorem 8.36.

by Theorem 8.13. Since $p(\alpha) = 0$, this means that

$$f(\alpha)s(\alpha) = 1$$

and that $(\alpha^2 + 3\alpha + 1)^{-1} = [f(\alpha)]^{-1} = s(\alpha)$. In order to find $s(x)$ and $t(x)$, we use the Euclidean Algorithm:

$$p(x) = f(x)(x + 4) + (x + 3)$$
$$f(x) = (x + 3)(x) + 1.$$

Thus,

$$
\begin{aligned}
1 &= f(x) - x(x + 3) \\
  &= f(x) - x[p(x) - f(x)(x + 4)] \\
  &= f(x)[1 + x(x + 4)] + p(x)(-x) \\
  &= f(x)(x^2 + 4x + 1) + p(x)(-x),
\end{aligned}
$$

so we have $s(x) = x^2 + 4x + 1$ and $t(x) = -x$. Therefore,

$$(\alpha^2 + 3\alpha + 1)^{-1} = s(\alpha) = \alpha^2 + 4\alpha + 1.$$

The result may be checked by computing the product

$$(\alpha^2 + 3\alpha + 1)(\alpha^2 + 4\alpha + 1)$$

in $\mathbf{Z}_5(\alpha)$. ■

It is of some interest to consider an example similar to Example 4 but in a more familiar setting.

Example 5   The polynomial $p(x) = x^2 - 2$ is irreducible over the field $\mathbf{Q}$ of rational numbers. In the field $\mathbf{Q}(\sqrt{2})$ obtained by adjoining a zero $\alpha = \sqrt{2}$ of $p(x)$ to $\mathbf{Q}$, let us find the multiplicative inverse of the element $4 + 3\sqrt{2}$ by the method employed in Example 4. The polynomials $f(x) = 3x + 4$ and $p(x) = x^2 - 2$ are relatively prime over $\mathbf{Q}$. To find $s(x)$ and $t(x)$ such that

$$f(x)s(x) + p(x)t(x) = 1,$$

we need only one step in the Euclidean Algorithm:

$$p(x) = f(x) \cdot \left(\frac{1}{3}x - \frac{4}{9}\right) + \left(-\frac{2}{9}\right).$$

Multiplying by 9/2 and rewriting this equation, we obtain

$$f(x) \cdot \left(\frac{3}{2}x - 2\right) + p(x)\left(-\frac{9}{2}\right) = 1.$$

Since $p(\sqrt{2}) = 0$, this gives

$$f(\sqrt{2}) \cdot \left(\frac{3}{2}\sqrt{2} - 2\right) = 1$$

and

$$(4 + 3\sqrt{2})^{-1} = [f(\sqrt{2})]^{-1} = \frac{3}{2}\sqrt{2} - 2.$$

This agrees with the result obtained by the usual procedure of rationalizing the denominator:

$$\frac{1}{4 + 3\sqrt{2}} = \frac{(1)(4 - 3\sqrt{2})}{(4 + 3\sqrt{2})(4 - 3\sqrt{2})} = \frac{4 - 3\sqrt{2}}{-2}$$

$$= \frac{3}{2}\sqrt{2} - 2. \qquad \blacksquare$$

The result in Theorem 8.36 generalizes to the following theorem.

Theorem 8.39 SPLITTING FIELD

If $p(x)$ is a polynomial of positive degree n over the field F, there exists an extension field E of F that contains n zeros of $p(x)$.

Induction **Proof** The proof is by induction on the degree n of $p(x)$. If $n = 1$, then $p(x)$ has the form $p(x) = ax + b$, with $a \neq 0$. Since $p(x)$ has the unique zero $-a^{-1}b$ in F, the theorem is true for $n = 1$.

Assume the theorem is true for all polynomials of degree less than k, and let $p(x)$ be a polynomial of degree k. We consider two cases, depending on whether $p(x)$ is irreducible.

If $p(x)$ is irreducible, then there exists an extension field E_1 of F that contains a zero α of $p(x)$, by Theorem 8.36. By the Factor Theorem,

$$p(x) = (x - \alpha)q(x),$$

where $q(x)$ must have degree $k - 1$, according to Theorem 8.7. Since $q(x)$ is a polynomial over E_1 that has degree less than k, the induction hypothesis applies to $q(x)$ over E_1, and there exists an extension field E of E_1 such that $q(x)$ has $k - 1$ zeros in E. By Exercise 10 of Section 8.3, the zeros of $p(x)$ in E consist of α and the zeros of $q(x)$ in E. Thus, $p(x)$ has k zeros in E.

If $p(x)$ is reducible, then $p(x)$ can be factored as a product $p(x) = g(x)h(x)$, where $n_1 = \deg g(x)$ and $n_2 = \deg h(x)$ are positive integers such that $n_1 + n_2 = k$. Since $n_1 < k$, the induction hypothesis applies to $g(x)$ over F, and there exists an extension field E_1 of F that contains n_1 zeros of $g(x)$. Now $h(x)$ is a polynomial of degree $n_2 < k$ over E_1, so the induction hypothesis applies again to $h(x)$ over E_1, and there exists an extension field E of E_1 such that $h(x)$ has n_2 zeros in E. By Exercise 11 of Section 8.3, the zeros of $p(x)$ in E consist of the zeros of $g(x)$ in E together with the zeros of $h(x)$ in E. There are altogether $n_1 + n_2 = k$ of these zeros in E.

In either case, we have proved the existence of an extension field of F that contains k zeros of $p(x)$, and the theorem follows by induction.

If E is a field that contains all the zeros of a polynomial $p(x)$, and if no proper subfield of E contains all of these zeros, then E is called the **splitting field** of $p(x)$ because it is the "smallest" field over which $p(x)$ "splits" into first-degree factors.

The basic facts about zeros of polynomials have been presented in this chapter. The two most important facts are found in Theorems 8.26 and 8.39. Theorem 8.26 asserts that, for any polynomial $p(x)$ of positive degree n over **C**, the field **C** contains n zeros of $p(x)$. Theorem 8.39 states that, for an arbitrary field F and any polynomial

$p(x)$ of positive degree n over F, there exists an extension field of F that contains n zeros of $p(x)$.

Important as it is, the material in this chapter is only a small part of the knowledge about extension fields. The study of extension fields leads into the area of mathematics known as *Galois theory*. Interesting results concerning some ancient problems lie in this direction. One of these results is that it is impossible to trisect an arbitrary angle using only a straightedge and a compass. Another is that it is impossible to express the zeros of the general equation of degree 5 or more by formulas that use only the four basic arithmetic operations and extraction of roots.

The end of this book is actually a beginning. It is a gateway to higher mathematics courses in several directions, especially those in abstract algebra and linear algebra. These higher-level courses are more theoretical and stimulating intellectually, and they might well lead to a lifelong interest in mathematics.

EXERCISES 8.5

1. Each of the following polynomials $p(x)$ is irreducible over $\mathbf{Z}_3$. For each of these polynomials, find all the elements of $\mathbf{Z}_3[x]/(p(x))$ and construct addition and multiplication tables for this field.
   a. $p(x) = x^2 + x + 2$ b. $p(x) = x^2 + 1$

2. In each of the following parts, a polynomial $p(x)$ over a field F is given. Construct addition and multiplication tables for the ring $F[x]/(p(x))$ in each case and decide whether this ring is a field.
   a. $p(x) = x^2 + x + 1$ over $F = \mathbf{Z}_2$
   b. $p(x) = x^3 + 1$ over $F = \mathbf{Z}_2$
   c. $p(x) = x^3 + x + 1$ over $F = \mathbf{Z}_2$
   d. $p(x) = x^3 + x^2 + 1$ over $F = \mathbf{Z}_2$
   e. $p(x) = x^2 + x + 1$ over $F = \mathbf{Z}_3$
   f. $p(x) = x^2 + 2$ over $F = \mathbf{Z}_3$

In Exercises 3–6, a field F, a polynomial $p(x)$ over F, and an element of the field $F(\alpha)$ obtained by adjoining a zero α of $p(x)$ to F are given. In each case:
   a. Verify that $p(x)$ is irreducible over F.
   b. Write out a formula for the product of two arbitrary elements $a_0 + a_1\alpha + a_2\alpha^2$ and $b_0 + b_1\alpha + b_2\alpha^2$ of $F(\alpha)$.
   c. Find the multiplicative inverse of the given element of $F(\alpha)$.

3. $F = \mathbf{Z}_3, p(x) = x^3 + 2x^2 + 1, \alpha^2 + \alpha + 2$

4. $F = \mathbf{Z}_3, p(x) = x^3 + x^2 + 2x + 1, \alpha^2 + 2\alpha + 1$

5. $F = \mathbf{Z}_5, p(x) = x^3 + x + 1, \alpha^2 + 4\alpha$

6. $F = \mathbf{Z}_5, p(x) = x^3 + x^2 + 1, \alpha^2 + 2\alpha + 3$

7. For the given irreducible polynomial $p(x)$ over $\mathbf{Z}_3$, list all elements of the field $\mathbf{Z}_3(\alpha)$ that is obtained by adjoining a zero α of $p(x)$ to $\mathbf{Z}_3$.
   a. $p(x) = x^3 + 2x^2 + 1$
   b. $p(x) = x^3 + x^2 + 2x + 1$

8. If F is a finite field with k elements, and $p(x)$ is a polynomial of positive degree n over F, find a formula for the number of elements in the ring $F[x]/(p(x))$.

9. Find the multiplicative inverse of $\sqrt[3]{4} - 2\sqrt[3]{2} - 2$ in $\mathbf{Q}(\sqrt[3]{2})$, where $\mathbf{Q}$ is the field of rational numbers.

10. An element u of a field F is a perfect square in F if there exists an element v in F such that $u = v^2$. The quadratic formula can be generalized in the following way: Suppose that $1 + 1 \neq 0$ in F and let $p(x) = ax^2 + bx + c, a \neq 0$, be a quadratic polynomial over F.
   a. Prove that $p(x)$ has a zero in F if and only if $b^2 - 4ac$ is a perfect square in F.
   b. If $b^2 - 4ac$ is a perfect square in F, show that the zeros of $p(x)$ in F are given by

$$r_1 = \frac{-b + \sqrt{b^2 - 4ac}}{2a} \quad \text{and} \quad r_2 = \frac{-b - \sqrt{b^2 - 4ac}}{2a}$$

   and that these zeros are distinct if $b^2 - 4ac \neq 0$.

11. Determine whether each of the following polynomials has a zero in the given field F. If a polynomial has zeros in the field, use the quadratic formula to find them.
   a. $x^2 + 3x + 2,$ $F = \mathbf{Z}_5$
   b. $x^2 + 3x + 3,$ $F = \mathbf{Z}_5$
   c. $x^2 + 2x + 6,$ $F = \mathbf{Z}_7$
   d. $x^2 + 3x + 1,$ $F = \mathbf{Z}_7$
   e. $2x^2 + x + 1,$ $F = \mathbf{Z}_7$
   f. $3x^2 + 2x - 1,$ $F = \mathbf{Z}_7$

12. **a.** Find the value of c that will cause the polynomial $f(x) = x^2 + 3x + c$ to have 3 as a zero in the field $\mathbf{Z}_7$.
   b. Find the other zero of $f(x)$ in $\mathbf{Z}_7$.

Each of the polynomials $p(x)$ in Exercises 13–16 is irreducible over the given field F. Find all zeros of $p(x)$ in the field $F(\alpha)$ obtained by adjoining a zero of $p(x)$ to F. (In Exercises 15 and 16, $p(x)$ has three zeros in $F(\alpha)$.)

13. $p(x) = x^2 + 2x + 2,$ $F = \mathbf{Z}_3$
14. $p(x) = x^2 + x + 2,$ $F = \mathbf{Z}_3$
15. $p(x) = x^3 + x^2 + 1,$ $F = \mathbf{Z}_5$
16. $p(x) = x^3 + 2x^2 + 4x + 2,$ $F = \mathbf{Z}_5$

KEY WORDS AND PHRASES

A Pioneer in Mathematics ||||||||||||||||||||||||||||||||||||||||||||||||||||||||||||||||||||||||||
Carl Friedrich Gauss (1777–1855)

Carl Friedrich Gauss was born in Brunswick, Germany, on April 30, 1777. He is regarded as the greatest mathematician of the 19th century and has been called the Prince of Mathematics. Part of Gauss's greatness is due to the fact that his interest spanned all mathematics known in his time. Since then, the volume of knowledge in mathematics has become so large that no one person could ever hope to master the whole field. In this sense, he may have been the last complete mathematician.

The world was almost deprived of Gauss's genius when, as a child, he fell into an overflowing canal near his home. It is said that he surely would have drowned had he not been rescued by a passerby.

His mathematical genius became evident early in his life. He often said that he could reckon before he could talk. In school, his precocity attracted the attention of the Duke of Brunswick. The Duke decided to finance the education of the young prodigy and granted him a fixed pension so that he could devote himself to work without financial considerations.

Gauss made some of the greatest contributions to mathematics when he was a young man. He developed the method of least squares while preparing for university studies at Collegium Carolinium. Two years later, he solved a 2000-year-old problem by proving that a regular 17-sided polygon can be constructed with only a straightedge and a compass. In his doctoral dissertation, Gauss proved the Fundamental Theorem of Algebra, a result that had been accepted without proof for many years. In 1801, at the age of 24, he published the monumental work *Disquisitiones Arithmeticae,* in which he laid the foundations of the area of mathematics called *number theory.*

In 1801, when Gauss turned his attention to astronomy, he accomplished an extraordinary achievement. Using a scanty amount of data, he was able to accurately predict the orbit of the asteroid Ceres. He gained international acclaim. In 1807, he was appointed director of the astronomical observatory of Göttingen.

APPENDIX
THE BASICS OF LOGIC

In any mathematical system, just as in any language, there must be some undefined terms. For example, the words *set* and *element* are undefined terms. We think of a set as a collection of objects, and the individual objects as elements of the set. We need to understand the word *set* to describe the word *element* and vice versa. Hence, we must rely on our intuition to understand these undefined terms and feel comfortable using them to define new terms.

A **statement**, or **proposition**, is a declarative sentence that is either true or false, but not both. **Postulates** are statements (often expressed using undefined terms) that are assumed to be true. Postulates and definitions are used to prove statements called **theorems**. Once a theorem is proved to be true it can be used to establish the truth of subsequent theorems. A **lemma** is itself a theorem whose major importance lies not in its own statement, but in its role as a stepping stone toward the statement or proof of a theorem. Finally, a **corollary** is also a theorem, but is not so named because it is usually either a direct consequence of or a special case of a preceding theorem. To avoid "stealing the thunder" of the more important theorem, it is labeled a corollary.

We now briefly discuss the basic concepts of logic that are essential to the mathematician for constructing proofs. We use the letters p, q, r, s, and so on to represent statements. Consider the following statements.

p : The sum of the angles in a triangle is 180°.
q : $2^2 + 3^2 = (2 + 3)^2$
r : $x^2 + 1 = 0$
s : Beckie is pretty.

The statement p is a true proposition from plane geometry. The statement q is a false proposition, when we consider the usual multiplication and addition in the set of real numbers. The statement r is not a proposition, since its truth or falsity cannot be determined unless the value of x is known. Also, s is not a proposition, since its truth or falsity "lies in the eyes of the beholder" and depends on which "Beckie" is under consideration.

The statement r in the preceding paragraph can be clarified by placing restrictions on the variable x, such as "for every x," "for each x," "for all x," "for some x," "for at least one x," or "there exists an x." The phrases "for every x," "for all x," and "for each x " mean the same thing and are often abbreviated by the symbol $\forall$, called the **universal quantifier**. Similarly, the phrases "for some x," "for at least one x," and "there exists an x" mean the same thing and are abbreviated by the symbol $\exists$, called the **existential quantifier**. Another commonly used symbol is $\ni$, read "**such that**." Thus, the statement

$$\forall x, x > 0$$

is read

$$\text{"For every } x, x > 0."$$

Similarly, the statement

$$\exists \; y \ni y^2 + 1 = 0$$

is read

$$\text{"There exists a } y \text{ such that } y^2 + 1 = 0."$$

A statement about the variable x may be true for some values of x and false for other values of x. Some such statements can be proved by furnishing an example, while others cannot. The quantifier used in the statement determines the type of proof required.

If the statement has an existential quantifier, then one example where the statement is true will establish the statement as a theorem. Consider the statement

$$\text{"There exists an integer } x \text{ such that } x^2 + 2x = 24."$$

If the value 4 is assigned to x and it is then verified that $4^2 + 2(4) = 16 + 8 = 24$, this proves that the statement is true. The phrase "there exists an integer x" requires only one value of x that works to make the statement true.

If the statement has a universal quantifier, a specific example does not make a proof. Consider the statement

$$\text{"For any integer } n, n - 1 \text{ is a factor of } n^2 - 4n + 3."$$

If the value 7 is assigned to n and it is then verified that $n - 1 = 6$ is indeed a factor of $7^2 - 4(7) + 3 = 24 = 6(4)$, this illustrates a case where the statement is true, but *it does not prove that the statement is true for any values of n other than 7* and thus does not constitute a proof. The phrase "for any integer n" requires an argument that can be applied independently of the value of n. In this case, a proof can be supplied by demonstrating that

$$(n - 1)(n - 3) = n^2 - 4n + 3,$$

since this shows that $n - 1$ is always a factor of $n^2 - 4n + 3$.

If a statement about x with a universal quantifier is not true for at least one value of x, the statement is declared to be false (and therefore is not a theorem). Consider the statement

$$\text{"} x^2 < 2^x \text{ for all real numbers } x."$$

For $x = 3$,

$$3^2 < 2^3$$

is false. Therefore, the statement

$$\text{"} x^2 < 2^x \text{ for all real numbers } x"$$

is false.

A demonstration in which a statement is shown to be false for a certain value of the variable is called a **counterexample**. A statement with a universal quantifier can be proved false by finding just one counterexample, as was done in the last paragraph.

If p is a proposition, then the **negation of p** is denoted by $\sim p$ and is read "not p." If p is a true proposition, then $\sim p$ must be false and vice versa. We illustrate the idea using a truth table (see Figure A-1), where T stands for true and F stands for false.

Figure A-1

Truth Table for $\sim p$

p	$\sim p$
T	F
F	T

The negation of statements involving the universal quantifier and the existential quantifier are given next. We use $p(x)$ to represent a statement involving the variable x. Then the statement

$$\sim(\forall x, p(x)) \text{ is } \exists x \ni \sim p(x)$$

is read

"The negation of 'For every x, $p(x)$ is true'

is

'There exists an x such that $p(x)$ is false.' "

We also write

$$\sim(\exists x \ni p(x)) \text{ is } \forall x, \sim p(x)$$

and read

"The negation of 'There exists an x such that $p(x)$ is true'

is

'For every x, $p(x)$ is false.' "

Example 1   The negation of the statement

"All the students in the class are female"

is

"There exists at least one student in the class who is not female."   ■

Example 2   The negation of the statement

"There is at least one student who passed the course"

is

"All the students failed the course."   ■

Connectives are used to join propositions to make compound statements. Propositions p and q can be joined with the connective "and," commonly symbolized by $\wedge$ and called **conjunction**. We define $p \wedge q$ to be true only when both p is true and q is true. The corresponding truth table for $p \wedge q$ is given in Figure A-2.

Figure A-2

Truth Table for $p \wedge q$

p	q	$p \wedge q$
T	T	T
T	F	F
F	T	F
F	F	F

Similarly, propositions p and q can be joined with the connective "or," symbolized by $\vee$ and called **disjunction**. We define $p \vee q$ to be true when either p is true or q is true, or both p and q are true. The truth table for $p \vee q$ is given in Figure A-3.

Figure A-3

Truth Table for $p \vee q$

p	q	$p \vee q$
T	T	T
T	F	T
F	T	T
F	F	F

Probably the most important connective is **implication**, denoted by $\Rightarrow$. Suppose p and q are propositions. Then

$$p \Rightarrow q$$

is read in several ways:

> "p implies q"
> "if p then q"
> "p only if q"
> "p is sufficient for q"
> "q is necessary for p."

In each of these statements, p is called by **hypothesis** and q is called the **conclusion**.

Let us consider the following situations. Algebra class meets only three days a week, on Monday, Wednesday, and Friday. Let p and q be the following propositions:

p: Today is Monday.
q: Algebra class meets today.

Consider the implication

$$p \Rightarrow q.$$

This implication is true if both p and q are true:

Today is Monday $\Rightarrow$ Algebra class meets today.

Suppose p is true and q is false. Then the implication

Today is Monday $\Rightarrow$ Algebra class meets today

is false. Next suppose that p is false. The falsity of p does not affect the truth or falsity of q. That is,

Today is not Monday

does not give any information about whether algebra class meets today. Thus, we conclude that

$$p \Rightarrow q$$

is false only when p is true and q is false. We record these results in the truth table in Figure A-4.

Truth Table for $p \Rightarrow q$

Figure A-4

p	q	$p \Rightarrow q$
T	T	T
T	F	F
F	T	T
F	F	T

Another prominent connective is the **biconditional**, denoted by

$$p \Leftrightarrow q$$

and read in any one of three ways:

"p if and only if q"
"p is necessary and sufficient for q"
"p is equivalent to q."

The biconditional statement

$$p \Leftrightarrow q$$

can be expressed as the conjunction of two statements:

$$(p \Rightarrow q) \wedge (q \Rightarrow p).$$

The truth table in Figure A-5 illustrates that the statement $p \Leftrightarrow q$ is true when p and q are both true or both false; otherwise, $p \Leftrightarrow q$ is false.

Truth Table for $p \Leftrightarrow q$

Figure A-5

p	q	$p \Rightarrow q$	$q \Rightarrow p$	$(p \Rightarrow q) \wedge (q \Rightarrow p)$ $p \Leftrightarrow q$
T	T	T	T	T
T	F	F	T	F
F	T	T	F	F
F	F	T	T	T

If the truth tables for two propositions are identical, then the two propositions are said to be **logically equivalent**, and we use the $\Leftrightarrow$ symbol to designate this.

Example 3 Show that

$$\sim(p \wedge q) \Leftrightarrow (\sim p) \vee (\sim q).$$

Solution We examine the two columns headed by $\sim(p \wedge q)$ and by $(\sim p) \vee (\sim q)$ in the truth table in Figure A-6 and note that they are identical.

Truth Table for $\sim(p \wedge q) \Leftrightarrow (\sim p) \vee (\sim q)$

Figure A-6

p	q	$p \wedge q$	$\sim(p \wedge q)$	$\sim p$	$\sim q$	$(\sim p) \vee (\sim q)$
T	T	T	F	F	F	F
T	F	F	T	F	T	T
F	T	F	T	T	F	T
F	F	F	T	T	T	T

The statement in Example 3 is the logical form of one of **De Morgan's Laws**. The corresponding form for sets is given at the end of Section 1.1. The next example illustrates a truth table involving three propositions.

Example 4 Show that

$$r \wedge (p \vee q) \Leftrightarrow (r \wedge p) \vee (r \wedge q).$$

Solution We need eight rows in our truth table, since there are 2^3 different ways to assign true and false to the three different statements (see Figure A-7).

Truth Table for $r \wedge (p \vee q) \Leftrightarrow (r \wedge p) \vee (r \wedge q)$

Figure A-7

r	p	q	$p \vee q$	$r \wedge (p \vee q)$	$r \wedge p$	$r \wedge q$	$(r \wedge p) \vee (r \wedge q)$
T	T	T	T	T	T	T	T
T	T	F	T	T	T	F	T
T	F	T	T	T	F	T	T
T	F	F	F	F	F	F	F
F	T	T	T	F	F	F	F
F	T	F	T	F	F	F	F
F	F	T	T	F	F	F	F
F	F	F	F	F	F	F	F

In this text, we see some theorems whose statements involve an implication

$$p \Rightarrow q.$$

In some instances, it is more convenient to prove a statement that is logically equivalent to the implication $p \Rightarrow q$. The truth table in Figure A-8 shows that the implication

$$p \Rightarrow q \quad \text{(implication)}$$

is logically equivalent to the statement

$$\sim q \Rightarrow \sim p \text{ (contrapositive)},$$

called the **contrapositive** of $p \Rightarrow q$.

Truth Table for $(p \Rightarrow q) \Leftrightarrow (\sim q \Rightarrow \sim p)$

Figure A-8

p	q	$p \Rightarrow q$	$\sim q$	$\sim p$	$\sim q \Rightarrow \sim p$
T	T	T	F	F	T
T	F	F	T	F	F
F	T	T	F	T	T
F	F	T	T	T	T

Two other variations of the implication $p \Rightarrow q$ are given special names. They are

$$q \Rightarrow p \quad \text{is the \textbf{converse} of} \quad p \Rightarrow q$$

and

$$\sim p \Rightarrow \sim q \quad \text{is the \textbf{inverse} of} \quad p \Rightarrow q.$$

We note that the converse and the inverse are logically equivalent; that is,

$$(q \Rightarrow p) \Leftrightarrow (\sim p \Rightarrow \sim q).$$

Example 5 Let p and q be the following statements:

p: x is an even integer
q: x is an integer.

In Figure A-9, we describe the implication $p \Rightarrow q$ and its variations.

Figure A-9

Logically equivalent		Logically equivalent	
Implication	Contrapositive	Converse	Inverse
$p \Rightarrow q$	$\sim q \Rightarrow \sim p$	$q \Rightarrow p$	$\sim p \Rightarrow \sim q$
x is an even integer	x is not an integer	x is an integer	x is not an even integer
$\Rightarrow$	$\Rightarrow$	$\Rightarrow$	$\Rightarrow$
x is an integer	x is not an even integer	x is an even integer	x is not an integer
TRUE	TRUE	FALSE	FALSE

■

Example 6 Suppose p and q are the following statements:

p: The Panthers win this week.
q: The Panthers are in the playoffs next week.

Suppose the only way the Panthers go to the playoffs is if they win this week. Hence, if they do not win this week, they will not go to the playoffs next week. In Figure A-10, we examine the implication $p \Rightarrow q$ and its variations.

Logically equivalent		Logically equivalent	
Implication	Contrapositive	Converse	Inverse
$p \Rightarrow q$	$\sim q \Rightarrow \sim p$	$q \Rightarrow p$	$\sim p \Rightarrow \sim q$
Panthers win this week	Panthers are not in the playoffs next week	Panthers are in the playoffs next week	Panthers do not win this week
$\Rightarrow$	$\Rightarrow$	$\Rightarrow$	$\Rightarrow$
Panthers are in the playoffs next week	Panthers do not win this week	Panthers win this week	Panthers are not in the playoffs next week
TRUE	TRUE	TRUE	TRUE

Figure A-10

Since the implication and its converse are true, we write

$$p \Leftrightarrow q.$$

The method of **proof by contradiction** is sometimes useful in proving statements of the form "p implies q." As shown in Figure A-4, the statement "p implies q" is true in all cases except when p is true and q is false. In a proof by contradiction, we assume that p is true and that q is false and then reach a contradiction (an impossible situation).

To provide a simple example, consider the following propositions:[†]

p : x is an integer and x^2 is even.

q : x is an even integer.

We shall use a proof by contradiction to prove that $p \Rightarrow q$.

Assume that p is true and q is false. Since x is not an even integer, x must be an odd integer. That is, $x = 2n + 1$ for some integer n. This implies that

$$x^2 = (2n + 1)(2n + 1)$$
$$= 4n^2 + 4n + 1$$
$$= 2(2n^2 + 2n) + 1,$$

and therefore x^2 is an odd integer. This directly contradicts proposition p. Therefore, q must be true when p is true, and this means that p implies q.

APPENDIX EXERCISES

Prove that each of the statements in Exercises 1–6 is false.

1. For every real number x, $x^2 > 0$.

2. For any real number x, $x^2 \geq x$.

3. For each real number a, there is a real number b such that $ab = 1$.

4. $2^x < 3^x$ for all real numbers x.

[†]An integer m is defined to be an *even integer* if $m = 2k$ for some integer k, and m is defined to be an *odd integer* if $m = 2q + 1$ for some integer q. More details may be found in Section 1.2.

5. $-x < |x|$ for all real numbers x.

6. If x is a real number such that $x < 1$, then $x^2 < x$.

Prove that each of the statements in Exercises 7–12 is true.

7. There is an integer n such that $n^2 + 2n = 48$.

8. There is a real number x such that $x + \dfrac{1}{x} = \dfrac{13}{6}$.

9. $n^2 < 2^n$ for some integer n.

10. $1 + 3n < 2^n$ for some integer n.

11. There exists an integer n such that $n^2 + n$ is an even integer.

12. There exists an integer n such that $n^2 + 2n$ is a multiple of 5.

Write the negation of each of the statements in Exercises 13–36.

13. All the children received a Valentine card.

14. Every house has a fireplace.

15. Every senior graduated and received a job offer.

16. All the cheerleaders are tall and athletic.

17. There is a rotten apple in the basket.

18. There is a snake that is nonpoisonous.

19. There is a politician who is honest and trustworthy.

20. There is a cold medication that is safe and effective.

21. For every $x \in A, x \in B$. (The notation $x \in A$ is defined in Section 1.1.)

22. For every real number r, the square of r is nonnegative.

23. For every right triangle with sides a and b and hypotenuse c, we have $c^2 = a^2 + b^2$.

24. For any two rational numbers r and s, there is an irrational number j between them.

25. Every complex number has a multiplicative inverse.

26. For all 2×2 matrices A and B over the real numbers, we have $AB = BA$. (The product of two matrices is given in Definition 1.30 of Section 1.5.)

27. For all sets A and B, their Cartesian products satisfy the equation $A \times B = B \times A$. (The Cartesian product is defined in Definition 1.8 of Section 1.2.)

28. For any real number $c, x < y \Rightarrow cx < cy$.

29. There exists a complex number x such that $x^2 + 1 = 0$.

30. There exists a 2×2 matrix A over the real numbers such that $A^2 = I$ where $I = \begin{bmatrix} 1 & 0 \\ 0 & 1 \end{bmatrix}$ and $A^2 = A \cdot A$. (The product of two matrices is given in Definition 1.30 of Section 1.5.)

31. There exists a set A such that $A \subseteq A \cap B$. (The notation $A \subseteq A \cap B$ is defined in Section 1.1.)

32. There exists a complex number z such that $\bar{z} = z$. (The notation $\bar{z}$ is given in Definition 7.7 of Section 7.2.)

33. There exists a triangle with angles α, β, and γ such that $\alpha + \beta + \gamma > 180°$.

34. There exists an angle θ such that $\sin \theta = 2.1$.

35. There exists a real number x such that $2^x \leq 0$.

36. There exists an even integer x such that x^2 is odd.

Construct truth tables for each of the statements in Exercises 37–52.

37. $p \Leftrightarrow \sim(\sim p)$

38. $p \vee (\sim p)$

39. $\sim(p \wedge (\sim p))$

40. $p \Rightarrow (p \vee q)$

41. $(p \wedge q) \Rightarrow p$

42. $\sim(p \vee q) \Leftrightarrow (\sim p) \wedge (\sim q)$

43. $(p \wedge (p \Rightarrow q)) \Rightarrow q$

44. $(p \Rightarrow q) \Leftrightarrow \sim(p \wedge \sim q)$

45. $(p \Rightarrow q) \Leftrightarrow ((\sim p) \vee q)$

46. $(\sim(p \Rightarrow q)) \Leftrightarrow (p \wedge (\sim q))$

47. $(p \Rightarrow q) \Leftrightarrow (p \wedge (\sim q) \Rightarrow (\sim p))$

48. $r \vee (p \wedge q) \Leftrightarrow (r \vee p) \wedge (r \vee q)$

49. $(p \wedge q \wedge r) \Rightarrow ((p \vee q) \wedge r)$

50. $((p \Rightarrow q) \wedge (q \Rightarrow r)) \Rightarrow (p \Rightarrow r)$

51. $(p \Rightarrow (q \wedge r)) \Leftrightarrow ((p \Rightarrow q) \wedge (p \Rightarrow r))$

52. $((p \wedge q) \Rightarrow r) \Leftrightarrow (p \Rightarrow (q \Rightarrow r))$

In Exercises 53–68, examine the implication $p \Rightarrow q$ and its variations (contrapositive, inverse, and converse) by writing each in English. Determine the truth or falsity of each.

53. p: My grade for this course is A.
q: I can enroll in the next course.

54. p: My car ran out of gas.
q: My car won't start.

55. p: The Saints win the Super Bowl.
q: The Saints are the champion football team.

56. p: I have completed all the requirements for a bachelor's degree.
q: I can graduate with a bachelor's degree.

57. p: My pet has four legs.
q: My pet is a dog.

58. p: I am within 30 miles of home.
q: I am within 20 miles of home.

59. p: Quadrilateral $ABCD$ is a square.
q: Quadrilateral $ABCD$ is a rectangle.

60. p: Triangle ABC is isosceles.
q: Triangle ABC is equilateral.

61. p: x is a positive real number.
q: x is a nonnegative real number.

62. p: x is a positive real number.
q: x^2 is a positive real number.

63. p: $5x$ is odd.
q: x is odd.

64. p: $5 + x$ is odd.
q: x is even.

65. p: xy is even.
q: x is even or y is even.

66. p: x is even and y is even.
   q: $x + y$ is even.

67. p: $x^2 > y^2$
   q: $x > y$

68. p: $\dfrac{x}{y} > 0$
   q: $xy > 0$

State the contrapositive, converse, and inverse of each of the implications in Exercises 69–74.

69. $p \Rightarrow (q \vee r)$ **70.** $p \Rightarrow (q \wedge r)$

71. $p \Rightarrow \sim q$ **72.** $(p \wedge \sim q) \Rightarrow \sim p$

73. $(p \vee q) \Rightarrow (r \wedge s)$ **74.** $(p \wedge q) \Rightarrow (r \wedge s)$

BIBLIOGRAPHY

Ames, Dennis B. *An Introduction to Abstract Algebra*. Scranton, PA: International Textbook, 1969.

Anderson, Marlow, and Todd Feil. *A First Course in Abstract Algebra*. Boston: PWS, 1995.

Ball, Richard W. *Principles of Abstract Algebra*. New York: Holt, Rinehart and Winston, 1963.

Ball, W. W. Rouse. *Mathematical Recreations & Essays*. 11th ed. New York: Macmillan, 1960.

Beker, Henry. *Cipher Systems: The Protection of Communications*. New York: Wiley, 1982.

Birkhoff, Garrett, and Saunders MacLane. *A Survey of Modern Algebra*. 4th ed. New York: A. K. Peters Limited, 1996.

Bloch, Norman J. *Abstract Algebra with Applications*. Englewood Cliffs, NJ: Prentice-Hall, 1986.

Bondi, Christine (editor). *New Applications of Mathematics*. New York: Penguin Books, 1991.

Buchthal, David C., and Douglas E. Cameron. *Modern Abstract Algebra*. Boston: PWS-KENT, 1987.

Bundrick, Charles M., and John J. Leeson. *Essentials of Abstract Algebra*. Monterey, CA: Brooks/Cole, 1972.

Burton, David M. *Abstract Algebra*. Dubuque, IA: Wm. C. Brown, 1988.

———. *The History of Mathematics*. 3rd ed. Dubuque, IA: Wm. C. Brown, 1995.

Clark, Allan. *Elements of Abstract Algebra*. New York: Dover, 1984.

Cohn, P. M. *Algebra*. 2nd ed. 2 vols. New York: Wiley, 1982, 1989.

Connell, I. *Modern Algebra: A Constructive Introduction*. New York: Elsevier, 1981.

Crown, G., M. Fenrick, and R. Valenza. *Abstract Algebra*. New York: Marcel Dekker, 1977.

Dean, R. A. *Elements of Abstract Algebra*. New York: Wiley, 1966.

Dubisch, Roy. *Introduction to Abstract Algebra*. New York: Wiley, 1985.

Dummit, David S., and Richard M. Foote. *Abstract Algebra*. Englewood Cliffs, NJ: Prentice-Hall, 1991.

Durbin, John R. *Modern Algebra*. 4th ed. New York: Wiley, 1999.

Eves, Howard. *Great Moments in Mathematics (Before 1650)*. Washington, DC: Mathematical Association of America, 1983.

———. *Great Moments in Mathematics (After 1650)*. Washington, DC: Mathematical Association of America, 1981.

———. *An Introduction to the History of Mathematics*. 5th ed. Troy, MO: Saunders, 1983.

———. *In Mathematical Circles, Quadrants I and II*. Boston: PWS, 1969.

———. *In Mathematical Circles, Quadrants III and IV*. Boston: PWS, 1969.

Fraleigh, John B. *A First Course in Abstract Algebra*. 6th ed. Reading: MA: Addison-Wesley, 1998.

Fuchs, Laszlo. *Infinite Abelian Groups*. 2 vols. New York: Academic Press, 1973.

Gallian, Joseph A. *Contemporary Abstract Algebra*. 3rd ed. Lexington, MA: D. C. Heath, 1994.

Gilbert, W. *Modern Algebra with Applications*. New York: Wiley, 1976.

Goldstein, L. J. *Abstract Algebra: A First Course*. Englewood Cliffs, NJ: Prentice-Hall, 1973.

Goodman, Frederick M. *Algebra*. Englewood Cliffs, NJ: Prentice-Hall, 1998.

Hall, F. M. *Introduction to Abstract Algebra*. 2nd ed. Vol. 1. New Rochelle, NY: Cambridge University Press, 1980.

———. *Introduction to Abstract Algebra*. Vol. 2. New Rochelle, NY: Cambridge University Press, 1980.

Hall, Marshall, Jr. *The Theory of Groups*. 2nd ed. New York: Chelsea, 1961.

Herstein, I. N. *Abstract Algebra*. 2nd ed. New York: Macmillan, 1988.

Hillman, Abraham P., and Gerald L. Alexanderson. *A First Undergraduate Course in Abstract Algebra.* 5th ed. Boston: PWS, 1994.

Hungerford, T. W. *Algebra.* New York: Springer-Verlag, 1989.

Jacobson, N. *Basic Algebra I.* 2nd ed. San Francisco: Freeman, 1985.

———. *Lectures in Abstract Algebra.* 3 vols. New York: Springer-Verlag, 1981, 1984, 1997.

Jones, Burton W. *An Introduction to Modern Algebra.* New York: Macmillan, 1975.

Kahn, David. *The Codebreakers: The Story of Secret Writing.* New York: Macmillan, 1967.

———. *Kahn on Codes: Secrets of the New Cryptology.* New York: Macmillan, 1983.

Keesee, John W. *Elementary Abstract Algebra.* Lexington, MA: D.C. Heath, 1965.

Kline, Morris. *Mathematical Thought from Ancient to Modern Times.* New York: Oxford University Press, 1972.

Koblitz, Neal. *A Course in Number Theory and Cryptography.* 2nd ed. New York: Springer-Verlag, 1994.

Konheim, Alan G. *Cryptography, A Primer.* New York: Wiley, 1981.

Kuczkowski, J., and J. Gersting. *Abstract Algebra: A First Look.* New York: Marcel Dekker, 1977.

Kurosh, A. *Theory of Groups.* 2 vols. Translated by K. A. Hirsch. New York: Chelsea, 1979.

Lang, S. *Algebra.* 2nd ed. Menlo Park, CA: Addison-Wesley, 1984.

Larney, V. C. *Abstract Algebra: A First Course.* Boston: PWS, 1975.

Larsen, Max D. *Introduction to Modern Algebraic Concepts.* Reading, MA: Addison-Wesley, 1969.

Lederman, Walter. *Introduction to the Theory of Finite Groups.* 4th ed. New York: Interscience, 1961.

McCoy, Neal H. *Fundamentals of Abstract Algebra.* Boston: Allyn and Bacon, 1972.

———. *Rings and Ideals* (Carus Mathematical Monograph No. 8). Washington, DC: The Mathematical Association of America, 1968.

———. *The Theory of Rings.* New York: Chelsea, 1972.

McCoy, N. H., and T. R. Berger. *Algebra: Groups, Rings, and Other Topics.* Boston: Allyn and Bacon, 1977.

McCoy, Neal H., and Gerald Janusz. *Introduction to Modern Algebra.* 5th ed. New York: McGraw-Hill, 1992.

Mackiw, George. *Applications of Abstract Algebra.* New York: Wiley, 1985.

Marcus, M. *Introduction to Modern Algebra.* New York: Marcel Dekker, 1978.

Maxfield, John E., and Margaret W. Maxfield. *Abstract Algebra and Solution by Radicals.* New York: Dover, 1983.

Mitchell, A. Richard, and Roger W. Mitchell. *An Introduction to Abstract Algebra.* Monterey, CA: Brooks/Cole, 1970.

Moore, J. T. *Introduction to Abstract Algebra.* New York: Academic Press, 1975.

Mostow, George D., Joseph H. Sampson, and Jean-Pierre Meyer. *Fundamental Structures of Algebra.* New York: McGraw-Hill, 1963.

Newman, James R. *The World of Mathematics.* Vol. 1. Scranton, PA: Harper and Row, 1988.

Nicholson, W. Keith. *Introduction to Abstract Algebra.* Boston: PWS-Kent, 1993.

Niven, Ivan, and Herbert S. Zuckerman. *An Introduction to the Theory of Numbers.* 4th ed. New York: Wiley, 1980.

Paley, H., and P. Weichsel. *A First Course in Abstract Algebra.* New York: Holt, Rinehart and Winston, 1966.

Pinter, C. C. *A Book of Abstract Algebra.* 2nd ed. New York: McGraw-Hill, 1989.

Rotman, Joseph J. *The Theory of Groups: An Introduction.* 3rd ed. Dubuque, IA: Wm. C. Brown, 1984.

Saracino, Dan. *Abstract Algebra: A First Course.* Reading, MA: Addison-Wesley, 1980.

Schilling, Otto F. G., and W. Stephen Piper. *Basic Abstract Algebra.* Boston: Allyn and Bacon, 1975.

Schneier, Bruce. *Applied Cryptography: Protocols, Algorithms, and Source Code in C.* 2nd ed. New York: Wiley, 1996.

Scott, W. R. *Group Theory.* Englewood Cliffs, NJ: Prentice-Hall, 1964.

Seberry, Jennifer. *Cryptography: An Introduction to Computer Security.* Englewood Cliffs, NJ: Prentice-Hall, 1989.

Shapiro, Louis. *Introduction to Abstract Algebra.* New York: McGraw-Hill, 1975.

Sierpinski, W., and A. Schinzel. *Elementary Theory of Numbers.* New York: Elsevier, 1988.

Smith, Laurence Dwight. *Cryptography: The Science of Secret Writing.* New York: Dover Publications, 1955.

Spence, Lawrence E., and Charles Vanden Eynden. *Elementary Abstract Algebra.* New York: Harper Collins, 1993.

Tannenbaum, Peter, and Robert Arnold. *Excursions in Modern Mathematics.* 3rd ed. Englewood Cliffs, NJ: Prentice-Hall, 1998.

Walker, Elbert A. *Introduction to Abstract Algebra.* New York: Random House, 1987.

Weiss, Marie, J., and Roy Dubisch. *Higher Algebra for the Undergraduate.* 2nd ed. New York: Wiley, 1962.

Welsh, Dominic. *Codes and Cryptography.* New York: Oxford University Press, 1988.

ANSWERS TO SELECTED COMPUTATIONAL EXERCISES

EXERCISES 1.1, pages 10–11

1. a. $A = \{x \mid x \text{ is a nonnegative even integer less than 12}\}$

   c. $A = \{x \mid x \text{ is a negative integer }\}$

2. a. False   **c.** False   **e.** False

3. a. False   **c.** True   **e.** True   **g.** False   **i.** False

4. a. True   **c.** False   **e.** False   **g.** False

5. a. $\{0, 1, 2, 3, 4, 5, 6, 8, 10\}$ **c.** $\{0, 2, 4, 6, 7, 8, 9, 10\}$ **e.** $\varnothing$

   g. $\{0, 2, 3, 4, 5\}$ **i.** $\{1, 3, 5\}$ **k.** $\{1, 2, 3, 5\}$ **m.** $\{3, 5\}$

6. a. A **c.** $\varnothing$ **e.** A **g.** A **i.** U **k.** U **m.** A

7. a. $\{\varnothing, A\}$ **c.** $\{\varnothing, \{a\}, \{b\}, \{c\}, \{a, b\}, \{a, c\}, \{b, c\}, A\}$

8. a. One possible partition is $X_1 = \{x \mid x \text{ is a negative integer}\}$ and $X_2 = \{x \mid x \text{ is a nonnegative integer}\}$. Another partition is $X_1 = \{x \mid x \text{ is a negative integer}\}$, $X_2 = \{0\}$, $X_3 = \{x \mid x \text{ is a positive integer}\}$.

   c. One partition is $X_1 = \{1, 5, 9\}$ and $X_2 = \{11, 15\}$. Another partition is $X_1 = \{1, 15\}$, $X_2 = \{11\}$ and $X_3 = \{5, 9\}$.

9. a. $X_1 = \{1\}, X_2 = \{2\}, X_3 = \{3\}; X_1 = \{1\}, X_2 = \{2, 3\}; X_1 = \{2\}, X_2 = \{1, 3\}; X_1 = \{3\}, X_2 = \{1, 2\}$

11. $A = \{a, b\}, B = \{a, c\}$

12. a. $A \subseteq B$ **c.** $B \subseteq A$ **e.** $A = B = U$ **g.** $A = U$

33. $(A \cap B') \cup (A' \cap B) = (A \cup B) \cap (A' \cup B')$

34. a.

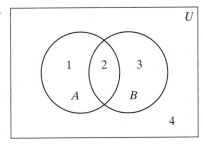

$A \cup B$:	Regions 1, 2, 3	$A - B$:	Region 1
$A \cap B$:	Region 2	$B - A$:	Region 3
$(A \cup B) - (A \cap B)$:	Regions 1, 3	$(A - B) \cup (B - A)$:	Regions 1, 3
$A + B$:	Regions 1, 3		

Each of $A + B$ and $(A - B) \cup (B - A)$ consists of Regions 1, 3.

c.

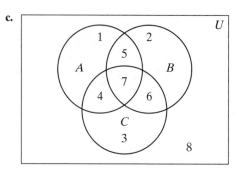

$$
\begin{array}{ll}
A:\ \text{Regions } 1, 4, 5, 7 & A \cap B:\ \text{Regions } 5, 7 \\
B + C:\ \text{Regions } 2, 3, 4, 5 & A \cap C:\ \text{Regions } 4, 7 \\
A \cap (B + C):\ \text{Regions } 4, 5 & (A \cap B) + (A \cap C):\ \text{Regions } 4, 5
\end{array}
$$

Each of $A \cap (B + C)$ and $(A \cap B) + (A \cap C)$ consists of Regions 4, 5.

Exercises 1.2, pages 19–22

1. a. $\{(a, 0), (a, 1), (b, 0), (b, 1)\}$ **c.** $\{(2, 2), (4, 2), (6, 2), (8, 2)\}$
    e. $\{(1, 1), (1, 2), (1, 3), (2, 1), (2, 2), (2, 3), (3, 1), (3, 2), (3, 3)\}$

2. a. Domain $= \mathbf{E}$, Codomain $= \mathbf{Z}$, Range $= \mathbf{Z}$
    c. Domain $= \mathbf{E}$, Codomain $= \mathbf{Z}$, Range $= \{y \mid y$ is a nonnegative even integer$\} =$
      $(\mathbf{Z}^{+} \cap \mathbf{E}) \cup \{0\}$

3. a. $f(S) = \{1, 3, 5, \dots\} = \mathbf{Z}^{+} - \mathbf{E}, f^{-1}(T) = \{-4, -3, -1, 1, 3, 4\}$
    c. $f(S) = \{0, 1, 4\}, f^{-1}(T) = \varnothing$

4. a. The mapping f is not onto, since there is no $x \in \mathbf{Z}$ such that $f(x) = 1$. The mapping f is
      one-to-one.
    c. The mapping f is onto and one-to-one.
    e. The mapping f is not onto, since there is no $x \in \mathbf{Z}$ such that $f(x) = -1$. It is not one-to-
      one, since $f(1) = f(-1)$ and $1 \neq -1$.
    g. The mapping f is not onto, since there is no $x \in \mathbf{Z}$ such that $f(x) = 3$. It is one-to-one.
    i. The mapping f is onto. It is not one-to-one, since $f(9) = f(4)$ and $9 \neq 4$.

5. a. The mapping f is both onto and one-to-one.
    c. The mapping f is both onto and one-to-one.
    e. The mapping f is not onto, since there is no $x \in \mathbf{R}$ such that $f(x) = -1$. It is not one-to-
      one, since $f(1) = f(-1)$ and $1 \neq -1$.

6. a. The mapping f is onto and one-to-one.

7. a. The mapping f is onto. The mapping is not one-to-one, since $f(-1) = f(1)$ and $-1 \neq 1$.
    c. The mapping f is onto and one-to one.

8. a. Let $f \colon \mathbf{E} \to \mathbf{E}$ where $f(x) = x$.
    c. Let $f \colon \mathbf{E} \to \mathbf{E}$ where

$$
f(x) = \begin{cases} x/2 & \text{if } x \text{ is a multiple of } 4 \\ x & \text{if } x \text{ is not a multiple of } 4. \end{cases}
$$

9. a. For arbitrary $a \in \mathbf{Z}, 2a - 1$ is odd, and therefore

$$
f(2a - 1) = \frac{(2a - 1) + 1}{2} = a.
$$

Thus, f is onto. But f is not one-to one, since $f(2) = 5$ and also $f(9) = 5$.

c. The mapping f is not onto, because there is no $x \in \mathbf{Z}$ such that $f(x) = 4$. Since $f(2) = 6$ and $f(3) = 6$, then f is not one-to-one.

10. a. The mapping f is not onto, because there is no $x \in \mathbf{R} - \{0\}$ such that $f(x) = 1$.

If $a_1, a_2 \in \mathbf{R} - \{0\}$,

$$f(a_1) = f(a_2) \implies \frac{a_1 - 1}{a_1} = \frac{a_2 - 1}{a_2}$$
$$\implies a_2(a_1 - 1) = a_1(a_2 - 1)$$
$$\implies a_2 a_1 - a_2 = a_1 a_2 - a_1$$
$$\implies -a_2 = -a_1$$
$$\implies a_2 = a_1.$$

Thus, f is one-to-one.

c. The mapping f is not onto, since there is no $x \in \mathbf{R} - \{0\}$ such that $f(x) = 0$. It is not one-to-one, since $f(2) = \frac{2}{5}$ and $f(\frac{1}{2}) = \frac{2}{5}$.

11. a. The mapping f is onto, since for every $(y, x) \in B = \mathbf{Z} \times \mathbf{Z}$ there exists an $(x, y) \in A = \mathbf{Z} \times \mathbf{Z}$ such that $f(x, y) = (y, x)$.

To show that f is one-to-one, we assume $(a, b) \in A = \mathbf{Z} \times \mathbf{Z}$ and $(c, d) \in A$ and

$$f(a, b) = f(c, d)$$

or

$$(b, a) = (d, c).$$

This means $b = d$ and $a = c$ and

$$(a, b) = (c, d).$$

c. Since for every $x \in B = \mathbf{Z}$ there exists an $(x, y) \in A = \mathbf{Z} \times \mathbf{Z}$ such that $f(x, y) = x$, the mapping f is onto. However, f is not one-to-one, since $f(1, 0) = f(1, 1)$ and $(1, 0) \neq (1, 1)$.

12. a. The mapping f is not onto, since there is no $a \in A$ such that $f(a) = 9 \in B$. It is not one-to-one, since $f(-2) = f(2)$ and $-2 \neq 2$.

c. With $T = \{4, 9\}$, $f^{-1}(T) = \{-2, 2\}$, and $f(f^{-1}(T)) = f(\{-2, 2\}) = \{4\} \neq T$.

13. a. $g(S) = \{2, 4\}$, $g^{-1}(g(S)) = \{2, 3, 4, 7\}$

14. a. $f(S) = \{-1, 2, 3\}$, $f^{-1}(f(S)) = S$

15. a. $(f \circ g)(x) = \begin{cases} 2x & \text{if } x \text{ is even} \\ 2(2x - 1) & \text{if } x \text{ is odd} \end{cases}$

c. $(f \circ g)(x) = \begin{cases} \dfrac{x + |x|}{2} & \text{if } x \text{ is even} \\ |x| - x & \text{if } x \text{ is odd} \end{cases}$

e. $(f \circ g)(x) = (x - |x|)^2$

16. a. $(g \circ f)(x) = 2x$ **c.** $(g \circ f)(x) = \dfrac{x + |x|}{2}$ **e.** $(g \circ f)(x) = 0$

17. n^m

19. $n(n - 1)(n - 2) \cdots (n - m + 1) = \dfrac{n!}{(n - m)!}$

■ EXERCISES 1.3,   pages 25–26

1. a. The mapping $f \circ g$ is not onto, since there is no $x \in \mathbf{Z}$ such that $(f \circ g)(x) = 1$. The mapping $f \circ g$ is one-to-one.

   c. The mapping $f \circ g$ is not onto, since there is no $x \in \mathbf{Z}$ such that $(f \circ g)(x) = 1$. It is not one-to-one, since $(f \circ g)(-2) = (f \circ g)(0)$ and $-2 \neq 0$.

   e. The mapping $f \circ g$ is not onto, since there is no $x \in \mathbf{Z}$ such that $(f \circ g)(x) = -1$. It is not one-to-one, since $(f \circ g)(1) = (f \circ g)(2)$ and $1 \neq 2$.

2. a. The mapping $g \circ f$ is not onto, since there is no $x \in \mathbf{Z}$ such that $(g \circ f)(x) = 1$. The mapping $g \circ f$ is one-to-one.

   c. The mapping $g \circ f$ is not onto, since there is no $x \in \mathbf{Z}$ such that $(g \circ f)(x) = -1$. It is not one-to-one, since $(g \circ f)(-1) = (g \circ f)(-2)$ and $-1 \neq -2$.

   e. The mapping $g \circ f$ is not onto, since there is no $x \in \mathbf{Z}$ such that $(g \circ f)(x) = 1$. It is not one-to-one, since $(g \circ f)(0) = (g \circ f)(1)$ and $0 \neq 1$.

3. Let $A = \{0, 1\}$, $B = \{-2, 1, 2\}$, $C = \{1, 4\}$. Let $g: A \to B$ be defined by $g(x) = x + 1$ and $f: B \to C$ be defined by $f(x) = x^2$. Then g is not onto, since $-2 \notin g(A)$. The mapping f is onto. Also, $f \circ g$ is onto, since $(f \circ g)(0) = f(1) = 1$ and $(f \circ g)(1) = f(2) = 4$.

5. a. Let $f: \mathbf{Z} \to \mathbf{Z}$ and $g: \mathbf{Z} \to \mathbf{Z}$ be defined by

$$f(x) = x \quad g(x) = \begin{cases} \dfrac{x}{2} & \text{if } x \text{ is even} \\ x & \text{if } x \text{ is odd.} \end{cases}$$

The mapping f is one-to-one and the mapping g is onto, but the composition $f \circ g = g$ is not one-to-one, since $(f \circ g)(1) = (f \circ g)(2)$ and $1 \neq 2$.

6. a. Let $f: \mathbf{Z} \to \mathbf{Z}$ and $g: \mathbf{Z} \to \mathbf{Z}$ be defined by

$$f(x) = \begin{cases} \dfrac{x}{2} & \text{if } x \text{ is even} \\ x & \text{if } x \text{ is odd} \end{cases} \quad g(x) = x.$$

The mapping f is onto and the mapping g is one-to-one, but the composition $f \circ g = f$ is not one-to-one, since $(f \circ g)(1) = (f \circ g)(2)$ and $1 \neq 2$.

8. a. Let $f(x) = x$, $g(x) = x^2$, and $h(x) = |x|$, for all $x \in \mathbf{Z}$.

■ EXERCISES 1.4,   pages 33–35

1. a. The set B is not closed, since $-1 \in B$ and $-1 * -1 = 1 \notin B$.

   c. The set B is closed.

2. a. Not commutative; not associative; no identity element

   c. Not commutative; not associative; no identity element

   e. Not commutative; not associative; no identity element

3. a. The binary operation $*$ is not commutative, since $D * A \neq A * D$.

   c. The elements A and B are inverses of each other and C is its own inverse.

10. a. A right inverse does not exist, since f is not onto.

   c. A right inverse $g: \mathbf{Z} \to \mathbf{Z}$ is $g(x) = x - 2$.

   e. A right inverse does not exist, since f is not onto.

   g. A right inverse does not exist, since f is not onto.

   i. A right inverse does not exist, since f is not onto.

k. A right inverse $g: \mathbf{Z} \to \mathbf{Z}$ is $g(x) = \begin{cases} x & \text{if } x \text{ is even} \\ 2x + 1 & \text{if } x \text{ is odd.} \end{cases}$

m. A right inverse $g: \mathbf{Z} \to \mathbf{Z}$ is $g(x) = \begin{cases} 2x & \text{if } x \text{ is even} \\ x - 2 & \text{if } x \text{ is odd.} \end{cases}$

11. a. A left inverse $g: \mathbf{Z} \to \mathbf{Z}$ is $g(x) = \begin{cases} \dfrac{x}{2} & \text{if } x \text{ is even} \\ 1 & \text{if } x \text{ is odd.} \end{cases}$

c. A left inverse $g: \mathbf{Z} \to \mathbf{Z}$ is $g(x) = x - 2$.

e. A left inverse $g: \mathbf{Z} \to \mathbf{Z}$ is $g(x) = \begin{cases} y & \text{if } x = y^3 \text{ for some } y \in \mathbf{Z} \\ 0 & \text{if } x \neq y^3 \text{ for some } y \in \mathbf{Z}. \end{cases}$

g. A left inverse $g: \mathbf{Z} \to \mathbf{Z}$ is $g(x) = \begin{cases} x & \text{if } x \text{ is even} \\ \dfrac{x + 1}{2} & \text{if } x \text{ is odd.} \end{cases}$

i. There is no left inverse, since f is not one-to-one.

k. There is no left inverse, since f is not one-to-one.

m. There is no left inverse, since f is not one-to-one.

13. Let $f: A \to A$, where A is nonempty.

$$f \text{ has a left inverse } \Leftrightarrow f \text{ is one-to-one, by Lemma 1.23}$$
$$\Leftrightarrow f^{-1}(f(S)) = S \text{ for every subset } S \text{ of } A, \text{ by}$$
$$\text{Exercise 25 of Section 1.2}$$

EXERCISES 1.5, **pages 44–46**

1. a. $A = \begin{bmatrix} 1 & 0 \\ 3 & 2 \\ 5 & 4 \end{bmatrix}$ **c.** $B = \begin{bmatrix} 1 & -1 & 1 & -1 \\ -1 & 1 & -1 & 1 \end{bmatrix}$

e. $C = \begin{bmatrix} 2 & 0 & 0 \\ 3 & 4 & 0 \\ 4 & 5 & 6 \\ 5 & 6 & 7 \end{bmatrix}$

2. a. $\begin{bmatrix} 3 & 0 & -4 \\ 8 & -8 & 6 \end{bmatrix}$ **c.** Not possible

3. a. $\begin{bmatrix} -5 & 7 \\ 8 & -1 \end{bmatrix}$ **c.** Not possible   **e.** $\begin{bmatrix} 4 & 2 \\ 3 & 7 \end{bmatrix}$ **g.** Not possible

i. $[4]$

4. $c_{ij} = \displaystyle\sum_{k=1}^{3} (i + k)(2k - j)$
$= (i + 1)(2 - j) + (i + 2)(4 - j) + (i + 3)(6 - j)$
$= 12i - 6j - 3ij + 28$

7. a. n **c.** 12

8.

·	I	A	B	C
I	I	A	B	C
A	A	B	C	I
B	B	C	I	A
C	C	I	A	B

9. (Answer not unique) $A = \begin{bmatrix} 1 & 2 \\ 3 & 4 \end{bmatrix}$, $B = \begin{bmatrix} 1 & 1 \\ 1 & 1 \end{bmatrix}$

11. (Answer not unique) $A = \begin{bmatrix} 1 & 2 \\ 1 & 2 \end{bmatrix}$, $B = \begin{bmatrix} -6 & -6 \\ 3 & 3 \end{bmatrix}$

13. $(A - B)(A + B) = \begin{bmatrix} 10 & 1 \\ 2 & 1 \end{bmatrix}$ and $A^2 - B^2 = \begin{bmatrix} 2 & 6 \\ -4 & 9 \end{bmatrix}$, $(A - B)(A + B) \neq A^2 - B^2$

15. $X = A^{-1}B$

20. b. For each x in G of the form $\begin{bmatrix} a & a \\ 0 & 0 \end{bmatrix}$, then $y = \begin{bmatrix} 1 & 1 \\ 0 & 0 \end{bmatrix}$. For each x in G of the form $\begin{bmatrix} 0 & 0 \\ a & a \end{bmatrix}$, then $y = \begin{bmatrix} 0 & 0 \\ 1 & 1 \end{bmatrix}$.

EXERCISES 1.6, pages 50–51

1. a. This is a mapping, since for every $a \in A$ there is a unique $b \in A$ such that (a, b) is an element of the relation.

   c. This is not a mapping, since the element 1 is related to three different values; $1R1$, $1R3$, and $1R5$.

   e. This is a mapping, since for every $a \in A$ there is a unique $b \in A$ such that (a, b) is an element of the relation.

2. a. The relation R is not reflexive, since $x \neq 2x$ for $x \neq 0$, $x \in \mathbf{Z}$. It is not symmetric, since $x = 2y \not\Rightarrow y = 2x$ for nonzero x and $y \in \mathbf{Z}$. It is not transitive, since $x = 2y$ and $y = 2z$ do not imply that $x = 2z$, for nonzero x, y, and z in $\mathbf{Z}$.

   c. The relation R is not reflexive and not symmetric, but it is transitive, since for arbitrary x, y, and $z \in \mathbf{Z}$ we have:

    (1) $x \not< x$

    (2) $x < y \not\Rightarrow y < x$

    (3) $x < y$ and $y < z \Rightarrow x < z$.

   e. The relation R is reflexive, symmetric, and transitive, since for arbitrary x, y, and z in $\mathbf{Z}$ we have:

    (1) $x - x = 5 \cdot 0$ and 0 is in $\mathbf{Z}$.

    (2) $x - y = 5 \cdot k$ for some k in $\mathbf{Z}$ implies $y - x = 5(-k)$ with $-k \in \mathbf{Z}$.

    (3) $x - y = 5k_1$ for some k_1 in $\mathbf{Z}$ and $y - z = 5k_2$ for some k_2 in $\mathbf{Z}$ imply $x - z = x - y + y - z = 5k_1 + 5k_2 = 5(k_1 + k_2)$ with $k_1 + k_2$ in $\mathbf{Z}$.

   g. The relation R is not reflexive, since $|-6| \not\leq |-6 + 1|$. It is not symmetric, since $|3| \leq |5 + 1|$, but $|5| \not\leq |3 + 1|$. It is not transitive, since $|4| \leq |3 + 1|$ and $|3| \leq |2 + 1|$, but $|4| \not\leq |2 + 1|$.

   i. The relation R is reflexive and transitive, but not symmetric, since for arbitrary x, y, and z in $\mathbf{Z}$ we have:

(1) $x = x \cdot 1$ with $1 \in \mathbf{Z}$
(2) $6 = 3(2)$ with $2 \in \mathbf{Z}$ but $3 \neq 6k$ where $k \in \mathbf{Z}$
(3) $y = xk_1$ for some $k_1 \in \mathbf{Z}$ and $z = yk_2$ for some $k_2 \in \mathbf{Z}$ imply $z = yk_2 = x(k_1k_2)$ with $k_1k_2 \in \mathbf{Z}$.

3. a. $\{-3, 3\}$

4. a. The relation R is reflexive and transitive but not symmetric, since for arbitrary non-empty subsets x, y, and z of A we have:

(1) x is a subset of x,
(2) x is a subset of y does not imply that y is a subset of x,
(3) x is a subset of y and y is a subset of z imply that x is a subset of z.

  c. The relation R is reflexive, symmetric and transitive, since for arbitrary nonempty subsets x, y, and z of A we have:

(1) x and x have the same number of elements.
(2) If x and y have the same number of elements, then y and x have the same number of elements.
(3) If x and y have the same number of elements and y and z have the same number of elements, then x and z have the same number of elements.

5. a. The relation is reflexive and symmetric but not transitive, since if x, y, and z are human beings, we have:

(1) x lives within 400 miles of x.
(2) x lives within 400 miles of y implies that y lives within 400 miles of x.
(3) x lives within 400 miles of y and y lives within 400 miles of z do not imply that x lives within 400 miles of z.

  c. The relation is symmetric but not reflexive and not transitive. Let x, y, and z be human beings, and we have:

(1) x is a first cousin of x is not a true statement.
(2) x is a first cousin of y implies that y is a first cousin of x.
(3) x is a first cousin of y and y is a first cousin of z do not imply that x is a first cousin of z.

  e. The relation is reflexive, symmetric, and transitive, since if x, y, and z are human beings, we have:

(1) x and x have the same mother.
(2) x and y have the same mother implies y and x have the same mother.
(3) x and y have the same mother and y and z have the same mother imply that x and z have the same mother.

6. a. The relation R is a equivalence relation on $A \times A$. Let a, b, c, d, p, and q be arbitrary elements of A.

(1) $(a, b)R(a, b)$ since $ab = ba$.
(2) $(a, b)R(c, d) \Rightarrow ad = bc \Rightarrow cb = da \Rightarrow (c, d)R(a, b)$
(3) $(a, b)R(c, d)$ and $(c, d)R(p, q) \Rightarrow ad = bc$ and $cq = dp$
$$\Rightarrow adcq = bcdp$$
$$\Rightarrow aq = bp \text{ since } c \neq 0 \text{ and } d \neq 0$$
$$\Rightarrow (a, b)R(p, q)$$

  c. The relation R is an equivalence relation on $A \times A$. Let a, b, c, d, p, and q be arbitrary elements of A.

(1) $(a, b)R(a, b)$ since $a^2 + b^2 = a^2 + b^2$.
(2) $(a, b)R(c, d) \Rightarrow a^2 + b^2 = c^2 + d^2 \Rightarrow c^2 + d^2 = a^2 + b^2 \Rightarrow (c, d)R(a, b)$

(3) $(a, b)R(c, d)$ and $(c, d)R(p, q) \Rightarrow a^2 + b^2 = c^2 + d^2$ and $c^2 + d^2 = p^2 + q^2$
$$\Rightarrow a^2 + b^2 = p^2 + q^2$$
$$\Rightarrow (a, b)R(p, q)$$

7. The relation R is reflexive and symmetric but not transitive.

8. a. The relation is symmetric but not reflexive and not transitive. Let x, y, and z be arbitrary elements of the power set $\mathcal{P}(A)$ of the nonempty set A.

(1) $x \cap x \neq \varnothing$ is not true if $x = \varnothing$.
(2) $x \cap y \neq \varnothing$ implies that $y \cap x \neq \varnothing$.
(3) $x \cap y \neq \varnothing$ and $y \cap z \neq \varnothing$ do not imply that $x \cap z \neq \varnothing$. For example, let $A = \{a, b, c, d\}$, $x = \{b, c\}$, $y = \{c, d\}$, and $z = \{d, a\}$. Then $x \cap y = \{c\} \neq \varnothing$, $y \cap z = \{d\} \neq \varnothing$ but $x \cap z = \varnothing$.

9. The relation is reflexive, symmetric, and transitive. Let x, y, and z be arbitrary elements of the power set $\mathcal{P}(A)$ and C a fixed subset of A.

(1) xRx since $x \cap C = x \cap C$
(2) $xRy \Rightarrow x \cap C = y \cap C \Rightarrow y \cap C = x \cap C \Rightarrow yRx$
(3) xRy and $yRz \Rightarrow x \cap C = y \cap C$ and $y \cap C = z \cap C$
$$\Rightarrow x \cap C = z \cap C$$
$$\Rightarrow xRz$$

Thus, R is an equivalence relation on $\mathcal{P}(A)$.

10. a. The relation is reflexive, symmetric, and transitive. Let a, b, and c represent arbitrary triangles in the plane. Then

(1) a is similar to a is true.
(2) a is similar to b implies that b is similar to a.
(3) a is similar to b and b is similar to c imply that a is similar to c.

11. c, j

13. a, c, d

15. $\bigcup_{\lambda \in L} A_\lambda = A_1 \cup A_2 \cup A_3 = \{a, b, c, d, e, f, g\}$, $\bigcap_{\lambda \in L} A_\lambda = A_1 \cap A_2 \cap A_3 = \{c\}$

EXERCISES 2.1, pages 57–58

27. All the addition postulates and all the multiplication postulates except 2c are satisfied. Postulate 2c is not satisfied, since $\{0\}$ does not contain an element different from 0. The set $\{0\}$ has the properties required in postulate 4, and postulate 5 is satisfied vacuously (that is, there is no counterexample). Thus, all postulates except 2c are satisfied.

EXERCISES 2.3, pages 69–70

1. a. $\pm1, \pm2, \pm3, \pm5, \pm6, \pm10, \pm15, \pm30$ **c.** $\pm1, \pm2, \pm4, \pm7, \pm14, \pm28$
e. $\pm1, \pm2, \pm3, \pm4, \pm6, \pm8, \pm12, \pm24$ **g.** $\pm1, \pm2, \pm4, \pm8, \pm16, \pm32$

3. $q = 34, r = 2$ **5.** $q = 32, r = 21$ **7.** $q = -4, r = 1$

9. $q = -52, r = 15$ **11.** $q = 0, r = 15$ **13.** $q = -32, r = 156$

15. $q = 0, r = 0$

19. If $a = 0$, then $n = -1$ makes $a - bn = 0 - b(-1) = b > 0$, and we have a positive element of S in this case. If $a \neq 0$, the choice $n = -2|a|$ gives $a - bn = a + 2b|a|$ as a specific example of a positive element of S. The problem does not explicitly require a proof that our element is positive, but this can be done as follows.

Since $b > 0$, we have $b \geq 1$ by Theorem 2.6. This implies $b|a| \geq |a|$ by Exercise 14 of Section 2.1. It follows from the definition of absolute value that $|a| \geq -a$. Now

$$b|a| \geq |a| \quad \text{and} \quad |a| \geq -a \Rightarrow b|a| \geq -a.$$

Since $a \neq 0$, $|a| > 0$, and therefore $|a| \geq 1$ by Theorem 2.6. Hence, $b|a| \geq b$ by Exercise 14 of Section 2.1.

$$b|a| \geq b \quad \text{and} \quad b > 0 \Rightarrow b|a| > 0$$

We have $b|a| \geq -a$ and $b|a| > 0$. By Exercise 12 of Section 2.1,

$$b|a| + b|a| > -a + 0,$$
$$2b|a| > -a, \text{ and}$$
$$a + 2b|a| > 0.$$

This shows that $a + 2b|a|$ is positive.

EXERCISES 2.4,   pages 76–77

1. $2, 3, 5, 7, 11, 13, 17, 19, 23, 29, 31, 37, 41, 43, 47, 53, 59, 61, 67, 71, 73, 79, 83, 89, 97$

2. a. $1400 = 2^3 \cdot 5^2 \cdot 7$; $980 = 2^2 \cdot 5 \cdot 7^2$; $(1400, 980) = 2^2 \cdot 5 \cdot 7 = 140$
  c. $3780 = 2^2 \cdot 3^3 \cdot 5 \cdot 7$; $16200 = 2^3 \cdot 3^4 \cdot 5^2$; $(3780, 16200) = 2^2 \cdot 3^3 \cdot 5 = 540$

3. a. $(a, b) = 3, m = 0, n = -1$ **c.** $(a, b) = 6, m = 2, n = -3$
  e. $(a, b) = 3, m = 2, n = 25$ **g.** $(a, b) = 9, m = -5, n = 3$
  i. $(a, b) = 3, m = -49, n = 188$ **k.** $(a, b) = 12, m = -3, n = 146$
  m. $(a, b) = 12, m = 5, n = 163$

4. a. $(4, 6) = 2$ **7.** $a = 6, b = 8, c = 9$

23. After a and b are written in their standard forms, the least common multiple of a and b can be found by forming the product of all the distinct prime factors that appear in the standard form of either a or b, with each factor raised to the greatest power to which it appears in either standard form.

24. a. The least common multiple of $1400 = 2^3 \cdot 5^2 \cdot 7$ and $980 = 2^2 \cdot 5 \cdot 7^2$ is $2^3 \cdot 5^2 \cdot 7^2 = 9800$.
  c. The least common multiple of $3780 = 2^2 \cdot 3^3 \cdot 5 \cdot 7$ and $16,200 = 2^3 \cdot 3^4 \cdot 5^2$ is $2^3 \cdot 3^4 \cdot 5^2 \cdot 7 = 113,400$.

25. a. An integer d is a **greatest common divisor** of a, b, and c if these conditions are satisfied:
  (1) d is a positive integer;
  (2) $d|a, d|b$, and $d|c$;
  (3) If $n|a, n|b$, and $n|c$, then $n|d$.

EXERCISES 2.5,   pages 82–83

1. $[0] = \{\ldots, -5, 0, 5, \ldots\}, [1] = \{\ldots, -4, 1, 6, \ldots\}$,
  $[2] = \{\ldots, -3, 2, 7, \ldots\}, [3] = \{\ldots, -2, 3, 8, \ldots\}$,
  $[4] = \{\ldots, -1, 4, 9, \ldots\}$

3. $x = 5$ **5.** $x = 11$ **7.** $x = 8$ **9.** $x = 173$ **11.** $x = 28$

13. $x = 7$ **15.** $x = 6$ **17.** $x = 2$ **23. a.** 3

31. $d = (6, 27) = 3$, and 3 divides 33; $x = 1, x = 10, x = 19$ are solutions.

33. $d = (8, 78) = 2$, and 2 divides 66; $x = 18$ and $x = 57$ are solutions.

35. $d = (34, 20) = 2$ and 2 divides 18; $x = 7$ and $x = 17$ are solutions.

37. $d = (24, 348) = 12$, and 12 does not divide 45; therefore, there are no solutions.

39. $d = (42, 74) = 2$, and 2 divides 30; $x = 6$ and $x = 43$ are solutions.

44. a. $x = 27$ or $x \equiv 27 \pmod{40}$

EXERCISES 2.6, pages 88–90

1. a. [3]   **c.** [4]   **e.** [6][4] = [0]   **g.** [6] + [6] = [0]

2. a. [1][2][3][4] = [24] = [4]   **c.** [1][2][3] = [6] = [2]

3. a.

+	[0]	[1]
[0]	[0]	[1]
[1]	[1]	[0]

c.

+	[0]	[1]	[2]	[3]	[4]
[0]	[0]	[1]	[2]	[3]	[4]
[1]	[1]	[2]	[3]	[4]	[0]
[2]	[2]	[3]	[4]	[0]	[1]
[3]	[3]	[4]	[0]	[1]	[2]
[4]	[4]	[0]	[1]	[2]	[3]

e.

+	[0]	[1]	[2]	[3]	[4]	[5]	[6]
[0]	[0]	[1]	[2]	[3]	[4]	[5]	[6]
[1]	[1]	[2]	[3]	[4]	[5]	[6]	[0]
[2]	[2]	[3]	[4]	[5]	[6]	[0]	[1]
[3]	[3]	[4]	[5]	[6]	[0]	[1]	[2]
[4]	[4]	[5]	[6]	[0]	[1]	[2]	[3]
[5]	[5]	[6]	[0]	[1]	[2]	[3]	[4]
[6]	[6]	[0]	[1]	[2]	[3]	[4]	[5]

4. a.

×	[0]	[1]
[0]	[0]	[0]
[1]	[0]	[1]

c.

×	[0]	[1]	[2]	[3]	[4]
[0]	[0]	[0]	[0]	[0]	[0]
[1]	[0]	[1]	[2]	[3]	[4]
[2]	[0]	[2]	[4]	[1]	[3]
[3]	[0]	[3]	[1]	[4]	[2]
[4]	[0]	[4]	[3]	[2]	[1]

e.

×	[0]	[1]	[2]	[3]	[4]	[5]	[6]
[0]	[0]	[0]	[0]	[0]	[0]	[0]	[0]
[1]	[0]	[1]	[2]	[3]	[4]	[5]	[6]
[2]	[0]	[2]	[4]	[6]	[1]	[3]	[5]
[3]	[0]	[3]	[6]	[2]	[5]	[1]	[4]
[4]	[0]	[4]	[1]	[5]	[2]	[6]	[3]
[5]	[0]	[5]	[3]	[1]	[6]	[4]	[2]
[6]	[0]	[6]	[5]	[4]	[3]	[2]	[1]

5. a. [9]   **c.** [13]   **e.** [173]
6. a. [1], [5]   **c.** [1], [3], [7], [9]   **e.** [1], [5], [7], [11], [13], [17]
7. a. [2], [3], [4]   **c.** [2], [4], [5], [6], [8]
  e. [2], [3], [4], [6], [8], [9], [10], [12], [14], [15], [16]
8. a. $[x] = [2]$ or $[x] = [5]$ **c.** $[x] = [2]$ or $[x] = [6]$
  e. No solution exists.   **g.** $[x] = [4]$ or $[x] = [10]$
10. a. $[x] = [4]^{-1}[5] = [10][5] = [11]$ **c.** $[x] = [7]^{-1}[11] = [7][11] = [5]$
  e. $[x] = [9]^{-1}[14] = [9][14] = [6]$ **g.** $[x] = [6]^{-1}[5] = [266][5] = [54]$
11. $[x] = [3]$, $[y] = [5]$
13. $[x] = [3]$, $[y] = [3]$
19. a. $[x] = [4]$ or $[x] = [5]$ **c.** $[x] = [1]$ or $[x] = [5]$

EXERCISES 2.7,   pages 95–97

1. Errors occur in 00010 and 11100.
3. Correct coded message:
  101101101   110110110   100100100   101101101   010010010   011011011
  Decoded message: 101   110   100   101   010   011
5. a. $\frac{3}{4}$ **c.** $\frac{2}{6} = \frac{1}{3}$
6. a. $(0.97)^4 + 4(0.97)^3(0.03) = 0.9948136$
7. a. $(0.9999)^8 = 0.9992003$
  c. $(0.9999)^8 + 8(0.9999)^7(0.0001) = 0.9999997$
  e. 1.000000
9. 1   **14. a.** 7   **c.** 1   **17. a.** Valid   **c.** Not valid
18. a. No error is detected.   **c.** An error is detected.
19. $y = -(10, 9, 8, 7, 6, 5, 4, 3, 2)$ **20. a.** 3   **c.** 3
22. a. 3   **c.** 3   **25.** 3
27. 00000   01010   10011   11001   00101   01111   10110   11100

■ **EXERCISES 2.8,**   pages 104–107

1. Ciphertext: APMHKPMKSHQ   HQVHAPMHUIQT

$f^{-1}(x) = x + 19 \bmod 27$

3. Plaintext: "tiger, do you read me?"

$f^{-1}(x) = x + 20 \bmod 31$

5. Ciphertext: FBBZXLXDGIXZUW

$f^{-1}(x) = 4x + 7 \bmod 27$

7. Plaintext: www.brookscole.com

$f^{-1}(x) = 19x + 2 \bmod 28$

9. Plaintext: mathematics

$f(x) = 9x + 13 \bmod 26$
$f^{-1}(x) = 3x + 13 \bmod 26$

11. Plaintext: there are 25 primes less than 100

$f(x) = 12x + 17 \bmod 37$
$f^{-1}(x) = 34x + 14 \bmod 37$

15. a. $n - 1$ **b.** $(n - 1)n - 1 = n^2 - n - 1$

17. Ciphertext: 62   49   75   26   49   73   75   50   61 $d = 37$

19. a. Ciphertext: 000   132   085   082   001   030   000
    b. Ciphertext: 001   050   105   039   000
    c. $d = 103$

21. Plaintext: quaternions

23. a. $\phi(5) = 4$;   1, 2, 3, 4   **c.** $\phi(15) = 8$;   1, 2, 4, 7, 8, 11, 13, 14
    e. $\phi(12) = 4$;   1, 5, 7, 11

25. The positive integers less than or equal to p^j that are not relatively prime to p^j are multiples of p, that is, elements of the set

$$\{1p, 2p, 3p, \ldots, (p^{j-1} - 1)p, p^{j-1}p\}.$$

Since this set contains p^{j-1} elements, then

$$\phi(p^j) = p^j - p^{j-1} = p^{j-1}(p - 1).$$

■ **EXERCISES 3.1,**   pages 118–122

1. Group

3. The set of all positive irrational numbers with the operation of multiplication does not form a group. The set is not closed with respect to multiplication. For example, $\sqrt{2}$ is a positive irrational number, but $\sqrt{2}\sqrt{2} = 2$ is not. Also, there is no identity element.

5. The set of all real numbers x such that $0 < x \le 1$ is not a group with respect to multiplication because not all elements have inverses.

7. Group   **9.** Group

11. The operation $\times$ is not associative, since

$$a \times (c \times a) = a \times e = a,$$

whereas

$$(a \times c) \times a = b \times a = c.$$

Also, there are no inverses for the elements a and b.

13.

$\times$	a	b	c	d
a	c	d	a	b
b	d	c	b	a
c	a	b	c	d
d	b	a	d	c

15. The set $\mathbf{Z}$ is an abelian group with respect to $*$. The identity element is -1. The element $-x - 2$ is the inverse of the element $x \in \mathbf{Z}$.

17. The set $\mathbf{Z}$ is not a group and hence not an abelian group with respect to the operation $*$. The operation is not associative. There is no identity element and hence no inverse elements.

19. The set $\mathbf{Z}$ is not a group and hence not an abelian group with respect to $*$. The identity element is 0, but 1 does not have an inverse in $\mathbf{Z}$.

21. Group

23. The set is not a group with respect to multiplication, since it does not have an identity element and hence no inverse elements.

25. Group

27. b.

$\times$	[1]	[2]	[3]	[4]	[5]	[6]
[1]	[1]	[2]	[3]	[4]	[5]	[6]
[2]	[2]	[4]	[6]	[1]	[3]	[5]
[3]	[3]	[6]	[2]	[5]	[1]	[4]
[4]	[4]	[1]	[5]	[2]	[6]	[3]
[5]	[5]	[3]	[1]	[6]	[4]	[2]
[6]	[6]	[5]	[4]	[3]	[2]	[1]

$[1]^{-1} = [1]$, $[2]$ and $[4]$ are inverses of each other, $[3]$ and $[5]$ are inverses of each other, and $[6]^{-1} = [6]$.

29.

×	I_3	P_1	P_2	P_3	P_4	P_5
I_3	I_3	P_1	P_2	P_3	P_4	P_5
P_1	P_1	I_3	P_3	P_2	P_5	P_4
P_2	P_2	P_5	I_3	P_4	P_3	P_1
P_3	P_3	P_4	P_1	P_5	P_2	I_3
P_4	P_4	P_3	P_5	P_1	I_3	P_2
P_5	P_5	P_2	P_4	I_3	P_1	P_3

35. The set G is not a group with respect to addition, since it does not contain an identity with respect to addition.

41. One possible choice for the elements, a, b, and c is the following: $a = \rho$, $b = \sigma$, and $c = \rho^2$. Then we have $\rho \circ \sigma = \gamma = \sigma \circ \rho^2$, but $\rho \neq \rho^2$.

51. $\mathcal{P}(A) = \{\varnothing, \{a\}, \{b\}, \{c\}, \{a, b\}, \{a, c\}, \{b, c\}, A\}$

+	$\varnothing$	$\{a\}$	$\{b\}$	$\{c\}$	$\{a, b\}$	$\{a, c\}$	$\{b, c\}$	A
$\varnothing$	$\varnothing$	$\{a\}$	$\{b\}$	$\{c\}$	$\{a, b\}$	$\{a, c\}$	$\{b, c\}$	A
$\{a\}$	$\{a\}$	$\varnothing$	$\{a, b\}$	$\{a, c\}$	$\{b\}$	$\{c\}$	A	$\{b, c\}$
$\{b\}$	$\{b\}$	$\{a, b\}$	$\varnothing$	$\{b, c\}$	$\{a\}$	A	$\{c\}$	$\{a, c\}$
$\{c\}$	$\{c\}$	$\{a, c\}$	$\{b, c\}$	$\varnothing$	A	$\{a\}$	$\{b\}$	$\{a, b\}$
$\{a, b\}$	$\{a, b\}$	$\{b\}$	$\{a\}$	A	$\varnothing$	$\{b, c\}$	$\{a, c\}$	$\{c\}$
$\{a, c\}$	$\{a, c\}$	$\{c\}$	A	$\{a\}$	$\{b, c\}$	$\varnothing$	$\{a, b\}$	$\{b\}$
$\{b, c\}$	$\{b, c\}$	A	$\{c\}$	$\{b\}$	$\{a, c\}$	$\{a, b\}$	$\varnothing$	$\{a\}$
A	A	$\{b, c\}$	$\{a, c\}$	$\{a, b\}$	$\{c\}$	$\{b\}$	$\{a\}$	$\varnothing$

53. The set A is an identity element. But the set $\mathcal{P}(A)$ is not a group with respect to the operation of intersection, since A is the only element that has an inverse.

55. Let n be a positive integer, $n \geq 2$. For elements $a_1, a_2, \ldots, a_n$ in a group G, the expression $a_1 + a_2 + \cdots + a_n$ is defined recursively by

$$a_1 + a_2 + \cdots + a_k + a_{k+1} = (a_1 + a_2 + \cdots + a_k) + a_{k+1}, \quad \text{for } k \geq 1.$$

EXERCISES 3.2,   pages 129–131

1. a. The set $\{e, \sigma\}$ is a subgroup of $S(A)$. The multiplication table is:

$\circ$	e	ρ
e	e	ρ
ρ	ρ	ρ^2

c. The set $\{e, \rho\}$ is not a subgroup of $S(A)$, since it is not closed. We have $\rho \circ \rho = \rho^2 \notin \{e, \rho\}$. The multiplication table is:

$\circ$	e	σ
e	e	σ
σ	σ	e

e. The set $\{e, \rho, \rho^2\}$ is a subgroup of $S(A)$. The multiplication table is:

$\circ$	e	ρ	ρ^2
e	e	ρ	ρ^2
ρ	ρ	ρ^2	e
ρ^2	ρ^2	e	ρ

g. The set $\{e, \sigma, \gamma\}$ is not a subgroup of $S(A)$, since it is not closed. We have $\gamma \circ \sigma = \rho \notin \{e, \sigma, \gamma\}$. The multiplication table is:

$\circ$	e	σ	γ
e	e	σ	γ
σ	σ	e	ρ^2
γ	γ	ρ	e

2. a. Subgroup

    c. The set $\{i, -i\}$ is not a subgroup of G, since it is not closed. We have $i \cdot i = -1 \notin \{i, -i\}$.

3. $\langle[6]\rangle = \{[0], [2], [4], [6], [8], [10], [12], [14]\}$

5. a. $\{[1], [3], [4], [9], [10], [12]\}$

6. a. $\langle A \rangle = \left\{ \begin{bmatrix} 0 & -1 \\ 1 & 0 \end{bmatrix}, \begin{bmatrix} -1 & 0 \\ 0 & -1 \end{bmatrix}, \begin{bmatrix} 0 & 1 \\ -1 & 0 \end{bmatrix}, \begin{bmatrix} 1 & 0 \\ 0 & 1 \end{bmatrix} \right\}$

7. a. $\langle A \rangle = \left\{ \begin{bmatrix} [2] & [0] \\ [0] & [3] \end{bmatrix}, \begin{bmatrix} [4] & [0] \\ [0] & [1] \end{bmatrix}, \begin{bmatrix} [1] & [0] \\ [0] & [4] \end{bmatrix}, \begin{bmatrix} [3] & [0] \\ [0] & [2] \end{bmatrix}, \begin{bmatrix} [0] & [0] \\ [0] & [0] \end{bmatrix} \right\}$

9. The set of all real numbers that are greater than 1 is closed under multiplication but is not a subgroup of G, since it does not contain inverses. (If $x > 1$, then $x^{-1} < 1$.)

18. a. $\{1, -1\}$

    c. $\{I_3\}$

25. The subgroup $\langle m \rangle \cap \langle n \rangle$ is the set of all multiples of the least common multiple of m and n.

27. Let $H_1 = \{e, \sigma\}$ and $H_2 = \{e, \gamma\}$.

29. Let $H_1 = \{e, \sigma\}$ and $H_2 = \{e, \gamma\}$.

EXERCISES 3.3, pages 138–140

1. $\langle e \rangle = \{e\}, \langle \rho \rangle = \{e, \rho, \rho^2\}, \langle \sigma \rangle = \{e, \sigma\}, \langle \gamma \rangle = \{e, \gamma\}, \langle \delta \rangle = \{e, \delta\}$

3. The element e has order 1. Each of the elements σ, γ, and δ has order 2. Each of the elements ρ and ρ^2 has order 3.

5. $o(I_3) = 1, o(P_1) = o(P_2) = o(P_4) = 2, o(P_3) = o(P_5) = 3$

6. **a.** $o(A) = 2$

7. **a.** $[1], [3], [5], [7]$

   c. $[1], [3], [7], [9]$

   e. $[1], [3], [5], [7], [9], [11], [13], [15]$

8. **a.** $\{[0]\}, \{[0], [6]\}, \{[0], [4], [8]\}, \{[0], [3], [6], [9]\}, \{[0], [2], [4], [6], [8], [10]\}, \mathbf{Z}_{12}$

   c. $\{[0]\}, \{[0], [5]\}, \{[0], [2], [4], [6], [8]\}, \mathbf{Z}_{10}$

   e. $\{[0]\}, \{[0], [8]\}, \{[0], [4], [8], [12]\}, \{[0], [2], [4], [6], [8], [10], [12], [14]\}, \mathbf{Z}_{16}$

9. **a.** $G = \langle [3] \rangle = \langle [5] \rangle$

   c. $G = \langle [2] \rangle = \langle [6] \rangle = \langle [7] \rangle = \langle [8] \rangle$

   e. $G = \langle [3] \rangle = \langle [5] \rangle = \langle [6] \rangle = \langle [7] \rangle = \langle [10] \rangle = \langle [11] \rangle = \langle [12] \rangle = \langle [14] \rangle$

10. **a.** $[3], [5]$

   c. $[2], [6], [7], [8]$

   e. $[3], [5], [6], [7], [10], [11], [12], [14]$

11. **a.** $\{[1]\}, \{[1], [6]\}, \{[1], [2], [4]\}, G$

   c. $\{[1]\}, \{[1], [10]\}, \{[1], [3], [4], [5], [9]\}, G$

   e. $\{[1]\}, \{[1], [16]\}, \{[1], [4], [13], [16]\}, \{[1], [2], [4], [8], [9], [13], [15], [16]\}, G$

13. **c.** $H = \left\{ \begin{bmatrix} 1 & 0 \\ 0 & 1 \end{bmatrix}, \begin{bmatrix} -\frac{1}{2} & -\frac{\sqrt{3}}{2} \\ \frac{\sqrt{3}}{2} & -\frac{1}{2} \end{bmatrix}, \begin{bmatrix} -\frac{1}{2} & \frac{\sqrt{3}}{2} \\ -\frac{\sqrt{3}}{2} & -\frac{1}{2} \end{bmatrix} \right\}$

16. **a.** $G = \{[1], [3], [7], [9], [11], [13], [17], [19]\}$

$\cdot$	$[1]$	$[3]$	$[7]$	$[9]$	$[11]$	$[13]$	$[17]$	$[19]$
$[1]$	$[1]$	$[3]$	$[7]$	$[9]$	$[11]$	$[13]$	$[17]$	$[19]$
$[3]$	$[3]$	$[9]$	$[1]$	$[7]$	$[13]$	$[19]$	$[11]$	$[17]$
$[7]$	$[7]$	$[1]$	$[9]$	$[3]$	$[17]$	$[11]$	$[19]$	$[13]$
$[9]$	$[9]$	$[7]$	$[3]$	$[1]$	$[19]$	$[17]$	$[13]$	$[11]$
$[11]$	$[11]$	$[13]$	$[17]$	$[19]$	$[1]$	$[3]$	$[7]$	$[9]$
$[13]$	$[13]$	$[19]$	$[11]$	$[17]$	$[3]$	$[9]$	$[1]$	$[7]$
$[17]$	$[17]$	$[11]$	$[19]$	$[13]$	$[7]$	$[1]$	$[9]$	$[3]$
$[19]$	$[19]$	$[17]$	$[13]$	$[11]$	$[9]$	$[7]$	$[3]$	$[1]$

c. $G = \{[1], [5], [7], [11], [13], [17], [19], [23]\}$

·	[1]	[5]	[7]	[11]	[13]	[17]	[19]	[23]
[1]	[1]	[5]	[7]	[11]	[13]	[17]	[19]	[23]
[5]	[5]	[1]	[11]	[7]	[17]	[13]	[23]	[19]
[7]	[7]	[11]	[1]	[5]	[19]	[23]	[13]	[17]
[11]	[11]	[7]	[5]	[1]	[23]	[19]	[17]	[13]
[13]	[13]	[17]	[19]	[23]	[1]	[5]	[7]	[11]
[17]	[17]	[13]	[23]	[19]	[5]	[1]	[11]	[7]
[19]	[19]	[23]	[13]	[17]	[7]	[11]	[1]	[5]
[23]	[23]	[19]	[17]	[13]	[11]	[7]	[5]	[1]

17. a. Not cyclic   **c.** Not cyclic

19. $a, a^5, a^7, a^{11}, a^{13}, a^{17}, a^{19}, a^{23}$

21. All subgroups of $\mathbf{Z}$ are of the form $\langle n \rangle$, n a fixed integer.

29. $p - 1$

EXERCISES 3.4,   pages 146–147

3. Let $\phi \colon \mathbf{Z}_4 \to G$ be defined by

$$\phi([0]_4) = [1]_5, \quad \phi([1]_4) = [2]_5, \quad \phi([2]_4) = [4]_5, \quad \phi([3]_4) = [3]_5.$$

5. Let $\phi \colon H \to S(A)$ be defined by

$$\phi(I_2) = I_A, \quad \phi(M_1) = \sigma, \quad \phi(M_2) = \rho, \quad \phi(M_3) = \rho^2, \quad \phi(M_4) = \gamma, \quad \phi(M_5) = \delta.$$

7. Let $\phi \colon \mathbf{Z} \to H$ be defined by $\phi(n) = \begin{bmatrix} 1 & n \\ 0 & 1 \end{bmatrix}$, $n \in \mathbf{Z}$. Then

$$\phi(n + m) = \begin{bmatrix} 1 & n + m \\ 0 & 1 \end{bmatrix} = \begin{bmatrix} 1 & n \\ 0 & 1 \end{bmatrix}\begin{bmatrix} 1 & m \\ 0 & 1 \end{bmatrix} = \phi(n) \cdot \phi(m)$$

for all $n, m \in \mathbf{Z}$.

9. Define $\phi \colon G \to H$ by $\phi(a + bi) = \begin{bmatrix} a & -b \\ b & a \end{bmatrix}$ for $a + bi \in G$. Let $x = a + bi \in G$ and $y = c + di \in G$. Then

$$\phi(xy) = \phi((a + bi)(c + di))$$
$$= \phi((ac - bd) + (bc + ad)i)$$

$$= \begin{bmatrix} ac - bd & -bc - ad \\ bc + ad & ac - bd \end{bmatrix}$$

$$= \begin{bmatrix} a & -b \\ b & a \end{bmatrix}\begin{bmatrix} c & -d \\ d & c \end{bmatrix}$$

$$= \phi(a + bi)\phi(c + di)$$

$$= \phi(x)\phi(y).$$

13. If G is nonabelian, then ϕ is not an isomorphism. If a and b are elements of G such that $ab \neq ba$, then $(ab)^{-1} \neq (ba)^{-1}$ and therefore $\phi(ab) = (ab)^{-1} = b^{-1}a^{-1} \neq a^{-1}b^{-1} = \phi(a)\phi(b)$.

19. For notational convenience we let a represent $[a]$. The elements 3 and $3^5 = 5$ are generators of G. The automorphisms of G are ϕ_1 and ϕ_2 defined by:

$$\phi_1: \begin{cases} \phi_1(1) = 1 \\ \phi_1(2) = 2 \\ \phi_1(3) = 3 \\ \phi_1(4) = 4 \\ \phi_1(5) = 5 \\ \phi_1(6) = 6 \end{cases} \qquad \phi_2: \begin{cases} \phi_2(3) = 5 \\ \phi_2(3^2) = \phi_2(2) = 5^2 = 4 \\ \phi_2(3^3) = \phi_2(6) = 5^3 = 6 \\ \phi_2(3^4) = \phi_2(4) = 5^4 = 2 \\ \phi_2(3^5) = \phi_2(5) = 5^5 = 3 \\ \phi_2(3^6) = \phi_2(1) = 5^6 = 1 \end{cases}$$

20. a. 2   **c.** 4

EXERCISES 3.5,   pages 150–151

1. a. ϕ is a homomorphism, and ker $\phi = \{1, -1\}$.
    c. ϕ is not a homomorphism.

EXERCISES 4.1,   pages 164–165

1. a. $(1, 4)(2, 5)$ **2. a.** $(1, 4, 8, 7, 2, 3)(5, 9, 6)$
    c. $(1, 4, 5, 2)$ **c.** $(1, 4, 8, 7)(2, 6, 5, 3)$
    e. $(1, 3, 5)(2, 4, 6)$ **e.** $(1, 2)(3, 4, 5)$
    g. $(1, 4)(2, 3, 5)$ **g.** $(1, 7, 6, 4, 3, 5, 2)$

3. a. even   **c.** odd   **e.** even   **g.** odd

4. a. odd   **c.** even   **e.** odd   **g.** even

5. a. two   **c.** four   **e.** three   **g.** six

6. a. six   **c.** four   **e.** six   **g.** seven

7. a. $(1, 4)(2, 5)$ **8. a.** $(1, 3)(1, 2)(1, 7)(1, 8)(1, 4)(5, 6)(5, 9)$
    c. $(1, 2)(1, 5)(1, 4)$ **c.** $(1, 7)(1, 8)(1, 4)(2, 3)(2, 5)(2, 6)$
    e. $(1, 5)(1, 3)(2, 6)(2, 4)$ **e.** $(1, 2)(3, 5)(3, 4)$
    g. $(1, 4)(2, 5)(2, 3)$ **g.** $(1, 2)(1, 5)(1, 3)(1, 4)(1, 6)(1, 7)$

9. a. $(3, 1, 4, 2) = (1, 4, 2, 3)$ **10. a.** $(1, 2)(4, 9)(5, 6)$
    c. $(1, 2, 4, 5)$ **c.** $(1, 2)(3, 4, 5)$
    e. $(1, 4, 2)(5, 3) = (1, 4, 2)(3, 5)$ **e.** $(3, 7, 4, 5)(6, 8)$

11. $g = f^3 = (1, 4)(2, 5)(3, 6), h = f^4 = (1, 5, 3)(2, 6, 4)$

13.

$(1, 2, 3, 4)$	$(1, 2, 3)$	$(1, 2)$	$(1, 2)(3, 4)$
$(1, 2, 4, 3)$	$(1, 3, 2)$	$(1, 3)$	$(1, 3)(2, 4)$
$(1, 3, 2, 4)$	$(1, 2, 4)$	$(1, 4)$	$(1, 4)(2, 3)$
$(1, 3, 4, 2)$	$(1, 4, 2)$	$(2, 3)$	(1)
$(1, 4, 2, 3)$	$(1, 3, 4)$	$(2, 4)$	
$(1, 4, 3, 2)$	$(1, 4, 3)$	$(3, 4)$	
	$(2, 3, 4)$		
	$(2, 4, 3)$		

15. $\langle(1, 2)\rangle = \{(1), (1, 2)\}$ has order 2.

$\langle(1, 2, 3)\rangle = \{(1), (1, 2, 3), (1, 3, 2)\}$ has order 3.

$\langle(1, 2, 3, 4)\rangle = \{(1), (1, 2, 3, 4), (1, 3)(2, 4), (1, 4, 3, 2)\}$ has order 4.

17. $\{e, \beta\}, \{e, \gamma\}, \{e, \Delta\}, \{e, \theta\}, \{e, \alpha^2\}, \{e, \alpha, \alpha^2, \alpha^3\}$

EXERCISES 4.2, pages 167–168

1. With $f_g \colon G \to G$ defined by $f_g(x) = gx$ for each $g \in G$, we obtain the following permutations on the set of elements of G:

$$f_e \colon \begin{cases} f_e(e) = e \\ f_e(a) = a \\ f_e(b) = b \\ f_e(ab) = ab \end{cases} \quad f_a \colon \begin{cases} f_a(e) = a \\ f_a(a) = e \\ f_a(b) = ab \\ f_a(ab) = b \end{cases} \quad f_b \colon \begin{cases} f_b(e) = b \\ f_b(a) = ab \\ f_b(b) = e \\ f_b(ab) = a \end{cases} \quad f_{ab} \colon \begin{cases} f_{ab}(e) = ab \\ f_{ab}(a) = b \\ f_{ab}(b) = a \\ f_{ab}(ab) = e. \end{cases}$$

The set $G' = \{f_e, f_a, f_b, f_{ab}\}$ is a group of permutations and the mapping $\phi \colon G \to G'$ defined by

$$\phi \colon \begin{cases} \phi(e) = f_e \\ \phi(a) = f_a \\ \phi(b) = f_b \\ \phi(ab) = f_{ab} \end{cases}$$

is an isomorphism from G to G'.

3. For notational convenience, we let a represent $[a]$ in this solution.

Let $f_a \colon G \to G$ be defined by $f_a(x) = ax$ for each $x \in G$. Then we have the following permutations:

$$f_2 \colon \begin{cases} f_2(2) = 4 \\ f_2(4) = 8 \\ f_2(6) = 2 \\ f_2(8) = 6 \end{cases} \quad f_4 \colon \begin{cases} f_4(2) = 8 \\ f_4(4) = 6 \\ f_4(6) = 4 \\ f_4(8) = 2 \end{cases} \quad f_6 \colon \begin{cases} f_6(2) = 2 \\ f_6(4) = 4 \\ f_6(6) = 6 \\ f_6(8) = 8 \end{cases} \quad f_8 \colon \begin{cases} f_8(2) = 6 \\ f_8(4) = 2 \\ f_8(6) = 8 \\ f_8(8) = 4. \end{cases}$$

The set $G' = \{f_2, f_4, f_6, f_8\}$ is a group of permutations, and the mapping $\phi \colon G \to G'$ defined by

$$\phi \colon \begin{cases} \phi(2) = f_2 \\ \phi(4) = f_4 \\ \phi(6) = f_6 \\ \phi(8) = f_8 \end{cases}$$

is an isomorphism from G to G'.

5. c. The mapping ϕ is not an isomorphism when G is not abelian.

7. c. The mapping ϕ is not an isomorphism when G is not abelian.

■ **EXERCISES 4.3,**   pages 172–173

1. $\{I, V\}$, where I is the identity mapping and V is the reflection about the vertical axis of symmetry

3. $\{I, R\}$, where I is the identity mapping and R is the counterclockwise rotation through $180°$ about the center of symmetry

5. Rotational symmetry only

7. Reflective symmetry only

9. Both rotational symmetry and reflective symmetry

11. $\{R, R^2, R^3 = I\}$, where I is the identity mapping and R is the counterclockwise rotation through $120°$ about the center of the triangle determined by the arrow tips

13. Let the vertices of the ellipses be numbered as in the following figure.

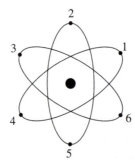

Then any symmetry of the figure can be identified with the corresponding permutation on $\{1, 2, 3, 4, 5, 6\}$, and the group G of symmetries of the figure can be described with the notation

$$G = \{R, R^2, R^3, R^4, R^5, R^6 = I, L, LR, LR^2, LR^3, LR^4, LR^5\},$$

where

$$
\begin{array}{ll}
I = (1) & L = (2, 6)(3, 5) \\
R = (1, 2, 3, 4, 5, 6) & LR = (1, 6)(2, 5)(3, 4) \\
R^2 = (1, 3, 5)(2, 4, 6) & LR^2 = (1, 5)(2, 4) \\
R^3 = (1, 4)(2, 5)(3, 6) & LR^3 = (1, 4)(2, 3)(5, 6) \\
R^4 = (1, 5, 3)(2, 6, 4) & LR^4 = (1, 3)(4, 6) \\
R^5 = (1, 6, 5, 4, 3, 2) & LR^5 = (1, 2)(3, 6)(4, 5).
\end{array}
$$

This is the same permutation group as the one in the answer to Exercise 24 of this exercise set.

15. Let the axes of symmetry be labeled as in the following figure.

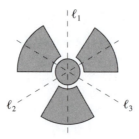

Then the group G of symmetries of the figure can be described as

$$G = \{I, R, R^2, L, LR, LR^2\},$$

where

> I is the identity mapping,
> R is the rotation through $120°$ counterclockwise about the center,
> R^2 is the rotation through $240°$ counterclockwise about the center,
> L is the reflection about the vertical axis ℓ_1,
> LR is the reflection about the axis ℓ_2, and
> LR^2 is the reflection about the axis ℓ_3.

17. Let I denote the identity mapping, and let t denote a translation of the set of E's one unit to the right. Then t^{-1} is a translation of the set of E's one unit to the left, and the collection

$$\{\ldots, t^{-2}, t^{-1}, t^0 = I, t, t^2, \ldots\}$$

are elements of the (infinite) group of symmetries of the figure. Let r denote the reflection of the figure about the horizontal axis of symmetry through the E's. Then $r^2 = I = r^0$, $rt = tr$, and the group of symmetries consists of all products of the form $r^i t^j$, where i is either 0 or 1 and j is an integer.

19. Let I denote the identity mapping, and let t denote a translation of the set of T's one unit to the right. Then t^{-1} is a translation of the set of T's one unit to the left. There is a vertical axis of symmetry through each copy of the letter **T**, and a corresponding reflection of the figure about that vertical axis. Each of these reflections is its own inverse. The group of symmetries consists of this infinite collection of reflections (one for each copy of the letter **T**) together with the identity I and all the integral powers of the translation t.

23. Using the same notational convention as in Example 11 of Section 4.1, the elements of G are as follows:

$$e = (1) \qquad\qquad \beta = (2,5)(3,4)$$
$$\alpha = (1,2,3,4,5) \qquad \gamma = \alpha\beta = \beta\alpha^4 = (1,2)(3,5)$$
$$\alpha^2 = (1,3,5,2,4) \qquad \Delta = \alpha^2\beta = \beta\alpha^3 = (1,3)(4,5)$$
$$\alpha^3 = (1,4,2,5,3) \qquad \theta = \alpha^3\beta = \beta\alpha^2 = (1,4)(2,3)$$
$$\alpha^4 = (1,5,4,3,2) \qquad \sigma = \alpha^4\beta = \beta\alpha = (1,5)(2,4).$$

With this notation, we obtain the following multiplication table for G.

$\circ$	e	α	α^2	α^3	α^4	β	γ	Δ	θ	σ
e	e	α	α^2	α^3	α^4	β	γ	Δ	θ	σ
α	α	α^2	α^3	α^4	e	γ	Δ	θ	σ	β
α^2	α^2	α^3	α^4	e	α	Δ	θ	σ	β	γ
α^3	α^3	α^4	e	α	α^2	θ	σ	β	γ	Δ
α^4	α^4	e	α	α^2	α^3	σ	β	γ	Δ	θ
β	β	σ	θ	Δ	γ	e	α^4	α^3	α^2	α
γ	γ	β	σ	θ	Δ	α	e	α^4	α^3	α^2
Δ	Δ	γ	β	σ	θ	α^2	α	e	α^4	α^3
θ	θ	Δ	γ	β	σ	α^3	α^2	α	e	α^4
σ	σ	θ	Δ	γ	β	α^4	α^3	α^2	α	e

25. 48

27. Using the same notational convention as in Example 11 of Section 4.1, the elements of $G = \{e, \alpha, \beta, \Delta\}$ are given by $e = (1), \alpha = (1, 3)(2, 4), \beta = (1, 4)(2, 3), \Delta = (1, 2)(3, 4)$. Let $\phi: G \to H$ be defined by

$$\phi(e) = \begin{bmatrix} 1 & 0 \\ 0 & 1 \end{bmatrix}, \qquad \phi(\alpha) = \begin{bmatrix} 1 & 0 \\ 0 & -1 \end{bmatrix},$$

$$\phi(\beta) = \begin{bmatrix} -1 & 0 \\ 0 & 1 \end{bmatrix}, \qquad \phi(\Delta) = \begin{bmatrix} -1 & 0 \\ 0 & -1 \end{bmatrix}.$$

■ EXERCISES 4.4, pages 182–184

1. a. $eH = \beta H = \{e, \beta\};$ $\alpha H = \gamma H = \{\alpha, \gamma\};$ $\alpha^2 H = \Delta H = \{\alpha^2, \Delta\};$
  $\alpha^3 H = \theta H = \{\alpha^3, \theta\}$
  b. $He = H\beta = \{e, \beta\};$ $H\alpha = H\theta = \{\alpha, \theta\};$ $H\alpha^2 = H\Delta = \{\alpha^2, \Delta\};$
  $H\alpha^3 = H\gamma = \{\alpha^3, \gamma\}$

3. a. $I_3 H = P_4 H = \{I_3, P_4\};$ $P_1 H = P_3^2 H = \{P_1, P_3^2\};$ $P_2 H = P_3 H = \{P_2, P_3\}$
  b. $HI_3 = HP_4 = \{I_3, P_4\};$ $HP_1 = HP_3 = \{P_1, P_3\};$ $HP_2 = HP_3^2 = \{P_2, P_3^2\}$

12. a. $\{e, \alpha^2\}$

15. Order 1: $\{(1)\}$
  Order 2: $\{(1), (1, 2)(3, 4)\},$ $\{(1), (1, 3)(2, 4)\},$ $\{(1), (1, 4)(2, 3)\}$
  Order 3: $\{(1), (1, 2, 3), (1, 3, 2)\},$ $\{(1), (1, 2, 4), (1, 4, 2)\},$ $\{(1), (1, 4, 3), (1, 3, 4)\},$
    $\{(1), (2, 3, 4), (2, 4, 3)\}$
  Order 4: $\{(1), (1, 2)(3, 4), (1, 3)(2, 4), (1, 4)(2, 3)\}$
  Order 12: A_4, as given in Example 8 of Section 4.1

17. The normal subgroups of the octic group G are $H_1 = \{e\}, H_2, = \{e, \alpha^2\}, H_3 = \{e, \alpha, \alpha^2, \alpha^3\},$
  $H_4 = \{e, \beta, \Delta, \alpha^2\}, H_5 = \{e, \gamma, \theta, \alpha^2\},$ and $H_6 = G$.

19. The normal subgroups of the quaternion group G are $H_1 = \{1\}, H_2 = \{-1, 1\},$
  $H_3 = \{i, -1, -i, 1\}, H_4 = \{j, -1, -j, 1\}, H_5 = \{k, -1, -k, 1\},$ and $H_6 = G$.

21. $H = \{e, \Delta\}, K = \{e, \beta, \Delta, \alpha^2\}$

33. $\{e, \alpha^2\}$

37. $\{(1), (1, 4)(2, 3), (1, 2)(3, 4), (1, 3)(2, 4)\}$

39. The equivalence classes are the same as the left cosets of H in G. That is, each equivalence
  class is a left coset of H in G, and each coset of H in G is an equivalence class.

45. a. $\{(1), (1, 3), (1, 2, 3, 4), (1, 3)(2, 4), (1, 4, 3, 2), (1, 2)(3, 4), (2, 4), (1, 4)(2, 3)\}$
  c. S_4

■ EXERCISES 4.5, pages 190–192

1. $G/H = \{H, \alpha H, \beta H, \gamma H\}$ where $H = \{e, \alpha^2\}, \alpha H = \{\alpha, \alpha^3\}, \beta H = \{\beta, \Delta\}, \gamma H = \{\gamma, \theta\}$.

$\cdot$	H	αH	βH	γH
H	H	αH	βH	γH
αH	αH	H	γH	βH
βH	βH	γH	H	αH
γH	γH	βH	αH	H

3. $G/H = \{H, iH, jH, kH\}$ where $H = \{1, -1\}, iH = \{i, -i\}, jH = \{j, -j\}, kH = \{k, -k\}$.

$\cdot$	H	iH	jH	kH
H	H	iH	jH	kH
iH	iH	H	kH	jH
jH	jH	kH	H	iH
kH	kH	jH	iH	H

5. The normal subgroups of the octic group G are $H_1 = \{e\}, H_2 = \{e, \alpha^2\}, H_3 = \{e, \alpha, \alpha^2, \alpha^3\}$, $H_4 = \{e, \beta, \Delta, \alpha^2\}, H_5 = \{e, \gamma, \theta, \alpha^2\}$, and $H_6 = G$. We consider the possible quotient groups.
(1) G/H_1 is isomorphic to G.
(2) $G/H_2 = \{H_2, \alpha H_2, \beta H_2, \gamma H_2\}$ is isomorphic to the Klein four group. (See Exercise 1 of Section 4.2.)
(3) Each of $G/H_3, G/H_4$, and G/H_5 is a cyclic group of order 2.
(4) $G/G = \{G\}$ is a group of order 1.

Thus, the homomorphic images of the octic group G are G itself, a Klein four group, a cyclic group of order 2, and a group with only the identity element.

7. The normal subgroups of the quaternion group G are $H_1 = \{1\}, H_2 = \{-1, 1\}, H_3 = \{i, -1, -i, 1\}$, $H_4 = \{j, -1, -j, 1\}, H_5 = \{k, -1, -k, 1\}$, and $H_6 = G$. We consider the quotient groups.
(1) G/H_1 is isomorphic to G.
(2) $G/H_2 = \{H_2, iH_2, jH_2, kH_2\}$ is isomorphic to the Klein four group. (See Exercise 1 of Section 4.2.)
(3) Each of $G/H_3, G/H_4$, and G/H_5 is the cyclic group of order 2.
(4) $G/G = \{G\}$ is a group of order 1.

Thus, the homomorphic images of the quaternion group G are G itself, a Klein four group, a cyclic group of order 2, and a group with only the identity element.

8. a. We have $G = \{[1], [3], [7], [9], [11], [13], [17], [19]\}$. The normal subgroups of G are

$$H_1 = \{[1]\},$$
$$H_2 = \{[1], [9]\},$$
$$H_3 = \{[1], [11]\},$$
$$H_4 = \{[1], [19]\},$$
$$H_5 = \{[1], [3], [9], [7]\},$$
$$H_6 = \{[1], [13], [9], [17]\},$$
$$H_7 = G.$$

The homomorphic images are the possible quotient groups.
(1) G/H_1 is isomorphic to G.
(2) Each of G/H_3 and G/H_4 is a cyclic group of order 4.
(3) G/H_2 is isomorphic to the Klein four group. (See Exercise 4 of this section.)
(4) Each of G/H_5 and G/H_6 is cyclic group of order 2.
(5) $G/G = \{G\}$ is a group of order 1.

Thus, the homomorphic images of G are the group G itself, a cyclic group of order 4, a Klein four group, a cyclic group of order 2, and a group with only the identity element.

c. The group G is given by $G = \{[1], [5], [7], [11], [13], [17], [19], [23]\}$. The trivial subgroups of G are $H_1 = \{[1]\}$ and G itself. Every element of G except $[1]$ has order 2, so there are seven cyclic subgroups $H_2, H_3, \ldots, H_8$ of order 2 in G. There are seven subgroups of G that are isomorphic to the Klein four group:

$$H_9 = \{[1],[5],[7],[11]\}, \qquad H_{10} = \{[1],[5],[13],[17]\},$$
$$H_{11} = \{[1],[5],[19],[23]\}, \qquad H_{12} = \{[1],[7],[13],[19]\},$$
$$H_{13} = \{[1],[7],[17],[23]\}, \qquad H_{14} = \{[1],[11],[13],[23]\},$$
$$H_{15} = \{[1],[11],[17],[19]\}.$$

The homomorphic images of G are the possible quotient groups.
(1) G/H_1 is isomorphic to G.
(2) Each of $G/H_2, G/H_3, \ldots, G/H_8$ is isomorphic to the Klein four group.
(3) Each of $G/H_9, G/H_{10}, \ldots, G/H_{15}$ is a cyclic group of order 2.
(4) $G/G = \{G\}$ is a group of order 1.

Thus, the homomorphic images of G are the group G itself, a four group, a cyclic group of order 2, and a group with only the identity element.

9. a. The left cosets of $H = \{(1), (1, 2)\}$ in $G = S_3$ are given by

$$(1)H = (1, 2)H = \{(1), (1, 2)\}$$
$$(1, 3)H = (1, 2, 3)H = \{(1, 3), (1, 2, 3)\}$$
$$(2, 3)H = (1, 3, 2)H = \{(2, 3), (1, 3, 2)\}.$$

The rule $aHbH = abH$ leads to

$$(1, 3)H(2, 3)H = (1, 3)(2, 3)H = (1, 3, 2)H$$

and also to

$$(1, 2, 3)H(1, 3, 2)H = (1, 2, 3)(1, 3, 2)H = (1)H.$$

We have $(1, 3)H = (1, 2, 3)H$ and $(2, 3)H = (1, 3, 2)H$, but

$$(1, 3)H(2, 3)H \neq (1, 2, 3)H(1, 3, 2)H.$$

Thus, the rule $aHbH = abH$ does not define a binary operation on the left cosets of H in G. (That is, the result is not well-defined.)

c. The left cosets of $H = \{(1), (2, 3)\}$ in $G = S_3$ are given by

$$(1)H = (2, 3)H = \{(1), (2, 3)\}$$
$$(1, 2)H = (1, 2, 3)H = \{(1, 2), (1, 2, 3)\}$$
$$(1, 3)H = (1, 3, 2)H = \{(1, 3), (1, 3, 2)\}.$$

The rule $aHbH = abH$ leads to

$$(1, 2)H(1, 3)H = (1, 2)(1, 3)H = (1, 3, 2)H$$

and also to

$$(1, 2, 3)H(1, 3, 2)H = (1, 2, 3)(1, 3, 2)H = (1)H.$$

We have $(1, 2)H = (1, 2, 3)H$ and $(1, 3)H = (1, 3, 2)H$, but

$$(1, 2)H(1, 3)H \neq (1, 2, 3)H(1, 3, 2)H.$$

Thus, the rule $aHbH = abH$ does not define a binary operation on the left cosets of H in G. (That is, the result is not well-defined.)

11. The mapping ϕ is a homomorphism, and ker $\phi = A_n$.

17. a. Let $G = \{a, a^2, a^3, a^4, a^5, a^6, a^7, a^8 = e\}$ be a cyclic group of order 8. The subgroup $H = \{a^2, a^4, a^6, a^8 = e\}$ of G is a cyclic group of order 4, and the mapping $\phi: G \to H$ defined by $\phi(x) = x^2$ is a homomorphism, since

$$\phi(xy) = (xy)^2$$
$$= x^2 y^2 \quad \text{since } G \text{ is abelian}$$
$$= \phi(x)\phi(y).$$

The mapping ϕ is an epimorphism, since
$$\phi(G) = \{\phi(a), \phi(a^2), \phi(a^3), \phi(a^4), \phi(a^5), \phi(a^6), \phi(a^7), \phi(e)\}$$
$$= \{a^2, a^4, a^6, a^8 = e, a^{10} = a^2, a^{12} = a^4, a^{14} = a^6, e\}$$
$$= \{a^2, a^4, a^6, a^8 = e\}$$
$$= H.$$
Thus, G has H as a homomorphic image.

EXERCISES 4.7,   pages 204–205

1. The cyclic group $C_9 = \langle a \rangle$ of order 9 is a p-group with $p = 3$.

3. a. $\langle (1, 2, 3) \rangle, \langle (1, 2, 4) \rangle, \langle (1, 3, 4) \rangle, \langle (2, 3, 4) \rangle$

5. a. $\mathbf{Z}_{10} = \langle [5] \rangle \oplus \langle [2] \rangle$ **c.** $\mathbf{Z}_{12} = \langle [3] \rangle \oplus \langle [4] \rangle$
$= \{[5], [0]\} \oplus \{[2], [4], [6], [8], [0]\}$ $= \{[3], [6], [9], [0]\} \oplus \{[4], [8], [0]\}$
$= C_2 \oplus C_5$ $= C_4 \oplus C_3$

6. a. Any abelian group of order 6 is isomorphic to $C_3 \oplus C_2$, where C_n is a cyclic group of order n.
   c. Any abelian group of order 12 is isomorphic to either $C_4 \oplus C_3$ or $C_2 \oplus C_2 \oplus C_3$.
   e. Any abelian group of order 36 is isomorphic to one of the direct sums $C_4 \oplus C_9$, $C_2 \oplus C_2 \oplus C_9$, $C_4 \oplus C_3 \oplus C_3$, $C_2 \oplus C_2 \oplus C_3 \oplus C_3$.

11. b. There are 24 distinct elements of G that have order 6.

EXERCISES 5.1,   pages 215–218

2. a. Ring
   c. Not a ring. The set is not closed with respect to multiplication. For example, $\sqrt[3]{5}$ is in the set, but the product $\sqrt[3]{5} \cdot \sqrt[3]{5} = \sqrt[3]{25}$ is not in the set.
   e. Not a ring. The set of positive real numbers does not contain an additive identity.
   g. Ring

3.

+	$\varnothing$	A	B	U
$\varnothing$	$\varnothing$	A	B	U
A	A	$\varnothing$	U	B
B	B	U	$\varnothing$	A
U	U	B	A	$\varnothing$

$\cdot$	$\varnothing$	A	B	U
$\varnothing$	$\varnothing$	$\varnothing$	$\varnothing$	$\varnothing$
A	$\varnothing$	A	$\varnothing$	A
B	$\varnothing$	$\varnothing$	B	B
U	$\varnothing$	A	B	U

5. The set $\mathcal{P}(A)$ is not a ring with respect to the operations of addition and multiplication as defined, since the set does not contain additive inverse elements.

7. a. $[2], [3], [4]$
   c. $[2], [4], [5], [6], [8]$
   e. $[2], [4], [6], [7], [8], [10], [12]$

8. a. $[1]^{-1} = [1], [5]^{-1} = [5]$
   c. $[1]^{-1} = [1], [3]^{-1} = [11], [5]^{-1} = [13], [7]^{-1} = [7], [9]^{-1} = [9], [11]^{-1} = [3],$
   $[13]^{-1} = [5], [15]^{-1} = [15]$
   e. $[1]^{-1} = [1], [3]^{-1} = [5], [5]^{-1} = [3], [9]^{-1} = [11], [11]^{-1} = [9], [13]^{-1} = [13]$

17. Let $R_1 = \mathbf{E}$ and R_2 be the set of all multiples of five. Each of R_1 and R_2 are subrings of $\mathbf{Z}$, but $R_1 \cup R_2$ is not, since it is not closed with respect to addition.

25. a. Yes

    b. The set S is a commutative ring, and it contains the unity [10].

    c. Yes

    d. Yes, [6] and [12]

    e. [2], [4], [8], [10], [14], [16]

27.

·	a	b	c	d
a	a	a	a	a
b	a	c	a	c
c	a	a	a	a
d	a	c	a	c

29. $\begin{bmatrix} 1 & 0 \\ 0 & 1 \end{bmatrix}, \begin{bmatrix} 1 & 0 \\ 0 & 0 \end{bmatrix}$

31. a. S is a subring of $M_2(\mathbf{Z})$.

    c. S is a subring of $M_2(\mathbf{Z})$.

32. b. All $\begin{bmatrix} a & b \\ 0 & c \end{bmatrix}$ with $a = \pm 1$ and $c = \pm 1$.

39. $\mathbf{Z}_2 \oplus \mathbf{Z}_2 = \{(0,0), (0,1), (1,0), (1,1)\}$

+	$(0,0)$	$(0,1)$	$(1,0)$	$(1,1)$
$(0,0)$	$(0,0)$	$(0,1)$	$(1,0)$	$(1,1)$
$(0,1)$	$(0,1)$	$(0,0)$	$(1,1)$	$(1,0)$
$(1,0)$	$(1,0)$	$(1,1)$	$(0,0)$	$(0,1)$
$(1,1)$	$(1,1)$	$(1,0)$	$(0,1)$	$(0,0)$

·	$(0,0)$	$(0,1)$	$(1,0)$	$(1,1)$
$(0,0)$	$(0,0)$	$(0,0)$	$(0,0)$	$(0,0)$
$(0,1)$	$(0,0)$	$(0,1)$	$(0,0)$	$(0,1)$
$(1,0)$	$(0,0)$	$(0,0)$	$(1,0)$	$(1,0)$
$(1,1)$	$(0,0)$	$(0,1)$	$(1,0)$	$(1,1)$

EXERCISES 5.2,   pages 221–223

1. a. The set of real numbers of the form $m + n\sqrt{2}$ where m and n are integers is an integral domain. It is not a field, since not every element (for example, $2 + 0\sqrt{2}$) has a multiplicative inverse.

    c. The set of real numbers of the form $a + b\sqrt[3]{2}$ where a and b are rational numbers is neither an integral domain nor a field, since it is not a ring. The set is not closed with respect to multiplication. For example: $\sqrt[3]{2} \cdot \sqrt[3]{2} = \sqrt[3]{4}$ is not in the set.

    e. The set of all complex numbers of the form $m + ni$ where $m \in \mathbf{Z}$ and $n \in \mathbf{Z}$ is an integral domain. It is not a field, since not every element (for example, $2 + 0i$) has a multiplicative inverse.

g. The set of all complex numbers of the form $a + bi$ where a and b are rational numbers is both an integral domain and field.

3. a. The set S is not an integral domain, since the elements [6] and [12] are zero divisors.

   b. The set S is not a field, since [6] and [12] do not have multiplicative inverses.

5. The ring W is commutative, since if (x, y) and (z, w) are elements of W, we have

$$(x, y) \cdot (z, w) = (xz - yw, xw + yz)$$
$$= (zx - wy, zy + wx)$$
$$= (z, w) \cdot (x, y).$$

The element $(1, 0)$ in W is the unity element, since for (x, y) in W we have

$$(x, y) \cdot (1, 0) = (1, 0) \cdot (x, y)$$
$$= (1x - 0y, 1y + 0x)$$
$$= (x, y).$$

7. a. S is a commutative ring.

   b. S has the unity element $\begin{bmatrix} 1 & 1 \\ 0 & 0 \end{bmatrix}$.

   c. S is an integral domain.

   d. S is a field.

9. a. R is a commutative ring.

   b. R has the unity element $\begin{bmatrix} 1 & 0 \\ 0 & 1 \end{bmatrix}$.

   c. R is an integral domain.

   d. R is a field.

15. a. [173]

   c. [27]

■ **EXERCISES 5.3,**   pages 299–230

9. Define $\phi: W \to R$ by

$$\phi((x, y)) = \begin{bmatrix} x & -y \\ y & x \end{bmatrix}.$$

The mapping ϕ is clearly a one-to-one correspondence from W to R.

$$\phi((x, y) + (z, w)) = \phi((x + z, y + w))$$
$$= \begin{bmatrix} x + z & -y - w \\ y + w & x + z \end{bmatrix} = \begin{bmatrix} x & -y \\ y & x \end{bmatrix} + \begin{bmatrix} z & -w \\ w & z \end{bmatrix}$$
$$= \phi((x, y)) + \phi((z, w))$$

$$\phi((x, y) \cdot (z, w)) = \phi((xz - yw, xw + yz))$$
$$= \begin{bmatrix} xz - yw & -xw - yz \\ xw + yz & xz - yw \end{bmatrix} = \begin{bmatrix} x & -y \\ y & x \end{bmatrix} \cdot \begin{bmatrix} z & -w \\ w & z \end{bmatrix}$$
$$= \phi((x, y)) \cdot \phi((z, w))$$

Thus, ϕ is an isomorphism.

11. a. For notational convenience in this solution, we write 0 for [0], 1 for [1], and 2 for [2] in $\mathbf{Z}_3$. Then

$$S = \{(0, 1), (0, 2), (1, 1), (1, 2), (2, 1), (2, 2)\}$$

Since $(0, 1) \sim (0, 2), (1, 1) \sim (2, 2)$ and $(1, 2) \sim (2, 1)$ in S, the distinct elements of Q are [0, 1], [1, 1], and [2, 1].

   b. Define $\phi: D \to Q$ by

$$\phi(0) = [0, 1]$$
$$\phi(1) = [1, 1]$$
$$\phi(2) = [2, 1].$$

15. The set of all quotients for D is the set Q of all equivalence classes $[m + ni, r + si]$, where $m + ni \in D$ and $r + si \in D$ with not both r and s equal 0. To show that Q is isomorphic to the set C of all complex numbers of the form $a + bi$ where a and b are rational numbers, we define $\phi: Q \to C$ by

$$\phi([m + ni, r + si]) = \frac{m + ni}{r + si}.$$

This rule does define a mapping from Q into C, since for $[m + ni, r + si] \in Q$ we can write

$$\frac{m + ni}{r + si} = \frac{mr + ns}{r^2 + s^2} + \frac{nr - ms}{r^2 + s^2}i,$$

which is an element in C.

To show that ϕ is onto, let $a + bi$ be an arbitrary element in C. Since a and b are both rational numbers, there exist integers p, q, t, and u such that

$$a = \frac{p}{q} \quad \text{and} \quad b = \frac{t}{u}.$$

Then the element $[pu + qti, qu + 0i]$ is in Q, and

$$\phi([pu + qti, qu + 0i]) = \frac{pu + qti}{qu + 0i}$$
$$= \frac{p}{q} + \frac{t}{u}i$$
$$= a + bi.$$

To show that ϕ is one-to-one, let $[m + ni, r + si]$ and $[x + yi, z + wi]$ be elements of Q such that

$$\phi([m + ni, r + si]) = \phi([x + yi, z + wi]).$$

Then

$$\frac{m + ni}{r + si} = \frac{x + yi}{z + wi},$$

and this implies that

$$(m + ni)(z + wi) = (r + si)(x + yi).$$

By the definition of equality in Q, we have

$$[m + ni, r + si] = [x + yi, z + wi],$$

and therefore ϕ is one-to-one. Since

$$\phi([m + ni, r + si] + [x + yi, z + wi])$$
$$= \phi([m + ni)(z + wi) + (r + si)(x + yi), (r + si)(z + wi)])$$
$$= \frac{(m + ni)(z + wi) + (r + si)(x + yi)}{(r + si)(z + wi)}$$
$$= \frac{m + ni}{r + si} + \frac{x + yi}{z + wi}$$
$$= \phi([m + ni, r + si]) + \phi([x + yi, z + wi])$$

and
$$\phi([m + ni, r + si] \cdot [x + yi, z + wi])$$
$$= \phi([(m + ni)(x + yi), (r + si)(z + wi)])$$
$$= \frac{(m + ni)(x + yi)}{(r + si)(z + wi)}$$
$$= \frac{m + ni}{r + si} \cdot \frac{x + yi}{z + wi}$$
$$= \phi([m + ni, r + si]) \cdot \phi([x + yi, z + wi]),$$

ϕ is an isomorphism from Q to C.

EXERCISES 6.1,   pages 245–246

5. a. Let $I_1 = (2)$ and $I_2 = (3)$. Then 2 and 3 are in $I_1 \cup I_2$, but the sum $2 + 3 = 5$ is not in $I_1 \cup I_2$. Hence, $I_1 \cup I_2$ is not an ideal of $\mathbf{Z}$.

13. a. $\{[0]\}, \mathbf{Z}_7$

   c. $\{[0]\}$

      $([6]) = \{[0], [6]\}$

      $([4]) = \{[0], [4], [8]\}$

      $([3]) = \{[0], [3], [6], [9]\}$

      $([2]) = \{[0], [2], [4], [6], [8], [10]\}$

      $\mathbf{Z}_{12}$

   e. $\{[0]\}$

      $([10]) = \{[0], [10]\}$

      $([5]) = \{[0], [5], [10], [15]\}$

      $([4]) = \{[0], [4], [8], [12], [16]\}$

      $([2]) = \{[0], [2], [4], [6], [8], [10], [12], [14], [16], [18]\}$

      $\mathbf{Z}_{20}$

17. The set U is not an ideal of S. $X = \begin{bmatrix} 1 & 2 \\ 0 & 1 \end{bmatrix}$ is in U, and $R = \begin{bmatrix} 1 & 2 \\ 0 & 3 \end{bmatrix}$ is in S, but $XR = \begin{bmatrix} 1 & 8 \\ 0 & 3 \end{bmatrix}$ is not in U.

EXERCISES 6.2,   pages 253–255

5. b. $\ker \theta = \left\{ \begin{bmatrix} 0 & 0 \\ y & 0 \end{bmatrix} \middle| y \in \mathbf{Z} \right\}$, $\phi\left(\begin{bmatrix} x & 0 \\ y & 0 \end{bmatrix} + \ker \theta \right) = \theta\left(\begin{bmatrix} x & 0 \\ y & 0 \end{bmatrix} \right) = x$

9. $\ker \theta = \left\{ \begin{bmatrix} 2m & 2n \\ 2p & 2q \end{bmatrix} \middle| m, n, p, q \in \mathbf{Z} \right\}$

11. θ is a homomorphism, since
$$\theta([a] + [b]) = 4([a] + [b])$$
$$= 4[a] + 4[b]$$
$$= \theta([a]) + \theta([b])$$

and
$$\theta([a])\theta([b]) = (4[a])(4[b])$$
$$= 16[ab]$$
$$= [16ab]$$
$$= [4ab]$$
$$= \theta([ab])$$
$$= \theta([a][b]).$$

13.

+	[0]	[2]	[4]	[6]
[0]	[0]	[2]	[4]	[6]
[2]	[2]	[4]	[6]	[0]
[4]	[4]	[6]	[0]	[2]
[6]	[6]	[0]	[2]	[4]

·	[0]	[2]	[4]	[6]
[0]	[0]	[0]	[0]	[0]
[2]	[0]	[4]	[0]	[4]
[4]	[0]	[0]	[0]	[0]
[6]	[0]	[4]	[0]	[4]

The mapping $\phi: R \to R'$ given by

$$\phi(a) = [0], \quad \phi(b) = [2], \quad \phi(c) = [4], \quad \phi(d) = [6]$$

is an isomorphism.

15. a. θ does not preserve addition, θ preserves multiplication, θ is not a homomorphism.

   c. θ preserves addition, θ does not preserve multiplication, θ is not a homomorphism.

   e. θ does not preserve addition, θ preserves multiplication, θ is not a homomorphism.

16. a. The ideals of $\mathbf{Z}_6$ are $I_1 = \{[0]\}$, $I_2 = \{[0], [3]\}$, $I_3 = \{[0], [2], [4]\}$, and $I_4 = \mathbf{Z}_6$. We consider the quotient rings:

   (1) $\mathbf{Z}_6/I_1$ is isomorphic to $\mathbf{Z}_6$.

   (2) $\mathbf{Z}_6/I_2 = \{I_2, [1] + I_2, [2] + I_2\}$ is isomorphic to $\mathbf{Z}_3$.

   (3) $\mathbf{Z}_6/I_3 = \{I_3, [1] + I_3\}$ is isomorphic to $\mathbf{Z}_2$.

   (4) $\mathbf{Z}_6/\mathbf{Z}_6 = \{\mathbf{Z}_6\}$ is a ring with only the zero element.

Thus, the homomorphic images of $\mathbf{Z}_6$ are (isomorphic to) $\mathbf{Z}_6$, $\mathbf{Z}_3$, $\mathbf{Z}_2$, and $\{0\}$.

   c. The ideals of $\mathbf{Z}_{12}$ are $I_1 = \{[0]\}$, $I_2 = \{[0], [6]\}$, $I_3 = \{[0], [4], [8]\}$, $I_4 = \{[0], [3], [6], [9]\}$, $I_5 = \{[0], [2], [4], [6], [8], [10]\}$, and $I_6 = \mathbf{Z}_{12}$. The quotient rings are as follows:

   (1) $\mathbf{Z}_{12}/I_1$ is isomorphic to $\mathbf{Z}_{12}$.

   (2) $\mathbf{Z}_{12}/I_2 = \{I_2, [1] + I_2, [2] + I_2, [3] + I_2, [4] + I_2, [5] + I_2\}$ is isomorphic to $\mathbf{Z}_6$.

   (3) $\mathbf{Z}_{12}/I_3 = \{I_3, [1] + I_3, [2] + I_3, [3] + I_3\}$ is isomorphic to $\mathbf{Z}_4$.

   (4) $\mathbf{Z}_{12}/I_4 = \{I_4, [1] + I_4, [2] + I_4\}$ is isomorphic to $\mathbf{Z}_3$.

   (5) $\mathbf{Z}_{12}/I_5 = \{I_5, [1] + I_5\}$ is isomorphic to $\mathbf{Z}_2$.

   (6) $\mathbf{Z}_{12}/\mathbf{Z}_{12} = \{\mathbf{Z}_{12}\}$ is a ring with only the zero element.

The homomorphic images of $\mathbf{Z}_{12}$ are (isomorphic to) $\mathbf{Z}_{12}$, $\mathbf{Z}_6$, $\mathbf{Z}_4$, $\mathbf{Z}_3$, $\mathbf{Z}_2$, and $\{0\}$.

   e. The ideals of $\mathbf{Z}_8$ are $I_1 = \{[0]\}$, $I_2 = \{[0], [4]\}$, $I_3 = \{[0], [2], [4], [6]\}$ and $I_4 = \mathbf{Z}_8$. The quotient rings are as follows:

   (1) $\mathbf{Z}_8/I_1$ is isomorphic to $\mathbf{Z}_8$.

   (2) $\mathbf{Z}_8/I_2 = \{I_2, [1] + I_2, [2] + I_2, [3] + I_2\}$ is isomorphic to $\mathbf{Z}_4$.

   (3) $\mathbf{Z}_8/I_3 = \{I_3, [1] + I_3\}$ is isomorphic to $\mathbf{Z}_2$.

   (4) $\mathbf{Z}_8/\mathbf{Z}_8 = \{\mathbf{Z}_8\}$ is a ring with only the zero element.

The homomorphic images of $\mathbf{Z}_8$ are (isomorphic to) $\mathbf{Z}_8$, $\mathbf{Z}_4$, $\mathbf{Z}_2$, and $\{0\}$.

19. b. -1 is the zero element, and 0 is the unity of R'.

EXERCISES 6.3, pages 259–260

1. a. 2 **c.** 6 **e.** 12

3. In $\mathbf{Z}_6$, [1] has additive order 6, and [2] has additive order 3.

7. b. Exercise 2 assures us that e, a, and b all have additive order 2. The other entries in the table can be determined by using the fact that D forms a group with respect to addition. For example, $e + a = a$ would imply $e = 0$, so $e + a = b$ must be true.

+	0	e	a	b
0	0	e	a	b
e	e	0	b	a
a	a	b	0	e
b	b	a	e	0

EXERCISES 6.4, pages 262–263

5. $R/I = \{I, 1 + I, \sqrt{2} + I, 1 + \sqrt{2} + I\}$ **7.** $\mathbf{E}/I = \{I, 2 + I, 4 + I\}$

9. $\{[0], [3], [6], [9]\}$ and $\{[0], [2], [4], [6], [8], [10]\}$

19. $\{[0], [3], [6], [9]\}$ and $\{[0], [2], [4], [6], [8], [10]\}$

EXERCISE 7.1, pages 272–273

1. $0.\overline{5}$ **3.** $0.\overline{987654320}$ **5.** $3.\overline{142857}$

7. $31/9$ **9.** $4/33$ **11.** $83/33$

20. a. $a = \sqrt{2}$ and $b = -\sqrt{2}$ are irrational, but $a + b = 0$ is rational.

21. a. An element v of F is a *lower bound* of S if $v \leq x$ for all $x \in S$. An element v of F is a *greatest lower bound* of S if these conditions are satisfied:

(1) v is a lower bound of S.

(2) If $b \in F$ is a lower bound of S, then $b \leq v$.

EXERCISES 7.2, pages 279–280

1. $10 + 11i$ **3.** $-i$ **5.** $2 - 11i$

7. $\frac{2}{5} + \frac{1}{5}i$ **9.** $\frac{11}{50} + \frac{1}{25}i$ **11.** $\frac{21}{29} + \frac{20}{29}i$

13. a. $3i, -3i$ **c.** $5i, -5i$ **e.** $\sqrt{13}\,i, -\sqrt{13}\,i$

EXERCISES 7.3, pages 287–289

1. a. $-2 + 2\sqrt{3}i = 4(\cos\frac{2\pi}{3} + i\sin\frac{2\pi}{3})$ **c.** $3 - 3i = 3\sqrt{2}(\cos\frac{7\pi}{4} + i\sin\frac{7\pi}{4})$

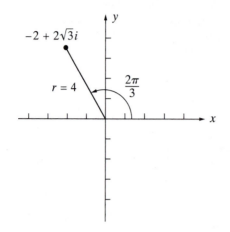

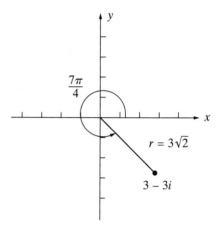

e. $1 + \sqrt{3}i = 2(\cos\frac{\pi}{3} + i\sin\frac{\pi}{3})$ **g.** $-4 = 4(\cos\pi + i\sin\pi)$

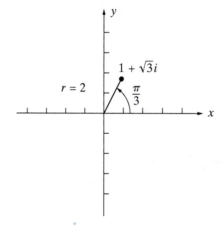

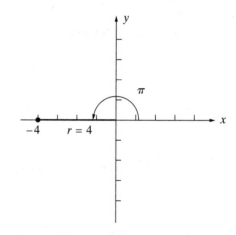

2. a. $4(\cos\frac{3\pi}{4} + i\sin\frac{3\pi}{4}) = -2\sqrt{2} + 2\sqrt{2}i$

   c. $6(\cos\frac{2\pi}{3} + i\sin\frac{2\pi}{3}) = -3 + 3\sqrt{3}i$

3. a. $-64\sqrt{3} - 64i$

   c. $512 + 512\sqrt{3}i$

   e. 1

   g. $-128 - 128\sqrt{3}i$

6. a.

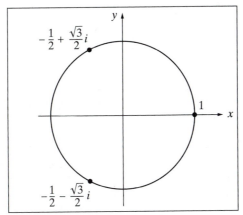

c.

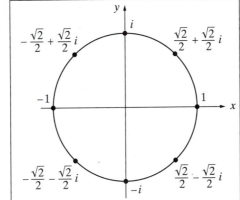

7. a. $\cos\frac{\pi}{18} + i\sin\frac{\pi}{18}, \cos\frac{13\pi}{18} + i\sin\frac{13\pi}{18}, \cos\frac{25\pi}{18} + i\sin\frac{25\pi}{18}$

  c. $\cos\frac{5\pi}{24} + i\sin\frac{5\pi}{24}, \cos\frac{17\pi}{24} + i\sin\frac{17\pi}{24}, \cos\frac{29\pi}{24} + i\sin\frac{29\pi}{24}, \cos\frac{41\pi}{24} + i\sin\frac{41\pi}{24}$

  e. $2(\cos\frac{\pi}{4} + i\sin\frac{\pi}{4}), 2(\cos\frac{13\pi}{20} + i\sin\frac{13\pi}{20}), 2(\cos\frac{21\pi}{20} + i\sin\frac{21\pi}{20}),$
  $2(\cos\frac{29\pi}{20} + i\sin\frac{29\pi}{20}), 2(\cos\frac{37\pi}{20} + i\sin\frac{37\pi}{20})$

8. a. $\frac{3}{2} + \frac{3\sqrt{3}}{2}i, -3, \frac{3}{2} - \frac{3\sqrt{3}}{2}i$

  c. $\frac{\sqrt{3}}{2} + \frac{1}{2}i, -\frac{\sqrt{3}}{2} + \frac{1}{2}i, -i$

  e. $\frac{\sqrt{3}}{2} + \frac{1}{2}i, -\frac{1}{2} + \frac{\sqrt{3}}{2}i, -\frac{\sqrt{3}}{2} - \frac{1}{2}i, \frac{1}{2} - \frac{\sqrt{3}}{2}i$

  g. $\frac{1}{2} + \frac{\sqrt{3}}{2}i, -\frac{\sqrt{3}}{2} + \frac{1}{2}i, -\frac{1}{2} - \frac{\sqrt{3}}{2}i, \frac{\sqrt{3}}{2} - \frac{1}{2}i$

11. $\langle\cos\frac{2\pi}{3} + i\sin\frac{2\pi}{3}\rangle = \{\cos\frac{2\pi}{3} + i\sin\frac{2\pi}{3}, \cos\frac{4\pi}{3} + i\sin\frac{4\pi}{3}, \cos 0 + i\sin 0\}$

13. $\langle\cos\frac{5\pi}{3} + i\sin\frac{5\pi}{3}\rangle = \{\cos\frac{5\pi}{3} + i\sin\frac{5\pi}{3}, \cos\frac{4\pi}{3} + i\sin\frac{4\pi}{3}, \cos\pi + i\sin\pi, \cos\frac{2\pi}{3} + i\sin\frac{2\pi}{3},$
  $\cos\frac{\pi}{3} + i\sin\frac{\pi}{3}, \cos 0 + i\sin 0\}$

17. a. $\cos\frac{\pi}{3} + i\sin\frac{\pi}{3} = \frac{1}{2} + \frac{\sqrt{3}}{2}i, \cos\frac{5\pi}{3} + i\sin\frac{5\pi}{3} = \frac{1}{2} - \frac{\sqrt{3}}{2}i$

EXERCISES 8.1, pages 299–300

1. a. $c_0 x^0 + c_1 x^1 + c_2 x^2 + c_3 x^3$, or $c_0 + c_1 x + c_2 x^2 + c_3 x^3$
   c. $a_1 x^1 + a_2 x^2 + a_3 x^3$, or $a_1 x + a_2 x^2 + a_3 x^3$

2. a. $\displaystyle\sum_{j=0}^{2} c_j x^j$ **c.** $\displaystyle\sum_{i=1}^{4} x^i$

3. a. $2x^3 + 4x^2 + 3x + 2$ **c.** $4x^2 + 2x$
   e. $2x^2 + 2x + 3$ **g.** $4x^5 + 4x^2 + 7x + 4$

4. a. $2x^3 + 4x^2 + 2x + 1$ **c.** $8x^5 + 8x^4 + 4x^3 + 8x^2 + 4x + 6$
   e. $8x^5 + 8x^4 + 4x^3 + 5x^2 + 4x$ **g.** $2x^5 + 8x^4 + 7x^3 + 5x^2 + 7x$

5. a. The set S of all polynomials with zero constant term is nonempty, since it contains the zero polynomial. Both the sum and the product of polynomials with zero constant terms are again polynomials with zero constant terms, so S is closed under addition and multiplication. The additive inverse of a polynomial with zero constant term is also a polynomial with zero constant term, so S is a subring of $R[x]$.

   c. Let S be the set of all polynomials that have zero coefficients for all odd powers of x. Then x^2 is in S, so S is nonempty. For arbitrary

$$f(x) = \sum_{i=0}^{n} a_{2i} x^{2i} \quad \text{and} \quad g(x) = \sum_{i=0}^{m} b_{2i} x^{2i}$$

in S, let k be the larger of n and m. Then

$$f(x) + g(x) = \sum_{i=0}^{k} (a_{2i} + b_{2i}) x^{2i}$$

has zero coefficients for all odd powers of x, and therefore is in S. Also,

$$f(x)g(x) = \sum_{i=0}^{m+n} \left(\sum_{p+q=i} a_{2p} b_{2q} \right) x^{2i}$$

is in S, and

$$-f(x) = \sum_{i=0}^{n} (-a_{2i}) x^{2i}$$

is in S. Thus, S is a subring of $R[x]$.

6. a. The set S described in 5a is a subring of $R[x]$. Since a product of a polynomial with zero constant term and any other polynomial always has zero constant term, S is an ideal of $R[x]$.

   c. The set S described in 5c is not an ideal of $R[x]$. The polynomial x^2 is in S, but the product $x(x^2) = x^3$ is not in S.

9. a. $n^2(n-1)$ **b.** $n^m(n-1)$

EXERCISES 8.2, pages 306–307

1. $q(x) = 4x^2 + 3x + 2, r(x) = 4$
3. $q(x) = x + 2, r(x) = x^2 + x$
5. $q(x) = 5x^2 + 3, r(x) = 2x + 3$
7. $d(x) = x + 1$
9. $d(x) = x + 5$
11. $s(x) = x^2 + 2x + 1, t(x) = x$
13. $s(x) = x^2 + 2, t(x) = 4$

EXERCISES 8.3, pages 313–314

1. a. $x^2 - 2$ is irreducible over **Q**, reducible over **R** and **C**, since $x^2 - 2 = (x - \sqrt{2})(x + \sqrt{2})$.
   c. $x^2 + x - 2 = (x + 2)(x - 1)$ is reducible over the fields **Q**, **R**, and **C**.
   e. $x^2 + x + 2$ is irreducible over $\mathbf{Z}_3$ and $\mathbf{Z}_5$, $x^2 + x + 2 = (x + 4)^2$ is reducible over $\mathbf{Z}_7$.
   g. $x^3 - x^2 + 2x + 2$ is irreducible over $\mathbf{Z}_3$, $x^3 - x^2 + 2x + 2 = (x + 3)^3$ is reducible over $\mathbf{Z}_5$, $x^3 - x^2 + 2x + 2 = (x + 2)(x^2 + 4x + 1)$ is reducible over $\mathbf{Z}_7$.

3. a. $2x^3 + 1 = 2(x + 2)(x^2 + 3x + 4)$
   c. $3x^3 + x^2 + 2x + 4 = 3(x + 1)(x + 2)(x + 4)$
   e. $2(x + 1)(x + 2)(x^2 + 3)$
   g. $(x + 3)^2(x^2 + 2)$

7. $x^4 + 5x^2 + 4 = (x^2 + 1)(x^2 + 4)$ is reducible over **R** and has no zeros in the field of real numbers.

EXERCISES 8.4, page 322

1. a. $f(x) = x^2 - (3 + 2i)x + 6i, g(x) = x^3 - 3x^2 + 4x - 12$
   c. $f(x) = x^2 - (3 - i)x + (2 - 2i), g(x) = x^3 - 4x^2 + 6x - 4$
   e. $f(x) = x^2 - (1 + 5i)x - (6 - 3i), g(x) = x^4 - 2x^3 + 14x^2 - 18x + 45$
   g. $f(x) = x^3 - 3x^2 + (3 - 2i)x - (1 - 2i)$,
      $g(x) = x^5 - 5x^4 + 10x^3 - 10x^2 + 9x - 5$

2. a. $1 + i, 2$
   c. $i, (-1 + i\sqrt{3})/2, (-1 - i\sqrt{3})/2$

3. $5/2, -1$

5. $3/2, -1$

7. $-2, (1 + i\sqrt{3})/2, (1 - i\sqrt{3})/2$

9. $1, 1/3, -2$

11. $-1, 1/2, -4/3$

13. $x^4 - x^3 - 2x^2 + 6x - 4 = (x - 1)(x + 2)(x^2 - 2x + 2)$

15. $2x^4 + 5x^3 - 7x^2 - 10x + 6 = 2(x - \frac{1}{2})(x + 3)(x^2 - 2)$

17. a. Let $f(x) = 3 + 9x + x^3$. The prime integer 3 divides all the coefficients of $f(x)$ except the leading coefficient $a_n = 1$, and 3^2 does not divide $a_0 = 3$. Thus, $f(x)$ is irreducible, by Eisenstein's Criterion.
   c. Let $f(x) = 3 - 27x^2 + 2x^5$. The prime 3 divides all the coefficients of $f(x)$ except the leading coefficient $a_n = 2$, and 3^2 does not divide $a_0 = 3$. Thus, $f(x)$ is irreducible, by Eisenstein's Criterion.

EXERCISES 8.5, pages 332–333

1. a. Let $P = (p(x))$ and $\alpha = x + P$ in $\mathbf{Z}_3[x]/P$. The elements of $\mathbf{Z}_3[x]/P$ are

$$\{0, 1, 2, \alpha, \alpha + 1, \alpha + 2, 2\alpha, 2\alpha + 1, 2\alpha + 2\},$$

where $0 = 0 + P, 1 = 1 + P$, and $2 = 2 + P$. Addition and multiplication tables are as follows:

+	0	1	2	α	$\alpha + 1$	$\alpha + 2$	2α	$2\alpha + 1$	$2\alpha + 2$
0	0	1	2	α	$\alpha + 1$	$\alpha + 2$	2α	$2\alpha + 1$	$2\alpha + 2$
1	1	2	0	$\alpha + 1$	$\alpha + 2$	α	$2\alpha + 1$	$2\alpha + 2$	2α
2	2	0	1	$\alpha + 2$	α	$\alpha + 1$	$2\alpha + 2$	2α	$2\alpha + 1$
α	α	$\alpha + 1$	$\alpha + 2$	2α	$2\alpha + 1$	$2\alpha + 2$	0	1	2
$\alpha + 1$	$\alpha + 1$	$\alpha + 2$	α	$2\alpha + 1$	$2\alpha + 2$	2α	1	2	0
$\alpha + 2$	$\alpha + 2$	α	$\alpha + 1$	$2\alpha + 2$	2α	$2\alpha + 1$	2	0	1
2α	2α	$2\alpha + 1$	$2\alpha + 2$	0	1	2	α	$\alpha + 1$	$\alpha + 2$
$2\alpha + 1$	$2\alpha + 1$	$2\alpha + 2$	2α	1	2	0	$\alpha + 1$	$\alpha + 2$	α
$2\alpha + 2$	$2\alpha + 2$	2α	$2\alpha + 1$	2	0	1	$\alpha + 2$	α	$\alpha + 1$

$\cdot$	0	1	2	α	$\alpha + 1$	$\alpha + 2$	2α	$2\alpha + 1$	$2\alpha + 2$
0	0	0	0	0	0	0	0	0	0
1	0	1	2	α	$\alpha + 1$	$\alpha + 2$	2α	$2\alpha + 1$	$2\alpha + 2$
2	0	2	1	2α	$2\alpha + 2$	$2\alpha + 1$	α	$\alpha + 2$	$\alpha + 1$
α	0	α	2α	$2\alpha + 1$	1	$\alpha + 1$	$\alpha + 2$	$2\alpha + 2$	2
$\alpha + 1$	0	$\alpha + 1$	$2\alpha + 2$	1	$\alpha + 2$	2α	2	α	$2\alpha + 1$
$\alpha + 2$	0	$\alpha + 2$	$2\alpha + 1$	$\alpha + 1$	2α	2	$2\alpha + 2$	1	α
2α	0	2α	α	$\alpha + 2$	2	$2\alpha + 2$	$2\alpha + 1$	$\alpha + 1$	1
$2\alpha + 1$	0	$2\alpha + 1$	$\alpha + 2$	$2\alpha + 2$	α	1	$\alpha + 1$	2	2α
$2\alpha + 2$	0	$2\alpha + 2$	$\alpha + 1$	2	$2\alpha + 1$	α	1	2α	$\alpha + 2$

2. a. $\mathbf{Z}_2[x]/(p(x)) = \{0, 1, \alpha, \alpha + 1\}$ is a field.

+	0	1	α	$\alpha + 1$
0	0	1	α	$\alpha + 1$
1	1	0	$\alpha + 1$	α
α	α	$\alpha + 1$	0	1
$\alpha + 1$	$\alpha + 1$	α	1	0

$\cdot$	0	1	α	$\alpha + 1$
0	0	0	0	0
1	0	1	α	$\alpha + 1$
α	0	α	$\alpha + 1$	1
$\alpha + 1$	0	$\alpha + 1$	1	α

c. $\mathbf{Z}_2[x]/(p(x)) = \{0, 1, \alpha, \alpha + 1, \alpha^2, \alpha^2 + 1, \alpha^2 + \alpha, \alpha^2 + \alpha + 1\}$ is a field.

+	0	1	α	$\alpha + 1$	α^2	$\alpha^2 + 1$	$\alpha^2 + \alpha$	$\alpha^2 + \alpha + 1$
0	0	1	α	$\alpha + 1$	α^2	$\alpha^2 + 1$	$\alpha^2 + \alpha$	$\alpha^2 + \alpha + 1$
1	1	0	$\alpha + 1$	α	$\alpha^2 + 1$	α^2	$\alpha^2 + \alpha + 1$	$\alpha^2 + \alpha$
α	α	$\alpha + 1$	0	1	$\alpha^2 + \alpha$	$\alpha^2 + \alpha + 1$	α^2	$\alpha^2 + 1$
$\alpha + 1$	$\alpha + 1$	α	1	0	$\alpha^2 + \alpha + 1$	$\alpha^2 + \alpha$	$\alpha^2 + 1$	α^2
α^2	α^2	$\alpha^2 + 1$	$\alpha^2 + \alpha$	$\alpha^2 + \alpha + 1$	0	1	α	$\alpha + 1$
$\alpha^2 + 1$	$\alpha^2 + 1$	α^2	$\alpha^2 + \alpha + 1$	$\alpha^2 + \alpha$	1	0	$\alpha + 1$	α
$\alpha^2 + \alpha$	$\alpha^2 + \alpha$	$\alpha^2 + \alpha + 1$	α^2	$\alpha^2 + 1$	α	$\alpha + 1$	0	1
$\alpha^2 + \alpha + 1$	$\alpha^2 + \alpha + 1$	$\alpha^2 + \alpha$	$\alpha^2 + 1$	α^2	$\alpha + 1$	α	1	0

$\cdot$	0	1	α	$\alpha + 1$	α^2	$\alpha^2 + 1$	$\alpha^2 + \alpha$	$\alpha^2 + \alpha + 1$
0	0	0	0	0	0	0	0	0
1	0	1	α	$\alpha + 1$	α^2	$\alpha^2 + 1$	$\alpha^2 + \alpha$	$\alpha^2 + \alpha + 1$
α	0	α	α^2	$\alpha^2 + \alpha$	$\alpha + 1$	1	$\alpha^2 + \alpha + 1$	$\alpha^2 + 1$
$\alpha + 1$	0	$\alpha + 1$	$\alpha^2 + \alpha$	$\alpha^2 + 1$	$\alpha^2 + \alpha + 1$	α^2	1	α
α^2	0	α^2	$\alpha + 1$	$\alpha^2 + \alpha + 1$	$\alpha^2 + \alpha$	α	$\alpha^2 + 1$	1
$\alpha^2 + 1$	0	$\alpha^2 + 1$	1	α^2	α	$\alpha^2 + \alpha + 1$	$\alpha + 1$	$\alpha^2 + \alpha$
$\alpha^2 + \alpha$	0	$\alpha^2 + \alpha$	$\alpha^2 + \alpha + 1$	1	$\alpha^2 + 1$	$\alpha + 1$	α	α^2
$\alpha^2 + \alpha + 1$	0	$\alpha^2 + \alpha + 1$	$\alpha^2 + 1$	α	1	$\alpha^2 + \alpha$	α^2	$\alpha + 1$

e. The elements of $\mathbf{Z}_3[x]/(p(x))$ are given by

$$\{0, 1, 2, \alpha, \alpha + 1, \alpha + 2, 2\alpha, 2\alpha + 1, 2\alpha + 2\}.$$

This ring is not a field, since $\alpha + 2$ does not have a multiplicative inverse.

+	0	1	2	α	$\alpha + 1$	$\alpha + 2$	2α	$2\alpha + 1$	$2\alpha + 2$
0	0	1	2	α	$\alpha + 1$	$\alpha + 2$	2α	$2\alpha + 1$	$2\alpha + 2$
1	1	2	0	$\alpha + 1$	$\alpha + 2$	α	$2\alpha + 1$	$2\alpha + 2$	2α
2	2	0	1	$\alpha + 2$	α	$\alpha + 1$	$2\alpha + 2$	2α	$2\alpha + 1$
α	α	$\alpha + 1$	$\alpha + 2$	2α	$2\alpha + 1$	$2\alpha + 2$	0	1	2
$\alpha + 1$	$\alpha + 1$	$\alpha + 2$	α	$2\alpha + 1$	$2\alpha + 2$	2α	1	2	0
$\alpha + 2$	$\alpha + 2$	α	$\alpha + 1$	$2\alpha + 2$	2α	$2\alpha + 1$	2	0	1
2α	2α	$2\alpha + 1$	$2\alpha + 2$	0	1	2	α	$\alpha + 1$	$\alpha + 2$
$2\alpha + 1$	$2\alpha + 1$	$2\alpha + 2$	2α	1	2	0	$\alpha + 1$	$\alpha + 2$	α
$2\alpha + 2$	$2\alpha + 2$	2α	$2\alpha + 1$	2	0	1	$\alpha + 2$	α	$\alpha + 1$

$\cdot$	0	1	2	α	$\alpha + 1$	$\alpha + 2$	2α	$2\alpha + 1$	$2\alpha + 2$
0	0	0	0	0	0	0	0	0	0
1	0	1	2	α	$\alpha + 1$	$\alpha + 2$	2α	$2\alpha + 1$	$2\alpha + 2$
2	0	2	1	2α	$2\alpha + 2$	$2\alpha + 1$	α	$\alpha + 2$	$\alpha + 1$
α	0	α	2α	$2\alpha + 2$	2	$\alpha + 2$	$\alpha + 1$	$2\alpha + 1$	1
$\alpha + 1$	0	$\alpha + 1$	$2\alpha + 2$	2	α	$2\alpha + 1$	1	$\alpha + 2$	2α
$\alpha + 2$	0	$\alpha + 2$	$2\alpha + 1$	$\alpha + 2$	$2\alpha + 1$	0	$2\alpha + 1$	0	$\alpha + 2$
2α	0	2α	α	$\alpha + 1$	1	$2\alpha + 1$	$2\alpha + 2$	$\alpha + 2$	2
$2\alpha + 1$	0	$2\alpha + 1$	$\alpha + 2$	$2\alpha + 1$	$\alpha + 2$	0	$\alpha + 2$	0	$2\alpha + 1$
$2\alpha + 2$	0	$2\alpha + 2$	$\alpha + 1$	1	2α	$\alpha + 2$	2	$2\alpha + 1$	α

3. a. We have $p(0) = 1$, $p(1) = 1$, and $p(2) = 2$. Therefore, $p(x)$ is irreducible, by Theorem 8.20.

    b. $(a_0 + a_1\alpha + a_2\alpha^2)(b_0 + b_1\alpha + b_2\alpha^2)$
      $= (a_0b_0 + 2a_1b_2 + 2a_2b_1 + 2a_2b_2)$
      $+ (a_0b_1 + a_1b_0 + 2a_2b_2)\alpha$
      $+ (a_0b_2 + a_1b_1 + a_1b_2 + a_2b_0 + a_2b_1 + a_2b_2)\alpha^2$

    c. $(\alpha^2 + \alpha + 2)^{-1} = \alpha + 1$

5. a. Since $p(0) = 1$, $p(1) = 3$, $p(2) = 1$, $p(3) = 1$, and $p(4) = 4$, Theorem 8.20 assures us that $p(x)$ is irreducible.

b. $(a_0 + a_1\alpha + a_2\alpha^2)(b_0 + b_1\alpha + b_2\alpha^2)$

$= (a_0b_0 + 4a_1b_2 + 4a_2b_1)$

$\quad + (a_0b_1 + a_1b_0 + 4a_1b_2 + 4a_2b_1 + 4a_2b_2)\alpha$

$\quad + (a_0b_2 + a_1b_1 + a_2b_0 + 4a_2b_2)\alpha^2$

c. $(\alpha^2 + 4\alpha)^{-1} = 4\alpha^2 + 3\alpha + 2$

7. a. $0, 1, 2, \alpha, \alpha + 1, \alpha + 2, 2\alpha, 2\alpha + 1, 2\alpha + 2, \alpha^2, \alpha^2 + 1, \alpha^2 + 2, 2\alpha^2, 2\alpha^2 + 1, 2\alpha^2 + 2,$
$\alpha^2 + \alpha, \ \alpha^2 + \alpha + 1, \ \alpha^2 + \alpha + 2, \ 2\alpha^2 + \alpha, \ 2\alpha^2 + \alpha + 1, \ 2\alpha^2 + \alpha + 2, \ \alpha^2 + 2\alpha,$
$\alpha^2 + 2\alpha + 1, \alpha^2 + 2\alpha + 2, 2\alpha^2 + 2\alpha, 2\alpha^2 + 2\alpha + 1, 2\alpha^2 + 2\alpha + 2$

9. $(-4 + \sqrt[3]{2} - 3\sqrt[3]{4})/22$

11. a. $3, 4$ **c.** $2, 3$ **e.** $5, 5$

13. $\alpha, 2\alpha + 1$ **15.** $\alpha, 2\alpha^2 + 3\alpha, 3\alpha^2 + \alpha + 4$

APPENDIX EXERCISES,   pages 342–345

1. For $x = 0$, the statement $0^2 > 0$ is false.

3. For $a = 0$ and any real number b, the statement $0 \cdot b = 1$ is false.

5. For $x = -4$, the statement $-(-4) < |-4|$ is false.

7. For $n = 6$, the statement $6^2 + 2(6) = 48$ is true.

9. For $n = 5$, the integer $5^2 < 2^5$ is true.

11. For $n = 3$, the integer $3^2 + 3$ is an even integer.

13. There is at least one child who did not receive a Valentine card.

15. There is at least one senior who either did not graduate or did not receive a job offer.

17. All of the apples in the basket are not rotten.

19. All of the politicians are dishonest or untrustworthy.

21. There is at least one $x \in A$ such that $x \notin B$.

23. There exists a right triangle with sides a and b and hypotenuse c such that $c^2 \neq a^2 + b^2$.

25. Some complex number does not have a multiplicative inverse.

27. There are sets A and B such that the Cartesian products $A \times B$ and $B \times A$ are not equal.

29. For every complex number x, $x^2 + 1 \neq 0$.

31. For all sets A and B, the set A is not a subset of $A \cap B$.

33. For any triangle with angles α, β, and γ, the inequality $\alpha + \beta + \gamma \leq 180°$ holds.

35. For every real number x, $2^x > 0$.

37. TRUTH TABLE for $p \Leftrightarrow \sim(\sim p)$

p	$\sim p$	$\sim(\sim p)$
T	F	T
F	T	F

We examine the two columns headed by p and $\sim(\sim p)$ and note that they are identical.

39. TRUTH TABLE for $\sim(p \wedge (\sim p))$

p	$\sim p$	$p \wedge (\sim p)$	$\sim(p \wedge (\sim p))$
T	F	F	T
F	T	F	T

41. TRUTH TABLE for $(p \wedge q) \Rightarrow p$

p	q	$p \wedge q$	$(p \wedge q) \Rightarrow p$
T	T	T	T
T	F	F	T
F	T	F	T
F	F	F	T

43. TRUTH TABLE for $(p \wedge (p \Rightarrow q)) \Rightarrow q$

p	q	$p \Rightarrow q$	$p \wedge (p \Rightarrow q)$	$(p \wedge (p \Rightarrow q)) \Rightarrow q$
T	T	T	T	T
T	F	F	F	T
F	T	T	F	T
F	F	T	F	T

45. TRUTH TABLE for $(p \Rightarrow q) \Leftrightarrow ((\sim p) \vee q)$

p	q	$p \Rightarrow q$	$\sim p$	$(\sim p) \vee q$
T	T	T	F	T
T	F	F	F	F
F	T	T	T	T
F	F	T	T	T

We examine the two columns headed by $p \Rightarrow q$ and $(\sim p) \vee q$ and note that they are identical.

47. TRUTH TABLE for $(p \Rightarrow q) \Leftrightarrow ((p \wedge (\sim q)) \Rightarrow (\sim p))$

p	q	$p \Rightarrow q$	$\sim q$	$p \wedge (\sim q)$	$\sim p$	$(p \wedge (\sim q)) \Rightarrow (\sim p)$
T	T	T	F	F	F	T
T	F	F	T	T	F	F
F	T	T	F	F	T	T
F	F	T	T	F	T	T

We examine the two columns headed by $p \Rightarrow q$ and $(p \wedge (\sim q)) \Rightarrow (\sim p)$ and note that they are identical.

49. TRUTH TABLE for $(p \wedge q \wedge r) \Rightarrow ((p \vee q) \wedge r)$

p	q	r	$p \wedge q \wedge r$	$p \vee q$	$(p \vee q) \wedge r$	$(p \wedge q \wedge r) \Rightarrow ((p \vee q) \wedge r)$
T	T	T	T	T	T	T
T	T	F	F	T	F	T
T	F	T	F	T	T	T
T	F	F	F	T	F	T
F	T	T	F	T	T	T
F	T	F	F	T	F	T
F	F	T	F	F	F	T
F	F	F	F	F	F	T

51. TRUTH TABLE for $(p \Rightarrow (q \wedge r)) \Leftrightarrow ((p \Rightarrow q) \wedge (p \Rightarrow r))$

p	q	r	$q \wedge r$	$p \Rightarrow (q \wedge r)$	$p \Rightarrow q$	$p \Rightarrow r$	$(p \Rightarrow q) \wedge (p \Rightarrow r)$
T	T	T	T	T	T	T	T
T	T	F	F	F	T	F	F
T	F	T	F	F	F	T	F
T	F	F	F	F	F	F	F
F	T	T	T	T	T	T	T
F	T	F	F	T	T	T	T
F	F	T	F	T	T	T	T
F	F	F	F	T	T	T	T

We examine the two columns headed by $p \Rightarrow (q \wedge r)$ and $(p \Rightarrow q) \wedge (p \Rightarrow r)$ and note that they are identical.

53. The implication $(p \Rightarrow q)$ is true: My grade for this course is A implies that I can enroll in the next course.

The contrapositive $(\sim q \Rightarrow \sim p)$ is true: I cannot enroll in the next course implies that my grade for this course is not A.

The inverse $(\sim p \Rightarrow \sim q)$ is false: My grade for this course is not A implies that I cannot enroll in the next course.

The converse $(q \Rightarrow p)$ is false: I can enroll in the next course implies that my grade for this course is A.

55. The implication $(p \Rightarrow q)$ is true: The Saints win the Super Bowl implies that the Saints are the champion football team.

The contrapositive $(\sim q \Rightarrow \sim p)$ is true: The Saints are not the champion football team implies that the Saints did not win the Super Bowl.

The inverse $(\sim p \Rightarrow \sim q)$ is true: The Saints did not win the Super Bowl implies that the Saints are not the champion football team.

The converse $(q \Rightarrow p)$ is true: The Saints are the champion football team implies that the Saints did win the Super Bowl.

57. The implication $(p \Rightarrow q)$ is false: My pet has four legs implies that my pet is a dog.

The contrapositive $(\sim q \Rightarrow \sim p)$ is false: My pet is not a dog implies that my pet does not have four legs.

The inverse $(\sim p \Rightarrow \sim q)$ is true: My pet does not have four legs implies that my pet is not a dog.

The converse $(q \Rightarrow p)$ is true: My pet is a dog implies that my pet has four legs.

59. The implication $(p \Rightarrow q)$ is true: Quadrilateral $ABCD$ is a square implies that quadrilateral $ABCD$ is a rectangle.

The contrapositive $(\sim q \Rightarrow \sim p)$ is true: Quadrilateral $ABCD$ is not a rectangle implies that quadrilateral $ABCD$ is not a square.

The inverse $(\sim p \Rightarrow \sim q)$ is false: Quadrilateral $ABCD$ is not a square implies that quadrilateral $ABCD$ is not a rectangle.

The converse $(q \Rightarrow p)$ is false: Quadrilateral $ABCD$ is a rectangle implies that quadrilateral $ABCD$ is a square.

61. The implication $(p \Rightarrow q)$ is true: x is a positive real number implies that x is a nonnegative real number.

The contrapositive $(\sim q \Rightarrow \sim p)$ is true: x is a negative real number implies that x is a nonpositive real number.

The inverse $(\sim p \Rightarrow \sim q)$ is false: x is a nonpositive real number implies that x is a negative real number.

The converse $(q \Rightarrow p)$ is false: x is a nonnegative real number implies that x is a positive real number.

63. The implication $(p \Rightarrow q)$ is true: $5x$ is odd implies that x is odd.

The contrapositive $(\sim q \Rightarrow \sim p)$ is true: x is not odd implies that $5x$ is not odd.

The inverse $(\sim p \Rightarrow \sim q)$ is true: $5x$ is not odd implies that x is not odd.

The converse $(q \Rightarrow p)$ is true: x is odd implies that $5x$ is odd.

65. The implication $(p \Rightarrow q)$ is true: xy is even implies that x is even or y is even.

The contrapositive $(\sim q \Rightarrow \sim p)$ is true: x is odd and y is odd implies that xy is odd.

The inverse $(\sim p \Rightarrow \sim q)$ is true: xy is odd implies that x is odd and y is odd.

The converse $(q \Rightarrow p)$ is true: x is even or y is even implies that xy is even.

67. The implication $(p \Rightarrow q)$ is false: $x^2 > y^2$ implies that $x > y$.

The contrapositive $(\sim q \Rightarrow \sim p)$ is false: $x \le y$ implies that $x^2 \le y^2$.

The inverse $(\sim p \Rightarrow \sim q)$ is false: $x^2 \le y^2$ implies that $x \le y$.

The converse $(q \Rightarrow p)$ is false: $x > y$ implies that $x^2 > y^2$.

69. Contrapositive: $\sim(q \lor r) \Rightarrow \sim p$, or $((\sim q) \land (\sim r)) \Rightarrow \sim p$

Converse: $(q \lor r) \Rightarrow p$

Inverse: $\sim p \Rightarrow \sim(q \lor r)$, or $\sim p \Rightarrow ((\sim q) \land (\sim r))$

71. Contrapositive: $q \Rightarrow \sim p$

Converse: $\sim q \Rightarrow p$

Inverse: $\sim p \Rightarrow q$

73. Contrapositive: $\sim(r \land s) \Rightarrow \sim(p \lor q)$, or $((\sim r) \lor (\sim s)) \Rightarrow ((\sim p) \land (\sim q))$

Converse: $(r \land s) \Rightarrow (p \lor q)$

Inverse: $\sim(p \lor q) \Rightarrow \sim(r \land s)$, or $((\sim p) \land (\sim q)) \Rightarrow ((\sim r) \lor (\sim s))$

INDEX

392

Octic Group

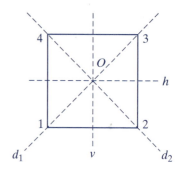

$\{e, \alpha, \alpha^2, \alpha^3, \beta, \gamma, \Delta, \theta\}$

$e = (1)$	$\beta = (1,4)(2,3)$
$\alpha = (1,2,3,4)$	$\gamma = (2,4)$
$\alpha^2 = (1,3)(2,4)$	$\Delta = (1,2)(3,4)$
$\alpha^3 = (1,4,3,2)$	$\theta = (1,3)$

Group Table

$\circ$	e	α	α^2	α^3	β	γ	Δ	θ
e	e	α	α^2	α^3	β	γ	Δ	θ
α	α	α^2	α^3	e	γ	Δ	θ	β
α^2	α^2	α^3	e	α	Δ	θ	β	γ
α^3	α^3	e	α	α^2	θ	β	γ	Δ
β	β	θ	Δ	γ	e	α^3	α^2	α
γ	γ	β	θ	Δ	α	e	α^3	α^2
Δ	Δ	γ	β	θ	α^2	α	e	α^3
θ	θ	Δ	γ	β	α^3	α^2	α	e

Klein Four Group

$G = \{e, a, b, ab\}$

Group Table

$\cdot$	e	a	b	ab
e	e	a	b	ab
a	a	e	ab	b
b	b	ab	e	a
ab	ab	b	a	e